Membrane Biogenesis

NATO ASI Series

Advanced Science Institutes Series

A series presenting the results of activities sponsored by the NATO Science Committee, which aims at the dissemination of advanced scientific and technological knowledge, with a view to strengthening links between scientific communities.

The Series is published by an international board of publishers in conjunction with the NATO Scientific Affairs Division

A Life Sciences B Physics	Plenum Publishing Corporation London and New York
C Mathematical and Physical Sciences D Behavioural and Social Sciences E Applied Sciences	Kluwer Academic Publishers Dordrecht, Boston and London
F Computer and Systems Sciences G Ecological Sciences H Cell Biology	Springer-Verlag Berlin Heidelberg New York London Paris Tokyo

Series H: Cell Biology Vol. 16

Membrane Biogenesis

Edited by

Jos A. F. Op den Kamp

Laboratory of Biochemistry
State University of Utrecht
Padualaan 8, P.O. Box 80054
3508 TB Utrecht, The Netherlands

Springer-Verlag
Berlin Heidelberg New York London Paris Tokyo
Published in cooperation with NATO Scientific Affairs Division

Proceedings of the NATO Advanced Study Institute on New Perspectives in the Dynamics of Assembly of Biomembranes held in Cargèse, Corsica, France, August 24 – September 4, 1987

ISBN 3-540-18566-6 Springer-Verlag Berlin Heidelberg New York
ISBN 0-387-18566-6 Springer-Verlag New York Berlin Heidelberg

Library of Congress Cataloging-in-Publication Data. Nato Advanced Study Institute on New Perspectives in the Dynamics of Assembly of Biomembranes (1987 : Cargèse, Corsica) Membrane biogenesis / edited by Jos A. F. Op den Kamp. p. cm.—(NATO ASI series. Series H, Cell biology ; vol. 16) "Proceedings of the NATO Advanced Study Institute on New Perspectives in the Dynamics of Assembly of Biomembranes held in Cargèse, Corsica, France, August 24 – September 4, 1987"—T.p. verso. "Published in cooperation with NATO Scientific Affairs Division." Includes index.
ISBN 0-387-18566-6 (U.S.)
1. Membranes (Biology)—Congresses. I. Kamp, Jos A. F. Op den (Jos Arnoldus Franciscus), 1939-. II. North Atlantic Treaty Organization. Scientific Affairs Division. III. Title. IV. Series. [DNLM: 1. Biogenesis—congresses. 2. Cell Membrane—congresses. 3. Membrane Lipids—congresses. QH 601 N2797n 1987]
QH601.N377 1987 574.87'5—dc 19 DNLM/DLC for Library of Congress 88-4471

Printed in Germany

Printing: Druckhaus Beltz, Hemsbach; Binding: J. Schäffer GmbH & Co. KG, Grünstadt
2131/3140-543210

PREFACE

Many individual aspects of the dynamics and assembly of biological membranes have been studied in great detail. Cell biological approaches, advanced genetics, biophysics and biochemistry have greatly contributed to an increase in our knowledge in this field.It is obvious however, that the three major membrane constituents - lipids, proteins and carbohydrates- are studied, in most cases separately and that a coherent overview of the various aspects of membrane biogenesis is not readily available.

The NATO Advanced Study Institute on "New Perspectives in the Dynamics of Assembly of Biomembranes" intended to provide such an overview: it was set up to teach students and specialists the achievements obtained in the various research areas and to try and integrate the numerous aspects of membrane assembly into a coherent framework. The articles in here reflect this. Starting with detailed contributions on phospholipid structure, dynamics, organization and biogenesis, an up to date overview of the basic, lipidic backbone of biomembranes is given. Extensive progress is made in the research on membrane protein biosynthesis. In particular the post- and co-translational modification processes of proteins, the mechanisms of protein translocation and the sorting mechanisms which are necessary to direct proteins to their final , intra - or extracellular destination have been characterized in detail. Modern genetic approaches were indispensable in this research area: gene cloning, hybrid protein construction, site directed mutagenesis and sequencing techniques elucidated many functional aspects of specific nucleic acid and amino acid sequences.

Sandwiched inbetween these articles dealing specifically with lipids or proteins are these contributions in which the coherence and mutual interactions of membrane constituents are most clearly expressed. The fact that many different interactions, including covalent bonding, between lipids, carbohydrates and proteins do occur as well as the increasing awareness of the effects that the membrane components can exert on each other with respect to structure, localization, mobility and function have been clearly described

Utrecht, November 1987

Jos A.F. Op den Kamp

CONTENTS

TECHNIQUES TO DETERMINE TRANSBILAYER ORGANIZATION AND DYNAMICS OF MEMBRANE PHOSPHOLIPIDS

Ben Roelofsen and Jos A.F. Op den Kamp

Department of Biochemistry,
University of Utrecht,
P.O. Box 80.054,
3508 TB Utrecht,
The Netherlands

INTRODUCTION

Nowadays, there are no more arguments opposing the view that the structural backbone of a biological membrane is provided by a lipid bilayer, the fatty acyl chains of the phospholipids forming the hydrophobic core of the membrane and their polar head groups facing the aqueous environments at either side of the bilayer. As a consequence of the primary task of a membrane, which is to separate two aqueous compartments from one another in which entirely different processes take place, it is conceivable that the chemical characteristics of one side of a membrane differ considerably from those of the other. Indeed, the absolute asymmetry in both transversal localization and orientation of, respectively the periferal and integral, proteins in a membrane has already been recognized a couple of decades ago. Since the early seventies, it is also known that the different classes of phospholipids may be distributed over both halves of the bilayer in a highly asymmetric fashion, which phenomenon particularly applies to plasma membranes. The first information on this point has been gained from studies on erythrocytes and, still, the membrane of those cells is the best characterized one with regards to the transbilayer organization and dynamics of its lipids (Op den Kamp, 1979).

NATO ASI Series, Vol. H16
Membrane Biogenesis. Edited by J. A. F. Op den Kamp

TECHNIQUES FOR PHOSPHOLIPID LOCALIZATION STUDIES

The three types of probes that are most commonly used to assess the transbilayer distribution of phospholipids in a membrane are: (i) group specific chemical reagents, (ii) phospholipases and (iii) phospholipid transfer proteins. The general strategy for their application in such studies is simple and straightforward. Intact, closed membranes are exposed to the action of the probe and the modification of the phospholipids thus obtained is compared with that after treatment of open membranes with the same reagent. The philosophy behind this approach will be obvious. When intact cells are treated with the reagent, only those phospholipids will be modified that are localized in the outer half of the bilayer. Treatment of open membranes serves as an essential control to make sure that, under the conditions applied, the probe can react with all of its potential target molecules when it has access to both sides of the membrane.

This general strategy implies a number of prerequisites which should be fulfilled to achieve conclusive results. These comprise the following: (i) the structural integrity of the membrane should be maintained during the experiment, so that the action of the probe remains restricted to that side of the membrane where it is supposed to explore its action; (ii) this also implies that, preferably, the membrane studied should be impermeable to the reagent; (iii) transbilayer movements of phospholipids should be absent, or at least very slow relative to the time scale of the experiment; and (iv) the treatment as such should not give rise to transbilayer rearrangements of phospholipids.

Each of the three above mentioned types of probes will be briefly discussed below.

Group specific chemical reagents

In the past, quite a number of chemical reagents have been developed which specifically react with free NH_2-groups (Op den Kamp, 1979; Etemadi, 1980), and therefore can provide information as to the localization of the amino-phospholipids, phosphatidylethanolamine (PE) and phosphatidylserine (PS). A reagent that has been most widely used for this purpose is tri-nitrobenzenesulphonic acid (TNBS). However, the reaction between TNBS and amino-phospholipids may be incomplete, particularly in case of a membrane containing relatively high amounts of these phospholipids, because of sterical hindrance as a consequence of the introduction of the bulky TNB-group, whereas the reactivity of the negatively charged TNBS towards PS will be impaired by electrostatic repulsion between this reagent and the -also negatively charged- polar head group of this phospholipid (Op den Kamp, 1979). Moverover, and in contrast to what has been believed until recently, TNBS can easily permeate through the red cell membrane (Haest *et al.*, 1981).

Another NH_2-group specific reagent is fluorescamine, which has some major advantages .

When applied at pH 8.0 (or higher), it reacts with free NH_2-groups within a second, the excess of the reagent being destroyed by aqueous hydrolysis within a minute. In contrast to this hydrolysis product, the product formed by reaction with a NH_2-group is a fluorophor, which enables a sensitive quantitative determination of the reaction product. That this probe, despite its ability to permeate through a membrane, can be satisfactorily used to determine the distribution of PE in the plasma membrane of a cell which also contains intracellular membrane systems, is shown in figure 1.

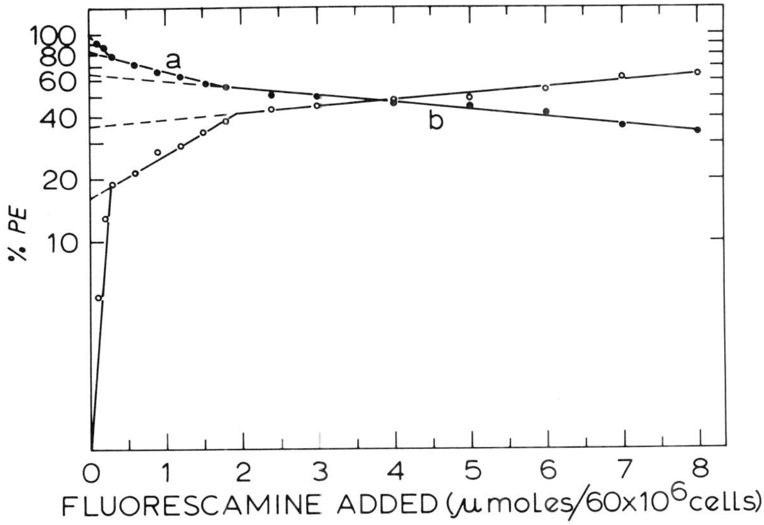

FIGURE 1
Semi-logarithmic plot of the loss in phosphatidylethanolamine (●) and of the appearance of its fluorescamine derivative (o) in Friend cells labeled at 0-4°C with increasing amounts of fluorescamine. Phosphatidylethanolamine pools attributable to the outer and inner plasma membrane monolayer and to the intracellular membranes were derived by extrapolating lines a and b, resp., of the curve representing the loss of phosphatidylethanolamine (●).

Intact Friend erythroleukaemic cells have been exposed to increasing concentrations of fluorescamine. The semi-logarithmic dose-response curves of the labeled, respectively unlabeled, PE in these treated cells, reveals three different pools (Rawyler *et al.*, 1984). The first and fast reacting pool represents the PE in the outer leaflet of the plasma membrane, the second pool represents that fraction of the PE in the inner monolayer of this membrane, and the third and slowly reacting pool reflects the PE located in the subcellular membranes. The distribution of PE in the plasma membrane of these cells, as determined by this technique, could be confirmed by experiments using phospholipases (Rawyler *et al.*, 1985).

Phospholipases

Phospholipases represent the most popular group of probes used in phospholipid localization studies, due to the differences in their mode of action and substrate specificity (Roelofsen, 1982).

Phospholipases A_2 (PLA_2), which can be isolated from pancreas, bee venom and a variety of snake venoms, can only attack diacyl *glycero*-phospholipids. These enzymes specifically split the ester bond at the 2-position of the glycerol back bone, thus producing a free fatty acid and a 1-acyl lyso-derivative. It is important to note that both these reaction products remain in the membrane.

Phospholipases C (PLC), which are usually from bacterial origin, cut off the complete polar head group from a phospholipid. This results in the production of diglycerides or ceramides, depending on whether the substrates are diacyl glycero-phospholipids or sphingomyelin (SM), respectively. These digestion products are left behind in the membrane. The polar head groups that are split off are readily water soluble and will, therefore, immediately leave the membrane. Sphingomyelinase C (SMase C) is a genuine phospholipase C, but it has an almost absolute specificity towards SM.

Phospholipases D are from vegetable origin and usually isolated from cabbage or peanuts. Those enzymes convert glycerophospholipids into the highly negatively charged phosphatidic acid.

The application of phospholipases in phospholipid localization studies requires some special prerequisites (Roelofsen, 1982), which concern: (i) the purity of the enzyme used, because any contamination that disrupts the structural stability of the membrane studied may largely affect the reliability of the results; (ii) a proper and complete inhibition of phospholipase activity after the incubation and prior to lipid extraction, because most of these enzymes will not be inactivated by organic solvents such as methanol and chloroform, which are most commonly used for such extraction procedures; and (iii) the selection of the proper procedure to extract the lipids from the phospholipase treated membranes, taking into account that this treatment may have caused extensive changes in lipid composition of the membranes which, in turn, may have important consequences for their extraction and final recovery.

The possibilities and limitations that are inherent to the use of phospholipases in phospholipid localization studies, can be illustrated by their use on erythrocytes. The strategy of such experiments is depicted in figure 2. In addition to the exposure of intact cells and open ghost membranes to the action of phospholipases, the erythrocyte provides the possibility of performing a third type of experiments (Fig. 2). To that end, a suspension of open ghosts (prepared by hypotonic lysis) is supplemented with a phospholipase in its inactive form, i.e. in the presence of a chelator such as EDTA to remove the divalent cations that are necessary for its activity. Subsequently, ghosts are resealed by restoring isotonicity, thereby trapping the (inactive)

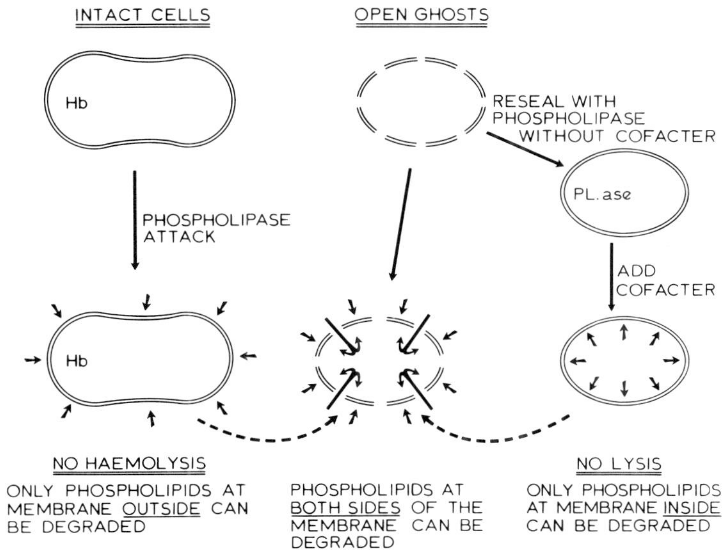

FIGURE 2
Scheme of phospholipase attack on intact red cells and open ghosts, as well as inside resealed ghosts. Hb, haemoglobin; PL.ase, phospholipase.

enzyme in their cytosol. After several washes of the resealed ghosts to remove the phospholipase from the medium, the trapped enzyme is activated by addition of its essential cofactor which, in case of phospholipase A_2, is Ca^{2+}. The enzyme then starts to degrade its substrates present in the inner membrane leaflet. The hydrolysis patterns thus obtained should, at least in the ideal situation, be complementary to the hydrolysis pattern of phospholipids achieved by treatment of the intact cells with the enzyme.

Treatment of intact human erythrocytes under non-lytic conditions with pure *Naja naja* PLA_2 ultimately results in the hydrolysis of 68% of the PC. When after 1 h, SMase C is added to this incubation mixture, upto 82% of the SM is rapidly degraded. It is of considerable interest to note that this degradation of SM in the outer monolayer enables the PLA_2, that is still present, to degrade an additional fraction of the PC as well as 20% of the PE (Zwaal *et al.*, 1975). Even under those conditions, any hydrolysis of PS is still not detectable which indicates that this phospholipid is exclusively located at the cytoplasmic side of the membrane. The degradation of the other three phospholipids comprises 48% of the total phospholipid complement of the membrane (Table I), suggesting that indeed all phospholipids in the outer monolayer have been degraded to completion. Hence, it is important to note that a complete picture of the phospholipids constituting the outer membrane leaflet, can only be obtained by using this combination of phospholipases. PLA_2 alone is not able to degrade all of its substrates in the outer monolayer. This is due to the fact that the action of the enzyme ceases as a consequence of an increase in lateral

TABLE I Percentage of nonlytic degradation of phospholipids by phospholipases in human erythrocytes. Incubations were started in the presence of phospholipase A_2 (*N.naja*), followed by the addition of sphingomyelinase C (*S. aureus*) after 1 h.

Incubation time (min)	Enzyme		% degradation				
	PLA_2 (*N.naja*)	SMase C (*S.aureus*)	Total phospholipid	SM	PC	PE	PS
0			0	0	0	0	0
60			20	0	68	0	0
120			48	82	76	20	0

surface pressure in that layer (Fujii and Tamura, 1979). This, in turn, is due to the production of split products (free fatty acids and lyso-PCs) in that layer, since these two split products occupy a surface area which is slightly larger than that occupied by the original diacyl-PC. In this situation, the enzyme can no longer reach its substrates; in other words, the reaction is self-quenching (Roelofsen *et al.*). Conversily, the removal of the polar head group from SM by SMase C has the opposite effect, namely a decrease in lateral surface pressure in the outer monolayer, which enables the PLA_2 that is still present to complete its job (Zwaal *et al.*, 1975).

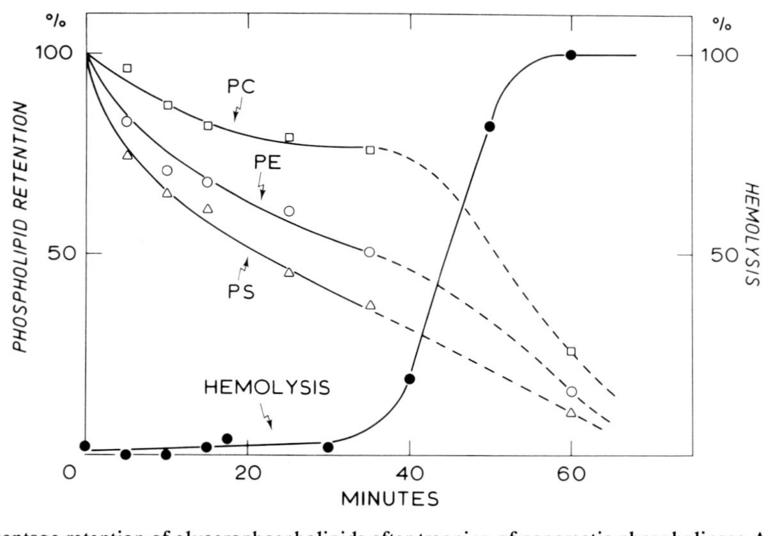

Percentage retention of glycerophospholipids after trapping of pancreatic phospholipase A_2 (without Ca^{2+}) inside resealed human erythrocyte ghosts followed by the addition of Ca^{2+} to start enzymatic breakdown at t=0. PC, phosphatidylcholine; PE, phospha- tidylethanolamine; PS, phosphatidylserine.

Figure 3 shows the time course for the hydrolysis of the phospholipids in the inner leaflet, as brought about by pancreatic PLA_2 trapped inside resealed ghosts. Unfortunately this treatment ultimately results in complete lysis of these ghosts. Nevertheless, it is of interest to note that -just before the onset of lysis- the degradation of PC tends to level off at 25%, which is exactly complementery to the 76% of the PC that can be degraded in the outer layer of the intact erythrocyte (Table I). On the other hand, the hydrolysis of PE and PS is already considerable and still in full swing at the moment lysis of the cells starts to take place, indicating that indeed the amino-phospholipids are predominantly located in the inner half of the bilayer (Zwaal *et al.*, 1975).

Phospholipid transfer proteins

A third type of probes to determine membrane phospholipid distributions has been introduced more recently, namely the phospholipid transfer proteins (Bloj and Zilversmit, 1981). Their major advantage is that, at least in principle, they do neither modify the phospholipid composition, nor the structure of the membrane studied.

A transfer protein that has been most widely used so far is the PC-specific one (PC-TP), purified from beef liver. This water soluble protein mediates a one-for-one exchange of PC molecules between two membrane systems (Bloj and Zilversmit, 1981). The intact cells to be studied, usually referred to as the "acceptor", are incubated in the presence of the PC-TP together with a suitable "donor" system, i.e. either small unilamellar vesicles or rat liver microsomes containing radiolabeled PC of known specific activity. During the incubation, part of the PC in the acceptor membrane will be replaced by radioactive species. After the incubation and separation of donors and acceptors, the lipids from both systems are extracted, chromatographically separated and the specific radioactivity of the PC in both fractions determined. By comparing both values, it is quite easy to calculate what fraction of the PC in the acceptor membrane is in equilibrium with that in the donor system.

Figure 4 shows the results of such experiments involving intact erythrocytes from man (panel A) and rat (panel B), respectively. In normal human erythrocytes, 74-76% of the PC appeared to be readily available for PC-TP mediated exchange (van Meer *et al.*, 1980). This is in perfect agreement with the fraction of PC, namely 76%, that is available in the intact cell for attack by PLA_2 in the presence of SMase C (compare Table I). Obviously, this pool represents the fraction of the PC in the outer membrane leaflet which, as can be deduced from the fact that the exchange profile ultimately reaches a horizontal plateau (Fig. 4, panel A), does not appear to equilibrate with the PC in the inner layer. The latter phenomenon indicates that, relative to the time scale of the experiment, transbilayer movements ("flip-flop") of PC in the human erythrocyte must be very slow.

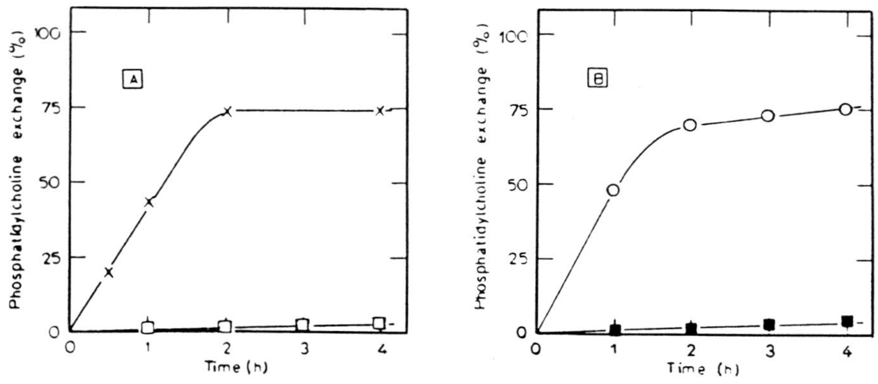

FIGURE 4
Exchange of PC between erythrocytes and rat liver microsomes. The amount of PC exchanged in the absence (◘) and presence (X,O) of PC-TP is plotted as a percentage of total erythrocyte PC, for human (A) and rat (B) erythrocytes.

Rat erythrocytes appeared to behave a bit differently in this respect, in that a fast exchangeable pool, comprising 50-60% of all the PC present, was followed by a slowly exchangeable one (Fig. 4, panel B). From this biphasic curve, it was estimated that PC equilibrates between the two halves of the bilayer with a half-time of 7 h (van Meer *et al.*, 1980). Both these results are in quite good agreement with data obtained by using a non-specific TP (Crain and Zilversmit, 1980) or phospholipases Renooij *et al.*, 1976), respectively showing 63 and 62% of the PC in the outer monolayer of the rat erythrocyte membrane and half-times for PC flip-flop of 7.2 and 4.5 h.
From the results gained from experiments involving phospholipases, chemical reagents and the PC-TP, it could be concluded that the four major phospholipid classes in the human erythrocyte show a marked asymmetry in their distribution over both halves of the membrane bilayer (Op den Kamp, 1979; Zwaal *et al.*, 1975). The two choline containing phospholipids dominate the outer layer, in which 82% of the SM and 76% of the PC is found, whereas the inner membrane leaflet mainly consists of the amino-phospholipids: 80% of the PE and all of the PS.

DETERMINATION OF PHOSPHOLIPID TRANSBILAYER MOBILITY

In the procedure described in the previous section, the first aim of the experiments was to determine the pool sizes of the PC present at either side of a membrane bilayer. In many cases, an estimate can be made of the rates of transbilayer movement of the molecules by analysis of the kinetics of biphasic exchange processes.

To determine the actual flip-flop rates of (even individual molecular species of) PC, a technique has been developed recently (van Meer and Op den Kamp, 1982) in which PC-TP is used to introduce radiolabeled PC into the outer membrane layer of the intact erythrocyte, after which its fate is followed by selectively degrading the outer monolayer by phospholipase treatment (Fig. 5).

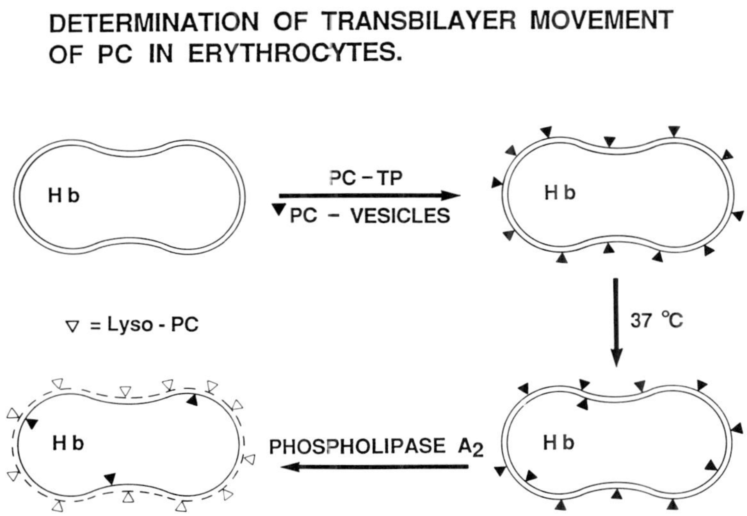

FIGURE 5
Principle of an experiment to determine the rate of transbilayer movement of PC in the erythrocyte membrane. Abbreviations: Hb, heamoglobin; PLEP, phospholipid exchange protein, i.e., PC-TP.

Intact erythrocytes are incubated for 1 h in the presence of PC-TP, either with [*methyl*-^{14}C]PC labeled rat liver microsomes or with sonicated vesicles prepared from egg PC mixed with an equimolar amount of cholesterol, PA (5mol%) and trace amounts of [*methyl*-^{14}C]egg PC and [^{14}C]cholesteryloleate or glycerol[^{3}H]trioleate, the latter compounds as non-exchangeable marker. After removal of the PC-TP and the [^{14}C]PC donor system by careful washing, the erythrocytes are subsequently incubated in isotonic buffer to enable the radioactive PC, introduced into the outer monolayer, to equilibrate between the two halves of the membrane bilayer. At timed intervals, samples are taken and the intact cells are subsequently incubated with a suitable phospholipase A_2, thereby selectively converting the PC in the outer monolayer into its lyso-derivative. It should be kept in mind, however, that the use of phospholipase A_2 alone does not result in a complete degradation of the PC in the outer monolayer. This can only be achieved when this enzyme is used in combination with sphingomyelinase C, which should be added after preincubation of the cells with the phospholipase A_2 (Roelofsen, B., 1982; Zwaal *et al.*, 1975). The lyso-PC thus produced, representing the PC in the outer monolayer, and the residual PC originating from the inner leaflet, can be easily separated from one another, which enables an accurate determination of their specific radioactivities. The rate of transbilayer movement of the PC and the corresponding half-time for this process can be easily calculated from the time-dependent decrease in specific radioactivity of the lyso-PC, which should be proportional to the increase in that of the PC. In practice, the half-time value of PC flip-flop is determined from the slope of the straight line that is achieved when the changes in relative specific radioactivity of the lyso-PC are plotted semilogarithmically versus the time of incubation.

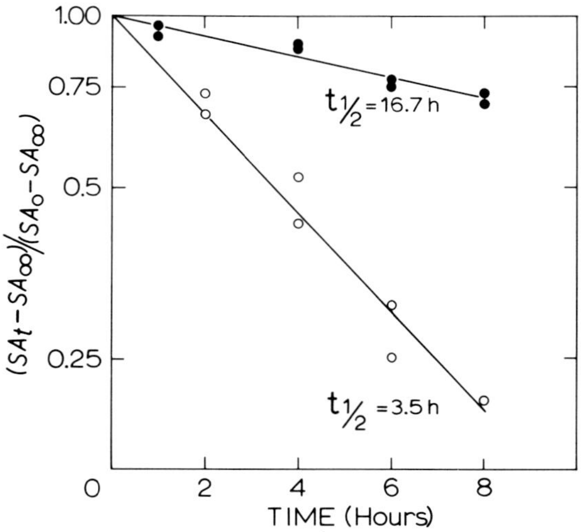

FIGURE 6
Transbilayer movement of 1,2-dioleoyl-PC (●) and 1-palmitoyl-2-linoleoyl-PC (O) in the human erythrocyte during incubation of the cells at 37°C.

Figure 6 shows the results of a typical experiment, involving 1,2-dioleoyl-PC and 1-palmitoyl-2-linoleoyl-PC, respectively. The half-time values ($t_{1/2}$) for the rate of their transbilayer movements, as indicated in Figure 6, have been calculated from the slopes of the corresponding lines.

The above method has been used to determine the rates of flip-flop of four individual molecular species of PC in the human erythrocyte membrane (Middelkoop *et al.*, 1986). The results are presented in Table II.

In agreement with earlier studies (van Meer and Op den Kamp, 1982) , unsaturated PC species experience transbilayer movements in the human red cell membrane which are considerably faster than that of a fully saturated one, such as 1,2-dipalmitoyl-PC (Table II).

Table II

Flip-flop of individual species of phosphatidylcholine in the human erythrocyte membrane. Figures represent means ± S.D. values.

PC species	Halftime (h)
1,2-Dipalmitoyl	26.3 ± 4.4
1,2-Dioleoyl	14.4 ± 3.5
1-Palmitoyl-2-linoleoyl	2.9 ± 1.7
1-Palmitoyl-2-arachidonoyl	9.7 ± 1.6

On the other hand, and in contrast to earlier suggestions (van Meer and Op den Kamp, 1982), there appears to exist no direct proportionality between the rate of flip-flop of a particular PC species and its total degree of unsaturation. 1-Palmitoyl-2-linoleoyl-PC experiences a transbilayer movement which is about four to five times as fast as that of 1,2-dioleoyl-PC (Table II; Fig. 6), although these two species have the same degree of unsaturation. This indicates that not only the total number of double bonds in a PC molecule, but more specifically their distribution over the two acyl chains, is an important determinant of its transbilayer mobility. Possibly even more surprising is the finding that the flip-flop rate of the higher unsaturated 1-Palmitoyl-2-arachidonoyl-PC is appreciably slower than that of the 1-palmitoyl-2-linoleoyl species (Table II).

At present, one may only speculate whether these differences are a consequence of differences in the molecular geometry of the various PC species which, in turn, may influence their packing in the bilayer as well as their interactions with adjacent membrane constituents. Alternatively, they may represent a reflection of a possible heterogeneity in their lateral distribution in the membrane, i.e., in different domains in which the "facilities" for transbilayer movements may differ from one another. Further studies will be necessary to gain insight into the factors which govern the transbilayer mobility of the PC molecules in the erythrocyte membrane, and, more specifically, that of the individual molecular species.

Using radio-labeled egg-PC, rather than individual molecular species, as probe molecules in this flip-flop assay procedure, one in fact determines the average half-time value of PC flip-flop. This value appears to be 15-20 h in the human erythrocyte membrane, when determined at 37°C (Franck *et al.*, 1983). Although this process is relatively slow, it is still fast when compared to the average lifetime of the human erythrocyte, which is 120 days. Hence, it is fascinating to realize that this transbilayer movement has no consequences as to the asymmetric distribution of the different phospholipid classes in the red cell membrane. The mechanisms that are involved in maintaining this marked phospholipid asymmetry will be dealt with in the following chapter.

REFERENCES

Bloj, B, Zilversmit, D B. Lipid transfer proteins in the study of artificial and natural membranes. Mol. Cell. Biochem. 40: 163-172, 1981.

Crain, R C, Zilversmit, D B. Two non-specific phospholipid exchange proteins from beef liver. 2. Use in studying the asymmetry and transbilayer movement of phosphatidylcholine, phosphatidylethanolamine and sphingomyelin in intact rat erythrocytes. Biochemistry 19: 1440-1447, 1980.

Etemadi, A H, Membrane asymmetry. A survey and critical appraisal of the methodology. II. Methods for assessing the unequal distribution of lipids. Biochim. Biophys. Acta, 604: 423-475, 1980.

Franck, P F H, Chiu, D T-Y, Op den Kamp, J A F, Lubin, B, van Deenen, L L M and Roelofsen, B. Accelerated transbilayer movement of phosphatidylcholine in sickled erythrocytes. A reversible process. J. Biol. Chem. 258: 8435-8442, 1983.

Fujii, T, Tamura, A. Asymmetric manipulation of the membrane lipid bilayer of intact human erythrocytes with phospholipases A, C and D induces a change in cell shape. J. Biochem. 86: 1345-1352, 1979.

Haest, C W M, Kamp, D, Deuticke, B, Penetration of 2,4,6-trinitrobenzene sulfonate into human erythrocytes. Consequences for studies on phospholipid asymmetry. Biochim. Biophys. Acta, 640: 535-543, 1981.

van Meer, G, Poorthuis, B J H M, Wirtz, K W A, Op den Kamp, J A F, van Deenen, L L M. Transbilayer distribution and mobility of phosphatidylcholine in intact erythrocyte membranes. A study with phosphatidylcholine exchange protein. Eur. J. Biochem. 103: 283-288, 1980.

van Meer, G, Op den Kamp, J A F. Transbilayer movement of various phosphatidylcholine species in intact human erythrocytes. J. Cell. Biochem. 19: 193-204, 1982.

Middelkoop, E, Lubin, B H, Op den Kamp, J A F, Roelofsen, B. Flip-flop rates of individual molecular species of phosphatidylcholine in the human red cell membrane. Biochim. Biophys. Acta, 855: 421-424, 1986.

Op den Kamp, J A F, Lipid asymmetry in membranes. Annu. Rev. Biochem. 48: 47-71, 1979.

Rawyler, A, Roelofsen, B, Op den Kamp, J A F, The use of fluorescamine as a permeant probe to localize phosphatidylethanolamine in intact Friend erythroleukaemic cells. Biochim. Biophys. Acta, 769: 330-336, 1984.

Rawyler, A, van der Schaft, P H, Roelofsen, B, Op den Kamp, J A F, Phospholipid localization in the plasma membrane of Friend erythroleukaemic cells and mouse erythrocytes. Biochemistry, 24: 1777-1783, 1985.

Renooij, W, van Golde, L M G, Zwaal, R F A, van Deenen, L L M. Topological asymmetry of phospholipid metabolism in rat erythrocyte membranes. Eur. J. Biochem. 61: 53-58, 1976.

Roelofsen, B. Phospholipases as tools to study the localization of phospholipids in biological membranes. A critical review. J. Toxicol. - Toxin Revs. 1: 87-197, 1982.

Roelofsen, B, Sibenius Trip, M, Verheij, H M, Zevenbergen, J L. The action of cobra venom phospholipase A_2 isoenzymes towards intact human erythrocytes. Biochim. Biophys. Acta, 600: 1012-1017.

Zwaal, R F A, Roelofsen, B, Comfurius, P, van Deenen, L L M. Organization of phospholipids in human red cell membranes as detected by the action of various purified phospholipases. Biochim. Biophys. Acta, 406: 83-96, 1975.

TRANSBILAYER ORGANIZATION AND MOBILITY OF PHOSPHOLIPIDS IN NORMAL AND PATHOLOGIC ERYTHROCYTES

Laurens L. M. van Deenen, Ben Roelofsen and Jos A.F.Op den Kamp.
Department of Biochemistry,
University of Utrecht,
PO Box 80.054,
3508 TB Utrecht, The Netherlands.

INTRODUCTION

Regarding the marked asymmetric distribution of phospholipids in the erythrocyte membrane, which is characterized by the two choline containing phospholipids, sphingomyelin (SM) and phosphatidyl- choline (PC), being mainly located in the outer monolayer, whereas most of the phosphatidylethanolamine (PE) and all of the phosphatidylserine (PS) is found in the cytoplasmic leaflet (Zwaal, et.al. 1975, Op den Kamp, 1984), some important questions may be raised. These concern: (i) its constitution during biogenesis of the cell, (ii) its maintenance during the life span of the erythrocyte, and last but not least, (iii) its physiologic significance.

NATO ASI Series, Vol. H16
Membrane Biogenesis. Edited by J. A. F. Op den Kamp

BIOGENESIS OF RED CELL PHOSPHOLIPID ASYMMETRY

In an attempt to answer the first of the above mentioned questions, comparative studies have been performed on mature murine erythrocytes and Friend erythroleukaemic cells (Rawyler et. al.1985). The latters are erythroid cells, derived from susceptible mouse spleens infected with the Friend virus complex, which are blocked at an early stage of their normal differentiation pathway, presumably between the burst-forming unit erythroid and colony-forming unit erythroid stage.These cells, which closely resemble pro-erythroblasts, can only proliferate under normal growth conditions.

The phospholipid distribution in the plasma membranes of both Friend cells and mouse erythrocytes has been assessed by using three independent techniques, i.e. treatment of intact cells with phospholipases A_2 and C, fluorescamine labeling and a PC transfer protein mediated exchange procedure (Rawyler et. al. 1985,Rawyler et. al. 1984). The results of these studies are depicted in figure 1. The very same extent of asymmetry

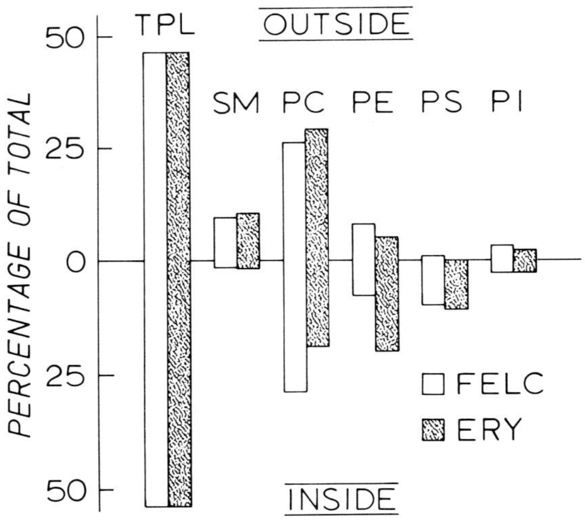

FIGURE 1
Phospholipid composition and transbilayer distribution in plasma membranes of Friend erythroleukaemic cells (open bars) and mature mouse erythrocytes (solid bars). Abbreviations: TPL, total phospholipid; SM, sphingomyelin; PC, PE, PS, and PI, phosphatidylcholine, -ethanolamine, -serine, and -inositol, respectively.

in distribution of SM in the mature erythrocyte membrane is already observed in the plasma membrane of the Friend cell. Though not yet as absolute as in the erythrocyte membrane, a highly asymmetric distribution in the plasma membrane of the precursor cell is also observed for PS. PC and PE, on the other hand, are still randomly distributed over both plasma membrane leaflets of the pro- erythroblast, so that these two phospholipids have to undergo considerable changes in their transbilayer orientation during further development of the cell.

Very recent studies (Nijhof et.al. 1986 ,Van der Schaft et. al. 1987) showed that this situation is maintained up to the normoblast stage, the last stage in development before the extrusion of the nucleus. In the reticulocyte, on the other hand, we find an asymmetric phospholipid distribution which is completely identical to that in the mature erythrocyte (Van der Schaft et. al. 1987). It has to be concluded, therefore, that abrupt changes in the transbilayer distribution of both PC and PE have to take place during the process of enucleation, or immediately thereafter. Further studies will be necessary to clearify the molecular mechanisms of these changes.

MAINTENANCE OF RED CELL PHOSPHOLIPID ASYMMETRY

At present, two independent mechanisms are believed to be responsible for maintaining phospholipid asymmetry in the erythrocyte membrane, i.e. (i) an interaction of both amino-phospholipids, PE and PS, with the membrane skeleton and (ii) an ATP-dependent system which mediates a translocation of these phospholipids from the outer to the inner membrane leaflet.

Interaction of amino-phospholipids with the membrane skeleton.

During the past decade, a great many of independent studies involving model systems, such as monomolecular lipid films and liposomes, as well as intact erythrocytes treated with oxidative reagents, such as diamide or tetrathionate, elucidated that amino-phospholipids (PE and PS) specifically interact with the membrane skeleton, most specifically its spectrin components. (See ref. 7 for a recent review). This membrane skeleton is composed of hetero-dimers of a and b spectrin molecules which, spontaneously, associate to form tetramers (but also hexa- and octamers) by head to head interactions. These complexes, in turn, are interconnected at their tail-ends by complexes of actin oligomers and the so-called band 4.1 protein. This two-dimensional protein network is coupled to the membrane via ankyrin, a protein which provides the link between the b spectrin chains and the cytoplasmic domain of band 3, one of the major integral proteins which is known to be responsible for the transport of anions through the membrane (Rothstein, 1984, Low, 1986). More recently, a second type of interaction between this skeletal network and the membrane has been proposed, namely an interaction of band 4.1 with the cytoplasmic domain of glycophorin (Anderson and Marchesi,1985), the major sialo-glycoprotein of the red cell.

Since this protein network closely underlies the cytoplasmic side of the membrane, it seems conceivable that interactions between this network and the amino-phospholipids play an important role in the stabilization of the lipid bilayer and the asymmetric localization of the amino-phospholipids herein. Indeed, perturbations of the membrane skeleton itself, or of its inter- action with the lipid bilayer, invariably result in a considerable destabilization of the latter as reflected by an accelerated phospholipid flip-flop (1Franck et. al .1982,1Franck et .al. 1983, Franck et. al. 1985).

ATP-dependent translocation of amino-phospholipids.

Very recently, another and original hypothesis regarding the maintenance of phospholipid asymmetry in the red cell membrane has been launched (Seigneuret and Devaux, 1984). Comparative studies on native and energy- deprived human erythrocytes have been performed, involving spin-labeled phospholipid analogs. These analogs, which had palmitic acid at the 1- and a 5-carbon short fatty acid, carrying a nitroxide group as spin label, at the 2-position, incorporate spontaneously into the outer membrane layer of intact human erythrocytes. Their equilibration over the two halves of the bilayer was monitored by the addition of ascorbate at timed intervals, which reduces the nitroxide group (thereby quenching its ESR signal) in those probe molecules still present in the outer monolayer. It was found that both the labeled PE and PS migrated rapidly to the inner monolayer when these lipids had been introduced into fresh erythrocytes. PC, on the other hand, largely resided in the outer leaflet; its translocation profile showing the characteristics of a (slow) passive diffusion process (Seigneuret and Devaux, 1984). Most interestingly, such a translocation profile was also observed for PE and PS when their spin labeled analogs were added to erythrocytes previously depleted of their ATP. Similar results were obtained in experiments with resealed ghosts. Appreciable translocations of PE and PS, but not of PC, in favor of the inner monolayer were observed only when the ghosts had been resealed in the presence of MgATP.

Similar experiments, involving radiolabeled species of naturally occurring glycerophospholipids and radiolabeled bovine brain SM, confirmed the above results (Tilley et. al. 1986). A non-specific lipid transfer protein from beef liver was used to introduce these long-chain diacyl glycerophospholipids and the SM into the outer membrane leaflet of intact human red cells. The transbilayer equilibration of those reporter molecules was monitored by treatment of the intact cells, at timed intervals, with phospholipase A_2 (PLA_2) to probe the glycerophospholipids and with sphingomyelinase C (SMase C) to follow the fate of the SM.

In fresh erythrocytes, about 95% of the PS accumulated in the inner monolayer within 1 h, whereas the translocation of 80% of the PE appeared to take a couple of hours more (Fig.2).The radiolabeled PC, however, reached an equilibrium after as much as 21 h, 75% of it still being present in the outer layer. Again in similarity with the earlier studies (Seigneuret and Devaux, 1984), all phospholipids largely resided in the outer membrane leaflet when inserted into cells previously depleted of their ATP. Furthermore, it is worth mentioning that, once the amino-phospholipids had reached the cytoplasmic membrane leaflet in fresh erythrocytes, subsequent depletion of the cells of their ATP did not result in their retranslocation back to the outer monolayer (Tilley et. al. 1986).

These observations do not seem to leave much room for any other conclusion than that the translocation of PE and PS from the outer to the inner leaflet of the red cell membrane is an energy dependent process, a process which plays a major role in maintaining the asymmetric distribution of phospholipids in this membrane (Seigenueret and Devaux, 1984, Tilley et. al. 1986, Zachowski et. al. ,1985, Daleke and Huestis, 1985, Zachowski et. al. 1986, Franck et. al. 1983).

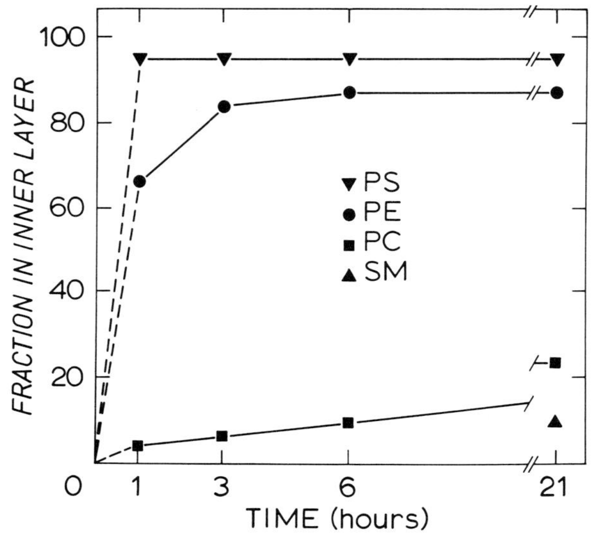

FIGURE 2
Transbilayer migration of radiolabeled dioleoyl glycero- phospholipids and bovine brain SM, previously inserted into the outer membrane layer of fresh intact human erythrocytes, using a non-specific lipid-transfer protein. Abbreviations: TPL, total phospholipid; SM, sphingomyelin; PC, PE, PS, and PI, phosphatidylcholine, -ethanolamine, -serine, and -inositol, respectively.

Phospholipid asymmetry and transbilayer mobility in sickle cells.

Reversibly sicklable cells (RSCs) provide a unique model system to gain further insight into the above discussed two potential mechanisms that control phospholipid asymmetry and lipid bilayer stability in the red cell membrane, since the effects of both ATP-depletion and uncoupling of the lipid bilayer from the membrane skeleton, can be studied in the same cell.

Using the technique outlined in the preceeding chapter, and involving the PC-specific transfer protein from beef liver to introduce trace amounts of radiolabeled egg PC into the outer membrane layer of intact cells, we determined the rate of PC flip-flop in RSCs under oxy- and deoxygenating conditions (Franck et. al. 1983). The results of those studies are depicted in figure 3, showing the decline in relative specific radioactivity of the PC in the outer

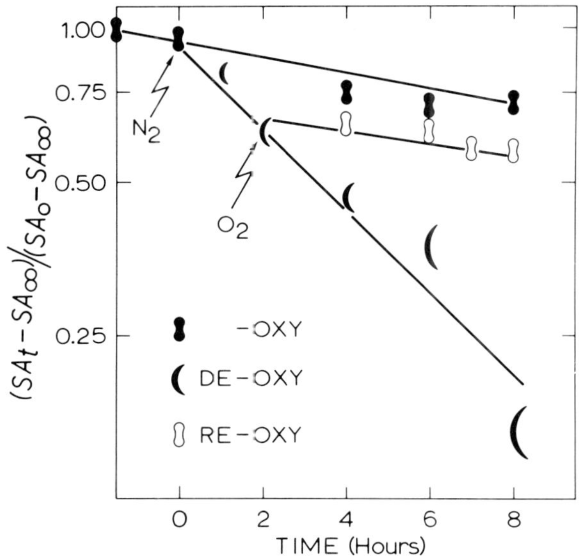

FIGURE 3
Time-dependent changes in relative specific radioactivity, calculated as $(SA_t\text{-}SA_{\bullet})/(SA_0\text{-}SA_{\bullet})$, of PC in the outer monolayer (detected as lyso-PC after non-lytic treatment of the cells with phospholipase A_2) of reversibly sickable cells (RCSs). Discoid RSCs were incubated under room air (-OXY). At zero time, part of this suspension was deoxygenated under nitrogen (N_2) which induced the cells to sickle (DE-OXY). After two hours under nitrogen, part of the deoxygenated suspension was briefly exposed to pure oxygen (O_2) which induced the cells to re-adopt their discoid shape immediately, and the incubation continued under room air (RE-OXY). Half-times for transbilayer movements of PC are calculated from the slopes of the lines depicted in this figure. SA_0, SA_t, and $SA_{\bullet}$, specific radio-activity of the lyso-PC at zero time, time t and infinite time, respectively.

monolayer, plotted semilogarithmically versus the time of incubation of the cells at 37°C. From the slope of the line for oxygenated (discoid) RSCs (Fig.3), a half-time value for the transbilayer movement of PC of approximately 20 h can be determined, which is very similar to the value found for normal erythrocytes under the same conditions (compare preceeding chapter). Incubation of the RSCs under nitrogen, which induces sickling of these cells, causes an accelerated decrease in relative specific radioactivity of the PC in the outer monolayer, revealing a half-time value of 3.5 h. This means that the process of PC flip-flop in deoxygenated (sickled) RSCs is six times as fast as that in oxygenated (discoid) RSCs. It is of particular interest to note that, when the deoxygenated cells are rapidly unsickled by a brief exposure to pure oxygen and subsequent incubation under room air, the accelerated flip-flop of PC is immediately reduced to its original (normal) low level (Fig.3). This clearly indicates that, not only the change in cell morphology, but also the change in transbilayer mobility of PC in the RSC membrane is completely reversible (Franck et. al. 1983).

It was recently observed that the characteristic elongated spicules, which protrude from the sickled RSC, can be released from the cell when it is subjected to a series of sickling and unsickling cycles by repetitive deoxygenation and reoxygenation (Allan et. al. 1982, Franck et. al 1985). Again using the above technique, it appeared that the rate of PC flip-flop in the membrane of the deoxygenated despiculated (remnant) RSC was much slower than that in the membrane of the deoxygenated native RSC, and in fact not appreciably different from the situation in oxygenated RSCs or normal erythrocytes (Franck et. al 1985). This suggests that the abnormality observed in the sickled RSC might be confined to those areas of the membrane that are in spicular form. As it was also known that the free spicules, that can be released from the RSCs, are essentially devoid of membrane skeletal proteins (Allan et. al. 1982, Franck et. al. 1985), the following hypothesis was developed (Fig.4). In those areas of the membrane that are in spicular form, the interaction between the phospholipids in the inner membrane leaflet and the membrane skeleton is broken by protrusion of the bundles of polymerized haemoglobin-S through gaps in this skeletal network (Opden Kamp et. al. 1986). Since in those spicules the amino-phospholipids no longer experience the constraints of the membrane skeleton, they may experience an enhanced freedom to move. It seems plausible that this is in fact the primary phenomenon that is sensed by the PC molecules which, in turn, gives rise to their accelerated transbilayer movement (Franck et. al. 1983,Franck et. al 1985).

The broken interaction of PS with the membrane skeleton may allow a translocation of this phospholipid to the outer monolayer. Indeed, the exposure of PS on the outer surface of the membrane of the sickled RSC has been detected by treatment of the intact cells with

PLA_2 (Lubin et. al. 1981). However, it was noted very recently that this probe is not very suitable to be used on cells which have a destabilized lipid bilayer (which obviously is the case in the sickled RSC), as its action as such may induce transbilayer reorientations of substrates under those conditions (Franck et. al 1986). As an alternative, the so-called prothrombinase assay was used. This assay system, which is based on the conversion of prothrombin into thrombin and comprises the purified clotting factors Xa, Va, prothrombin and Ca^{2+}, has been shown to be a very sensitive, highly specific and non-invasive method to detect the presence of PS in a lipid or membrane surface (Franck et. al. 1985, Franck et. al. 1986)

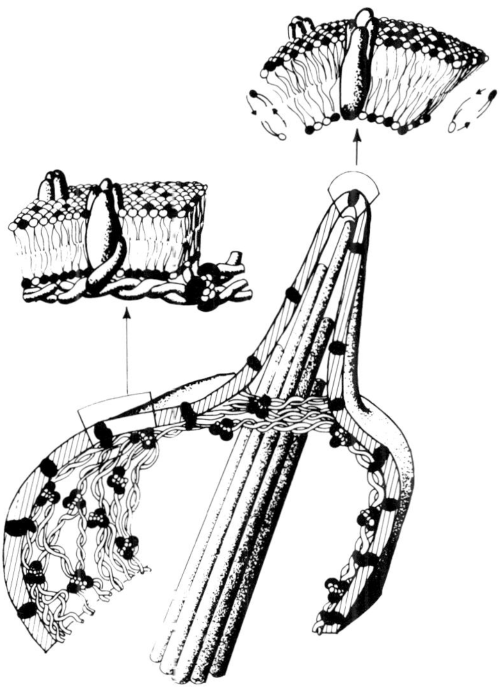

SPICULE FORMATION IN SICKLED ERYTHROCYTES

FIGURE 4
Spicule formation in sickled erythrocytes. During the process of sickling, haemoglobin S molecules polymerize into fibers and align with other fibres to bundles. These bundles of polymerized haemoglobin protrude through the cytoskeleton, thereby breaking the interaction between the membrane bilayer and the membrane cytoskeleton, which results in the formation of the long, terminal spicules of the sickled erythrocyte. In those areas of the membrane where the bilayer and cytoskeleton are uncoupled, the phospholipids probably no longer experience the constraints of the cytoskeleton and an enhanced transbilayer mobility may ultimately occur, eventually resulting in an altered distribution of the phospholipids over both halves of the bilayer.

This assay system was used to determine the presence of PS in the outer membrane leaflet of RSCs, incubated under nitrogen to induce their sickled morphology (Middelkoop et. al. 1987). The response of fresh (ATP containing) RSCs in this assay system was rather low and did not change during incubation of the cells under nitrogen for up to 7 h (Fig.5). This indicates that only a very small fraction of the PS

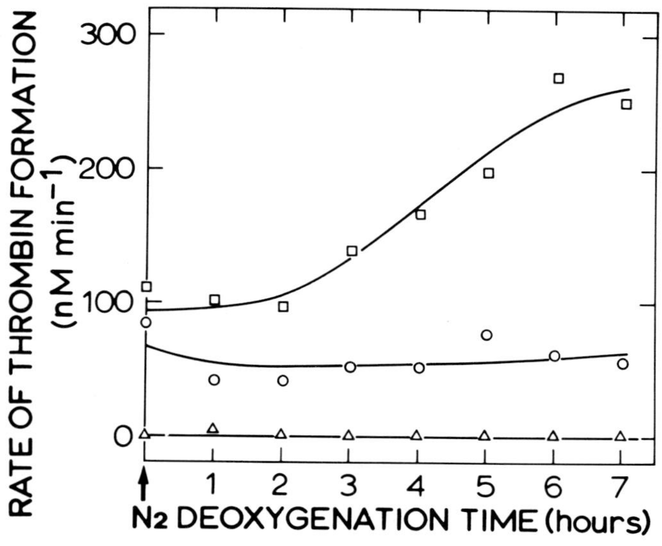

FIGURE 5
Effect of deoxygenation on the prothrombinase activity of fresh (o) and ATP-depleted (□) RSCs, as well as of ATP-depleted normal erythrocytes (Δ). Suspensions of cells were incubated under nitrogen and, at the time points indicated, samples were taken and the prothrombinase activity of the cells determined. Fresh and ATP-depleted RSCs have been derived from the same blood sample. The figure shows the results of a typical experiment.

was present in the outer layer of the sickled cell, and probably only in those areas of the membrane that are in spicular form. However, when those cells had been depleted of their ATP, prior to sickling, the rate of thrombin formation increased considerably with the time the cells had been incubated under nitrogen (Fig.5). This clearly points to a time-dependent increase in the amount of PS in the outer membrane leaflet of the ATP-depleted sickled cell. Incubation of ATP-depleted normal erythrocytes under nitrogen, on the other hand, did not lead to the formation of thrombin in this assay (Fig.5), indicating that all of the PS remained in the inner membrane leaflet.

The interpretation of these results is as follows: In those area's of the membrane of the sickled cell that are in spicular form, PS no longer experiences the constraints of the membrane skeleton and can be translocated to the outer monolayer. Due to rapid lateral movements, it will be spread out over the entire outer membrane leaflet. When the cell still contains ATP, PS will be retranslocated to the inner monolayer and will be retained in this layer in those areas of the membrane where the interaction with the membrane skeleton is preserved. The areas of the membrane that are in spicular form only account for a few % of the total membrane surface, which explains why in the ATP-containing sickled cell only a small fraction of the PS is exposed on the outer surface and that this fraction does not change in time (Fig. 5). In the ATP-depleted sickled cell, however, the ATP-dependent translocase cannot operate, which means that the PS once translocated to the outer monolayer will remain in that leaflet. Since the PS that has disappeared from the inner monolayer of the spicular membrane will be replenished from the surrounding areas of that layer, the translocation of PS from the inner to the outer layer will continue. This causes a time-dependent increase in the amount of PS in the outer monolayer, until the distribution reaches an equilibrium (Fig.5). This interpretation is in full agreement with the observation that ATP-depleted normal erythrocytes show an absolutely negative response in the prothrombinase assay upon deoxygenation. In those cells, the interaction between the lipid bilayer and the membrane skeleton is completely preserved, retaining all of the PS in the inner monolayer.

CONCLUSIONS

From all of the above studies, it can be concluded that the asymmetric distribution of phospholipids in the erythrocyte membrane is maintained by both an ATP-dependent translocase cannot operate, which means that the PS once translocated to the outer monolayer will remain in that leaflet. Since the PS that has disappeared from the inner monolayer of the spicular membrane will be replenished from the surrounding areas of that layer, the translocation of PS from the inner to the outer layer will continue. This causes a time-dependent increase in the amount of PS in the outer monolayer, until the distribution reaches an equilibrium (Fig.5). This interpretation is in full agreement with the observation that ATP-depleted normal erythrocytes show an absolutely negative response in the prothrombinase assay upon deoxygenation. In those cells, the interaction between the lipid bilayer and the membrane skeleton is completely preserved, retaining all of the PS in the inner monolayer.

CONCLUSIONS

From all of the above studies, it can be concluded that the asymmetric distribution of phospholipids in the erythrocyte membrane is maintained by both an ATP-dependent translocation of the amino-phospholipids towards the cytoplasmic half of the lipid bilayer and by an interaction of these phospholipids with the membrane skeleton. The latter provides a major contribution to the transversal stabilization of the lipid bilayer, as was also evidenced by recent studies on hereditary pyropoikilocytes (Franck et. al. 1985).

Regarding the procoagulant activity that the presence of PS generates in a membrane surface, the physiological significance of its exclusive localization in the cytoplasmic half of the membrane will be obvious.

REFERENCES

Allan, D, Limbrick, A R, Thomas, P, Westerman, M P. Release of spectrin-free spicules on reoxygenation of sickled erythrocytes. Nature, 295: 612-613, 1982.

Anderson, R A, Marchesi, VT. Regulation of the association of membrane skeletal protein 4.1 with glycophorin by a poly- phosphoinositide. Nature, 318: 295-298, 1985.

Bevers, E M, Comfurius, P, van Rijn, J L M L, Hemker, H C, Zwaal, R F A. Generation of prothrombin-converting activity and the exposure of phosphatidylserine at the outer surface of platelets. Eur. J. Biochem. 122: 429-436, 1982.

Daleke, D L, Huestis, W H. Incorporation and translocation of amino-phospholipids in human erythrocytes. Biochemistry, 24: 5406-5416, 1985.

Franck, P F H, Roelofsen, B, Op den Kamp, J A F. Complete exchange of phosphatidylcholine from intact erythrocytes after protein cross-linking. Biochim.Biophys. Acta, 687: 105-108, 1982.

Franck, P F H, Chui, D T-Y, Op den Kamp, J A F, Lubin, B, van Deenen, L L M, Roelofsen, B. Accelerated transbilayer movement of phospha- tidylcholine in sickled erythrocytes. A reversible process. J. Biol. Chem. 258: 8435-8442, 1983.

Franck, P F H, Bevers,EM, Lubin, B H, Comfurius, P, Chiu, D T Y, Op den Kamp, J A F, Zwaal, R F A, van Deenen,L L M, Roelofsen, B. Uncoupling of the membrane skeleton from the lipid bilayer. The cause of accelerated phospholipid flip-flop leading to an enhanced procoagulant activity of sickled cells. J. Clin. Invest. 75: 183-190, 1985.

Franck, P F H, Op den Kamp, J A F, Roelofsen, B, van Deenen, L L M. Does diamide treatment of intact human erythrocytes cause a loss of phospholipid asymmetry? Biochim. Biophys. Acta. 857: 75-84, 1986.

Franck, P F H, Op den Kamp, J A F, Lubin, B, Berendsen, W, Joosten, P, Briët, E, van Deenen, L L M, Roelofsen, B. Abnormal transbilayer mobility of phosphatidylcholine in hereditary pyropoikilocytosis reflects the increased heat sensitivity of the membrane skeleton.Biochim.Biophys. Acta, 815: 259-267, 1985.

Haest , C W M. Interactions between membrane skeletal proteins and the intrinsic domain of the erythrocyte membrane. Biochim. Biophys. Acta, 694: 331-352, 1982.

Low, P S. Structure and function of the cytoplasmic domain of band 3: center of erythrocyte membrane - peripheral protein interactions. Biochim. Biophys. Acta, 864: 145-168, 1986.

Lubin, B, Chui, D, Bastacky, J, Roelofsen, B, van Deenen, L L M. Abnormalities in membrane phospholipid organization in sickled erythrocytes. J. Clin. Invest, 67: 1643-1649, 1981.

Middelkoop, E, Lubin, B H, Bevers, E M, Op den Kamp, J A F, Comfurius, P, Chui, D T-Y, Zwaal, R F A, van Deenen, L L M, Roelofsen, B. Studies on sickled erythrocytes provide evidence that the asymmetric distribution of phosphatidylserine in the red cell membrane is maintained by both ATP-dependent translocation and interaction with membrane skeletal proteins. Blood, manuscript submitted, 1987.

Nijhof, W, van der Schaft, P H, Wierenga, P K, Roelofsen, B, Op den Kamp, J A F, van Deenen, L L M. The transbilayer distribution of phosphatidylethanolamine in erythroid plasma membranes during erythropoiesis. Biochim. Biophys.Acta, 862: 273 -277, 1986.

Op den Kamp, J A F. Lipid asymmetry in membranes. Annu.Rev. Biochem. 48: 47-71, 1979.

Op den Kamp, J A F, Roelofsen, B, van Deenen, L L M. Structural and dynamic aspects of phosphatidylcholine in the human erythrocyte membrane. Trends Biochem. Sci. 10: 320-323, 1985.

Rawyler , A, van der Schaft, P H, Roelofsen, B, Op den Kamp, J A F. Phospholipid localization in the plasma membrane of Friend erythroleukaemic cells and mouse erythrocytes, Biochemistry, 24: 1777-1783, 1985.

Rawyler, A, Roelofsen, B, Op den Kamp, J A F. The use of fluorescamine as a permeant probe to localize phosphatidyl- ethanolamine in intact Friend erythroleukaemic cells. Biochim. Biophys. Acta, 769: 330-336, 1984.273-277, 1986.

Rothstein, A. The functional architecture of band 3, the anion transport protein of the red cell membrane, Can. J. Biochem. Cell Biol. 62: 1198-1204, 1984.

van der Schaft, P H, Roelofsen, B, Op den Kamp, J A F, van Deenen, L L M. Phospholipid asymmetry during erythropoiesis. A study on Friend erythroleukaemic cells and mouse reticulocytes. Biochim. Biophys. Acta, 900: 103-115, 1987

Seigneuret, M, Devaux, P F. ATP-dependent asymmetric distribution of spin-labeled phospholipids in the erythrocyte membrane: Relation to shape changes. Proc. Natl. Acad. Sci. USA, 81: 3751-3755, 1984.

Tilley, L, Cribier, S, Roelofsen, B, Op den Kamp, J A F, van Deenen, L L M. ATP-dependent translocation of amino phospholipids across the human erythrocyte membrane. FEBS Lett. 194: 21-27, 1986.

Zachowski ,A, Fellman, P, Devaux, P F. Absence of transbilayer diffusion of spin-labeled sphingomyelin in human erythrocytes. Comparison with the diffusion of several spin-labeled glycero- phospholipids. Biochim. Biophys. Acta, 815: 510-514, 1985.

Zachowski, A, Favre, E, Cribier, S, Hervé, P, Devaux, P F. Outside-inside translocation of amino-phospholipids in the human erythrocyte membrane is mediated by a specific enzyme. Biochemistry, 25: 2585-2590, 1986.

Zwaal,R F A, Roelofsen, B, Comfurius, P, van Deenen , L L M. Organization of phospholipids in human red cell membranes as detected by the action of various purified phospholipases. Biochim. Biophys. Acta, 406: 83-96, 1975.

LIPID LOCALIZATION AND MOBILITY IN THE PLASMALEMMA OF AORTIC ENDOTHELIAL CELLS ARE REVERSIBLY AFFECTED BY THE PRESENCE OF CELL JUNCTIONS

J.F. Tournier, A. Lopez, N. Gas, J.F. Tocanne
Centre de Recherche de Biochimie et de Génétique Cellulaires du CNRS
118, route de Narbonne
31062 Toulouse Cedex
FRANCE

ABSTRACT :

Bovine aortic endothelium can form, in vitro, a confluent monolayer of small and strictly non-overlapping cells which retain the differentiated properties of their in vivo counterparts (Vlodavsky et al. 1979). The localization and the dynamic properties of the lipophilic probe 5N-(hexadecanoyl)-aminofluorescein (HEDAF) was observed to be markedly influenced by the presence of cell junctions. The probe was found to be restricted to the apical plasmalemma by junctional complexes. The probe was only observed in the basal membrane after disruption of cell conctacts. Fluorescence Recovery After Photobleaching (FRAP) experiments indicated that the lateral diffusion coefficient D of the probe on apical membranes fell when the cell junctions were disrupted, but returned to its initial value when cell contacts were restored.

INTRODUCTION :

The endothelium of large vessels is composed of a single layer of closely connected, highly flattened cells. Their apical surfaces face the blood stream, while their basal poles face the tissues. Three main environmental factors can affect the lateral mobility of lipids in the membranes of these cells : the cellular substrate (i.e., the extracellular matrix), blood-borne factors which can interact directly with the apical poles of endothelial cells, and cell-cell interactions. Membrane phospholipid mobility in endothelial cells has been found to depend on the cell substrate (Nakache et al. 1985). However, little is known about the

NATO ASI Series, Vol. H16
Membrane Biogenesis. Edited by J.A.F. Op den Kamp

influence of other environmental factors. This study was designed to investigate the role of cell junctional complexes on the localization and lateral mobility of HEDAF in the apical and basal membranes of endothelial cells.

MATERIALS AND METHODS :

Culture conditions and fluorescence labelling of endothelial cells :

Vascular endothelial cell cultures were established and cloned from ox aortic arch according to Gospodarowicz et al. (1976). Cells were grown at 37°C in a 5 % CO_2 atmosphere in Dulbecco's modified Eagle medium (Gibco) supplemented with 10 % heat-inactivated calf serum (Seromed). Each week, the monolayers were dissociated with 0.05 % trypsin, 0.02 % EDTA in phosphate buffered saline (PBS, STV Seromed), and then subcultured at a split ratio of 1/64 on glass coverslips. 2 ng of basic fibroblast growth factor per ml of culture medium (Bolhen et al. 1984) were added every other day until the cells reached confluence. In all experiments, we used cells 4 days after confluence (10 days of subculture) between passages 4 to 28. At this stage, the cells in the monolayers are strictly contact-inhibited. They rest on their own extracellular matrix, and form tight junctions (Fig. 1a,b). HEDAF (Molecular Probes), dissolved in dimethyl formamide (5 mM), was added to the cell cultures in PBS at final probe and solvent concentrations of 3 uM and 0.6 % v/v respectively. Contact was maintained for 5 min at 20°C, the cells were then washed twice with PBS. This procedure did not affect the viability of endothelial cells (not shown), and led to a typical fluorescence pattern, characteristic of plasma membrane labelling, for the duration of the measurements (30 min).

Fluorescence recovery after photobleaching :

FRAP experiments were carried out using an experimental set-up described elsewhere (Dupou et al. 1987). Briefly, the apparatus consisted of an epifluorescence microscope (Leitz Ortholux II), equipped with a

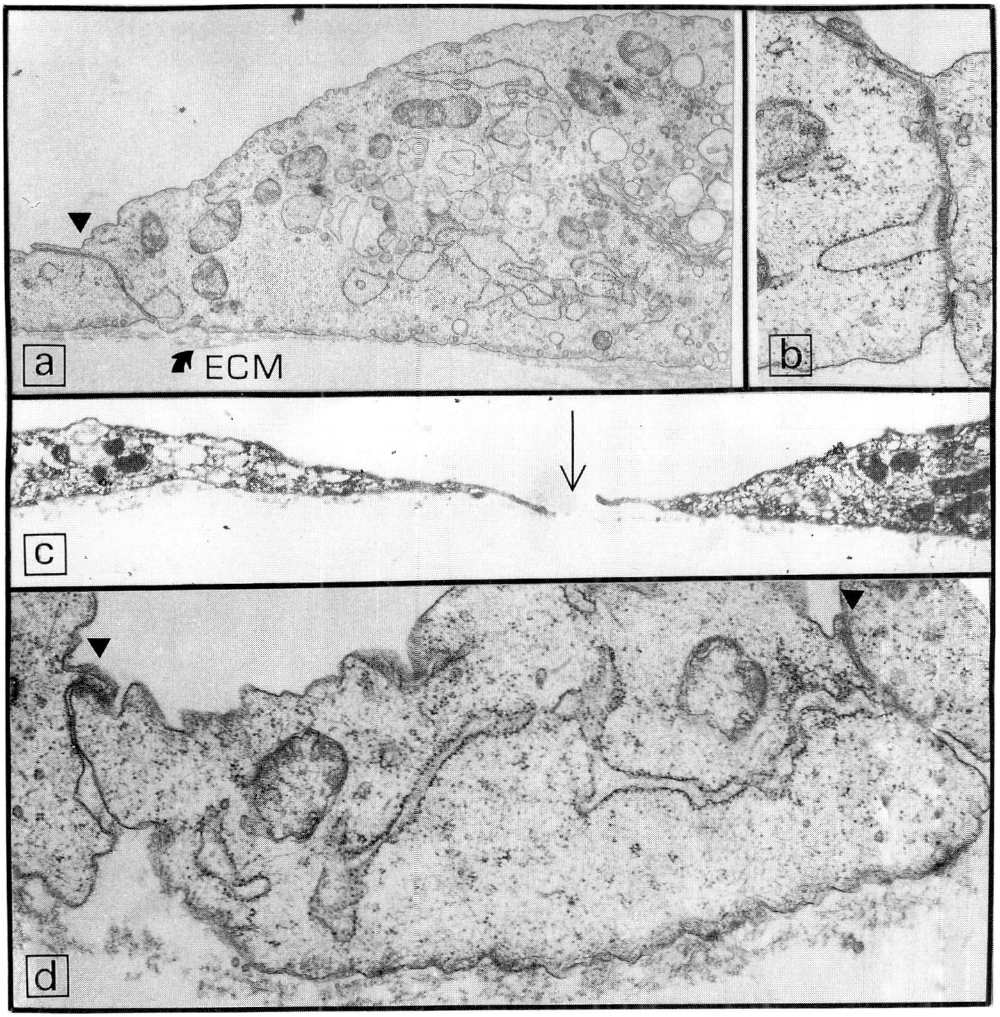

Fig. 1 : Electron micrographs of cross-sections of confluent bovine endothelial cells in the presence and absence of junctional complexes.

a) Vascular endothelium forms a single cell layer, sealed by tight junctions (arrowhead). Cells are thickened in the vicinity of the nucleus, and are hihly flattened at their peripheries (10000X). These polarized cells rest on their own extracellular matrix (ECM, arrow).

b) High magnification view of a junctional site (30000X).

c) The junctional complexes are disrupted by divalent cation depletion. Arrow indicates the free intercellular space (12000X).

d) After their disruption as in c, the typical junctional complexes (arrowheads) are restored by addition of PBS containing Ca^{++} and Mg^{++} (27000X).

conventional mercury arc lamp (Osram HBO 100 W/2). The excitation spot was 3.9 μm in diameter. In order to select the fluorescence signal originating from the illuminated area, a field diaphragm (calibrated pinhole) was interposed in the emission beam, in the conjugated image plane, just in front of the photomultiplier tube (EMI955B, USA). Photobleaching was carried out for a fixed time (approx. 400 ms) in order to control its extent (20 to 45 % of the initial fluorescence intensity). Fluorescence recovery was then monitored stepwise (6 steps) by measurement of the fluorescence intensity over short intervals (5 ms, controlled by a fast electronic shutter) after regular periods in the dark. The mean fluorescence intensity for each set of recovery times was used for the calculation of D, using the equation described by Soumpasis for $t/\tau > 0.1$, and the asymptotic value for $t/\tau > 0.1$ (Soumpasis, 1983).

RESULTS :

Localization of the probe on the surface of endothelial cells :

Measurement of the fluorescence intensity on labelled cells in various focal planes, showed that only the apical surfaces retained the stain. No fluorescence staining was observed at the basal poles (Fig. 2). In order to find out if this uneven distribution was dependent on junctional complexes, confluent cells were disjoined by exposing the cultures to PBS without divalent cations (PBS Ca^{++}, Mg^{++} free ; Gonzales-Mariscal et al. 1985 ; Pitelka et al. 1983) for 10 min at 20°C, and then labelled with HEDAF. As shown in Fig. 2, the fluorescence intensity appeared to be equally distributed on both apical and basal plasmalemma after disruption of junctional complexes (Fig. 1c). These results extend previous observations indicating that cell-cell contacts (probably tight junctions) act as a kind of fence. They appear to regionalize the diffusion of at least some membrane lipids to the apical plasmalemma of cells of epithelial origin (Dragsten et al. 1982 ; Spiegel et al. 1985 ; Van Meer et al. 1986). A further set of experiments was designed to find out whether the lateral mobility of the probe at the apical poles was affected by the tight contact of the cells in confluent cultures.

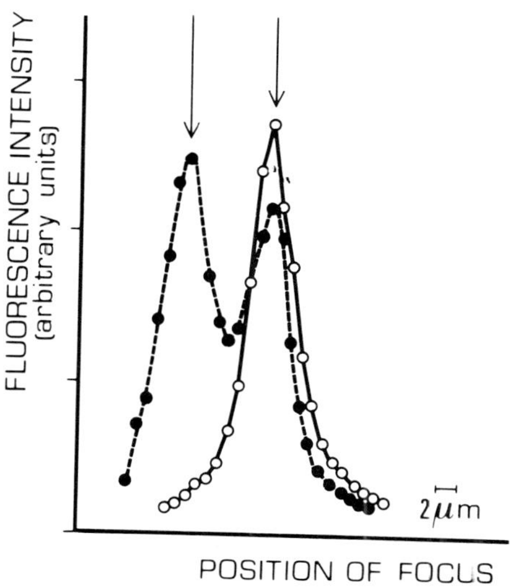

Fig. 2 : Fluorescence intensity profile from the plasma membrane as a function of the position of the focus through confluent cells, in the presence or absence of junctions.
The incident beam was used to excite fluorescence from a small spot (3.9 μm in diameter). The fluorescence intensity (arbitrary units) was recorded as the microscope focus was scanned in steps (1 μm) across the center of one cell.
(O——O) : confluent culture ; only the apical membrane was labelled.
(●– – –●) : confluent culture labelled after exposure to PBS without divalent cations ; the probe was distributed uniformly over the apical and basal plasmalemma.
The left peak is from the basal membrane, and the one on the right from the apical membrane (arrows).

Lateral diffusion coefficient of HEDAF in the plasma membranes of confluent cultures in the presence or absence of junctional complexes :

100 % fluorescence recovery was found in all the experiments, indicating that the probe was free to diffuse in the plane of the endothelial plasma membranes. The mean D values and their corresponding statistical distributions are shown in Table 1 and Fig. 3 respectively. A mean D value of 2.8 x 10^{-9} cm^2/sec (Fig. 3) was determined in FRAP experiments carried out on the apical membranes of a large number of confluent cells (Fig. 1a). When cell junctions were disrupted (Fig. 1c), the calculated diffusion coefficient fell to a mean value of 2 x 10^{-9} cm^2/s (Fig. 3). This value was identical to that measured when the probe was located in the basolateral membranes (Fig. 3). The restoration of cell-cell contacts by addition of divalent cations (Ca^{++} and Mg^{++}) induced the cells to regain their initial arrangement (Fig. 1d), with a mean D coefficient on apical membranes of 2.7 x 10^{-9} cm^2/s (Fig. 3). These effects on lateral mobility appeared to depend on cell-cell contacts, since differences in D coefficients were not observed in control cultures in the exponential growth phase (sparse cells without junctional complexes) in the presence or absence of divalent cations (not shown).

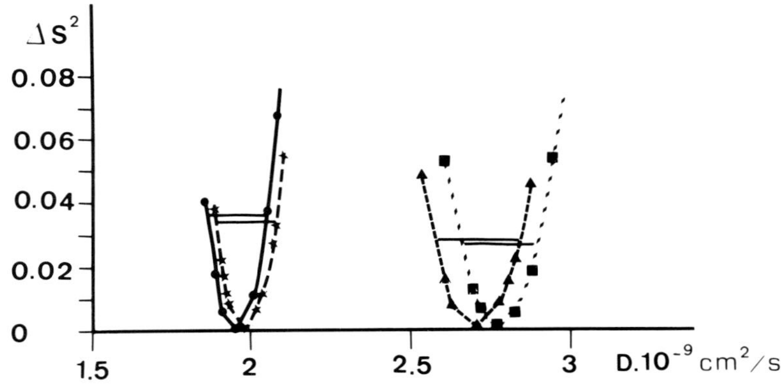

Fig. 3 : Plot of the normalized sum $\Delta S^2 = S^2Dj - S^2Dsol$ as a function of the lateral diffusion coefficient D for HEDAF :
a) on the apical plasmalemma of confluent endothelial cells (■-■).
b) on the apical plasmalemma after disruption of cell junctions by exposure of the cells to PBS without divalent cations (10 min at 20°C) (●—●).
c) on basal membranes after disruption of the junctional complexes as in b (★— —★).
d) on apical membranes after disruption (as in b) and restoration of cell contacts by incubation of the cells in PBS with Ca^{++} and Mg^{++}, 1 hour at 37°C (▲····▲).
In this expression, S^2Dj and S^2Dsol are defined as the sum of the quadratic deviations between the observed and calculated fluorescence recovery values for a given value Dj of the lateral diffusion coefficient D and the value for the solution Dsol. From these curves, and after computer calculation of Dsol values using a minimization program, the corresponding minimum (Dmin) and maximum (Dmax) values of Dsol were determined from Snedecar-Fisher's law, at a 5 % level of uncertainty (——).

DISCUSSION :

The results showed that a particular structural feature of these endothelial cells, i.e. the presence of cell junctions, influenced both the localization and the dynamics of a lipophilic probe in the plasma membrane. Cell junctions appeared to affect the localization of HEDAF on apical membranes of endothelial cells in a similar fashion to that described for epithelial cells (Dragsten et al. 1982 ; Spiegel et al. 1985 ; Van Meer et al. 1986). However, less is known about the effects of cell junctions on lateral diffusion. Interestingly, the HEDAF probe diffused faster in the presence of cell junctions than when they were disrupted. The basal and apical poles of isolated, confluent cells had the same D values. This value was similar to that found for the basal poles of non-confluent cells plated on extracellular matrix-coated dishes (unpublished results). Furthermore, the D value measured after resealing of the junctions was identical to that found prior to disruption, indicating that the process is reversible. The D value therefore appears to reflect the organization of the membrane with respect to cell contacts. This may be related to a redistribution of membrane proteins and/or lipids. For example, it has recently been shown that protein distribution over the cell surface is affected by cell-cell interactions (Vega-Salas et al. 1987). Moreover, the D value of lipids in cell membranes has been found to depend on the distribution, composition and relative mobility of membrane-bound proteins (O'Leary, 1987). In our experiments, we found similar D values on basal and apical poles of endothelial cells. This is in disagreement with the results of Makache et al. (1986) who observed a large difference in D values between the two poles. Further studies will be required to elucidate this discrepancy.

Table 1. Lateral diffusion coefficients of the lipophilic probe HEDAF in the plasma membrane of confluent endothelial cells in the presence and absence of junctional complexes.

Membrane studied	D (10^{-9} cm^2/s)		SD	No of recovery values
Apical on confluent cells	2.77	MIN 2,66 MAX 2,88	0.08	453
Apical on confluent cells without junctions	1.95	MIN 1,86 MAX 2,06	0.09	431
Basal on confluent cells without junctions	1.98	MIN 1,88 MAX 2,08	0.09	408
Apical on confluent cells after the resealing of junctional complexes	2.70	MIN 2,58 MAX 2,82	0.08	364

Dmin and Dmax were evaluated at a 95 % level of confidence (see caption to Fig. 3).

REFERENCES :

Bolhen P, Baird DA, Esch F, Ling GN, Gospodarowicz D (1986) Partial molecular characterization of pituitary fibroblast growth factor. Proc Natl Acad Sci USA 81 : 5364-5368.

Dragsten PR, Handler JS, Blumenthal R (1982) Fluorescent membrane probes and the mechanism of maintenance of cellular asymmetry in epithelia. Fed Proc 41 : 48-53.

Dupou L, Gualandris L, Lopez A, Duprat AM, Tocanne JF (1987) Alterations in lateral lipid mobility in the plasma membrane of urodelean ectodermal cells during gastrulation. Exp Cell Res 169 : 502-513.

Gonzales-Mariscal L, Chavez de Ramirez B, Cereijido M (1985) Tight junction formation in cultured epithelial cells (MDCK). J. Membrane Biol 86 : 113-125.

Gospodarowicz D, Moran J, Braun D, Birdwell C (1976) Clonal growth of bovine endothelial cells : fibroblast growth factor as a survival agent. Proc Natl Acad Sci USA 73 : 4120-4124.

Nakache M, Schreiber AB, Gaub H, Mc Connel HM (1985) Heterogeneity of membrane phospholipid mobility in endothelial cells depends on cell substrate. Nature 317 : 75-77.

O'Leary TJ (1987) Lateral diffusion of lipids in complex biological membranes. Proc Natl Acad Sci USA 84 : 429-433.

Pitelka DR, Taggart BN, Hamamoto ST (1983) Effects of extracellular calcium depletion on membrane topography and occluding junctions of mammary epithelial cells in culture. J Cell Biol 9 : 613-624.

Soumpasis DM (1983) Theoretical analysis of fluorescence recovery photobleaching experiments. Biophys J 41 : 95-97.

Spiegel S, Blumenthal R, Fishman PH, Handler JS (1985) Gangliosides do not move from apical to basolateral plasma membrane in cultured epithelial cells. Biochem Biophys Acta 821 : 310-318.

Van Meer G, Simons K (1986) The function of tight junctions in maintening differences in lipid composition between the apical and the basolateral cell surface domains of MDCK cells. EMBO J 5 : 1455-1464.

Vega-Salas DE, Salas PJI, Gundersen D, Rodriguez-Boulan E (1987) Formation of the apical pole of epithelial (Madin-Darby Canine Kidney) cells : polarity of an apical protein is independant of tight junctions while segregation of a basolateral marker requires cell-cell interactions. J Cell Biol 104 : 905-916.

Vlodavsky I, Gospodarowicz D (1979) Structural and functional alterations in the surface of vascular endothelial cells associated with the formation of a confluent cell monolayer and with the withdrawal of fibroblast growth factor. J Supramol Struct 12 : 73-114.

PHOSPHOLIPID DYNAMICS IN MEMBRANE BIOGENESIS IN HEPATOCYTES

Joan A.Higgins
Department of Biochemistry,
University of Sheffield,
Sheffield, S10 2 TN
Great Britain

The major site of synthesis of phospholipids in hepatocytes is the endoplasmic reticulum. Newly synthesised phospholipids may be retained in the endoplasmic reticulum membrane, sequestered in the cisternal space of the endoplasmic reticulum or Golgi membranes for secretion as plasma lipoproteins, or transfered to other membranes in the cell. Thus, hepatocytes have a major sorting problem in directing phospholipids to their final destination. This is comparable in concept to the sorting of proteins after synthesis by bound or free ribosomes. However, proteins are large molecules and carry information in their structure for their own glycosylation and for "address tags" for their final destination. The most well understood mechanism of protein sorting is the role of mannose-6-phosphate labelled oligosaccharides in directing the lysosomal proteins to the appropriate morphological site. In contrast, phospholipids are small molecules and they have no absolute distribution in cells. For example, phosphatidylcholine is a component of all membranes, occurs in each bilayer of membranes and is also a major component of the secreted plasma lipoproteins.

NATO ASI Series, Vol. H16
Membrane Biogenesis. Edited by J. A. F. Op den Kamp

Thus it is unlikely that information for the destination of phosphatidylcholine resides in the molecules themselves. However, there must exist in cells mechanisms which ensure that the composition of membranes and the two dimensional and transverse distribution of the phospholipids in the membrane bilayer is regulated.

As the endoplasmic reticulum is the site of synthesis of phospholipids in the hepatocyte understanding the mechanism of biogenesis of ths membrane is central to understanding the sorting of phospholipids in the cell.

DISTRIBUTION OF PHOSPHOLIPIDS IN THE ENDOPLASMIC RETICULUM MEMBRANE

Phospholipids are distributed asymmetrically in the endoplasmic reticulum membrane (Higgins and Dawson, 1977; Bollen and Higgins,1980; Higgins, 1982; and Higgins and Pigott, 1982). 70% of the phosphatidylcholine is in the outer leaflet and 60% of the phosphatidylethanolamine is in the inner leaflet of the bilayer. Sphingomyelin has a similar distribution to phosphatidylcholine and phosphatidylserine to phosphatidylethanolamine. This distribution has been demonstrated using phospholipase C, phospholipase D or using trinitrobenzene sulphonate (TNBS), which reacts with the aminophospholipids, to differentiate between the phospholipids of the outer and inner leaflets. The validity of such probes has been questioned because they produce changes in the bilayer components which might cause alterations in the bilayer structure (Op den Kamp, 1979;Van Meer, 1986). This has been discussed in more detail elsewhere but a number of pieces of evidence support the use of phospholipids or TNBS as probes (Higgins, 1985; Higgins and Hutson, 1986)

(i) Microsomal vesicles remain closed during experimental procedures and the bilayer structure is retained shown by electron microscopy using both sections or freeze fracture replicas.

(ii). When the vesicles are opened using a variety of physical or chemical methods the phospholipids behave as a single pool. Differential availability of phospholipids to probes in intact vesicles is therefore not a consequence of the specificity of the probe.

(iii) Similar results of the transverse distribution of the phospholipid are obtained using different probes. The pool of phosphatidylethanolamine available to TNBS or phospholipase C is the same in intact vesicles.

These observations suggest that the probes used do not cause a rearrangement of the membrane bilayer. This is also suggested by the studies of the biogenesis of the bilayer described below. These results would be difficult to explain if rearrangement of the bilayer occurred. It is possible that the bilayer structure is stabilized by peripheral membrane proteins. Such a stabilizing role has been suggested for the red blood cell cytoskeleton components spectrin and band 4.1 (Haest,et al 1978).

BIOGENESIS OF THE PHOSPHOLIPID BILAYER

The mechanism of biogenesis of the the endoplasmic reticulum has been investigated using two approaches.

(i) Investigation of the sites of enzymes involved in the synthesis of phospholipids using proteolytic enzymes under conditions in which the microsomal vesicles remain closed.

(ii) Determination of the site in the bilayer of newly synthesised phospholipids and their subsequent movement in the membrane bilayer using probes to differentiate between the inner and outer leaflets.

Phosphatidylethanolamine synthesis and translocation in microsomal membranes

De novo synthesis

Phosphatidylethanolamine is synthesised de novo by the CDP-ethanolamine pathway. The enzyme catalysing the final step in this synthesis is located at the cisternal surface of the endoplasmic reticulum membrane (Bell,Ballas and Coleman, 1981). Consistent with this using either phospholipase C or TNBS as probes we found that newly synthesised phosphatidylethanolamine labelled with CDP-^{14}C-ethanolamine *in vitro* is initially incorporated into the outer (cytosolic) leaflet of the microsomal membrane with a specific activity 3-4 times that of the inner (cisternal) leaflet (Hutson and Higgins, 1982). On continued incubation the labelled phosphatidylethanolamine rapidly increased in the inner leaflet and the specific activity of the two pools of phosphatidylethanolamine equilibrated (Fig.1).

In a pulse experiment microsomes were incubated for two minutes with CDP-^{14}C-ethanolamine, isolated by centrifugation, washed and reincubated (Hutson and

Higgins,1982) . The distribution and specific activities of the phosphatidylethanolamine of the inner and outer leaflets were probed using phospholipase C. Initially the specific activity of the outer leaflet phosphatidylethanolamine was three times that of the inner leaflet. On incubation, however, the specific activity of the outer leaflet fell and that of the inner leaflet rose. Phosphatidylethanolamine synthesised by the CDP-ethanolamine pathway is therefore incorporated into the outer leaflet of the membrane but rapidly moves across the bilayer to equilibrate with the inner leaflet. This equilibration does not affect the transverse distribution of the membrane which remains asymmetric with 60% of the phosphatidylethanolamine in the inner leaflet.

Base-exchange

Ethanolamine is also incorporated into phosphatidylethanolamine of microsomal membranes by base exchange. In this reaction ethanolamine in the incubation medium replaces a base on a pre-existing phospholipid in the membrane. The physiological role of base exchange is not understood: it may play a role in remodelling and maintaining the phospholipid bilayer. The base exchange reaction, however, provides a useful experimental system for incorporation of a radioactive label into the membrane phosphatidylethanolamine without *de novo* synthesis.

Incorporation of ^{14}C-ethanolamine into the phosphatidylethanolamine of microsomes was inhibited more than 90% by pretreatment of the microsomes with trypsin,

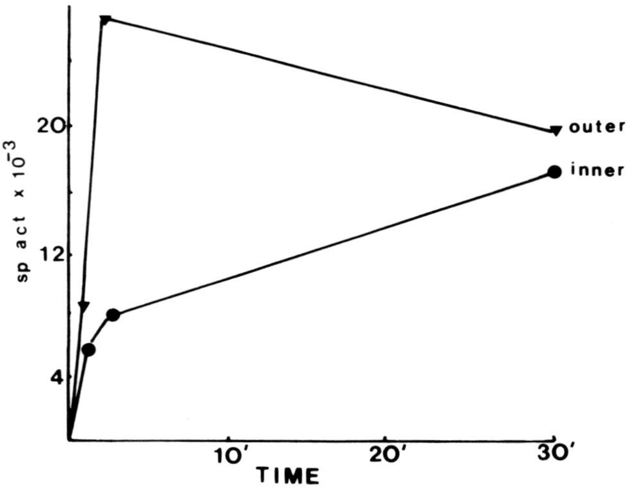

Fig, 1 Specific activities of phosphatidylethanolamine of the outer and inner leaflets of the microsomal membrane labelled by the CDP-ethanolamine pathway and probed by phospholpase C.

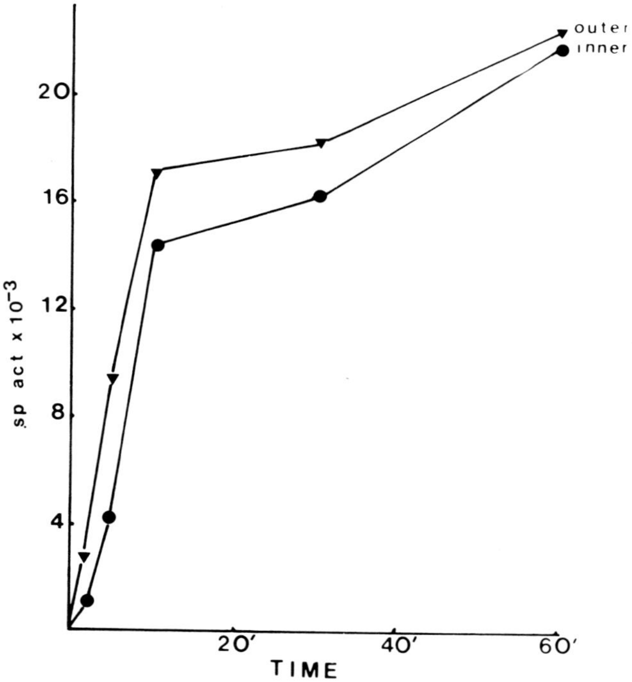

Fig.2 Specific activities of phosphatidylethanolamine of the inner and outer leaflets of the microsomal membrane labelled by base-exchange and probed with TNBS

suggesting that the enzyme is located at the cytosolic side of the endoplasmic reticulum membrane (Hutson and Higgins, 1985). Using TNBS as a probe we found that the ^{14}C-phosphatidylethanolamine is initially incorporated to the greatest extent into the outer leaflet of the membrane bilayer but rapidly appears in the inner leaflet (Fig.2). To determine whether labelled phosphatidylethanolamine is transferred from outer to inner leaflet microsomes, were labelled with ^{14}C-ethanolamine by base exchange for two minutes, isolated by centrifugation and reincubated. Phosphatidylethanolamine was transferred from the outer to the inner leaflet of the membrane. Translocation of labelled phosphatidylethanolamine was slowed by cooling to 0°, arrested by EDTA or EGTA and restored by addition of calcium ions (Hutson and Higgins, 1985).

Thus, phosphatidylethanolamine either synthesised *de novo* or labelled by base-exchange behaves in a similar way. The phospholipid is rapidly transferred from the outer leaflet to the inner leaflet of the membrane. The specific activities of the two pools are in equilibrium; however, the asymmetry of the membrane bilayer is retained.

Mechanisms of translocation of phosphatidylethanolamine across the membrane bilayer

Two general mechanisms may be involved in transmembrane movement of phospholipids.

(i) Changes in the conformation of the membrane lipids

producing transient non-bilayer structures thus allowing transmembrane equilibration of phospholipid at focal points in the membrane.

(ii) Protein mediated transmembrane movement of phospholipids.

Both mechanisms have been implicated in studies of the behaviour of phosphatidylethanolamine in membranes. In mixed phospholipid model membranes increased transmembrane movement of lipids takes place under conditions in which the bilayer structure is destabilized (Cullis et al 1985). This is dependent on the presence of phospholipids which prefer a hexagonal HII conformation. Phosphatidylethanolamine isolated from rat liver microsomes organizes in this non-bilayer arrangment at temperatures above 7° C. Isolated microsomal lipids form bilayers; however, NMR studies of such model membranes and microsomal vesicles have suggested that a part of the phospholipid may be in a non-bilayer conformation (Cullis et al, 1985; Van Duiju, 1986. Thus, in microsomal membranes or the endoplasmic reticulum in situ lateral segregation of phosphatidylethanolamine within the microsomal membrane might cause a phase change from bilayer to hexagonal HII conformation and hence allow transmembrane equilibration of the phosphatidylethanolamine.

In red blood cells several different investigations have suggested that transbilayer movement of phosphatidylethanolamine or phosphatidylserine is mediated by an ATP-dependent protein (Daleke and Huestis, 1985; Zachowski, Fellman and Devaux, 1985; Tilley et al, 1986;

Zachowski et al, 1986, 1987). From shape changes in whole red cells induced by partition of phospholipid analogues into the membrane Daleke and Huestis (1985) concluded that phosphatidylethanolamine and phosphatidylserine move across the bilayer. This translocation is dependent on ATP and inhibited by N-ethylmaleimide. Studies with spin-labelled analogues of phospholipids incorporated into red blood cell membranes suggested that phosphatidylethanolamine and phosphatidylserine move rapidly to the inner leaflet and become inaccessible to ascorbic acid in the incubation medium (Zachowski et al , 1985,1986). In contrast phosphatidylcholine and sphingomyelin analogues remain in the outer leaflet. Translocation of the aminophospholipids is dependent on ATP, inhibited by N-ethylmaleimide and unaffected by depletion of the red cell calcium. In other experiments labelled phospholipids were incorporated into the red cell membrane by phospholipid transfer proteins and the subsequent movement of the labelled phospholipids was followed using phospholipases to differentiate between the inner and outer leaflets of the membrane bilayer (Tilley et al, 1986). Results from these investigations also suggested that phosphatidylethanolamine and phosphatidylserine move rapidly from the outer leaflet to the inner leaflet of the bilayer. Phosphatidylcholine and sphingomyelin in contrast remain in the outer leaflet. Depletion of red cell ATP inhibits rapid translocation of the aminophospholipid. Recent investigations of the behaviour of spin labelled analogues of phospholipids have indicated that tranlocation of the aminophospholipids also occurs in lymphocytes

(Zachowski et al, 1987). Spin labelled phosphatidylserine and phosphatidylcholine incorporated into the plasma membrane are rapidly transferred to the inner leaflet at both 4° and 37°C. Analogues of the choline containing phospholipids move more slowly. Treatment of the lymphocytes with N-ethylmaleimide, inhibition of ATPase with vanadate ions and depletion of cell ATP inhibited rapid tranlocation of the aminophospholipids.

In microsomal membranes translocation of phosphatidylethanolamine from the outer to the inner leaflet is independent of de novo synthesis and therefore not linked to an increase of phosphatidylethanolamine in the outer leaflet. Translocation is slowed at 0° C, is calcium dependent and does not require added ATP. Thus, transmembrane movement of phosphatidylethanolamine in rat liver microsomes differs in a number of respects from that in red blood cells or lymphocytes. Calcium ions have been implicated in the production of the hexagonal HII conformation and low temperatures tend to stabilize the bilayer. At present therefore it is not possible to say which of the above general mechanisms are involved in the translocation of phosphatidylethanolamine in the endoplasmic reticulum.

Phosphatidylcholine synthesis and translocation in microsomal membranes

De novo synthesis

Phosphatidylcholine is synthesised de novo by the CDP-choline pathway. The enzyme catalysing the final step in this pathway is located at the cytosolic side of the endoplasmic reticulum membrane . Using phospholipase C as a probe we observed that more than 90% of the phosphatidylcholine synthesised by this pathway is retained in the outer leaflet of the membrane bilayer the specific activity of which was three times that of the inner leaflet (Higgins, 1979). When microsomes were pulsed with CDP-^{14}C-choline, isolated by centrifugation and chased with unlabelled CDP-choline, the labelled phosphatidylcholine remained in the outer leaflet of the bilayer (Higgins, 1979). ^{14}C-choline injected intraportally 60 minutes prior to removal of the liver and isolation of the microsomes was also incorporated preferentially into phosphatidylcholine of the outer leaflet of the membrane bilayer and remained at this site when the microsomes were incubated <u>in vitro</u>. 15 hours after injection of ^{14}C-choline the specific activities of the phosphatidylcholine of the inner and outer leaflets were the same. Therefore phoshatidylcholine does equilibrate across the bilayer <u>in vivo</u> but this is a slow process; <u>in vitro</u> equilibration does not occur during the period of our experiments.

Methylation pathway

Phosphatidylcholine is synthesised by methylation of phosphatidylethanolamine. Studies using trypsin to probe the site of the methyltransferases and phospholipase C to probe the site of the products of methylation have indicated that phosphatidylethanolamine at the inner leaflet of the microsomal membrane is methylated. During methylation the phosphatidylcholine is transferrred to the outer leaflet (Higgins, 1981). In vitro, phosphatidylcholine synthesised by this pathway is concentrated in the outer leaflet of the bilayer and does not equilibrate with the phosphatidylcholine of the inner leaflet.

^{14}C-methyl-methionone, ^{14}C-ethanolamine or ^{14}C-glycerol injected intraportally 60 minutes prior to isolation of the microsomes are all incorporated into phosphatidylcholine of the microsomal membranes. In each case using phospholipase C as a probe the specific activity of the labelled phosphatidylcholine of the inner and outer leaflets was the same. In vivo therefore phosphatidylcholine sythesised by the methylation pathway equilibrates across the bilayer.

Phospholipid-transfer proteins

Phospholipid transfer proteins provide a potentially valuble tool for studying the distribution of phospholipids in membranes. These proteins catalyse the exchange between phospholipid molecules in the outer leaflet of membranes and liposomes. Phospholipid transfer proteins have been used successfully in studies of the distribution of phospholipids in red blood cell membranes (Op den Kamp,1979); however, when these proteins were used to investigate the distribution of phosphatidylcholine in microsomal membranes it was observed that all of the phospholipid was available for exchange in a single pool

(Zilversmit and Hughes, 1977; Jackson, Westerman and Wirtz, 1978; Van den Besseler, et al 1978. It was not possible by this method to distinguish between the inner and outer leaflets of the membrane. These observations are difficult to reconcile with those described above in which phosphatidylcholine synthesised by the CDP-choline pathway remains in the outer leaflet of the bilayer. To examine this more closely we used phosphatidylcholine transfer protein and phospholipase C sequentially as probes of the distribution of phosphatidylcholine in microsomal membranes (Hutson, Higgins and Wirtz, 1985).

One explanation of the above observations is that phospholipid transfer proteins induce transmembrane equilibration of the phospholipids of the microsomal membrane. To test this possibility microsomes were incubated with and without phosphatidylacholine transfer protein with an excess of liposomes prepared from either rat liver microsomal phospholipid or phosphatidylcholine to allow complete exchange of the phoshatidylcholine. The microsomes were separated by centrifugation on a cushion of 0.75M sucrose and the distribution of phospholipids in the membranes determined using phospholipase C. The transverse distribution of the phospholipid was the same in the unincubated microsomes, in the microsomes incubated with liposomes in the absence of phosphatidycholine transfer protein, and with liposomes in microsomes incubated in the presence of phosphatidylcholine transfer protein (Hutson, Higgins and Wirtz, 1985).

In a second experiment microsomes were labelled with

^{14}C-choline injected intraportally so that the outer leaflet of the membrane was labelled. These were incubated with phosphatidylcholine transfer proein to allow exchange between the phospholipids of the outer leaflets of the vesicles. The specific activities of the phosphatidylcholine of the outer and inner leaflets was unaltered by this procedure.

To test whether the phosphatidylcholine pools of the inner and outer leaflets of the bilayer equilibrate during exchange, microsomes labelled with ^{14}C-choline injected intraportally 60 minutes prior to preparation were incubated with an excess of liposomes and phosphatidylcholine transfer protein. The microsomes and liposomes were separated and the specific activities of the phosphatidylcholine in the inner and outer leaflets of the membrane determined using phospholipase C (Hutson, Higgins and Wirtz, 1985). The specific activity of the outer leaflet was initally more that twice that of the inner leaflet and more than 90% of the radioactive lipid was in this leaflet. On incubation the specific activities of the phosphatidylcholine of the outer and inner leaflets fell, while that of the liposomes rose (Fig. 3). The specific activities of the two pools of phosphatidylcholine did not equilibrate during incubation and at the end of the experiment the specific activity of the outer leaflet was slightly higher than that of the inner leaflet.

Microsomes were also labelled with ^{14}C-glycerol injected intraportally 60 minutes before preparation so that the specific activities of the phosphatidylcholine in the

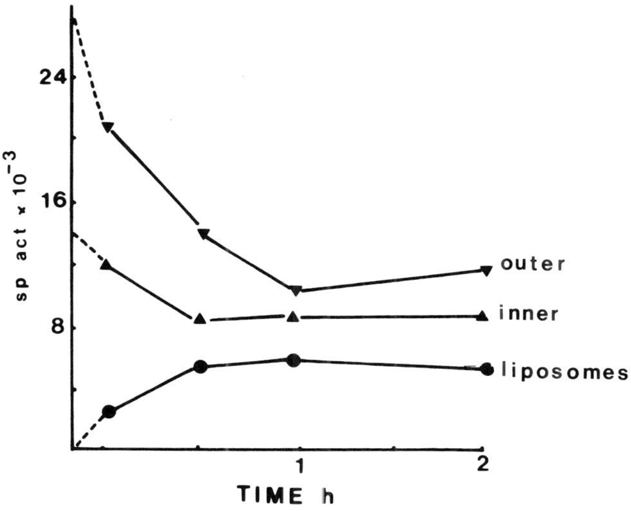

Fig. 3. Specific activities of phosphatidylcholine of the outer and inner leaflet of the microsomal membrane after incubation with liposomes in the presence of phospholipid transfer protein. See text for details.

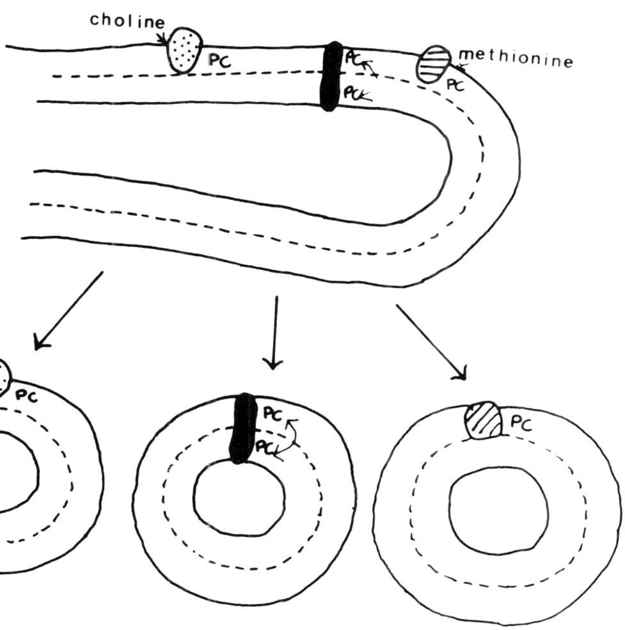

Fig. 4. A model for the tranlocation of phosphatidylcholine in the microsomal membrane. For details see text.

inner and outer leaflets of the bilayer were similar. When these microsomes were incuated with liposomes and phosphatidylcholine transfer protein followed by phospholipase C treatment both inner and outer pools of phosphatidylcholine were found to exchange in parallel (Hutson, Higgins and Wirtz, 1985). Phosphatidylcholine transfer protein therefore exchanges phosphatidylcholine in both the inner and outer leaflets of the bilayer. However, during this exchange the two pools do not equilibrate with each other and nor is transmembrane movement of the membrane phospholipids or the labelled phosphatidylcholine induced.

A model for translocation of microsomal phosphatidylcholine

One possible mechanism for these conflicting observations is shown diagrammatically in Fig.4. This suggests that pools of phosphatidylcholine synthesised by the CDP-choline pathway or the methylation pathway are sequestered topographically in the endoplasmic reticulum membrane and separated from the site of translocation of phosphatidylcholine (translocase). Phosphatidylcholine synthesised by the CDP-choline pathway equilibrates slowly with the rest of the phosphatidylcholine in the outer leaflet of the membrane bilayer. Phosphatidylcholine synthesised by the methylation pathway equilibrates more quickly. Thus *in vivo* phosphatidylcholine synthesised by the CDP-choline pathway remains in the outer leaflet while phosphatidylcholine synthesised by the methylation pathway moves in the two dimensional plane of the membrane outer leaflet and is transfered across the bilayer. When the liver is homogenised, however, the topographically distinct enzymes systems would tend to be located in different microsomal vesicles. Thus, phosphatidylcholine synthesised by the CDP-choline pathway or the methylation pathway in isolated microsomes would remain in the outer leaflet because the phospholipid cannot gain access to the translocase.

One way to test this hypothesis would be to label the phosphatidylcholine of the outer leaflet of the mebrane bilayer and to fuse the microsomal vesicles. If the translocase and synthetic enzymes are located in a separate subpopulation of vesicles then fusion of the microsomes should allow the phosphatidylcholine in the outer leaflet to gain access to the translocase and hence to move to the inner leaflet of the bilayer.

Paiment and co-workers have reported that incubation of ribosome-stripped microsomes with GTP, ATP, manganese and magnesium ions causes fusion to produce large vesicles (Paiment et al, 1980;1987). We have made use of this observation to test the effect of vesicle fusion on the behaviour of phosphatidylcholine in the bilayer. Microsomes were prepared with labelled phosphatidylcholine in the outer leaflet by injection of ^{14}C-choline in vivo. These were stripped of ribosomes and incubated under the fusion conditions reported by Paiement et al (1987). The specific activity of the phosphatidylcholine of the outer leaflet was initially 2.8 times that of the inner leaflet. When the ribosomes were stripped of ribosomes and incubated in buffer alone the specific activities of the two leaflets were unaltered. When the microsomes were incubated under fusion conditions the specific activities of the outer and inner leaflets equilibrated. Omission of either ATP or manganese and magnesium ions prevented this equilibrium.

It is possible that translocation of phosphatidylcholine from the outer to the inner leaflet is activated by fusion conditions rather than fusion dependent. Paiement et al (1987) reported that smooth microsomes do not fuse and suggested that the fusion-mechanism is restricted to the rough microsome. Similarly we found that only the phosphatidylcholine of rough microsomes equilibrated; the distribution of labelled phosphatidylcholine in smooth microsomes remained asymmetric on incubation of these under fusion conditions. We consider therefore that fusion is necessary for translocation and that this suggests that phosphatidylcholine is translocated across the microsomal membrane by a localized mechanism. It is probable that this is a protein-mediated process. Bishop and Bell (1985) have reported that a water soluble analogue of phosphatidylcholine is transported from the incubation medium into the microsomal vesicle content. This tranport is inhibited by N-ethylmaleimide or TNBS, is energy-endependent, sensitive to proteases from the outside of the vesicle, and exhibits stereospecificity. Transport of a water soluble phospholipid analogue may not necessarily use the same mechanism as tranlocation of a hydrophobic membrane phospholipid. However, our observations also favour the existence of a phosphatidylcholine translocase. The characteristics of this translocase are under further investigation in this laboratory.

TABLE ONE

Ratio of specific activity of phosphatidylcholine of outer to inner leaflet of microsomes fused after labelling with ^{3}H-choline in vivo

Original microsomes	2.78
Incubated microsomes	
A. Incubated alone	2.64
B. Complete fusion medium	1.01
C. Fusion medium minus ions	2.74
D. Fusion medium minus ATP	2.23

Microsomes were isolated from rat liver 30 minutes after intraportal injection of ^{3}H-choline (25 μc/100 g body weight. An aliquot was retained and the remained treated with pyrophosphate (5mM) in 0.25M sucrose, 40mM tris,pH 7.4 for 15 minutes, isolated by centrifugation and washed. The pellet was resuspended in 40 mM tris, 30 mM KCl, pH 7.4. In A, microsomes were incubated alone; in B, they were incubated with GTP (0.5 mM), ATP (1.0 mM), $MnCl_2$ (2.5 mM) and $MgCl_2$ (7.5 mM); in C they were incubated as B but without ions, and in D they were incubated as B without ATP. Incubation was for 120 min at 37° after which microsomes were isolated by centrifugation resuspended in 0.25 M sucrose 40 mM tris pH 7.4 containing 1.0 mM $CaCl_2$. Aliquots were incubated with and without phospholipase C for 10 min. Lipids were extracted and the specific activities of the phosphatidylcholine of the outer (hydrolysed) leaflet and the inner (unhydrolysed) leaflet determined. Results are the average of two experiments performed in triplicate.

CONCLUSIONS

The transverse distribution of the phospholipid bilayer

of the endoplasmic reticulum is asymmetric. This asymmetry is produced by synthesis of phospholipids at the cytosolic side of the endoplasmic reticulum followed by a regulated translocation of the phospholipid to produce an asymmetric distribution. Phosphatidylethaolamine rapidly equilibrates across the bialyer by a process which is calcium dependent, temperature dependent and energy independent. Phosphatidylcholine synthesised by the CDP-choline pathway in vitro remains in the outer leaflet of the membrane bilayer. This appears to be a consequence of lateral segregation of a phosphatidylcholine translocase and choline phosphotransferase. After fusion of microsomal vesicles labelled phosphatidylcholine in the outer leaflet equilibrates across the bilayer. During equilibration of either phosphatidylethanolamine or phosphatidylcholine the membrane retains compositional asymmetry.

This research was supported by a grant from the Wellcome Trust.

REFERENCES

Bell, R.M., Ballas, L.M., Coleman, R.A. (1981) Lipid Topogenesis. J Lipid Res 22:391-403

Bishop,W.R., Bell,R.M. (1985) Assembly of the endoplasmic reticulum phospholipid bilayer: the phosphatidylcholine transporter. Cell 42: 51-60

Bollen, I.C., Higgins,J.A. (1980) Phospholipid asymmetry in rough and smooth endoplasmic reticulum membranes of untreated and phenobarbital treated rat liver. Biochem J 189: 475-480

Cullis, P.R., Hope,M.J., de Kruiff,B., Verkleij, A.J., Tilcock,C.P.S. (1985) Structural properties and functional roles of phospholipids in biological membranes. in: Kuo,J.F.(ed) Phospholipids and cellular recognition Vol I. CRC Press, Florida,1-59

Daleke, D.L. Huestis, W.H. (1985) Incorporation and translocation of aminophospholipids in human erythrocytes. Biochemistry 24: 5406-5416

Higgins, J.A. (1979) Asymmetry of the site of choline incorporation into phosphatidylcholine of rat liver microsomes. Biochim Biophys Acta 558: 48-57

Higgins, J.A. (1981) Biogenesis of endoplasmic reticulum phosphatidylcholine. Translocation of intermediates across the membrane bilayer during methylation of phosphatidylethanolamine. Biochim Biophys Acta 640: 1-15

Higgins, J.A. (1982) Fine structural changes in rat liver microsomes treated with phospholipase C. J Cell Sci 53: 211-225

Higgins, J.A. (1985) Biogenesis of the phospholipid bilayer of endoplasmic reticulum membranes. in Phospholipids in the nervous system. Horrocks, L.A., Kanfer, J.N., Porcellati, G. (eds) Raven Press, New York.267-278

Higgins, J.A., Dawson, R.M.C. (1977) Asymmetry of the phospholipid bilayer of rat liver endoplasmic reticulum. Biochim Biophys Acta 470:342-356

Higgins, J.A.,Hutson, J.L. (1986) Probes of the distribution of phospholipids in endoplasmic reticulum membranes. TIBS 11: 401

Higgins, J.A. Piggott, C.A.(1982) Asymmetric distribution of phosphatidylethanolamine in the endoplasmic reticulum demonstrated using trinitrobenzenesulphonate as a probe. Biochim Biophys Acta 693: 151-158

Hutson, J.L. Higgins, J.A. (1982) Asymmetric synthesis followed by transmembrane movement of phosphatidylethanolamine in rat liver endoplasmic reticulum. Biochim Biophys Acta 687: 247-256

Hutson, J.L.,Higgins,J.A. (1985) Asymmetric synthesis and transmembrane movement of phosphatidylethanolamine synthesised by base-exchange in rat liver endoplasmic reticulum. Biochim Biophys Acta 835: 236-243

Hutson, J.L., Higgins,J.A., Wirtz, K.W.A. (1985) Microsomal membranes contain phosphatidylcholine that equilibrates across the bilayer and phosphatidylcholine that does not. FEBS letters 183: 145-149

Jackson, R.L., Westerman, J., Wirtz, K.W.A. (1978) Complete exchange of phospholipids between microsomes and plasma lipoproteins mediated by liver phospholipid-exchange proteins. FEBS letters 94:38-42

Op den Kamp, J.A.F. (1979) Lipid asymmetry in membranes. Ann Rev Biochem 48: 47-71

Paiement, J., Beaufay, H., Godelaine, D. (1980) Coalescence of microsomal vesicles from rat liver a phenomenon occurring in parallel with enhancement of glycosylation activity during incubation of stripped rough microsomes with GTP. J Cell Biol 86:29-37

Paiement, J.,Rindress, D., Smith, C.E., Poliquin, L., Bergeron, J.J.M. (1987) Properties of a GTP-sensitive microdomain in rough microsomes. 898: 6-22

Tilley, L., Criber,S., Roelofsen,B., Op den Kamp, J.A.F., Van Deenen, L.L.L. ATP-dependent translocation of amino phospholipids across the human erythrocyte membrane. FEBS letters 194: 21-27

Van Den Besseler, A.M.H.P., De Kruijff, B., Van Den Bosch, H., Van Deenen, L.L.M. (1978) Phosphatidylcholine mobility in liver microsomal membranes Biochim Biophys Acta 510: 242-255

Van Duijn, G., Luiker, J.,Verklij, A.J., DeKruijff, B., (1986) Relation between lipid polymorpism and transbilayer movement of lipids in rat liver microsomes. Biochim Biophys Acta 863: 193-204

Van Meer, G.,(1986) The lipid bilayer of the endoplasmic reticulum TIBS 11: 194-195

Zachowski, R., Fellman, P., Devaux, P.F., (1985) Absence of transbilayer diffusiom of spin labelled sphingomyelin in human erythrocytes. Biochim Biophys Acta 897: 197-200

Zachowski, A., Favre, E., Cribier, S., Herve, P., Devaux, P. (1986) Outside inside tranlocation of aminophospholipids in the human erythrocyte membrane is mediated by a specific enzyme. Biochemistry 25:2585-2590

Zachowski, A., Herrmann, A., Paraf, A., Devaux, P.F. (1987) Phospholipid outside-inside translocation in lymphocyte plasma membranes is a protein mediated phenomenon. Biochim Biophys Acta 897: 197-200

Zilversmit, D.B. and Hughes, M.E. (1977) Extensive exchange of rat liver microsomal phospholipids. Biochim Biophys Acta 469: 99-110

THE REGULATION OF PHOSPHATIDYLCHOLINE SYNTHESIS AT THE SUBCELLULAR LEVEL IN KREBS II ASCITE CELLS

F. Tercé, M. Record, G. Ribbes, H. Chap and L. Douste-Blazy
INSERM Unité 101, Biochimie des Lipides
Hôpital Purpan, 31059 Toulouse cédex, France

It is a quite general feature that phosphatidylcholine is the major phospholipid in eucaryotic cells. Therefore, the regulation of phosphatidylcholine has been intensively studied in the past decade (Pelech, 1984a). The synthesis of this phospholipid appears induced by very different stimuli such as phorbol esters (Pelech, 1984b), premature delivery (Weinhold, 1981) or fibroblast pneumocyte factor (Post, 1986). It is now well established that all the stimuli trigger cytidylyltransferase (EC 2.7.7.15) activity, which catalyzes the limiting step of phosphatidylcholine biosynthesis through the de novo pathway. Stimulation of enzyme activity occurs by the translocation of soluble cytidylyltransferase from the cytosol to cell membranes with a subsequent activation of the enzyme by membranous lipids (Pelech, 1984a). Fatty acid interaction or phosphorylation-dephosphorylation (Pelech, 1984c; Pelech, 1982) have been pointed out to explain the enzyme translocation. However it is not known wether the cytidylyltransferase is translocated to specific membrane sites and wether the nature of the membrane bearing the enzyme could be of some importance.

To study this point, we have taken advantage of a model described a few years ago by Sleight and Kent (Sleight, 1983). They found that CHO cells treated by exogenous phospholipase C in non-lytic conditions respond by an increase in phosphatidylcholine biosynthesis involving the classical translocation of soluble cytidylyltransferase. Diglycerides have been shown to interact in vitro directly with the enzyme (Choy, 1979) and the authors assumed that phosphatidylcholine-depleted plasma membrane attracted the soluble enzyme, but failed to isolate the plasma membrane (Kent, 1979).

NATO ASI Series, Vol. H16
Membrane Biogenesis. Edited by J. A. F. Op den Kamp

Krebs II cells as a model for studying phosphatidylcholine synthesis regulation

Krebs II cell is a tumor strain growing in the peritoneal cavity of Swiss mice and maintained by weekly injection (Record, 1977). This model has been used for many years in our laboratory to study phospholipid metabolism at the subcellular level (El Tamer, 1985). In Krebs cells, phosphatidylcholine is by far the major phospholipid class and represents 47.3% of total phospholipids (Record, 1984). When cells were treated in non-lytic conditions by C. Welchii phospholipase C (specific for phosphatidylcholine and sphingomyelin), the incorporation of (^{3}H)-choline was increased by about two times (Tercé, 1984). Only little amount of (^{3}H)-choline was incorporated into sphingomyeline and (^{14}C)-ethanolamine incorporation was not modified by cell treatment. Cell appeared to respond rapidly to non lytic phospholipase C attack since no lag period was observed (Tercé, 1984) during the time course of (^{3}H)-choline incorporation. In addition, by phosphate measurement, we found no difference in phospholipid composition between normal and treated cells, indicating they regulate very finely their phosphatidylcholine content. This observation suggested that diglycerides generated in plasma membrane outer layer were resynthesized into phosphatidylcholine (Tercé, 1984). We have previously established the composition of the two plasma membrane leaflets of Krebs cells (Record, 1984). Among the three phosphatidylcholine subclasses (diacyl-, alkenyl-acyl- and alkyl-acyl-), the outer leaflet contains only diacyl-glycerophosphorylcholine (diacyl-GPC) and alkenyl-acyl-glycerophosphorylcholine (alkenylacyl-GPC). Therefore we analysed the radioactivity of phosphatidylcholine subclasses after phospholipase C treatment of cells prelabelled with (^{3}H)-choline. Effectively only diacyl-GPC and alkenylacyl-GPC exhibited an increased labelling (Tercé, 1984). This result allowed us to conclude that diglycerides generated in the cell surface were not stored by plasma membrane but participated to phosphatidylcholine resynthesis. At this step of our work, the results already obtained indicated that diglyceride enrichment of plasma membrane might be not the clue of cytidylyltransferase activation. We have further investigated this point by checking the subcellular distribution of the enzyme.

Subcellular distribution of cytidylyltransferase following stimulation of phosphatidylcholine synthesis

Results reported in the next two paragraphs are the object of a future publication (Tercé et al., in preparation), where more detailed experimental data will be available. We summarize herein the most striking features of the study.

In order our model matched that of CHO cells, cytidylyltransferase had to fulfill two characteristics : a) this enzyme must catalyze the limiting step of the _de novo_ pathway for phosphatidylcholine biosynthesis; b) the enzyme must be translocated from the cytosol to membranes.

The first point was investigated by incubating Krebs cells with (^{3}H)-choline for 3 hours and measuring the radioactivity of phosphocholine, CDP-choline and phosphatidylcholine. We obtained 50 times less label in CDP-choline as compared with its precursor, phosphocholine. In the same way, radioactivity of CDP-choline represented 30 times less than that of its direct product, phosphatidylcholine. Therefore, CDP-choline produced by cytidylyltransferase is the limiting choline compound in the cell. We also performed another type of experiments in which cells incubated with (^{3}H)-choline for 3 hours were further treated with phospholipase C. At variance with control cells, those treated with phospholipase C displayed a decrease in the phosphocholine radioactivity with a concomitant increase in phosphatidylcholine labelling. Choline and CDP-choline pools did not change upon cell treatment indicating they were in a steady state throughout the experiment. Therefore, conversion of phosphocholine is obviously the limiting reaction in the phosphatidylcholine _de novo_ pathway of Krebs cells. This was assessed by the stimulation of cytidylyltransferase specific activity, rising from 1.14 $nmol.min^{-1}.mg^{-1}$ in homogenate of resting cells to 1.89 nmol.min^{-1}.mg^{-1} in treated cells. This change in enzyme activity was in the same order of magnitude as the increase of (^{3}H)-choline incorporation previously reported in treated cells.

The subcellular distribution was first investigated by differential centrifugation after cell disruption by nitrogen cavitation. In control cells, the total cytidylyltransferase activity was distributed for 15.7% in the 1000xg pellet (nuclei and debris), 50.8% in the 150,000xg pellet (microsomes) and 33.5% in the cytosol. After 3 hours cell treatment with phospholipase C, the percentage of total activity in the nuclei pellet was unchanged. Instead, the activity distribution rose from 50.8% to 65.3% in the microsomes at the expense of the cytosol.

Nature of the target cell membrane for cytidylyltransferase translocation

It is well established that microsomes correspond to a mixture of cell membranes (Molnar, 1976). To answer the question wether cytidylyltransferase was translocated to a specific membrane or randomly distributed, we had to isolate plasma membrane from endoplasmic reticulum. This was achieved using a procedure we developped a few years ago on Krebs cell (Record, 1982) and which has been extended to a larger variety of cells such as neutrophil (Ribbes, 1985), platelets (Fauvel, 1986), HL60 cells (Record, 1986), neuroblastoma cells (Chakravarthy, 1985). The rationale of the method is to use an alkaline pH all along cell fractionation to induce a selective density shift of the endoplasmic reticulum vesicles which then nicely separate from plasma membrane vesicles. By means of this method, we found that membrane-bound cytidylyltransferase in control cells strictly located in the endoplasmic reticulum. After cell treatment with phospholipase C, we were still able to isolate plasma membrane on the gradient, at variance with what was claimed for embryonic muscle cells (Kent, 1979). Surprisingly, we observed that cytidylyltransferase was not translocated from cytosol to plasma membrane, but specially to endoplasmic reticulum. In addition, Golgi apparatus, mitochondria and lysosomes were not involved in the translocation process. Therefore, endoplasmic reticulun appears as a "catalytic surface" where enzymes (cytidylyltransferase and cholinephosphotransferase) and substrates (diradylglycerol and CDP-choline) would be closely associated for a better efficiency in phosphatidylcholine resynthesis. This process is also enhanced in our model by a concomitant stimulation of cholinephosphotransferase in the same range as cytidylyltransferase.

Our results rule out the hypothesis suggesting that phospholipase C-altered plasma membrane would be the preferential site for cytosolic cytidylyltransferase translocation (Sleight, 1980). Anyway, the mechanism(s) which direct the enzyme specifically to the endoplasmic reticulum still remain to be understood. A general scheme summarizing the results reported in the paper is presented below.

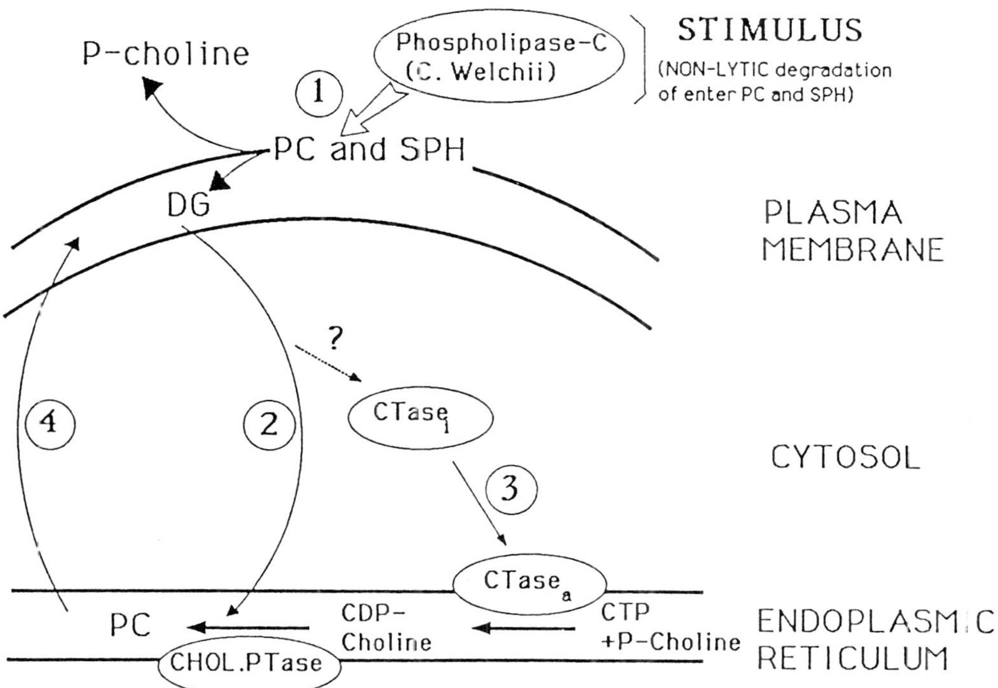

CTase= Cytidylyltransferase   Chol PTase= Choline phosphotransferase

Legend : Proposed regulation pathways for phosphatidylcholine synthesis in Krebs II cells.

Phosphatidylcholine synthesis stimulation was induced by non lytic degradation (1) of plasma membrane phosphatidylcholine (PC) and sphingomyelin (SPH). Ceramide did not appear to be resynthesized. Instead, diradylglycerol (DG) were internalized (2) to the endoplasmic reticulum and resynthesized into PC (see Tercé, 1984) which probably are transported back to the plasma membrane by exchange protein (4). Simultaneously, inactive cytosolic cytidylyltransferase ($CTase_i$) was translocated to the endoplasmic reticulum (3) where it was activated by membranous lipids ($CTase_a$), increasing the amount of CDP-choline available for PC resynthesis. Wether diglycerides interact with cytosolic cytidylyltransferase (broken line) to induce its translocation is not known.

REFERENCES

Chakravarthy BR, Spence MW, Clarke JTR and Cook HW (1986) Rapid isolation of neuroblastoma plasma membrane on percoll gradients. Characterization and lipid composition. Biochim. Biophys. Acta 812:223-233

Choy PC, Farrens SB and Vance DE (1979) Lipid requirements for the aggregation of CTP:phosphocholine cytidylyltransferase in rat liver cytosol Can. J. Biochem. 57:605-612

El Tamer A, Record M, Chap H and Douste-Blazy L (1985) Turnover of phosphocholine and phosphoethanolamine in ether-phospholipids of Krebs II ascite cells Lipids 20:699-703

Fauvel J, Chap H, Roques V, Levy-Toledano S and Douste-Blazy L (1986) Biochemical characterization of plasma membranes and intracellular membranes isolated from human platelets using percoll gradient Biochim. Biophys. Acta 856: 155-164

Kent C (1979) Stimulation of phospholipid metabolism in embryonic muscle cells treated with phospholipase C Proc. Natl. Acad. Sci. USA 76:4474-4478

Molnar J (1976) Role of endoplasmic reticulum and Golgi apparatus in the biosynthesis of plasma glycoproteins in: The Enzymes of Biological Membranes (Martonosi A ed.) vol. 2 Plenum Press, New York and London, pp.385-419

Pelech SL and Vance DE (1982) Regulation of rat liver cytosolic CTP:phosphocholine cytidylyltransferase by phosphorylation and dephosphorylation J. Biol. Chem. 257:14198-14202

Pelech SL and Vance DE (1984a) Regulation of phosphatidylcholine biosynthesis Biochim. Biophys. Acta 779:217-251

Pelech SL, Paddon HB and Vance DE (1984b) Phorbol esters stimulate phosphatidylcholine biosynthesis by translocation of CTP:phosphocholine cytidylyltransferase from cytosol to microsqmes Biochim. Biophys. Acta 795:447-451

Pelech SL, Cook HW, Paddon HB and Vance DE (1984c) Membrane-bound CTP:phosphocholine cytidylyltransferase regulates the rate of phosphatidylcholine synthesis in HeLa cells treated with unsatturated fatty acids Biochim. Biophys. Acta 795:433-440

Post M, Barsoumian A and Smith BT (1986) The cellular mechanism of glucocorticoid accelaration of fetal lung maturation. Fibroblast-pneumocyte factor stimulates choline-phosphate cytidylyltransferase activity J. Biol. Chem. 261: 2179-2184

Record M, Lloveras J, Ribbes G and Douste-Blazy L (1977) Phospholipases A_1 and A_2 in subcellular fractions and plasma membranes of Krebs II ascites cells Cancer Res. 37:4372 -4377

Record M, Bes JC, Chap H and Douste-Blazy L (1982) Isolation and characterization of plasma membranes from Krebs II ascite cells using percoll gradient Biochim. Biophys. Acta 688:57-65

Record M, El Tamer A, Chap H and Douste-Blazy L (1984) Evidence for a highly assymetric arrangement of ether- and diacyl -phospholipid subclasses in the plasma membrane of Krebs II ascites cells Biochim. Biophys. Acta 778:449-456

Record M, Hoffman D and Snyder F (1986) Comparative subcellular distribution of platelet-activating factor (PAF) and its antitumor methoxy analog in HL60 cells Proceedings of American Association for Cancer Research, abstract n°56, p. 14

Ribbes G, Ninio E, Fontan P, Record M, Chap H, Benveniste J and Douste-Blazy L (1985) Evidence that biosynthesis of platelet-activating factor (PAF- acether) by human neutrophils occurs in an intracellular membrane FEBS Lett. 191:195-199

Sleight R and Kent C (1980) Regulation of phosphatidylcholine biosynthesis in cultured chick embryonic muscle treated with phospholipase C J. Biol. Chem. 255:10644-10650

Sleight R and Kent C (1983) Regulation of phosphatidylcholine biosynthesis in mammalian cells. I. Effects of phospholipase C treatment on phosphatidylcholine metabolism in Chinese Hamster Ovary cells and LM mouse fibroblasts J. Biol. Chem. 258:824-830

Tercé F, Record M, Chap H and Douste-blazy L (1984) Different susceptibility of alkylacyl- versus diacyl- and alkenylacyl- phosphatidylcholine subclasses to stimulation of biosynthesis by phospholipase C Biochem. Biophys. Res. Commun. 125:413-419

Weinhold PA, Feldman DA, Quade MM, Miller JC and Brooks BL (1981) Evidence for a regulatory role of CTP:phosphocholine cytidylyltransferase in the synthesis of phosphatidylcholine in fetal lung following premature birth biochim. Biophys. Acta 665:134-144

PHOSPHOLIPID ASYMMETRY DURING ERYTHROPOIESIS

Jos A.F. Op den Kamp, Andre Rawyler*, Peter van der Schaft+ and Ben Roelofsen

Department of Biochemistry
State University of Utrecht
Padualaan 8, Transitorium III
NL-3584 CH Utrecht
The Netherlands

The membrane of the human erythrocyte is one of the best studied biological membranes. Its composition is known and many important aspects of its structure and function have been elucidated. A detailed overview of the phospholipid composition and organization of this membrane has been presented in chapter 1.

A prominent feature of the phospholipid bilayer is its asymmetry. In all the species investigated sofar the choline containing phospholipids, phosphatidylcholine* (PC) and sphingomyelin (SPH) are preferentially located in the outer membrane layer whereas phosphatidylethanolamine (PE) and phosphatidylserine (PS) are preferentially resp. completely present in the inner layer (Op den Kamp, 1979). Several questions regarding this phospholipid asymmetry have been studied recently. A possible function of the asymmetry could be the coupling of lipid bilayer to the underlying skeleton of spectrin-actin.

* Present address: Laboratoire de Physiologie Végétale, Université de Neuchatel, Chantemerle 20, CH-2000 Neuchatel, Switzerland
+ Present address: PFW Nederland bv., Amersfoort, The Netherlands

NATO ASI Series, Vol. H16
Membrane Biogenesis. Edited by J.A.F. Op den Kamp

The aminogroup containing phospholipids PE and PS should be the major determinants in this (Haest *et al.*, 1978; Mombers *et al.*, 1979; Franck *et al.*, 1983). The same interactions could be responsible for the maintenance of lipid asymmetry, although it became evident recently that metabolic energy is also required (Seigneuret and Devaux, 1984; Tilley *et al.*, 1986).

This paper deals with some questions concerning the biogenesis of phospholipid asymmetry in the membranes of mammalian erythrocytes.
The erythrocyte is the endproduct of a complex differentiation pathway which starts in the bone marrow by the specific commitments of stem cells.
The plasma membrane of these cells is the precursor of the plasma membrane of the mature erythrocyte. The obvious questions are:

- What is the lipid composition and organization in the plasma membrane of these early progenitor cells?
- How do these parameters change during differentiation and which mechanisms are involved?
- At what stage in the differentiation pathway is the final phospholipid asymmetry obtained?

Relevant to mention beforehand is the observation that the phospholipid asymmetry of the erythrocyte is stable and, as far as we know now, does not change when the cells grow older. Attempts to find differences in phospholipid asymmetry between various populations, enriched in young or old cells via centrifugation techniques, were not successful. (Roelofsen *et al.*, unpublished observations).

* Abbreviations used: phosphatidylcholine: PC; phosphatidylethanolamine: PE; phosphatidylserine: PS; phosphatidylinositol: PE; sphingomyelin: SPH; Friend eythroleukemic cells: FELC; dimethylsulfoxide: DMSO; colonyforming erythroid: CFU-E; phosphatidylcholine transfer protein: PC-TP.

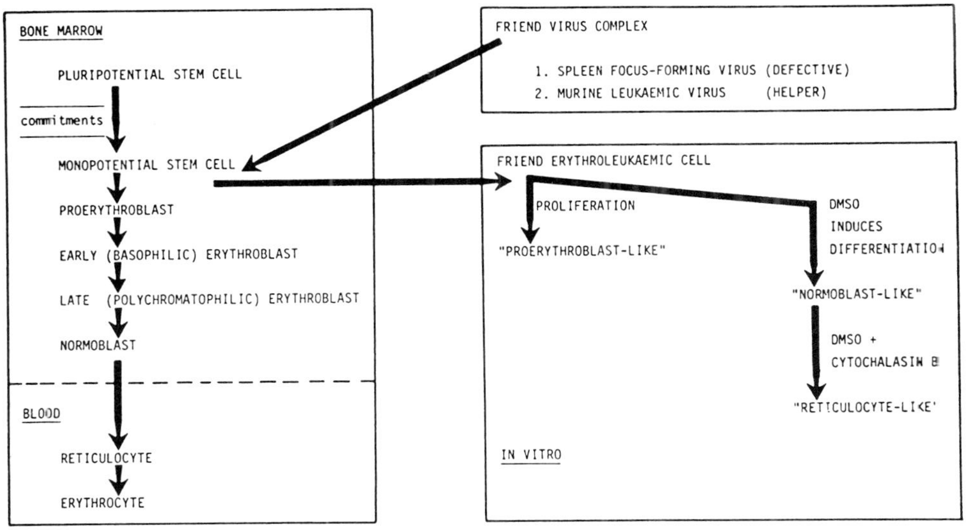

Fig. 1 Schematic overview of bone marrow stem cell differentiation and the origin of FELC.

Erythropoiesis

Pluripotential stem cells, the progenitors of the erythrocytes and of other cells which occur in the blood stream reside in the bone marrow. Specific commitment yields the monopotential stem cells which can differentiate into a more mature cell type, the proerythroblast. Evolution proceeds through a series of differentiation steps to the normoblast stage. At this point the cells are about to leave the bone marrow: the release into the bloodstream is preceeded by the extrusion of intracellular material such as the nucleus, mitochondria etc.. The resulting reticulocyte contains one membrane only - the plasma membrane.

The investigations summarized below were concentrated on the following cell types: (see Fig. 1).

i *The proerythroblast-like Friend erythroleukemic cell*

Experimentally, bone marrow cells are complicated: their isolation is difficult and the available amounts are much too low. Therefore a derivative of the proerythroblast was studied. The Friend virus complex has been used frequently to arrest differentiation and to transform the committed stem cell into proliferating proerythroblast-like cells which can be easily cultured *in vitro* (Friend *et al.*, 1971).

ii *Partially differentiated Friend erythroleukemic cells*

Although the normal differentiating ability of the Friend cells (FELC) is blocked by the transforming virus complex, non-specific agents can, at least partly, overcome this inhibition . Dimethylsulfoxide (DMSO) is one of those compounds and growth of the cells in the presence of this agent can lead to the synthesis of hemoglobin and result in a cell type resembling the normoblast (Fig. 1).

iii *In vitro differentiating early progenitor cells*

The so-called colony forming erythroid (CFU-E) cells can be isolated following thiamphenicol treatment of mice (Nijhof and Wierenga, 1983). The different stages up to the reticulocyte stage can be studied by *in vitro* culture of these cells (Nijhof and Wierenga, 1984). The isolated CFU-E and their differentiated forms resemble most closely, as far as known now, the *in vivo* occurring erythroid progenitor cells.

iv *Reticulocytes*

Phenylhydrazine treatment of mice results in an accumulation of reticulocytes which can be easily isolated and purified. These cells are not fully comparable to the *in vivo* occurring reticulocytes because they accumulate under stress. Nevertheless they are useful replacements especially because the abundance of reticulocytes is low.

v *Mature erythrocytes isolated from mice*

Phospholipid localization techniques

Standard techniques to localize phospholipids are phospholipase treatments, coupling of lipid polar headgroups to probe molecules and phospholipid exchange. Detailed overviews of merits and drawbacks of these approaches have been discussed elsewhere (Op den Kamp, 1979; Etemadi 1980; Roelofsen, 1982). In our experiments the following specific tools were used.

i *Phospholipase A_2:* purified from *Naja naja* or bee venom. Its action is illustrated in Fig. 2. The enzyme was used under non-lytic conditions so as to hydrolyse only the phospholipids in the outer layer of the plasma membrane. Intactness of cells was determined by measuring hemoglobin retention (for reticulocytes and erythrocytes) and retention of the cytosolic lactate dehydrogenase (for the progenitor cells).

ii *Phospholipase C:* purified from *Bacillus cereus*. This enzyme (see Fig. 2) has been used to study the phospholipid distribution in the plasma membrane of the Friend cells, again under non-lytic conditions. The phospholipids in the membranes of mature erythrocytes and the reticulocytes are not accessible for this enzyme.

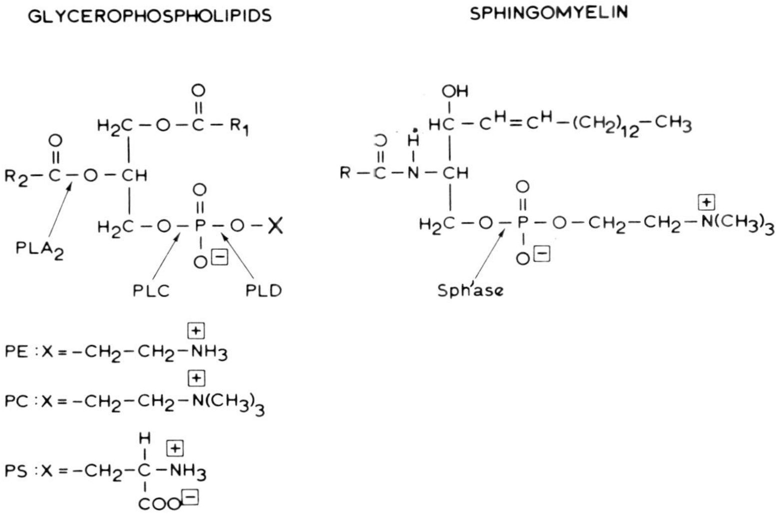

Fig. 2 Action of phospholipases.

iii *Sphingomyelinase C:* purified from *Staphylococcus aureus*. Its activity is similar to that of the phospholipase C but the enzyme is highly specific for sphingomyelin (Fig. 2). The enzyme has been used to determine the localization of SPH in the plasma membrane. Furthermore it appeared to be a useful tool in the procedure to determine the ratio of plasma membrane phospholipid to total cellular phospholipid (see page). More information on the various enzymes and their applications are described by Roelofsen (1982).

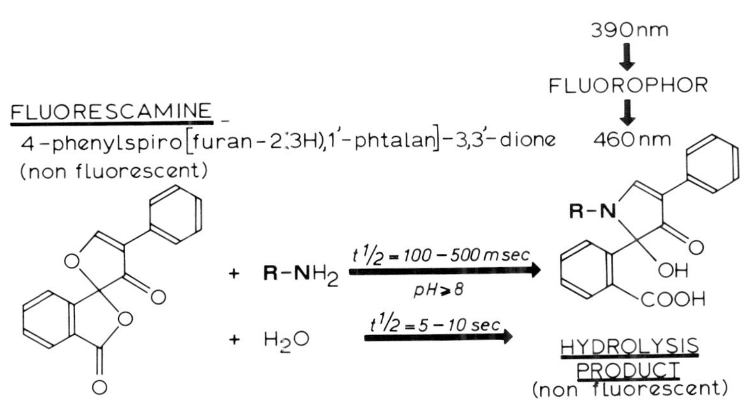

Fig. 3 Fluorescamine reaction with free amino groups.

iv *Fluorescamine.* A permeable probe which reacts quickly with free NH_2 groups (Fig. 3). The technique to use this probe for the localization of PE has been worked out in details (Rawyler *et al.*, 1984). Cells are incubated for 15 sec with increasing amounts of the probe followed by quenching of excess probe, lipid extraction and lipid analysis. A semilogarithmic plot of residual (non-reacted)PE versus probe concentration (Fig. 4) may elucidate the various pools of PE which are present in the cell. In the example shown in Fig. 4 the pools present in the outer layer of the plasma membrane, the inner layer of the plasma membrane and the intracellular membranes are labelled at different concentrations. Making use of the fluorescent properties of the fluorescamine-PE adduct, allows the detection of small amounts of PE which facilitates studies in which the amount of material is limited.

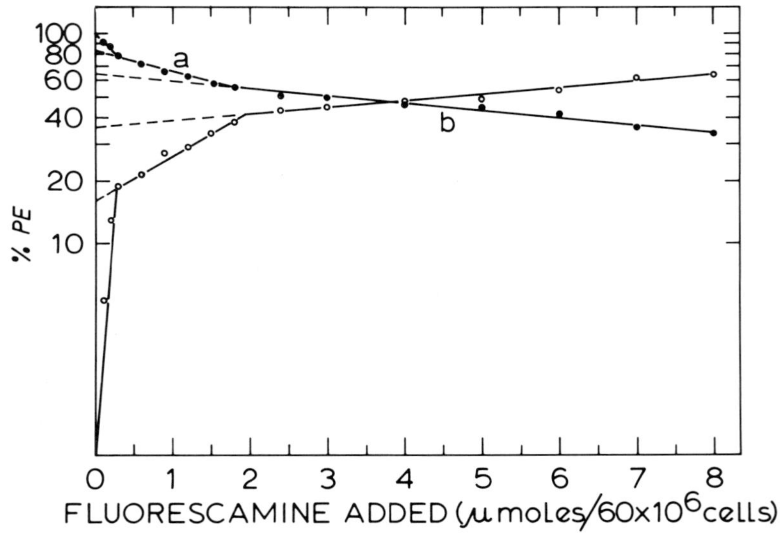

Fig. 4 Labelling of PE by fluorescamine in intact FELC revealing three pools of PE. Extrapolation of line (a) to (o) shows the amount of PE present in the outer layer of the plasma membrane; extrapolation of line (6) shows how much PE is present in the inner layer of the plasma membrane. The residual PE is present in intracellular membrane systems.

v *Phospholipid exchange.* A protein purified from beef liver is able to mediate the transfer of single PC molecules between various membranes. This PC-TP gives a one for one exchange and provides, when a suitable radioactive PC donor system is available, a non-destructive and reliable way to determine PC distribution (Fig. 5). The essential prerequisite for a correct interpretation of the data is that PC transbilayer movement is slow in comparison with the exchange procedure (see also chapter 1).

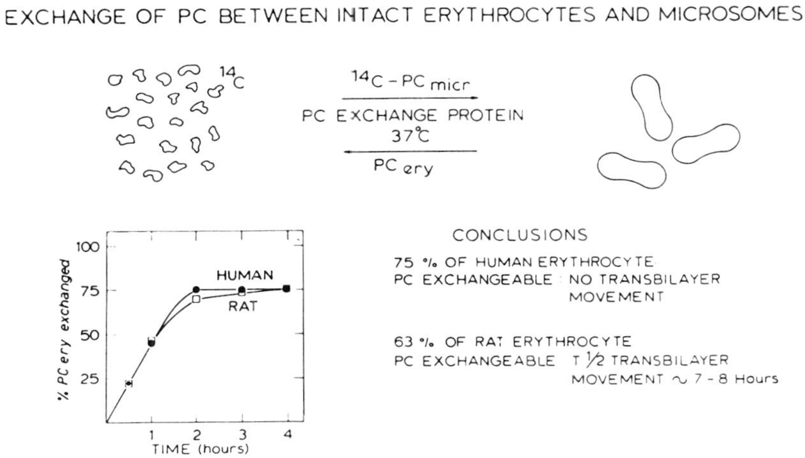

Fig. 5 Exchange of PC between intact erythrocytes and microsomal membranes.

Phospholipid localization in the plasma membrane of cells containing intracellular membrane systems

In case all of the phospholipid of a cell is present in one membrane only, phospholipid localization studies are relatively easy to perform and the data can be interpreted readily. In case of the FELC's and other complex cells which contain a variety of intracellular membrane systems the interpretation of the data obtained with the regular techniques can be complicated. Several direct and indirect approaches have been pursued. In the indirect approach plasma membrane derivatives, free of intracellular material are studied. An inside-out derivative can be obtained by a procedure developed by Sandra and Pagano (1978). Through the uptake of latex beads by mouse LM cells (a fibroblast cell line) phagolysomes are formed which can be readily isolated and purified. Rightside-out plasma membrane derivatives can be obtained by making use of specific virusses. Enveloped virusses such as the influenza virus (Lenard and Rothman, 1976) and Semliki Forest Virus (Van Meer *et al.*, 1981) bud through the plasma membrane of the host cell and obtain this way their lipid bilayer directly from the host cell plasma membrane. The major drawback of these indirect approaches stems from the very way by which plasma membranes are obtained: indeed, the risk of obtaining specialized regions of plasma membranes, which are not representative of the whole membrane, cannot be overlooked. The direct approach is based on the treatment of whole cells with the tools summarized above followed by phospholipid extraction and analysis. To achieve a correct interpretation of the data, it is essential to know the phospholipid composition of the plasma membrane and the percentage of total cellular phospholipid which is present in the plasma membrane. The first information can be obtained by the isolation and purification of plasma membrane. For FELC this procedure was carried out following classical centrifugation techniques (Fig. 6).

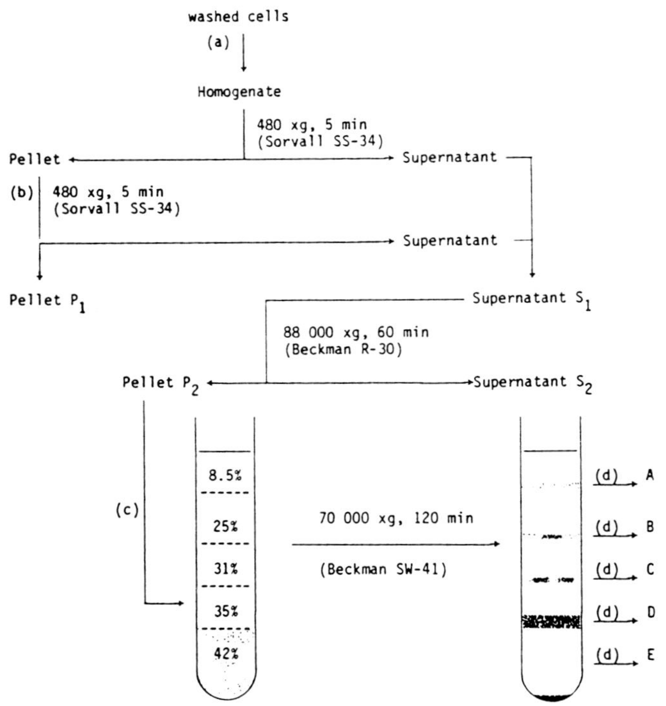

Fig. 6 Isolation and purification of plasma membranes from FELC and DMSO-FELC. Fractions are: A, unknown; B, mainly lysosomal membranes; C, plasma membranes; D, microsomes; E, mitochondria.

The purity of all fractions was assessed by measuring standard marker enzyme activities (Table I) (Rawyler *et al.*, 1983). In addition to these regular procedures a method was developed to assay plasma membrane purity on the basis of a phospholipid marker.

Table I.
Distrubution of subcellular marker enzymes in the fractionation of differentiated Friend cells. For each constituent, relative specific activities expressed as recovered activity/mg protein in each fraction divided by activity/mg protein in the homogenate are given. The amount of each constituent in each fraction, expressed as percentage of the amount found in the homogenate, is given in parentheses. Recovery is the sum of recovered activities in all fractions expressed as percentage of the activity present in the homogenate. Markers for plasma membranes were constituents 1, 2 and 3; for endoplasmic reticulum (microsomal fraction) 4 and 5; for mitochondria 6 and for lysosomes 7 and 8.

Constituent	Fraction							Recovery (%)
	P1	S2	A	B	C	D	E	
1.5'-nucleotidase	0.50 (6.7)	1.05 (23.2)	13.98 (1.3)	20.14 (5.4)	16.01 (6.7)	2.49 (8.6)	1.83 (2.9)	54.8
2.alkaline phospha-tase	0.30 (19.7)	0.53 (28.0)	24.63 (3.2)	50.71) (14.3	41.93 (21.7)	4.12 (15.3)	7.67 (4.3)	106.5
3.alkaline phospho-diesterase I	0 (0)	0.38 (20.0)	61.70 (8.0)	50.99 (14.3)	26.98 (14.0)	2.33 (8.7)	1.66 (9.3)	74.3
4.NADH cytochro-me c-reductase	0.31 (30.0)	0.01 (7.8)	1.41 (0.05)	1.82 (0.30)	2.81 (1.3)	8.45 (46.2)	3.97 (3.5)	88.9
5.NADH diaphorase	0.61 (18.9)	0.33 (20.3)	2.68 (0.1)	1.53 (0.1)	1.37 (0.2)	9.56 (53.1)	1.9 7(5.2)	97.9
6.succinate dehydrogenase	2.03 (33.0)	0.90 (37.1)	0 (0)	0 (0)	0 (0)	4.44 (12.7)	5.00 (16.0)	98.8
7.ß-D-glucuroni-dase	0.96 (6.0)	0.10 (61.0)	3.35 (0.4)	4.40 (1.2)	12.50 (6.5)	5.65 (21.0)	1.33 (0.7)	96.8
8.ß-N-acetyl-D-glucosaminidase	0.40 (19.6)	0.28 (12.1)	4.18 (0.5)	6.49 (1.8)	13.67 (5.5)	5.17 (19.2)	1.26 (1.4)	60.1
Recovered protein	43.1	46.1	0.1	0.3	0.7	3.7	5.7	99.9

An overwhelming amount of evidence has been presented to show that phosphatidylinositol (PI) and its phosphorylated derivatives PIP and PIP_2 play an essential role in the communication between the cell and its environment (for a review see Berridge and Irvine, 1984). In the socalled "PI-cycle", PI is phosphorylated to PIP and PIP_2 and external signals induce the hydrolysis of these compounds giving rise to the formation of the second messengers diacylglycerol and the inositolphosphates. We have tried to use the PI phosphorylation steps as a marker system for the plasma membrane (Fig. 7).

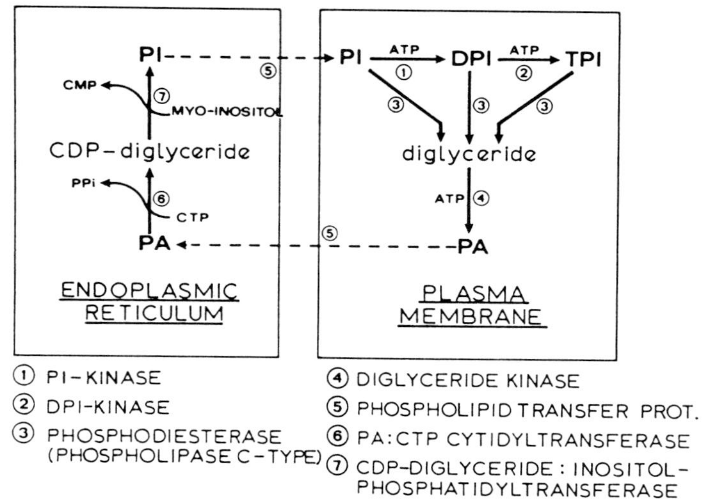

Fig. 7 Schematic representation of PI metabolism.

Radiolabelled ATP was added for a short period of time to intact FELC just before cell homogenization and plasma membrane isolation was started. Phospholipid analysis showed that all of the ^{32}P was incorporated in that short incubation period in PIP and PIP_2 and that this had occurred exclusively in the plasma membrane (Fig. 8) (Rawyler *et al.*, 1982).

Having purified the plasma membrane, it was possible to determine its phospholipid composition. The data are presented in Table II. The final problem which had to be solved, before actual phospholipid localization studies could be performed, was to determine the relative amount of phospholipid present in the plasma membrane. Of course it is possible to derive this value from the measurements of the absolute amounts of phospholipid in the purified plasma membrane and in the whole cells in combination with a value for plasma membrane recovery during the purification. On the other hand, a specific approach measuring direct lipid recovery seems more elegant. Chap *et al.* (1979) have worked out such a method and the rationale is illustrated in Fig. 9.

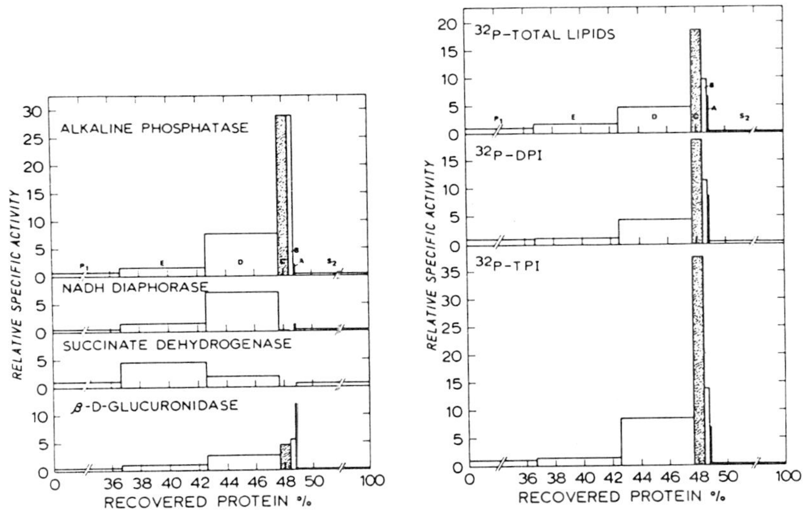

Fig. 8 Distribution of various marker enzymes in the different membrane fractions of FELC homogenates (left: see Fig. 6 and table I for explanation). At the right side the ^{32}P radioactivity in total lipid extracts DPI (=PIP) and TPI (=PIP_2) in the same membrane fractions is given.

Table II Phospholipid composition of the plasmamembranes of Friend cells, DMSO-treated Friend cells, reticulocytes and mouse erythrocytes

	PC	PE	SPH	PS	PI	LPC
Friend cells	55.2	16.5	10.2	10.6	5.3	1.4
DMSO-Friend cells	55.5	17.1	9.9	10.5	2.3	1.7
Reticylocytes	47.0	25.9	12.4	8.8	3.4	2.5
Erythrocytes	45.5	24.3	11.5	10.8	4.1	4

ESTIMATION OF THE AMOUNT OF PHOSPHOLIPID PRESENT IN THE PLASMA MEMBRANE

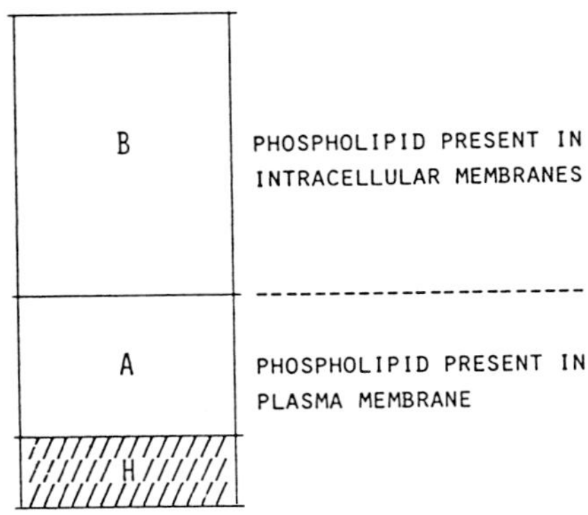

H = AMOUNT OF PHOSPHOLIPID HYDROLYZED UNDER NON-LYTIC CONDITIONS

(1) DETERMINE HYDROLYSIS IN WHOLE CELL EXTRACT $= \frac{H}{A+B} \times 100\%$

(2) DETERMINE HYDROLYSIS IN PLASMA MEMBRANE EXTRACT $= \frac{H}{A} \times 100\%$

RATIO $\frac{(1)}{(2)} = \frac{A}{A+B}$ = AMOUNT OF PHOSPHOLIPID PRESENT IN PLASMA MEMBRANE

Fig. 9 Estimation of the amount of phospholipid present in the plasma membranes of FELC.

The lipid marker -sphingomyelin- which is present in the outer layer of the plasma membrane is hydrolyzed under non-lytic conditions by sphingomyelinase C. Lipid extracts from the whole cell and from a purified plasma membrane fraction are prepared and analysed for both the SPH content and total phospholipid. A simple calculation, shown in Fig. 9, learns how much of the total phospholipid is present in the plasma membrane (Table III).

Table III Relative abundance of each phospholipid class in the plasma membranes of native Friend cells (F) and DMSO-FELC (D)

	composition (mole %) in				amount of each phospholipid class in plasma membrane			
	cells		plasma membranes		% of total phospholipid		% of phospho-lipid class	
PL	F	D	F	D	F	D	F	D
PC	59.9	60.0	55.2	55.5	21.2	22.8	35.8	38.0
PE	18.5	18.0	16.5	17.1	6.3	7.0	34.3	39.1
SPH	5.6	5.7	10.2	9.9	4.0	4.1	70.7	71.4
PI	8.3	8.0	5.3	2.3	2.3	2.3	27.6	28.0
PS	4.0	4.2	10.6	10.5	4.1	4.3	103.0	103.0
LPC	0.7	0.6	1.4	1.7	0.5	0.6	77.6	90.1
DPG	3.0	3.5	n.d.	n.d.				
Total	100.0	100.0	100.0	100.0	38.4	41.1		

With this information at hand, whole cells can be treated with phospholipases, fluorescamine and exchange proteins without the need of purifying the plasma membrane in each individual experiment.

Phospholipid composition and localization

The detailed description of experiments, procedures and results is published elsewhere (Rawyler *et al.*, 1983, 1984, 1985; Nijhof *et al.*, 1986; Van der Schaft *et al.*, 1987). At this place a short synopsis will be given.

Plasma membranes of Friend cells and DMSO induced FELC do not differ in phospholipid composition nor in phospholipid distribution over inner- and outer layer. As it was shown before already that also phospholipid metabolism is not affected by DMSO treatment of Friend erythroleukemic cells (Storm *et al.*, 1982) it seems reasonable to conclude that the early stages of erythropoiesis proceed without drastic effects on the phospholipid complement of the plasma membrane. An interesting observation is the absolute confinement of the cell with PS within the plasma membrane of both FELC and DMSO-FELC (Table III), suggesting that the fate of PS molecules is precauciously determined in this particular cell type. On the other hand, reticulocytes and erythrocytes which represent the final stages of the differentiation pathway, show considerable differences in their phospholipid composition when compared to that in the plasma membrane of the precursor cells (Table II).In particular the PC and PE contents differ considerably between the early stages of development and the final mature membranes.

A similar trend is seen in Table IV and Fig. 10, which present the data of phospholipid localization experiments. Again no difference between FELC's and DMSO-FELC on the one side and reticulocytes and erythrocytes on the other side but considerable differences

Table IV.
Relative amounts of FELC (F) and DMSO-FELC (D) plasma membrane phospholipids present in the outer monolayer. Data are expressed as the percent of the total amount of each phospho-lipid class present in the plasma membrane.

Probe	PC(%)		PE(%)		PS(%)		PI(%)		SPH(%)	
	F	D	F	D	F	D	F	D	F	D
SPH									79-85	78-83
PLA_2	45-47	42-47	50-55	44-49			47-51	54-61		
PL C	42-45	39-45	35-41	33-38	15-20	15-20				
PC-TP	45-50	47-53								
fluorescamine			47-52	41-46	9-10	9-10				

when these two "groups" of cells are compared to one another (Fig. 10). A few remarks about specific phospholipids should be made.

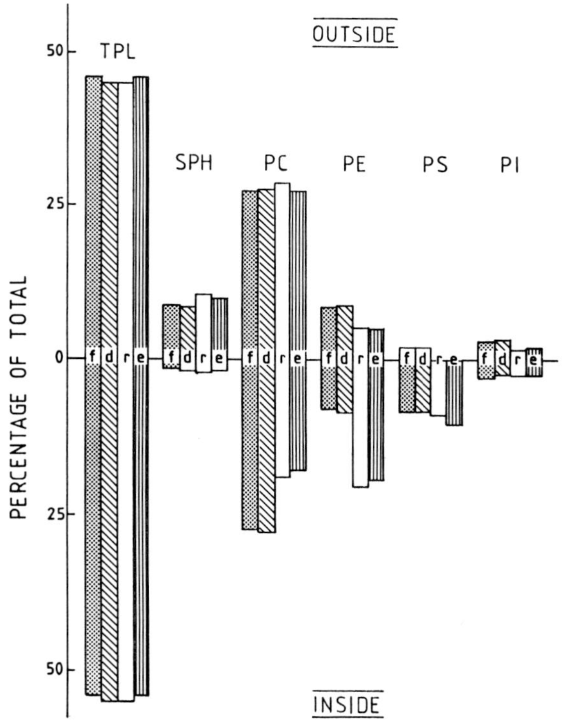

Fig. 10 Distribution of total phospholipids (TPL as well as various phospholipid species over the inner and outer leaflet of the plasma membrane of Friend cells (f), DMSO-FELC (d), reticulocytes (r) and mature mouse erythrocytes (e).

- Sphingomyelin is already at an early stage of erythropoiesis predominantly located in the outer layer. The mechanism by which this specific location is brought about and maintained is completely unknown. Relevant to mention here is that in all (abnormal) erythrocytes investigated sofar the SPH is mostly outside (Haest *et al.*, 1978; Lubin *et al.*, 1981) and that transbilayer movement of this lipid is, most likely, very slow (Boegheim *et al.*, 1983).
- Phosphatidylserine, at least the majority of it, is already located in the inner layer of the FELC's plasma membrane. From studies as described in chapter 1 it is obvious that at least two processes are responsible for the internal PS localization in the erythrocyte membrane. An ATP-dependent PS translocation, bringing PS inward, and an interaction of the PS with the intracellular membrane skeleton. However, the latter structure has not been yet formed at the early stages of erythropoiesis. It is only at the enucleation step and the subsequent maturation of the reticulocytes that the spectrin/actin network bcomes fully developed (Zweig *et al.* 1981; Geiduschek and Singer, 1979). It is possible, therefore, that the ATP-driven translocation system is the only process involved in maintaining PS asymmetry in the plasma membrane of the nucleated precursor cell. Further experiments are necessary to confirm this.

PC and PE are distributed randomly at an early stage of erythropoiesis and clearly asymmetric at the end of this process. A rearrangement is therefore required. Noteworthy, in this respect, is the fact that reticulocytes are much smaller than the Friend cells which means that only part of the original phospholipid complement of the plasma membrane of the erythroblast will end up in the reticulocyte. It is most likely that at the enucleation stage, when not only intracellular but also plasma membrane material is lost, the definite rearrangement of phospholipids occurs.

This idea is substantiated by studies on the CFU-E system discussed above (Nijhof *et al.*, 1986). These *in vitro* differentiating cells can be followed up to the late normoblast stage with the fluorescamine labelling technique. It is obvious that the distribution of PE over inner and outer layer of the plasma membrane does not change during this part of the erythropoiesis. Unfortunately, not enough cells can be obtained to follow the fate of the PC. Therefore, the actual stage at which the final phospholipid asymmetry is formed cannot be denominated with certainty. It is most likely, however, that the enucleation is the most crucial step, but further experiments are necessary to prove this and to elucidate possible mechanisms responsible for the ultimate establishement of asymmetry.

Berridge M J, Irvine R F, (1984) Inositol triphosphate, a novel second messenger in cellular signal transduction. Nature 312:315-321

Boegheim J P J, van Linde M, Op den Kamp J A F, Roelofsen B (1983) The sphingomyelin pools in the outer and inner layer of the human erythrocyte membrane are composed of different molecular species. Biochim Biophys Acta 735:438-442

Chap H J, Zwaal R F A, van Deenen L L M (1979) Action of highly purified phospholipases on blood platelets. Evidence for an asymmetric distribution of phospholipids in the surface membranes. Biochim Biophys Acta 467:146-164

Etemadi A H (1980) Membrane asymmetry. A survey and critical appraisal of the methodology II. Methods for assessing the unequal distribution of lipids. Biochim Biophys Acta 604:423-475

Franck P F H, Chiu D T Y, Op den Kamp J A F, Lubin B, van Deenen L L M, Roelofsen B (1983) Accelerated transbilayer movement of phosphatidylcholine in sickled erythrocytes. A reversible process. J Biol Chem 258:8434-8442

Friend C, Scher W, Holland J G, Sato T (1971) Hemoglobin synthesis in murine virus-induced leukemic cells in vitro: stimulation of erythroid differentiation by dimethylsulfoxide. Proc Nat Acad Sci USA 68:378-382

Geiduschek J B, Singer S J, (1979) Molecular changes in the membranes of mouse erythroid cells accompanying differentiation. Cell 16:149-63

Haest C W M, Plasa G, Kamp D, Deuticke B (1978) Spectrin as stabilizer of the phospholipid asymmetry in the human erythrocyte membrane. Biochim Biophys Acta 509:21-32

Lenard J, Rothman J E, (1976) Transbilayer distribution and movement of cholesterol and phospholipid in the membrane of influenza virus. Proc Natl Aca Sci USA 73:391-395

Lubin B, Chiu D, Bastacky J, Roelofsen B, Van Deenen L L M (1981) Abnormalities in membrane phospholipid organization in sickled erythrocytes. J Clin Invest 67:1643

Mombers C, Verkleij A J, de Gier J, van Deenen L L M (1979) The interaction of spectrin/actin and synthetic phospholipids. Biochim Biophys Acta 551:27-281

Nijhof W, van der Schaft P M, Wierenga P K, Roelofsen B, Op den Kamp J A F, van Deenen L L M (1986) The transbilayer distribution of phosphatidylethanolamine in erythroid plasma membranes during erythropoiesis. Biochim Biophy Acta 862:273-277

Nijhof W, Wierenga P, (1983) Isolation and characterization of the erythroid progenitor cell: CFU-E J Cell Biol 96:386-392

Nijhof W, Wierenga P, Staal G E J, Jansen G, (1984) Changes in activities and isozyme patterns of glycolytic enzymes during erythroid differentiation in vitro. Blood 64:607-613

Op den Kamp J A F (1979) Lipid asymmetry in membranes. Annu Rev Biochem 48:47-71

Rawyler A J, Roelofsen B, Wirtz K W A, Op den Kamp J A F, (1982) (Poly) Phosphoinositide phosphorylation is a marker for membrane in Friend erythroleukaemic cells. FEBS lett 148:140-144

Rawyler A J, Roelofsen B, Op den Kamp J A , van Deenen L L M (1983) Isolation and characterization of plasma membranes from Friend erythroleukaemic cells. A study with sphingomyelinase C. Biochim Biophys Acta 730:130-138

Rawyler A, Roelofsen B, Op den Kamp J A F (1984) The use of fluorescamine as a permeant probe to localize phosphatidylethanolamine in intact Friend erythroleukeamic cells. Biochim Biophys Acta 769:330

Rawyler A J, van der Schaft P H, Roelofsen B, Op den Kamp J A F (1985) Phospholipid localization in the plasma membrane of Friend erythroleukaemic cells and mouse erythrocytes. Biochem 24:1777-1783

Roelofsen B (1982) Phospholipases as tools to study the localization of phospholipids in biological membranes. J Toxicol Rev 1:8

Sandra A, Pagano R E (1978) Phospholipid asymmetry in LM cell plasma membrane derivatives: polar head group and acylchain distributions. Biochem 17:332-338

Seigneuret M, Devaux P F (1984) ATP-dependent asymmetric distribution of spin-labelled phospholipids in the erythrocyte membrane: Relation to shape changes. Proc Natl Acad Sci USA 81:3751-3755

Storm G, Bosman G I C G M, Boer , Roelofsen B, Op den Kamp J A F, van Deenen L L M (1982) Studies on phospholipid turnover in Friend erythroleukaemic cells. Biochem Intern 5:169-176

Tilley L, Cribier S, Roelofsen B, Op den Kamp J A F, van Deenen L L M (1986) ATP-dependent translocation of aminophospholipids across the human erythrocyte membrane FEBS Lett 194:21-27

van der Schaft P M, Roelofsen B., Op den Kamp J A F, van Deenen L L M (1987) Phospholipid asymmetry during erythropoiesis. A study on Friend erythroleukaemic cells and mouse reticulocytes. Biochim Biophys Acta 900:103-115

van Meer G, Simons K, Op den Kamp J A F, van Deenen L L M (1981) Phospholipid asymmetry in Semliki Forest Virus grown on baby hamster kidney (BHK-21) cells. Biochem 20:1974-1981

Zweig S E, Tokuyasu K T and Singer S J (1981) Member-associated changes during erythropoiesis on the mechanism of maturation of reticulocytes to erythrocytes. J Supramol Struct Cell Biochem 17:163-181

FACTORS WHICH MAY ALTER THE ASSEMBLY OF BIOMEMBRANES SO AS TO INFLUENCE THEIR STRUCTURE OR FUNCTION

Fred A. Kummerow
Burnsides Research Laboratory
University of Illinois
1208 West Pennsylvania Avenue
Urbana, Illinois 61801 USA

We are currently using both simple liposomes and mammalian cells to study factors possibly governing the dynamics of assembly of biomembranes. Some factors we found to be important to the properties of simple liposomes as well as biomembranes are 25-hydroxycholesterol, 25-hydroxycholecalciferol, 26-hydroxycholesterol and the oxidation products of highly unsaturated fatty acids. They effect the permeability of membranes to Ca^{+} ions and the phospholipid head group composition of biomembranes. Their influence on membranes is concentration dependent. It is, in some cases, dependent on the presence of antioxidants which counteract their influence on biomembranes, the presence of surface active agents and yet unknown factors.

<u>The Influence of Hydroxylated Sterols on the Permeability of Liposomes</u>

An important property of biomembranes is their impermeability to Ca^{2+} and other cations. This has been most clearly demonstrated in studies with phosphatidylcholine liposomes (Serhan et. al., 1981). Studies in our laboratory (Holmes et. al., 1983 and Holmes and Kummerow, 1985) have shown that hydroxylated sterols increased the permeability of phospholipid liposomes. Either 25-hydroxy vitamin D or 25-hydroxycholesterol at 10 mole % substantially increased the permeability of phosphatidylcholine liposomes to Ca^{2+}, whereas with vitamin D_3 (cholecalciferol) and cholesterol at 10 mole %, the liposomes were impermeable. The phospholipid liposomes (MLV) consisted of 70 mole % egg yolk phosphatidylcholine, 20 mole % dicetyl phosphate and 10 mole % of the indicated sterol. They were prepared by pressure filtration (Olson et. al., 1979) and encapsulating the Ca^{2+} sensitive dye, arsenazo III (Weissmann et. al., 1980). Free arsenazo III was

NATO ASI Series, Vol. H16
Membrane Biogenesis. Edited by J. A. F. Op den Kamp

separated from the liposome-entrapped dye by Sephadex G-50 chromatography. Two to twelve minutes after the addition of Ca^{2+}, to give a final concentration of 1 mM Ca^{2+}, uptake was monitored at 650 nm using a calculated molar extinction coefficient of 1.89 x 10^4 for the Ca^{2+} arsenazo III complex.

The Influence of Peroxidized Fatty Acids on the Permeability of Liposomes

Peroxidized fatty acids also increased the permeability of phosphatidyl choline liposomes to Ca^{2+}. Large unilamellar vesicles (LUVs) consisting of dipalmitoylphosphatidylcholine (DPPC) and dipalmitoylphosphatidylglycerol (DPPG) in a 80:20 mole % ratio were prepared by the solvent vaporization method of Deamer and Bangham (1976). When peroxidized methyl arachidonate and/or cholesterol were incorporated into vesicles, they replaced a corresponding mole % of the DPPC utilized for vesicle synthesis. The negatively charged lipid DPPG was incorporated into all vesicles to prevent vesicle clumping.

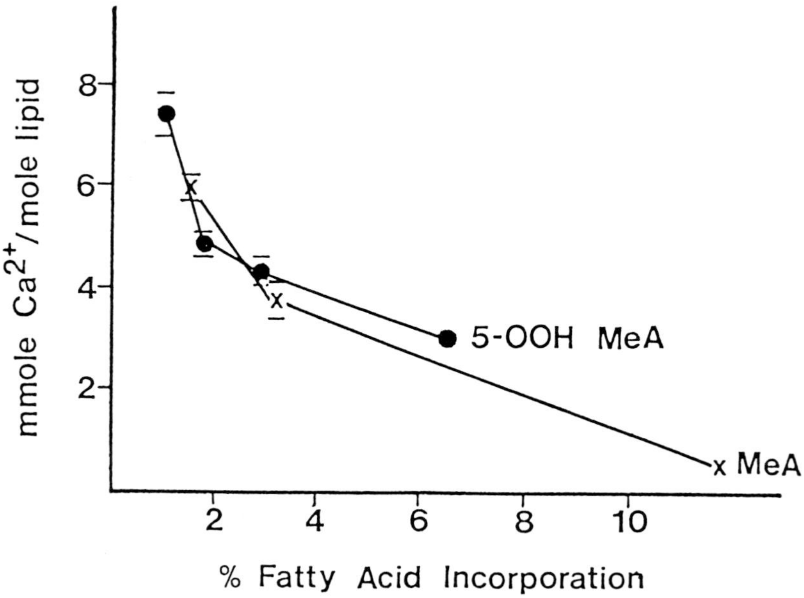

Figure 1. Plot of Vesicle Ca^{2+} Uptake Related to % Fatty Acid Incorporation for Methyl Arachidonate and 5-Hydroxy-peroxy-Methyl Arachidonate

Arsenaro III was trapped within the LUVs for detection of Ca^{2+} uptake (Bauer, 1981).

The results (Moriarity, 1987) indicated that vesicles made only of DPPC and DPPG exhibited the greatest Ca^{2+} uptake (Fig. 1). The incorporation of methyl arachidonate decreased Ca^{2+} uptake. However in comparison with methyl arachidonate the incorporation of a hydroperoxide of methyl arachidonate increased Ca^{2+} uptake. The 5-hydroperoxymethyl arachidonate was more effective in increasing Ca^{2+} uptake than the 12-hydroxperoxymethyl arachidonate (Fig. 2).

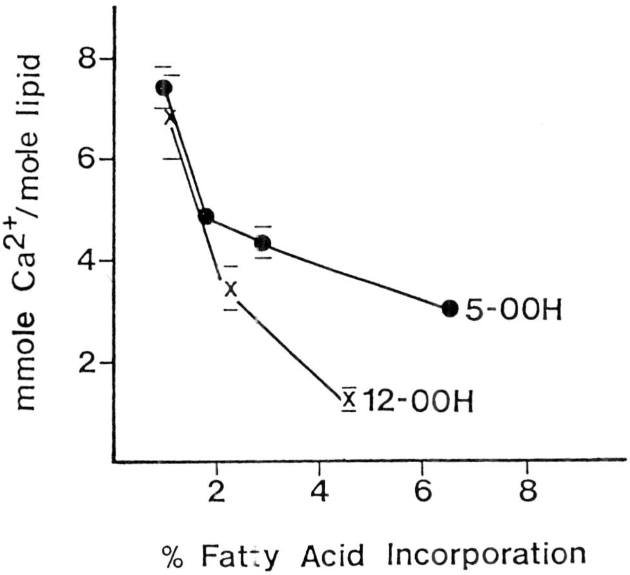

Figure 2. Plot of Vesicle Ca^{2+} Uptake Related to % Fatty Acid Incorporation for 5-Hydroxperoxy-Methyl Arachidonate and 12-Hydroperoxy-Methyl Arachidonate

Cholesterol had an ordering influence on the other membrane components (Table 1). The presence of 20 mole % cholesterol in DPPC:DPPG membranes decreased their uptake of Ca^{2+} to half its level when cholesterol was absent. The incorporation of both cholesterol (20 mole %) and methyl arachidonate (2 mole %) further decreased Ca^{2+} uptake to a fraction of that apparent in DPPC:DPPG vesicles. Both hydroxylated sterols and a peroxidized polyunsaturated fatty acid therefore cause liposomes to be more permeable (leaky) to Ca^{2+}.

TABLE 1 - Ca^{2+} Uptake of Vesicles Containing Cholesterol and Methyl Arachidonate or 15-Hydroperoxy Methyl Arachidonate

	Vesicle Ca^{2+} Uptake (mmol Ca^{2+}/mol lipid)	
Lipid Incorporated	$\bar{x}$	SEM
DPPC:DPPG	7.35	0.16
DPPC:DPPG:Cholesterol	3.51	0.10
DPPC:DPPG:Cholesterol:MeA (2.0%)	0.52	0.10
DPPC:DPPG:Cholesterol:15-OOH MeA (2.0%)	2.88	0.23

The Influence of Hydroxylated Sterols on the Permeability of Biomembranes

We studied the influence of 25-hydroxycholecalciferol and 26-hydroxycholesterol on the permeability of biomembranes with the use of arterial cells and blood platelets.

a) Hydroxycholecalciferol

Rasmussen et. al. (1982) provided one of the clearest demonstrations that a biological function can be regulated by an alteration in the degree of membrane fluidity in "target tissue" sensitive to 25-hydroxycholecalciferol (25-hydroxy vitamin D). We have found that "nontarget tissue", such as the coronary artery is also sensitive to 25-hydroxycholecalciferol (Kamio et. al., 1977 and Taura et. al., 1978, 1979). The incorporation of cholecalciferol (Vitamin D_3) in the diet of swine increased intimal thickening of the coronary arteries and increased the concentration of 25-hydroxycholecalciferol in the serum. An increase in the number of smooth muscle cells, globular elastic fiber, collagen fibrils and glycosaminoglycans contributed to the intimal thickening in the coronary arteries from swine fed increasing amounts of vitamin D (Table 2).

A level of vitamin D_3 less than the dietary intake of vitamin D by Americans increased intimal thickening and serum 25-hydroxy vitamin D levels from 5.6 u and 15 ng/ml at 2 months to 19.2 u and 27 ng/ml at 6 months of age. An increase in dietary vitamin D intake increased the

TABLE 2 - Intimal Thickening and Serum 25 OH Vitamin D Level of Swine at 2 and 6 Months of Age

Level of Vitamin D in Diet	Int. Thickness	Serum 25 OH Vitamin D_3 level
300 I.U./Kg for 2 mo.	5.6 ± 2.1 u	15.0 ± 7.6 ng/ml
300 I.U./Kg for additional 4 mo.	19.2 ± 2.6 "	27.3 ± 4.9 "
2,500 I.U./Kg for additional 4 mo.	24.2 ± 2.6 "	55.6 ± 10.4 "
25,000 I.U./Kg for additional 4 mo.	28.0 ± 3.8 "	438.7 ± 100.7 "

serum 25-hydroxyvitamin D level and increased intimal thickness significantly. The serum 25-hydroxy vitamin D level of Americans is approximately 27 ng/ml; levels as high as 64 ng/ml have been reported (Holmes and Kummerow, 1983). The influence of vitamin D on the dynamics of assembly of biomembranes of cells in the intima remains unknown.

b) Hydroxylated Cholesterol

In mammalian models hydroxylation of the cholesterol side chain at carbon 26 and then hydroxylation of the cyclopentaphenathrene ring at carbon 7 and 12 is the pathway taken in the metabolism of cholesterol to cholic acid (Danielsson, 1973). We have recently found that 26-hydroxycholesterol increased the permeability of Ca^{2+} into gel filtered platelets prepared from the blood of rats (Hallam and Rink, 1985). Gel-filtered platelets were either incubated for one hour with various concentrations of 26-hydroxycholesterol or were incubated for different times with a constant concentration of 26-hydroxycholesterol. The cytoplasmic calcium was measured using quin 2, a compound whose fluorescence increases when it complexes with calcium (Ca^{2+}) (Sheu et. al., 1984).

The results showed that after an one hour incubation of platelets at 37°C with concentrations of 26-hydroxycholesterol of 0, 50, 100, 250, 500, 750, 1000, and 10000 ng/ml, the intracellular free calcium increased from 68 nM to 190 nM (Fig. 3). However, the incubation of platelets with concentrations of cholesterol of 0, 250, 500, and 1000 ng/ml, did not significantly increase intracellular free calcium. When

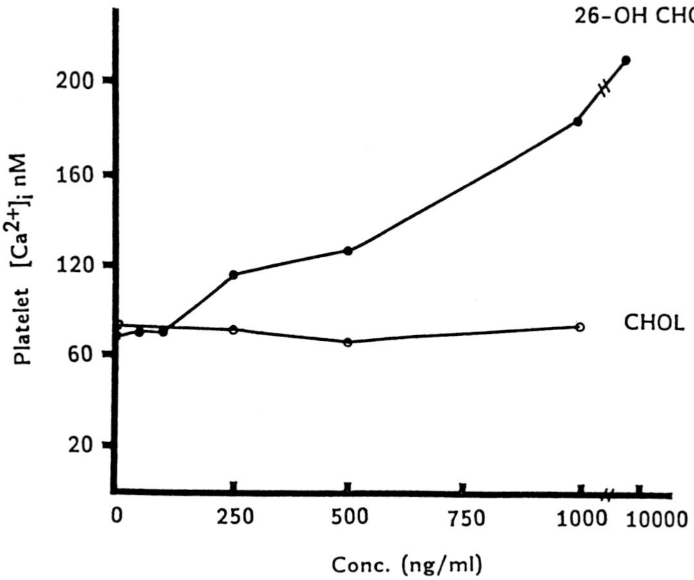

Figure 3. A comparison of the effects of 26-hydroxycholesterol and cholesterol on platelet intracellular free calcium

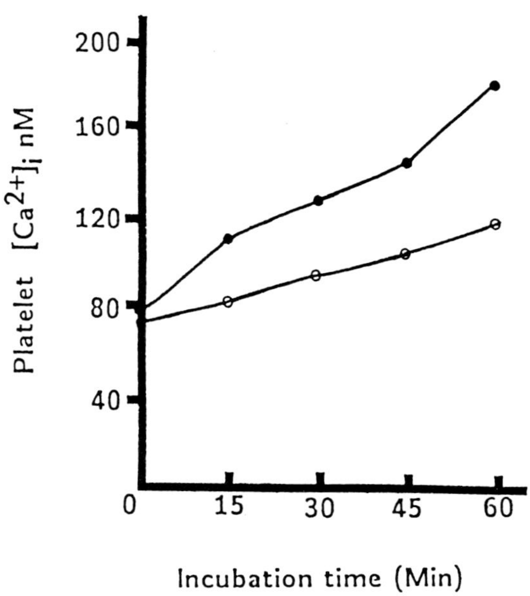

Figure 4. The time dependence on the effect of 26-hydroxycholesterol on platelet intracellular free calcium at two different concentrations

incubating platelets for 0, 15, 30, 45, or 60 min at a constant concentration of 26-hydroxycholesterol, (Fig. 4) the intracellular free calcium of platelets increased from 74 to 113 nm with 500 ng/ml 26-hydroxycholesterol. With 1000 ng/ml 26-hydroxycholesterol the platelet intracellular free calcium was raised from 80 nM to 182 nM. These results showed that the effect of 26-hydroxycholesterol on platelet calcium is concentration and time dependent. The greater the concentration of 26-hydroxycholesterol in the incubation media, the higher the platelet Ca^{2+}. The longer the platelets were exposed to the 26-hydroxycholesterol, the greater the influence on Ca^{2+} (Kou, 1987).

Platelets that were incubated for one hour at 37°C with cholesterol or different concentrations of mono- or di- esters of palmityl 26-hydroxycholesterol in gel-filtration buffer containing 1 mM Ca^{2+}, and 20 uM quin 2/AM indicated that the concentrations of cholesterol or mono- or di- esters of palmityl 26-hydroxycholesterol ranging from 50 to 1000 ng/ml had no significant influence on platelet intracellular free calcium (Fig. 5).

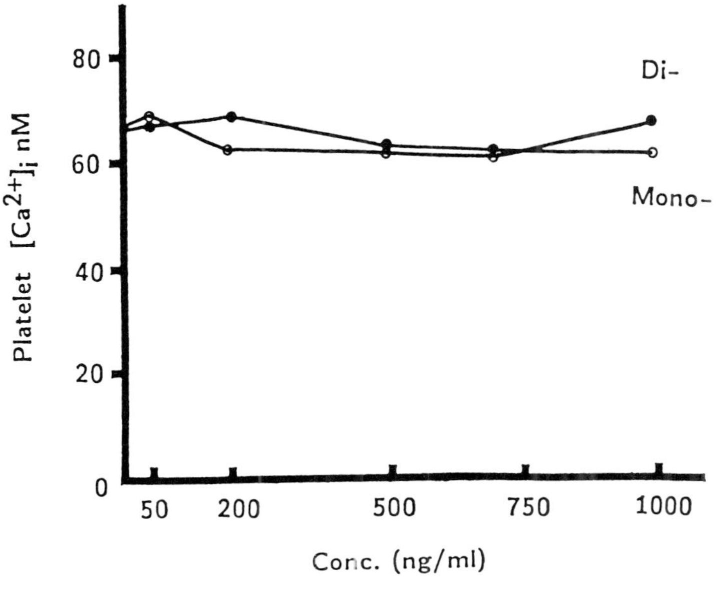

Figure 5. The effect of mono- and di- esters of palmityl 26-hydroxycholesterol on platelet intracellular free calcium

These results indicated that both the 3ß- and 26-hydroxy groups are necessary to perturbation of the platelet membrane by Ca^{2+} and to increase the concentration of intracellular free calcium (Holmes, 1985). 26-Hydroxycholesterol may incorporate into membranes, change membrane fatty acid composition (Yachnin et. al., 1979) or change membrane enzyme localization, thus inducing the membrane to be more "leaky" to cations.

The Influence of Peroxidized Fatty Acids on the Permeability of Biomembranes

Lipid peroxides can cause membrane protein changes (Sullivan and Stern, 1984), can bind to nucleic acids (Summerfield and Tappel, 1981) and can cause lipid peroxidation in human aortas (Glavind and Hartman, 1951). We have used the restricted ovulator (RO) chickens as a model to study the influence of lipid peroxidation on intimal thickening (Tokuyasu et. al. 1980, Toda et. al., 1980). The RO chicken carries a genetic defect which results in the development of hyperlipidemia, an elevation in serum lipid peroxidation (Smith et. al., 1985) and atherosclerosis. A comparison of plasma hormone levels in the non-egg laying RO hens with laying hens indicated that the plasma of the RO hen contained three times more estrogen than laying hens (Leszczynski et. al., 1984) (Table 3). Five month old hens injected with seven 10 mg/kilo of body wt. of 17ß-estradiol every other day developed a five-fold

TABLE 3 - Lipid Concentration of Plasma from Chickens Injected every 2nd day with 10 mg/Kg body weight 17ß-Estradiol for 14 days

Supplement to Diet	Triglyceride mg/dl	Cholesterol mg/dl	Malonaldehyde nM/ml
None	1734±2513	226±199	8.53±0.98
5% Corn oil	1003± 360	190± 39	9.87±1.17
Estradiol	13449±1401	1242±167	75.63±2.95
Estradiol + corn oil	13858±3360	1195±273	76.71±9.01

Leszczynski, D.E., Pikul, J. and Kummerow, F.A. (1987)

increase in plasma lipid and an eight-fold increase in malonaldehyde levels in 14 days. Five % corn oil supplementation of the diet significantly increased the malonaldehyde concentration in the lipids extracted from the liver and heart but not from plasma or skeletal muscle (Leszczynski et. al., 1987). Restricted ovulator chickens fed 1000 I.U. vitamin E per kilogram of diet maintained their hyperlipidemia, but serum peroxide levels were reduced to those of laying hens as indicated by thiobarbituric acid values (TBA), (Smith and Kummerow, 1987). The intimal thickness of the aorta was measured by light microscopy (Table 4).

TABLE 4 - Serum Triglycerides, Peroxidation and Intimal Thickening as a Function of Vitamin E Supplementation of Diet

	Layer Chicken		RO Chicken	
	Vit. E -	Vit. E +	Vit. E -	Vit. E +
Triglyceride mg/dl	570 ± 2.90	640 ± 14.0	8160 ± 2320	7160 ± 4850
TBA mmole/ml	12.2 ± 2.8	7.7 ± 1.4	40.9 ± 6.3	8.4 ± 2.0
Int. Thickening um	18 ± 1	15 ± 3	24 ± 1	15 ± 2

Smith, T. and Kummerow, F.A. (1987) 4th Int'l. Congress on O_2 Radicals, p.98, Held at Uni. of California at San Diego, La Jolla, CA 6/27-7/3/87

Hyperlipidemic hens which had their serum peroxide levels reduced by dietary vitamin E had intimal thicknesses equivalent to that of laying hens. Vitamin E may therefore have some influence on arterial cell assembly by counteracting peroxidized lipids.

The Influence of Detergent, Aging and Plaque Formation on Phospholipid Head Group Composition

a) *Rhodopseudomonas sphaeroides*

Rhodopseudomonas sphaeroides represents a model organism for the study of biomembrane structure. *R. sphaeroides* is capable of growth by

TABLE 5 - Whole Cell Phospholipid Composition of *Rhodopseudomonas sphaeroides* % of Total Phospholipids

	-TRIS	+TRIS
N-acylphosphatidylserine(NAPS)	0.9%	41.3%
Phosphatidylglycerol	37.0	25.5
Phosphatidylcholine	18.6	14.0
Phosphatidylethanolamine	45.2	19.6

Donahue, et. al. (1982) J. Bacteriology, 152:595-606

both aerobic and anaerobic respiration as well as fermentation. Depending upon the partial pressure of oxygen and light in its environment, *R. sphaeroides* undergoes morphologic and physiological changes and thus provides a simple system for a study of developmental regulation of membrane formation. The phospholipid composition of the intracytoplasmic membrane (ICM) is given in Table 5. The presence of a Tris buffer agent in the media changed the phospholipid head group composition in *Rhodopseudomonas sphaeroides*. Approximately 40% of N-acyl phosphatidylserine (NAPS) accumulated primarily at the expense of phosphatidylethanolamine. NAPS represented only 1-2 % of the total phospholipid when cells were grown in the absence of Tris buffer (Donahue et. al., 1982). Thus the presence of Tris altered the flow of metabolites so as to change the phospholipid head group composition of the intracytoplasmic membrane. It is possible that the presence of yet unknown factors in the substrate may also change the structure of biomembranes which contain cholesterol as well as phospholipids.

b) The phospholipid head group composition of mammalian biomembranes

The phospholipid head group composition of mammalian biomembranes is influenced by both age and a disease process (Kummerow, 1983). The phospholipid composition of the intimal layer of the coronary arteries of 3-week old swine contained a larger percentage of phosphatidylcholine and less sphingomyelin than the intimal layer from 2-year old swine (Table 6). The cholesterol/phospholipid ratio did not change

TABLE 6 - Phospholipid Composition of Lipid Extracted from the Corona Artery of Swine fed a Corn/Soybean Ration

Phospholipid	Age of Swine	
Phosphatidyl = (P)	3 weeks	2 years
P ethanolamine	25.18 ± 0.94%	24.76 ± 0.78%
P choline (PC)	44.74 ± 2.90	33.91 ± 0.97
Sphingomyelin	16.06 ± 3.53	23.72 ± 1.02
P inositol, P serine	11.35 ± 1.45	14.55 ± 1.19
Lyso P choline	trace	1.28 ± 0.52
Lyso P ethanolamine	trace	0.80 ± 0.40
Cholesterol/PC ratio	4.25:1 ± 0.9	4.36:1 ± 0.25

significantly. A similar trend was noted in the phospholipid composition of the intimal layer of human coronary arteries and the involved as compared to the uninvolved human aorta. The plaques (involved area) contained less phosphatidylethanolamine and less phosphatidylcholine and more sphingomyelin than the uninvovled area (Table 7). Intimal tissue from older subjects and the involved areas contained significantly more lysolecithin than that from younger subjects and the uninvolved area.

The precentage of phosphatidylcholine in human arterial tissue from the younger subjects was identical to that in two year old swine.

TABLE 7 - Phospholipid Composition of Human Arterial Tissue

Phospholipid	Intima tissue from younger subjects and uninvolved areas	Intima tissue from older subjects and involved areas
Phosphatidyl ethanolamine	8.8%	2.4%
Phosphatidyl choline (PC)	34.1	19.2
Sphingomyelin (S)	44.8	68.8
Phosphatidyl inositol	2.3	0.8
Phosphatidyl serine	2.7	0.8
Lysolecithin	3.9	8.0
Cardiolipin	1.8	1.4
Phosphatidic acid	1.0	0.6

However the percentage of phosphaditylethanolamine was significantly lower and the sphingomyelin content significantly higher than that in two year old swine.

Similar data on the percentage of phospholipid in the left coronary arteries of younger and older men were obtained by Ylä-Herttuala et. al. (1987). The increase from only traces of lysolecithin in 3 week old swine to 8% in older human subjects and the involved areas of the artery may be significant to the permeability of the intima by Ca^{2+}. Lysolecithin is a surface active agent. Its increasing presence may effect the concentration of intracellular free calcium in the intima (Taura et. al., 1979).

The phospholipid head group composition of mammalian membranes gradually seem to adjust to the changes that occur during "aging". The intimal layer of the left descending coronary artery in swine gradually thickened with age, from 5.6 u at 2 months of age to 85 u at 72 months of age (Table 8). The intimal layer of human coronary arteries from men below 27 yrs of age measures approximately 65 u in thickness compared to 160 u for men over 60 years of age (Tracy, 1987).

TABLE 8 - Intimal Thickness of the Proximal Region of the Left Anterior Desending Coronary Artery in Aging Swine Kept on a Basal Corn/Soy Ration

Age	Thickness
2 months	5.6 ± 2.1 um
6 months	22.5 ± 2.5 um
36 months	37.2 ± 5.8 um
72 months	85.3 ± 14.9 um

Adjustments in phospholipid composition during "aging" may be necessary to keep the lipids in the biomembranes in the proper fluid crystalline state. Brasitus et. al. (1986) have shown that an increase in the polyunsaturated fatty acid content of the phosphatidylcholine and phosphatidylethanolamine in intestinal brush border membrane "target tissue" increased membrane fluidity as measured by a membrane probe and an increased calcium flux through the biomembrane. It is apparent from the study of Brasitus et. al. (1986) and Rasmussen et. al. (1982) that vitamin D acts as a regulator, to some extent, of the fatty acid composition of intestinal brush border membranes in an attempt to control the fluidity of the membrane lipid. The regulators for cell membrane assembly of "non target" tissue are not as apparent.

Recent <u>in vitro</u> fluorescent polarization studies based on the use of n-9-anthroyloxy fatty acid as a fluorescent probe have shown differences in areas of fluidity or microviscosity in phosphatidylcholine of different fatty acid compositions. Studies with vesicles prepared from dipalmitoyl phosphatidylcholine or 1 palmitoyl 2-oleyl-sn-glycero-3-phosphocholine by Vincent and Gallay (1984) have shown that "micro-viscosity" or "fluidity" was highest in the seventh carbon region of <u>cis</u> oleic acid and that the addition of cholesterol enhanced or disordered the fatty acid chain at various points in the chain. Upon addition of cholesterol, Vincent and Gallay found no ordering effects occurring in the gel phase in the second carbon regions, i.e. near the aqueous interface. An enhancement of the ordering was evident in the seventh to ninth carbon regions. The disordering of the hydrocarbon chains in the 12th carbon region, observed in the absence of cholesterol, disappeared. In the 16th carbon regions, cholesterol had a strong disordering effect on the hydrocarbon chains. As the fatty acid composition of the phospholipids in arterial tissue (Emken, 1983) reflects dietary fatty acid composition, it is necessary to consider what kind of dietary fatty acid composition would provide for the most stable biomembranes (Kummerow, 1987).

The enzymes involved in biomembrane assembly normally determine the composition of phospholipids and the phospholipid/cholesterol ratio so that proper membrane fluidity is always maintained. However, this relationship may be perturbed by the presence of the hydrophilic hydroxy group in a polyunsaturated fatty acid chain, on a sterol side chain or on a sterol ring. The position of the hydroxy group in an oxidized fatty acid seems to be significant to the extent of membrane perturbation. The 5-hydroxyperoxy group caused a phosphatidylcholine liposome to be more "leaky" to calcium than a 12-hydroxyperoxy in methyl arachidonate. A sterol with only one hydroxy group, such as cholesterol and cholecalciferol did not increase the permeability of liposomes or biomembranes while the dihydroxy derivatives did. The presence of a hydroxy group at position 26 on the cholesterol side chain increased intracellular free calcium in blood platelets, the same phenomena may occur in the biomembranes in other tissues. For example, an increased concentration of 26-hydroxycholesterol in the plasma may increase intracellular free calcium in arterial cells and be significant to the development of atherosclerosis (Holmes & Kummerow, 1985). In fact, chemical structure or phenomenon which changes the hydrophilic properties of biomembranes may alter their function and result in pathology in biomembranes (Table 9).

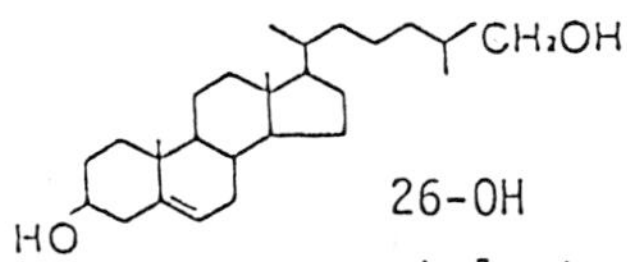

26-OH cholesterol

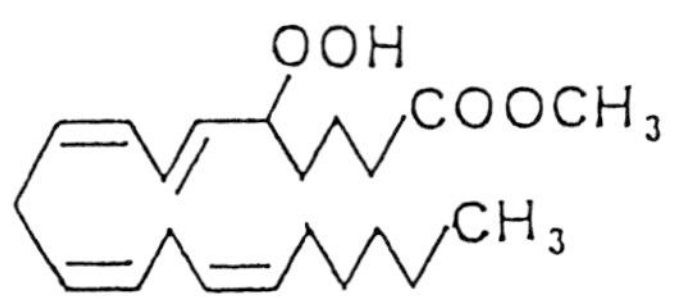

5-hydroxy peroxy me. arachidonate

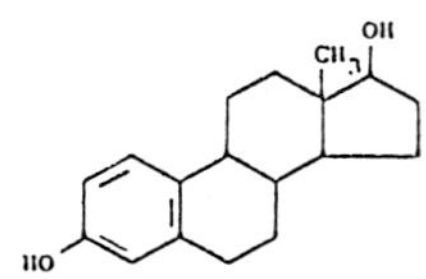

Estradiol

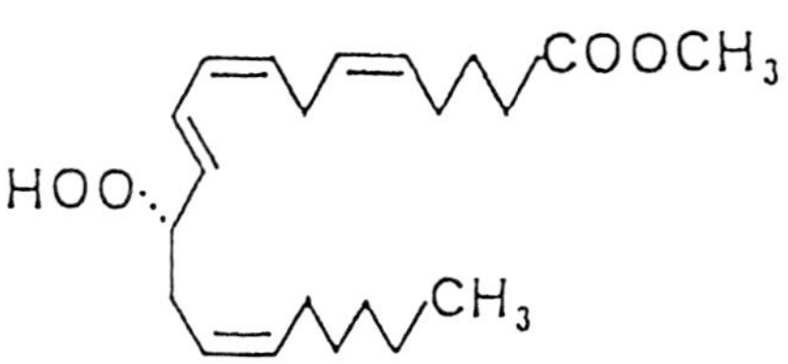

12-hydroxy me. arachidonate

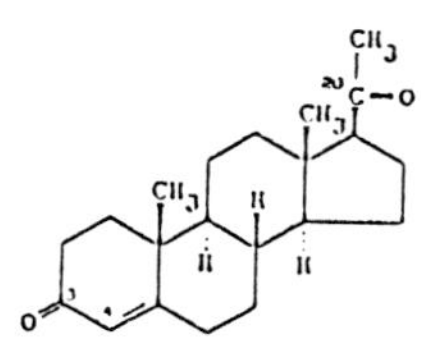

Progesterone

Table 9 - The structure of various oxidized lipids which may have significance for the hydrophilic properties of biomembranes

The five-fold increase in plasma lipid levels in only 14 days after estrogen treatment provides an example for the influence of a dihydroxylated sterol ring on lipid synthesis. Treatment with a non hydroxylated hormone such as progesterone failed to increase lipid synthesis (Hagan et. al., 1984). As an eight-fold increase in malonaldehyde level also occurred with estrogen supplementation, it is evident that high plasma lipid levels require additional antioxidants as shown in the RO chickens fed additional Vitamin E. Whether a change in a single amino acid in the protein sequence in biomembranes may cause a similar change in the hydrophilic properties of biomembranes has not been considered to date.

In summary, studies with both liposomes and biomembranes indicated that the presence of oxidized lipids increases the hydrophilic properties of their membranes and allows for an increased flow of calcium ions through the membrane leading to possible pathological consequences.

Acknowledgement: I am indebted to Drs. N. Moriarity for Fig. 1,2 and Table 1, I-Ling Kou for Figs 3-5, M. Honda for Tables 2,8, D. Leszczynski for Table 3, T. Smith for Table 4, M. Mahfouz for Tables 6,7, to D. McCulley and V. Yamamoto for preparation of the manuscript and to the Wallace Genetic Fd., the Illinois Corn Marketing Bd., the Illinois Beef Council and the National Science Foundation for funding support.

References

Bauer PJ (1981) Affinity and stoichimetry of calcium binding by arsenazo III. Analyt Biochem 110:61-72

Brasitus TA, Dudeja PK, Eby B, Lau K (1986) Correction by 1,25-dihydroxycholecalciferol of the abnormal fluidity and lipid composition of enterocyte brush border membranes in Vitamin D deprived rats. J Biol Chem 261:16404-16409

Danielsson H (1973) Mechanism of bile acid biosynthesis. In: The Bile Acids, vol. 2. Nair PP and Kritchevsky D (eds), Plenum Press, New York

Deamer DW, Bangham AD (1976) Large volume liposomes by an ether vaporization method, Biophys Acta 443:629-634

Donahue TJ, Cain BD, Kaplan S (1982) Alterations in the phospholipid composition of rhodopseudomonas sphaeroides and other bacteria induced by tris, J of Bacteriology, 152:595-606

Emken EA (1983) Biochemistry of unsaturated fatty acid isomers. J Am Oil Chem Soc 60:995-1004

Glavind J, Hartmann S (1951) The occurrence of peroxidized lipids in atheromatous human aortas. Experienta 7:464-474

Hagan RC, Leszczynski DE, Kummerow FA (1984) Comparative plasma lipid response of pullets and laying hens to estradiol and progesterone. Toxicology and Applied Pharmacology 76:483-489

Hallam TJ, Rink TJ (1985) Responses to adenosine diphosphate in human platelets loaded with the fluorescent calcium indicator quin 2. J Physiol 368:131-146

Holmes RP, Kummerow FA (1983) The relationship of adequate and excessive intake of vitamin D to health and disease. J Am Clin Nutr 2:173-199

Holmes RP, Mahfouz M, Travis BD, Yoss N, Keenan M (1983) The effect of membrane lipid composition on the permeability of membranes to Ca^{2+}, In: Biomembranes and Cell Function. Kummerow FA, Benga G, Holmes RP (eds). New York Acad of Sci, New York Vol. 144 p. 44-56

Holmes RP, Kummerow FA (1985) Membrane perturbations in atherosclerosis. In: Membrane Fluditiy in Biology, vol. 3. Aloia RC, Boggs JM, (eds). Academic Press Inc, NY pp. 281-305

Holmes RP (1985) The influence of membrane lipids on the permeability of membranes to Ca^{2+}. In: Structure and Properties of cell membranes. Vol. 2. Benga G, (ed). CRC Press Boca Raton, Florida

Kamio A, Kummerow FA, Imai H (1977) Degeneration of aortic smooth muscle cells in swine fed excess vitamin D_3. Arch Pathol Lab Med 101:378-381

Kou I-Ling (1987) The influence of dietary cholesterol on 26-hydroxycholesterol and the effect of 26-hydroxycholesterol on the intracellular free calcium level. Ph.D. thesis. The University of Illinois, Urbana, IL pp. 57-63

Kummerow FA (1983) Modification of cell membrane composition by dietary lipids and its implication for atherosclerosis. In: Biomembranes and Cell Function. Kummerow FA, Benga G, Holmes RP (eds). Ann New York Acad Sci New York Vol. 414 p. 29-43

Kummerow FA (1987) Nutritional aspects of isomeric fats. In: Lipids in Modern Nutrition. Vol. 13. Horisberger M and Bracco U (eds). Nestle Nutrition, Vevey/Raven Press pp. 213-221

Leszczynski DE, Hagan RC, Rowe SE, Kummerow FA (1984) Plasma sex horome and lipid patterns in normal and restricted-ovulator chicken hens. General and Comp Endocrinology 55:280-288

Leszczynski DE, Pikul J, Kummerow FA (1987) Lipid oxidation in plasma and tissue from estrogenized chicken hens. Biol Med and Metal Biol (in press)

Moriarity NJ (1987) The synthesis separation and identification of methyl arachidonate hydroperoxide positional isomers and their effect as components of large unilamellar vesicles on vesicle calcium ion uptake. Ph.D. thesis. The University of Illinois, Urbana, IL pp. 32-58

Olson FC, Hunt CA, Szoka FC, Vail WJ, Papahadjopoulos (1979) Preparation of liposomes of defied size distribution by extrusion through poly carbanate membranes. Biochim Biophys Acta 557:9-23

Rasmussen H, Matsumoto T, Fontaine O, Goodman DBP (1982) Role of changes in membrane lipid structure in the action of 1,25-dihydroxyvitamin D_3. Fed Proc 41:72-77

Serhan C, Anderson P, Goodman E, Dunham P, Weissmann G (1981) Phosphatidate and oxidized fatty acids on calcium ionophores-studies employing arsenazo III in liposomes. J Biol Chem 256:2736-2741

Sheu S, Sharma VK, Banerjec SP (1984) Measurement of cytoplasmic free calcium concentration in isolated rat ventricular myocytes with quin 2. Circulation Res 55:830-834

Smith TL, Toda T, Kummerow FA (1985) Plasma lipid peroxidation in hyperlipidemic chickens. Atherosclerosis 57:119-122

Smith TL, Kummerow FA (1987) Serum lipid peroxidation and athero-sclerosis in restricted ovulator chickens. 4th Int'l Congress on Oxygen Radicals, Nat'l Bureau of Standards, Univ. Cal. at San Diego, June 30 p. 98

Sullivan SG, Stern A (1984) Membrane protein changes induced by test-butyl hydroperoxide in red blood cells. Biochim Biophys Acta 774:215

Summerfield FW, Tappel AL (1976) Determination of maldialdehyde-DNA cross-links by fluorescence and incorporation of tritium. Analyt Biochem 111:77

Taura S, Taura M, Kummerow FA (1978) Human arterio- and athero-sclerosis: Identical to that in a 6 and 36 month old swine fed a corn diet free of cholesterol and saturated fat. Artery 4(2):100-106

Taura S, Taura M, Imai H, Kummerow FA, Tokuyasu K, Cho BHS (1979) Ultrastructure of cardiovascular lesions induced by hypervitaminosis D and its withdrawal. Arterial Wall 4(4):249-259.

Toda T, Leszczynski DE, McGibbon WH, Kummerow FA (1980) Coronary arterial lesions in sexually mature non-layers, layers and roosters. Virchows Arch A Path Anat and Hist 388:123-125

Tokuyasu K, Imai H, Taura S, Cho S, Kummerow FA (1980) Aortic lesions in nonlaying hens with endogenous hyperlipidemia. Arach Pathol Lab Med 104:41-45

Tracy R (1987) Louisiana State University Medical School, New Orleans LA personal communication

Vincent M, Galley J (1984) Time-resolved fluorescence anisotrophy study of effect of cis double bond on structure of lecithin and cholesterol-lecithin bilayers using n-(9-anthroxyloxy) fatty acids as probes. Biochem 23:6514-6522

Weissmann C, Anderson P, Serhan C, Samuelsson E, Goodman E (1980) A general method employing arsenazo III in liposomes for study of calcium ionophores. Results with A23187 and prostaglandins. Nat'l Acad Sci USA 77:1506-1510

Yachnin S, Streuli RA, Gordon LI, Hsu RC (1979) Alternation of peripheral blood cell membrane function and morphology by oxygenated sterol; a membrane insertion hypothesis. In: Current Topics in Hemotology Vol. 2. Alan R. Liss Inc. New York

Ylä-Herttuala S, Sumuvuori H, Karkola K, Möttönen M, Nikkari T (1987) Atherosclerosis and biochemical composition of coronary arteries in Finnish men. Atherosclerosis 65:109-115

MYOCARDIAL CELL DEATH AND THE POSSIBLE ROLE OF SARCOLEMMAL PHOSPHOLIPIDS (BASED ON MORPHOLOGICAL OBSERVATIONS).

J.A. Post, T.J.C. Ruigrok[1)], J.M.J. Lamers[2)], P.D. Verdouw[3)] and A.J. Verkleij[4)]

Institute of Molecular Biology and Biotechnology
State University Utrecht
Padualaan 8
3584 CH Utrecht
The Netherlands

INTRODUCTION

This article gives an overview of a series of articles (Post et al., 1985a,b; 1987; Verkleij and Post, 1987), which led to a new hypothesis concerning myocardial cell death and which explains the observed morphological events in the context of current knowledge of the molecular organization of biological membranes. Myocardial cell death secondary to several pathological conditions is accompanied by a calcium overload of the myocardial cell (Alto and Dhalla, 1979; Katz and Reuter, 1979) and seems to be triggered by this overload. Two of the most studied systems for myocardial failure are the calcium-paradox and the ischemia-reperfusion model.

1) Department of Cardiology, University Hospital, Interuniversity Cardiology Institute, Utrecht, The Netherlands. 2) Department of Biochemistry, Erasmus University Rotterdam, Rotterdam, The Netherlands. 3) Thoraxcenter, Erasmus University Rotterdam, Rotterdam, The Netherlands. 4) Molecular Cell Biology, State University Utrecht, Utrecht, The Netherlands.

NATO ASI Series, Vol. H16
Membrane Biogenesis. Edited by J. A. F. Op den Kamp

The calcium-paradox concerns the "paradoxical"-effect of Ca^{++} when calcium is reintroduced after a short period of calcium-free perfusion of the heart (Zimmerman and Hülsmann, 1966). Reintroduction of calcium results in a severe contracture of the myocardial cell, a swelling of the mitochondria and a destruction of the sarcolemma (Muir, 1968; Zimmerman et al., 1967), which is reflected by a massive enzyme release (Hearse et al., 1978; Zimmerman and Hülsmann, 1966). At the time of reperfusion an influx of Ca^{++} takes place, which is thought to have a biphasic character (Nayler et al., 1983). First a relatively small gain in cytosolic calcium occurs, followed by a massive influx.

Ischemia-reperfusion refers to the experimental protocol in which the coronary flow is stopped and, after a period of ischemia, restored. Severe myocardial ischemia is associated with a depletion of high energy phosphates, glycogen, an accumulation of metabolic intermediates (Jennings et al., 1983), a decrease in pH (Bailey et al., 1982), and changes in the ionic balance of the cell (Borgers and Piper, 1986; Jennings et al., 1964). Restoration of the coronary flow after a short period of ischemia might salvage the myocardium. However, if ischemia is prolonged, then reperfusion results in dramatic damage of the myocardium. An increase of intracellular calcium occurs (Shen and Jennings, 1972) and a loss of intracellular constituents is observed (Hearse et al., 1978). The myofibrils are contracted and severely damaged. The sarcolemma is disrupted (Kloner et al., 1974) and freeze fracturing shows severe aggregation of the intramembranous particles (IMP's) (Ashraf and Halverson, 1977).

Thus all these interventions, which result in myocardial failure, seem to be accompanied by a loss of the barrier function of the sarcolemma, resulting in an influx of Ca^{++}-ions during reperfusion. We have studied the ultrastructure of Langendorff perfused hearts, subjected to these experimental protocols, with the use of fast freezing devices (a physical fixation method which does not require cryoprotection and chemical fixation and therefore excludes artefacts possibly induced by these interventions). Alternatively we used thin section electron microscopy and conventional freeze-fracturing. In addition to the Langendorff system we studied the morphology of ischemic-reperfused pig hearts _in vivo_ and correlated the morphology with the long term recovery of the ischemic part of the heart.

MATERIALS AND METHODS

Langendorff perfusion technique

New Zealand white rabbits were anesthetized with Hypnorm (fluanison/fentanyl) and thiopental. The animals were heparinized, the hearts were excised and subsequently perfused at 37°C by the Langendorff technique at a constant pressure of 6.4 kPa (65 cm H_2O). The standard perfusate (+ Ca) had the following composition (mmol/l): NaCl, 124.0; KCl, 4.7; $MgCl_2$, 1.0; $NaHCO_3$, 24.0; Na_2HPO_4, 0.5; $CaCl_2$, 1.3; glucose, 11.0. The perfusion fluids were equilibrated with 95% O_2/5% CO_2, resulting in a pH of 7.4 at 37°C.

Calcium paradox: After a perfusion period using the standard solution the heart underwent a calcium free perfusion. During calcium-free perfusion (-Ca) calcium was omitted from the standard perfusate and no correction was made for the small change in osmolarity. The following types of experiments were performed: (1) 60 min +Ca (3 hearts); (2) 45 min +Ca, 15 min -Ca (3 hearts); (3) 40 min +Ca, 15 min -Ca, 5 min +Ca (3 hearts): (4) 15 min +Ca, 15 min -Ca, 30 min +Ca (3 hearts). All experiments, therefore, had a duration of 60 min. At the end of the experiments biopsy specimens of the left ventricular free wall were excised and processed for electron microscopy as described below.

Ischemia-reperfusion: After an equilibration period of 15 min control perfusion, the hearts were made totally ischemic by stopping the perfusion flow. The following types of experiments were performed: (i) 90 min perfusion (3 hearts); (ii) 15 min perfusion, 90 min ischemia (3 hearts); (iii) 15 min perfusion, 90 min ischemia, 30 min reperfusion (3 hearts). At the end of the experiments, biopsy specimens of the left ventricular free wall were excised and cut into 1 mm^3 cubes and processed for electron microscopy.

In vivo ischemia-reperfusion

Surgical preparations: Yorkshire pigs weighing 11-18 kg (group I animals) were anesthetized and catheterized as described previously (Verdouw et al., 1977). Subsequently, the heart was exposed via a midsternal thoracotomy and suspended in a pericardial cradle. Ischemia was induced by occluding the left circumflex coronary artery with a microsurgical clamp. After 30 or 60 min the myocardium was abruptly reperfused by releasing the clamp. Three experiments were performed with each occlusion. When ventricular fibrillation occurred, the animals were promptly defibrillated. If attempts to restore normal sinus rhythm were unsuccessful within 60 sec, the animals were excluded from further study. Therefore the final analysis was carried out with four animals. The ischemic myocardium was clearly defined as a cyanotic region, sharply demarcated from the non-ischemic area. At the end of the 2 h reperfusion period, several needle-biopsies were taken from the non-ischemic and ischemic areas of the left ventricle and processed for electron microscopy.

In a parallel series of experiments the long-term effect of the above episodes of coronary artery occlusion and subsequent reperfusion on left ventricular function was determined (group II animals). For this purpose a left lateral thoracotomy was performed in 12 pigs (11-16 kg) under sterile conditions. After opening of the pericardium the left circumflex coronary artery was prepared free at its proximal portion over a length of 5 nm. Then the left circumflex coronary artery was completely occluded for 30 or 60 min in 6 animals each. After the occlusion the myocardium was abruptly reperfused and the chest closed after 2 h of reperfusion.

Echocardiographic analysis: Before surgery and after 2 weeks of reperfusion, two-dimensional echographic recordings of left ventricular function were obtained in the intact animal (group II animals) and recorded on videotape. Quantitative analysis of the videoframes was performed using the left parasternal short axis view at mid-papillary muscle level. Images at end-diastole and end-systole were manually traced using a digitizer, connected to a Microsonics computer system. Epicardial and endocardial contours of the same videoframes were traced using the leading edge method (O'Rourke et al., 1984). The calculated centroid of the epicardial contour was taken as the center of contraction, and each short axis was subdivided in 12 equal sectors, with the anterior intercept of the papillary muscle with the interventricular septum as the first radius. Systolic wall thickening (SWT) of each sector the left ventricle was calculated as: (EDT-EST)/EDT x 100% in which EDT and EST are the wall thicknesses at end-diastole and end-systole of each sector, respectively.

Electron microscopy

Thin-section electron microscopy: At the end of the experiments, biopsy specimens of the left ventricle free wall were excised and cut into cubes < 1 mm^3. These were immersion-fixed for at least 2 h in glutaraldehyde (2.5%)/tannic acid (2%), which was added to the perfusate that was used during the last step of the experiment. In some experiments the samples were fixed in 2.5% glutaraldehyde only.

After this first fixation the samples were washed in a 0.1 mol/l phosphate buffer (pH 7.1) and post-fixed for 30 min in a combination of 1% OsO_4 plus 1.5% $K_4Fe(CN)_6$. The samples were stained en bloc with 1% uranyl acetate in 70% acetone, dehydrated with the use of acetone and embedded in araldite resin. The blocks were sectioned on a Reichert OM-U2 ultramicrotome using a diamond knife. The thin sections were mounted on copper grids, stained with uranyl acetate and lead citrate and examined in a Philips 301 or Philips 201 electron microscope.

Freeze-fracture electron microscopy - conventional method: After the glutaraldehyde/tannic acid fixation the samples were washed in the perfusate that was used during the last step of the experiment and placed in this perfusate containing 30% glycerol. Gold alloy hats containing tissue were frozen in a mixture of liquid and solid nitrogen. The fracturing and shadowing were performed in a Balzers BAF freeze-etch unit. Fracturing was done at a vacuum of 5 x 10^{-7} torr; specimen-table temperature was -120°C and knife-temperature was -196°C. Shadowing was done with a Balzers Control Unit EVM 052A. A thirty-ångstrom thick layer of platinum-carbon was deposited (45° angle), followed by 100 Å of carbon (90° angle).

Freeze-fracture electron microscopy - ultrafast method: Immediately after termination of the experiments the biopsies were physically fixed without the use of any chemical fixation or cryofixation. A one-sided propane jet was used for this ultra-fast cryofixation of the specimens (Knoll et al., 1982). This method excluded ultrastructural changes induced by chemical fixation and cryoprotectants. Fracturing and shadowing were done in the same way as described above. The replicas obtained from both techniques were cleaned in a sodium hypochlorite solution and distilled water before mounting on copper grids, and were examined in a Philips 301 or Philips 201 electron microscope.

RESULTS

The results of the three different systems used will be presented separately. Although many more morphological changes take place, we will focus on the sarcolemma.

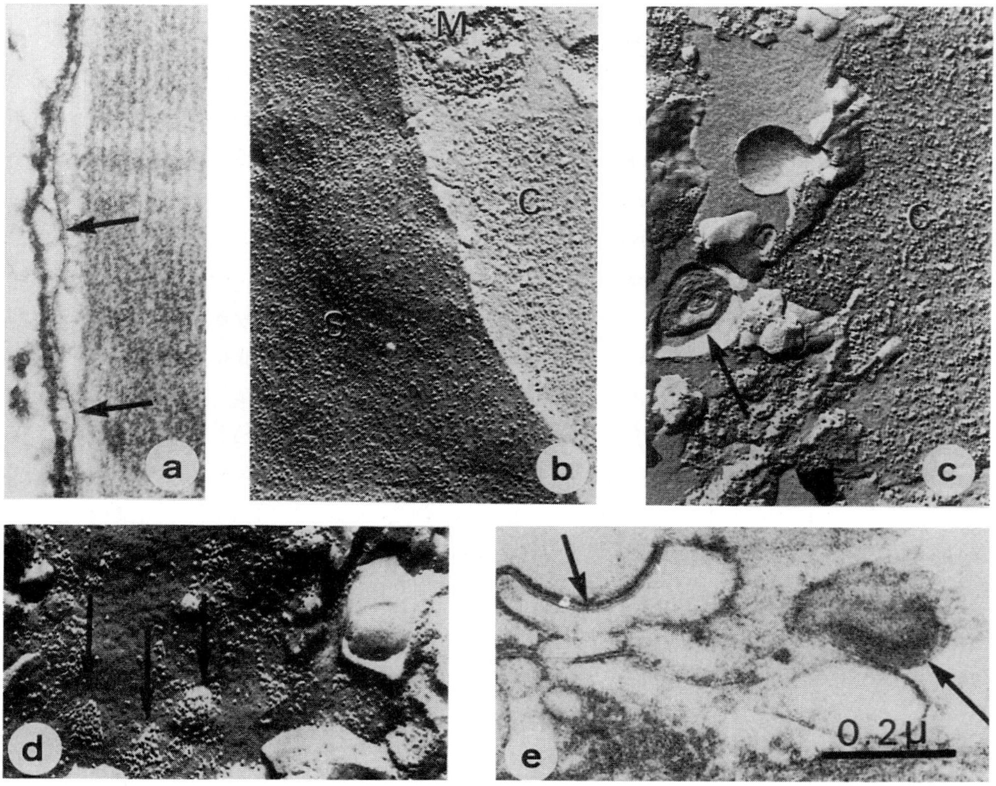

Fig. 1 Micrographs of isolated perfused rabbit hearts. a: after 15 min of calcium free perfusion. The glycocalyx is lifted from the lipid bilayer (49000x). b: 15 min of calcium free perfusion: no changes are observed by freeze fracturing (49000x). Subsequent reperfusion caused destruction of the sarcolemma, which is attended with aggregation of IMP's (d: 49000x) and the the extrusion of multilamellar vesicles (c: 49000x; e: 80500x).

Calcium-paradox

After 15 min of Ca^{++}-free perfusion the glycocalyx exhibited lifting from the bilayer (Fig. 1a), which has been reported previously (Frank, 1983). However, both freeze-fracture techniques showed no changes in the fractured sarcolemma (Fig. 1b). Five or 15 min of repletion of calcium after the calcium-free perfusion resulted in a severe aggregation of the IMP's and a disruption of the sarcolemma (Fig. 1c,d,e). This disruption was attended with the formation of multilamellar, lipidic vesicles.

Ischemia-reperfusion, Langendorff

Ninety min of ischemia resulted in an aggregation of the IMP's of the sarcolemma (Fig. 2b). No peeling of the glycocalyx was observed and the lipid bilayer remained continuous. Subsequent reperfusion of the heart (30 min) resulted in a severe aggregation of the IMP's (Fig. 2c). In addition both freeze-fracture techniques showed sarcolemmal formation of multilamellar vesicles. These contained no IMP's. The bilayer had lost its continuity. This was clearly seen when the sarcolemma was cross-fractured (Fig. 2e). When the fracture plane followed the membrane, smooth patches of overlying liposomes were observed on top of the bilayer (Fig. 2d,f). Thin sectioning also showed the disruption of sarcolemma after reperfusion. It has to be noted that the multilamellar structures could only be observed after tannic acid fixation (Fig. 2g). When the fixation was done with the use of glutaraldehyde alone the sarcolemma had also lost its integrity, but no multilamellar structures were observed (Fig. 2h). This indicates that the liposomal structures consist of pure lipid, since tannic acid is known to preserve lipid (Glauert, 1974), whereas glutaraldehyde fixation alone does not (Kalina and Pease, 1977). Next to the sarcolemmal changes we observed that during ischemia as well as during reperfusion the mitochondria extruded lipidic material. This lipidic material contained no IMP's (Fig. 3a) and again tannic acid was needed to preserve the structure with thin sectioning (Fig. 3b).

Ischemia-reperfusion, in vivo

Immediately following the coronary occlusion there was a complete loss of systolic wall thickening in the segment perfused by the left circumflex coronary artery assessed with a 5 MHz ultrasound crystal (Krautkamer-Branson, Lewistown, PA). When the needle-biopsies for electron microscopic investigation were taken 2 h after reperfusion, no sign of

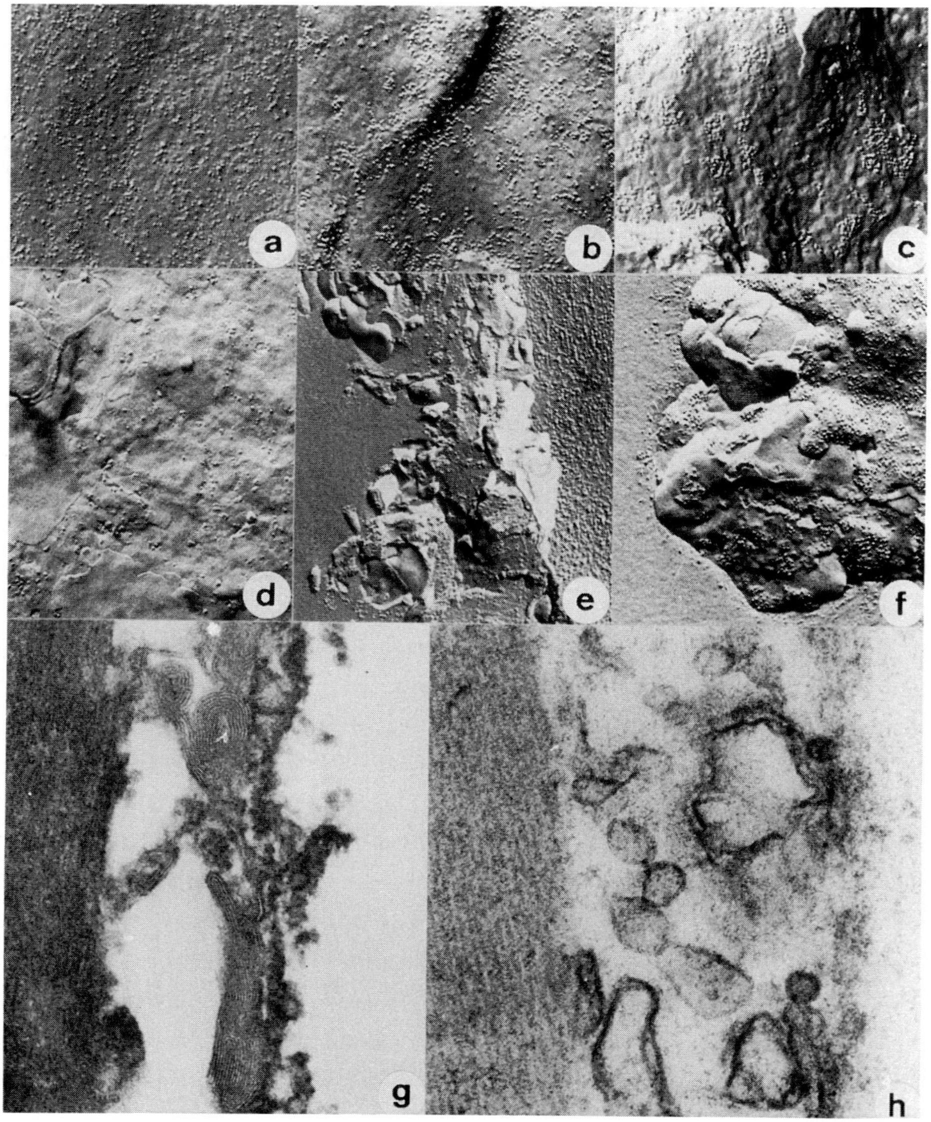

Fig. 2 Effects of ischemia and reperfusion on the ultrastructure of the sarcolemma. a-d: ultrafast freeze-fracture electron microscopy (43000x); e,f: conventional freeze-fracture electron microscopy (43000x); g,h: thin-section electron microscopy (71350x). a: control perfusion; b: ischemia, an aggregation of intramembranous particles can be observed; c-h: ischemia followed by reperfusion; c: severe aggregation of intramembranous particles is shown; d: particle-free plaques are present on top of the sarcolemma; e: cross-fractured sarcolemma, liposomal structures can be observed; f: multilamellar, particle-free structures can be observed on top of the sarcolemma; g: tannic acid preserved the liposomal structures; h: the use of only glutaraldehyde showed disruption of the sarcolemma, however, without the multilamellar structures.

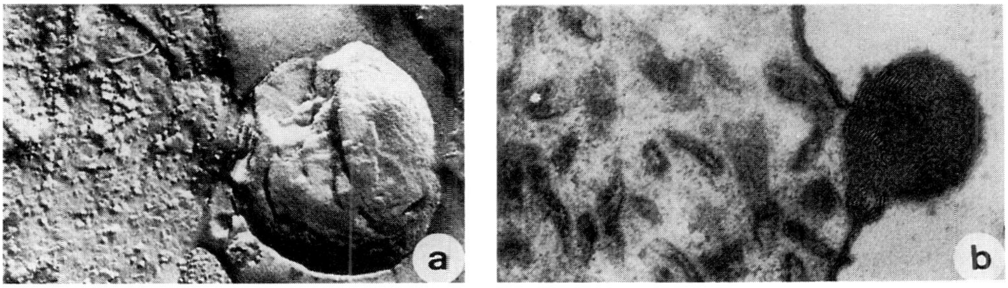

Fig. 3 Effects of ischemia on the ultrastructure of mitochondria. a: replica obtained by ultrafast freeze-fracturing, showing a particle-free, liposomal structure attached to a mitochondrion (64100x); b: thin-section micrograph (glutaraldehyde/tannic acid fixation), showing the extrusion of a liposomal structure (74900x).

recovery of function was observed, independent of the time of ischemia (30 or 60 min). Non-ischemic tissue samples of both the 30 and 60 min occluded hearts showed a normal appearance of the sarcolemma and were not different from samples taken from control hearts that had not been ischemic (not shown). The lipid bilayer was a continuous, trilamellar structure with the glycocalyx attached to it. Freeze-fracturing of these samples showed also an intact sarcolemma, with a random distribution of intramembranous particles. At the end of 2 h reperfusion thin-sectioning revealed that all samples collected from myocardium, that had been ischemic for 30 min only, showed intact sarcolemma. Yet, freeze-fracturing showed a change in distribution of the IMP's, which became aggregated in small clusters.

Two hours of reperfusion following 60 min of coronary artery occlusion caused dramatic changes at the sarcolemmal level. In these animals thin-sectioning demonstrated breakdown of the lipid-bilayer, with formation of multilamellar structures. Freeze-fracturing confirmed this sarcolemmal breakdown and clearly showed the presence of multilamellar, particle-free vesicles. Since the described morphological phenomena are identical to the ones shown in Fig. 2 they are not shown. In addition to the sarcolemmal changes the extrusion of lipidic material by the mitochondria was again observed, comparable to that observed in the Langendorff system. The two-dimensional echocardiography showed that the hearts, which had been ischemic for 30 min, had a good recovery of regional myocardial function after 2 weeks of reperfusion (Fig. 4b). No recovery was found in the animals which had been subjected to 60 min of ischemia (Fig. 4c).

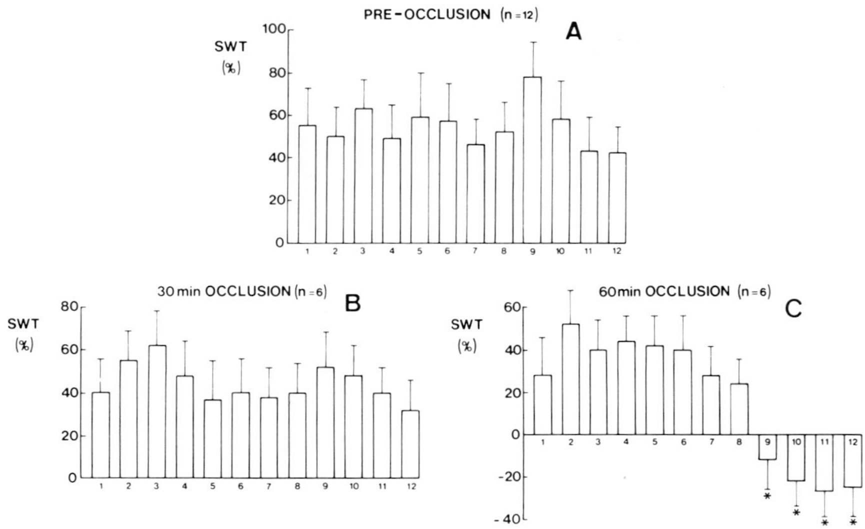

Fig. 4 Systolic wall thickening (%) in porcine hearts before left circumflex coronary artery occlusion (30 or 60 minutes) and after 2 weeks of recovery; (A) the pooled data (N=12) of the animals before occlusion; (B) 2 weeks recovery after 30 minutes of occlusion (N=6); (C) 2 weeks recovery after 60 minutes of occlusion (N=6). Notice the depressed function in the segments 9 12 (all perfused by the left circumflex coronary artery) after 2 weeks in the animals which have been subjected to 60 minutes of ischemia (*=P<0.05 versus comparable segment with 30 minutes occlusion).

DISCUSSION

The results obtained in both ischemia-reperfusion models showed that, under all the ischemic conditions studied, the experimental interventions were attended with the extrusion of lipidic material by part of the mitochondria. Of special interest is the fact that in the *in vivo* model the extrusion was observed in the hearts subjected to 30 or 60 min of ischemia. Two-dimensional echocardiography obtained after 2 weeks, however, showed recovery of myocardial function in the animals subjected to 30 min of ischemia, but not in those subjected to 60 min of ischemia. Apparently, a rather dramatic change in morphological appearance of part of the cellular mitochondria does not necessarily exclude long-term recovery of regional myocardial function.

A more reliable indicator of lethal injury seems to be the sarcolemmal destabilization and destruction. This phenomenon was not observed in tissue which had been ischemic for 30 min and reperfused, despite absence of function of the ischemic tissue at that time. However, two weeks later these hearts showed almost complete return of function. Sixty min of ischemia and 2 h of reperfusion resulted in a destruction of the sarcolemma and after 2 weeks no recovery was observed.

The destruction of the sarcolemma in all the models was attended with an aggregation of the IMP's and the extrusion of lipidic structures.

IMP-aggregation

IMP-aggregation has been observed in many membranes (for review see Verkleij and Ververgaert, 1978) and has been attributed to the formation of solid "islands" or domains of lipid. These occur when one or more of the lipid constituents of the membrane passes through its liquid crystalline to gel phase transition temperature and literally freezes, forming domains of solid lipid which exclude proteins (IMP's), leaving the circular "bald" patches visualized by freeze-fracture techniques. Such IMP aggregation has been reported in cases where the membrane has been cooled below its physiological temperature and is termed "thermotropic lateral phase separation" in reference to the lipids which are now present in both the liquid-crystalline and gel phases within the membrane. The aggregation of IMP's seen in myocardial sarcolemma after a period of ischemia or calcium deprivation both followed by reperfusion is not thermotropically induced and as such is thought to stem, upon reperfusion, from an increase of intracellular calcium. This, in turn, implies that calcium triggers this aggregation from the cytoplasmic side of the sarcolemma. How can we explain this proposition?

Particle aggregation, indicative of lateral phase separation, can be induced in erythrocyte ghosts in several ways; by increasing calcium concentration (mM range), by acdition of the positively charged polypeptide polylysine or by lowering the pH to below 5.5. This aggregation can only occur when the spectrin has been removed from the erythrocyte ghosts by washes with low ionic strength medium (Elgsaeter et al., 1976). After this treatment the erythrocyte membrane is orientated in an inside-out configuration and so the original cytoplasmic leaflet of the bilayer is now on the outside and is exposed to the calcium, polylysine and low pH. It has been postulated that the treatments mentioned above induce solidification of the negatively charged phospholipids phosphatidylserine (PS) and phosphatidylinositol (PI) resulting in IMP aggregation. These negatively charged phospholipids have been shown to be asymmetrically distributed across the bilayer of the erythrocyte membrane and to be almost exclusively confined to the inner cytoplasmic leaflet (Op den Kamp, 1979; Verkleij et al., 1973). The fact that IMP aggregation is caused by the lipid phase separation in the inner leaflet has been supported by the finding that cationic dyes, such as acridine orange, can only induce particle aggregation when they have penetrated intracellularly (Lelkes et al., 1983).

If the IMP aggregation observed in the sarcolemma is to be explained in the manner described for the erythrocyte then a prerequisite would be an asymmetric distribution of the membrane lipids with the negatively charged phospholipids present in the cytoplasmic leaflet. All the studies concerning lipid localization of lipids in membranes, so far investigated, indicate that lipids are asymmetrically distributed. The negatively charged phospholipids are predominantly present in the cytoplasmic leaflet (Op den Kamp, 1979). At this moment the lipid asymmetry of the sarcolemma is under investigation in our laboratory. Preliminary results indicate that the phospholipids are indeed asymmetrically distributed over the two monolayers of the sarcolemma and that this distribution very much resembles the phospholipid asymmetry of the erythrocyte.

<u>Extrusion of lipids from the sarcolemma</u>

In addition to IMP aggregation an extrusion of lipids occurs as a result of reperfusion following ischemia or calcium reintroduction after a period of Ca^{++}-free perfusion. This extrusion, in the form of blebbing of multilamellar, liposome-like structures, suggests that the stability of the membrane is lost during this period. This destabilization can be explained by consideration of the polymorphic behavior of membrane lipids. When isolated, certain membrane lipids prefer to adopt a non-bilayer configuration in aqueous dispersions, known as the hexagonal (H_{II}) phase, which consists of cylindrical tubes or micelles of lipid with an aqueous core. The lipid molecules are orientated in such a way that the headgroups are at the aqueous core of the cylinder or micelle and the acyl chains radiate outwards. Virtually all membrane phosphatidylethanolamines (PE's) will adopt such hexagonal H_{II} structures under physiological conditions (De Kruijff et al., 1985). In the biological membrane, however, this tendency is suppressed by the presence of other, bilayer preferring lipids, such as phosphatidylcholine (PC) and phosphatidylserine (PS) and proteins. In such a lipid mixture, where one of the components prefers the H_{II} phase, the bilayer can easily become unstable. Juxtaposition of such unstable bilayers will, in principle, result in contact sites and fusion which, it is proposed, proceed via inverted lipid structures. Such inverted lipid structures have been visualized by freeze-fracture as lipidic particles (Verkleij, 1984) and are indeed lipids in the H_{II} phase. Continued destabilization of such bilayers may eventually lead to severe breakdown of bilayer structure and gross formation of the H_{II} phase (Cullis et al., 1983; Verkleij, 1984). Later on, by mixing again with "bilayer-lipids", bilayer structures may be formed. This unstable equilibrium of bilayer and non-bilayer lipids can be disturbed by many factors, including calcium, via an effect on the phase transitions of negatively charged phospholipids. Addition of calcium to dispersions of lipid vesicles containing PE and PS results in fusion of those vesicles. This is because the calcium is able to induce a lateral phase separation of the PS in the bilayer, permitting fusion to proceed via H_{II} phase structures (Cullis et al., 1983). Lateral phase separation also occurs on addition of calcium to vesicles of phosphatidylinositol (PI) and PE (Nayar et al., 1982). If liposomes are made composed of the same phospholipid mixture present in the cytoplasmic monolayer of the inner leaflet of the erythrocyte (PE : PS : PC : SPH = 4 : 2 : 1 : 1) addition of calcium will again induce non-bilayer structures and fusion (Hope and Cullis, 1979). The calcium concentration required to induce such membrane destabilization in model systems is in the order of $\geq$ 2 mM. In the presence of physiological levels of other divalent cations, such as Mg^{2+}, this concentration is reduced to $\geq$ 0.2 mM (Cullis et al., 1983), concentrations that are not unlikely to occur in the myocardial cell during reperfusion following ischemia, or during the calcium paradox.

If we assume the existence of lipid asymmetry in the sarcolemma comparable to that of the erythrocyte, with PE, PS and PI preferentially in the cytoplasmic leaflet, then it is easy to envisage that an increase of cytosolic calcium during reperfusion will lead to the breakdown of the lipid-bilayer organization. This model is depicted in fig. 5.

The hypothesis of sarcolemmal lipid asymmetry as the root of membrane destabilization is currently under investigation along with a more detailed study of the physicochemical properties of sarcolemmal lipid species.

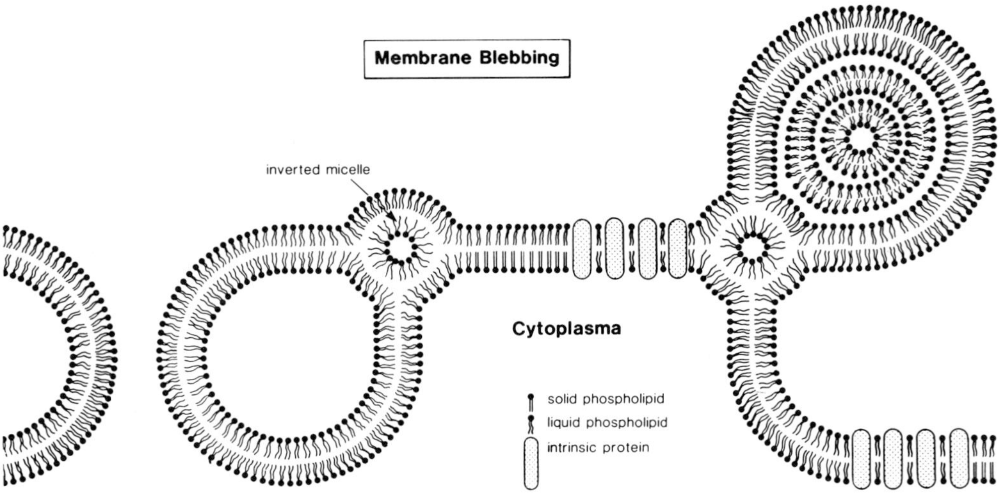

Fig. 5 Proposed model of the irreversible disruption of the sarcolemma. During the first seconds of reperfusion an increase of intracellular Ca^{++} takes place. This induces solidification of the negatively charged phospholipids in the inner monolayer of the sarcolemma, leading to a lateral phase separation of the lipids in the cytoplasmic monolayer and an aggregation of the intramembranous particles. Due to this lateral phase segregation a destabilization of the bilayer and fusion events take place by the formation of inverted micelles leading to the formation of blebs.

ACKNOWLEDGEMENT

This study is supported by the Dutch Heart Foundation.

REFERENCES

Alto LE, Dhalla NS (1979) Myocardial cation contents during induction of the calcium paradox. Am J Physiol 237: H713-H719

Ashraf M, Halverson CA (1977) Structural changes in the freeze-fractured sarcolemma of ischemic myocardium. Am J Pathol 88: 583-594

Bailey IA, Radda GK, Seymour AL, Williams SR (1982) Effects of insulin on myocardial metabolism and acidosis in normoxia and ischemia. Biochim Biophys Acta 720: 17-29

Borgers M, Piper HM (1986) Calcium shifts in anoxic cardiac myocytes. A cytochemical study. J Mol Cell Cardiol 18: 439-448

Cullis PR, de Kruijff B, Hope MJ, Verkleij AJ, Nayar R, Farren SB, Tilcock C, Madden TD, Bally MB (1983) Structural properties of lipids and their functional roles in biological membranes. In: Alia RC (ed) Membrane fluidity in biology. Concept of membrane structure. Academic Press New York vol I 39

De Kruijff B, Cullis PR, Verkleij AJ, Hope MJ, van Echteld CJA, Taraschi TF, van Hoogevest P, Killian JA, Rietveld A, van der Steen ATM (1985) Modulation of lipid polymorphism by lipid-protein interactions. In: Watts A, de Pont JJHHM (eds). Progress in protein-lipid interactions. Elsevier Science Publishers BV Amsterdam 89

Elgsaeter A, Shutton D, Branton D (1976) Intramembrane particle aggregation in erythrocyte ghosts. II. The influence on spectrin aggregation. Biochim Biophys Acta 426: 101-122

Frank JS (1983) Ca-depletion of the sarcolemma-ultrastructural changes. Eur Heart J 4 (suppl H): 23-27

Glauert AM (1974) Fixatives. In: Glauert AM (ed). Practical methods in electron microscopy. Elsevier/North Holland, Amsterdam Vol 3 p 30

Hearse DJ, Humphrey SM, Bullock GR (1978) The oxygen paradox and the calcium paradox: two facets of the same problem? J Mol Cell Cardiol 10: 641-668

Hope MJ, Cullis PR (1979) The bilayer stability of inner monolayer lipids from the human erythrocyte. FEBS Lett 107: 323-326

Jennings RB, Sommer HM, Kaltenbach JP, West JJ (1964) Electrolyte alterations in acute myocardial ischemic injury. Circ. Res. 14: 260-269

Jennings RB, Reimer KA, Jones RN, Peyton RB (1983) High energy phosphates, anaerobic glycolysis and irreversibility in ischemia. In: Spitzer JJ (ed). Myocardial ischemia. Plenum New York p 403

Kalina M, Pease DC (1977) The preservation of ultrastructure in saturated phosphatidyl cholines by tannic acid in model systems and type II pneumatocytes. J Cell Biol 74: 726-741

Katz AM, Reuter H (1979) Cellular calcium and cardiac cell death. Am J Cardiol 44: 188-190

Kloner RA, Ganote CE, Whalen DA, Jennings RB (1974) Effect of a transient period of ischemia on myocardial cells. II. Fine structures during the first minutes of reflow. Am J Pathol 74: 399-422

Knoll G, Oebel G, Plattner H (1982) A simple sandwich-cryogen-jet procedure with high cooling rates for cryofixation of biological materials in the native state. Protoplasm 111: 161-176

Lelkes G, Lelkes G, Szinyei Merse K, Hollan SR (1983) Intensive, reversible aggregation of intramembrane particles in non-hemolyzed human erythrocytes. A freeze-fracture study. Biochim Biophys Acta 732: 48-57

Muir AR (1968) A calcium-induced contracture of cardiac muscle cells. J Anat 102: 148-149

Nayar R, Schmidt S, Hope MJ and Cullis PR (1982) Structural preferences of phosphatidylinositol and phosphatidylinositol-phosphatidyletanolamine model membranes. Influence of Ca^{++} and Mg^{2+}. Biochim Biophys Acta 688: 169-176

Nayler WG, Elz JS, Perry SE, Dally MJ (1983) The biochemistry of uncontrolled calcium entry. Eur Heart J 4 (Suppl H): 29-41

Op den Kamp JAF (1979) Lipid asymmetry in membranes. Ann Rev Biochem 48: 47-71

O'Rourke RA, Hanrath P, Henry WN, Hugenholtz PG, Pisa Z, Roelandt J, Tanaka M (1984) Report of the joint international society and federation of cardiology / World Health Organization task force on recommendations for standardization of measurements from M-mode echocardiograms. Circulation 69: 854A-857A

Post JA, Leunissen-Bijvelt J, Ruigrok TJC, Verkleij AJ (1985a) Ultrastructural changes of sarcolemma and mitochondria in the isolated rabbit heart during ischemia and reperfusion. Biochim Biophys Acta 845: 119-123

Post JA, Nievelstein PFEM, Leunissen-Bijvelt J, Verkleij AJ and Ruigrok TJC (1985b) Sarcolemmal disruption during the calcium-paradox. J Mol Cell Cardiol 17: 265-273

Post JA, Lamers JMJ, Verdouw PD, Ten Cate FJ, Van der Giessen WJ, Verkleij AJ (1987) Sarcolemmal destabilization and destruction after ischemia and reperfusion and its relation with long-term recovery of regional left ventricular function in pigs. Eur Heart J 8: 423-430

Shen AC, Jennings RB (1972) Myocardial calcium and magnesium in acute ischemic injury. Am J Pathol 67: 417-440

Verdouw PD, Remme WJ, Hugenholtz PG (1977) Cardiovascular and antiarrhythmic effects of apridine (AC 1802) during partial occlusion of a coronary artery in the pig. Cardiovasc Res 11: 317-323

Verkleij AJ, Zwaal RFA, Roelofsen B, Comfurius P, Kastelyn D and van Deenen LLM (1973) The asymmetric distribution of phospholipids in the human red cell membrane. Biochim Biophys Acta 323: 178-193

Verkleij AJ, Ververgaert PHJTH (1978) Freeze-fracture morphology of biological membranes. Biochim Biophys Acta 515: 303-327

Verkleij AJ (1984) Lipidic intramembraneous particles. Biochim Biophys Acta 779: 43-63

Verkleij AJ, Post JA (1987) Physicochemical properties and organization of lipids in membranes. Its possible role in myocardial injury. Basic Res Cardiol: in press

Zimmerman ANE, Hülsmann WC (1966) Paradoxical influence of calcium ions on the permeability of the cell membranes of the isolated rat heart. Nature 211: 646-647

Zimmerman ANE, Daems W, Hülsmann WC, Snijder J, Wisse E, Durrer D (1967) Morphological changes of heart muscle caused by successive perfusion with calcium-free and calcium-containing solution (calcium paradox). Cardiovasc Res 1: 201-209

THE ROLE OF MITOCHONDRIAL MEMBRANE PHOSPHOLIPID POLAR HEADGROUPS IN YEAST CYTOCHROME *C* OXIDASE KINETICS.

AKHILESH TRIVEDI AND E. RENO TUSTANOFF,
Department of Biochemistry, The University of Western Ontario, London, Ontario, CANADA, N6A 4G5.

INTRODUCTION:

The availability of yeast lipid mutants is beginning to have an important impact in unravelling the relationship between mitochondrial membrane lipids and proteins. The use of sterol (Parks, 1978) and fatty acid auxotrophs (Lands and Graff, 1981, Trivedi *et al.*, 1986) has provided information on several aspects of the structural-functional relationships that exist in membranes of varying lipid composition. Recently we have resorted to using *Saccharomyces cerevisiae* mutants which are defective in various enzymes responsible for the synthesis of phospholipids to probe these relationships further (Trivedi *et al.*, 1987). These studies not only have emphasized the regulatory role of lipids, but have afforded an insight into their possible control of the assembly of mitochondrial membranes. Although we have been able to manipulate the phospholipid composition of yeast mitochondrial membranes significantly, the overall net charge of these membranes has been conserved due to biological compensation.

The charges on membrane lipid molecules may be an important factor in regulating the nature of biological membranes. The forces which are involved in maintaining the configuration of membranes, are quite likely to be influenced by the polar headgroups of phospholipids. To substantiate this premise, we have carried out a series of experiments using the yeast mutant *cho2* which has been shown by genetic and biochemical analysis to be defective in the methylation pathway of phosphatidylethanolamine (Hirsch and Henry, 1986), cf. Fig. 1. Employing this mutant we have been able to alter the nature of the phospholipid polar headgroups in mitochondrial membranes and have monitored the consequences of these manipulations on the electron transfer capability of the membrane-bound enzyme cytochrome *c* oxidase.

METHODS:

The *S. cerevisiae* mutant strain, *cho2* (MATa lys2 cho2-1) and its parental strain, *ade5a*, were obtained from Dr. S.A. Henry, Department of Biological Sciences, Carnegie-Mellon University, Pittsburgh, USA. These strains were

NATO ASI Series, Vol. H16
Membrane Biogenesis. Edited by J. A. F. Op den Kamp

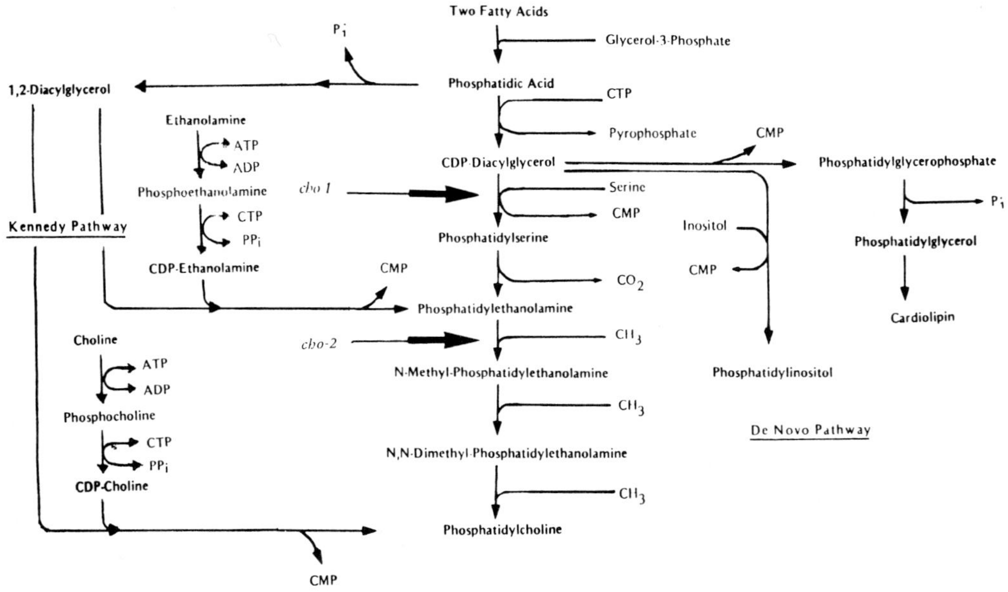

Fig. 1: Phospholipid Biosynthesis Pathways in Yeast.

maintained and cultured as previously reported by Hirsch and Henry (1986). For experiments, cells were pre-grown in YEPD liquid medium [1% Difco yeast extract, 2% glucose and 2% peptone (all w/v)] overnight and then transferred (ca. 5×10^6 cells/mL of medium) into synthetic minimal medium supplemented with either 1 mM choline or its analogues (Fantin, 1987). Cells were grown into their late logarithmic phase at 30°C. Auxotrophic integrity of mutant cells were monitored using replica plating techniques. Cell strains were also monitored for *petite* formation and if colonies contained more than 5% *petite* cells, they were discarded.

Mitochondria were isolated mechanically from harvested cells, cytochrome *c* -depleted, and analyzed for their phospholipid and fatty acyl composition as previously described (Trivedi *et al.*, 1986). The oxidase concentration (heme aa_3) were determined according to van Gelder (1966). The rate of oxidation of graded amounts of cytochrome *c* by various mitochondrial preparations was measured spectrophotometrically at 25°C (Trivedi *et al.*, 1986). Assays were carried out at high pH and low ionic strength (I=17 mM) as described by Errede *et al.* (1976). The data were depicted as Eadie-Hofstee plots of v/S versus v where v is the velocity ($M.s^{-1}$) and S is the substrate concentration (M). A series of biphasic curves were obtained as lines of best fit using a computer programme to plot the calculated data. Apparent K_m values were calculated from the negative slopes of the high and low affinity phases of these plots. The extension of the abscissa intercept of each phase of the curves gave the maximum velocity values.

RESULTS AND DISCUSSION:

It has been suggested by Rietveld and de Kruijff (1986) that there are two sites for the association of phospholipids with integral proteins during membrane biogenesis. In one, the phospholipid fatty acyl chains in the membrane bilayer interact with the core of the membrane-bound enzyme to provide a fluid environment which facilitates protein lateral movement. The other region of the molecule, the polar headgroups, intercalates and encompasses the protein complex at the bilayer surface and aids in forming a boundary. It was reported that the mitochondrial membrane bound-enzyme, cytochrome c oxidase, possesses both boundary and protein-bound phospholipids (Dennis, 1986). However, the role of these specific phospholipids has been questioned and their association with the oxidase is suspect as a result of recent findings (Fantin 1987). Results from our laboratory, indicate that the relationship between phospholipid and cytochrome c oxidase is not limited only in providing a suitable environment for the enzyme's activity but also contributes a more dynamic role in mitochondrial membrane biogenesis. This led us to ponder whether the assembly and/or function of proteins can be influenced by the phospholipid environment of the membrane. To assess this possibility, we varied the charge and the geometry of headgroups of specific phospholipids in situ and measured the effects of these alterations on the kinetics of cytochrome c oxidase activity as well as on the binding affinity of cytochrome c.

To this end, four different analogues of choline [HO-CH_2-CH_2-$^{+}N(CH_3)_3$] which potentially could be incorporated into the polar portion of the phospholipid glycerol backbone were either synthesized or obtained commercially. The mutant's ability to use these bases via the Kennedy pathway (Fig. 1) allowed us to change the nature of membrane phospholipid. The analogues used were: (i) 3-amino-1-propanol [HO-CH_2-CH_2-CH_2-$^{+}NH_3$] which would give rise to a phospholipid in which the -CH_2-CH_2- group linking the phosphoryl and quaternary ammonium group of the phosphorylethanolamine moiety is replaced by a propyl [-CH_2-CH_2-CH_2-] group; (ii) diethylethanolamine [HO-CH_2-CH_2-^{+}NH-$(C_2H_5)_2$] which would replace the three methyl constituents on the quaternary ammonium nucleus of the phospholipid base with two larger space filling ethyl groups; (iii) phosphoniumcholine [HO-CH_2-CH_2-$^+P(CH_3)_3$] would represent a phospholipid whose terminal nitrogen (atomic radius 0.53 A°) in the headgroup is replaced by a phosphorous atom (atomic radius 0.92 Å) and lastly (iv) sulfoniumcholine [HO-CH_2-CH_2-$^+S(CH_3)_2$] in which similar to phosphoniumcholine the nitrogen atom is replaced by a sulfur atom (atomic radius 1.06 A°). These choline analogues can be categorized into two distinct classes. Diethylethanolamine and 3-amino-1-propanol containing headgroups would be bulkier in size and may restrict the orientation of the bilayer. On the other hand, phosphonium- and sulfoniumcholine would alter the charges on the phospholipid head group as a result of the increased electronegativity and atomic radii.

To see whether these bases can be assimilated _via_ the Kennedy pathway and synthesize analogues of phosphatidylcholine, the mutant cells were grown in the presence of each base and their growth rates were observed (data not shown). The cells were able to utilize these bases and when isolated mitochondria from these cells were analyzed for their phospholipid content it was seen that these bases with their altered bulk and charge were incorporated into the mitochondrial membrane phospholipids (Table I). It can be seen, for example, when mutant cells were grown in the presence of diethylethanolamine, a major portion of the total phospholipid content was found to be phosphatidyldiethylethanolamine (30%). Similarly, by growing the _cho2_ mutant in the presence of the other analogues, noticeable membrane enrichment was observed in the newly synthesized phospholipid, e.g., phosphoniumphosphatidylcholine, sulfoniumphosphatidylcholine and phosphatidylpropanolamine.

On comparing the mitochondrial phospholipid composition of mutant cells grown in lipid base-supplemented medium with the wild-type, no change in the individual phospholipid content was noticed except that the level of phosphatidylserine was raised significantly in unsupplemented mutant cells. This was attributed to a lesion in the pathway which convert phosphatidylethanolamine to phosphatidylcholine. Since the synthesis of phosphatidylserine is a rate-limiting step in the biosynthesis of phospholipid in yeast (Atkinson _et al._, 1980), an enrichment in its level was expected in the mutant, however, the levels of this phospholipid dropped to that found in the parental strain on the addition of choline which circumvented the defect in the methylation pathway. The fatty acid composition of these analogue-supplemented membranes is shown in Table II. Overall, the ratio of unsaturated to saturated fatty acids changed marginally from 1.1 to 1.5 in these different preparations, however ,in contrast to its parental strain _ade5a_, the mitochondrial fatty

Table I. Mitochondrial Phospholipid Composition

Strain	Medium Supplemented (1mM)	Total Phospholipid (%)										
		PI[a]	PS	PC	PE	CL	SPC	PPC	PDE	PPA	UC	PC/PE
ade5a	none	16.7	7.4	43.3	15.2	11.5	-	-	-	-	5.9	2.8
cho2	none	10.2	33.5	14.8	26.2	10.2	-	-	-	-	5.1	0.6
cho2	choline	12.9	7.2	39.2	27.4	9.5	-	-	-	-	3.8	1.4
cho2	A.P.	11.5	24.2	4.9	12.6	11.8	-	-	-	30.2	4.8	0.4
cho2	D.E.	16.3	19.1	3.3	18.5	8.2	-	-	30.6	-	4.0	0.2
cho2	P.C.	13.8	17.5	5.3	15.6	11.6	-	30.4	-	-	5.8	0.4
cho2	S.C.	14.7	17.3	4.9	17.3	12.4	28.6	-	-	-	4.8	0.3

[a]PI (phosphatidylinositol); PS (phosphatidylserine); PC (phosphatidylcholine); PE (phosphatidylethanolamine); CL (cardiolipin); SPC (sulfoniumphosphatidylcholine); PPC (phosphoniumphosphatidylcholine); PDE (phosphatidyldiethanolamine); PPA (phosphatidylpropanolamine); UC (mainly mono and dimethylphosphatidylethanolamine); A.P. (3-amino-1-propanol); D.E. (diethylethanolamine); P.C. (phosphoniumcholine); S.C. (sulfoniumcholine).

Table II. Mitochondrial Fatty Acid Composition (%)

Strain	Medium addition (1mM)	12:0	14:0	16:0	16:1	18:0	18:1	U/S*
ade5a	none	0.9	0.9	36.4	27.5	6.4	27.9	1.2
cho2	none	1.5	2.4	27.4	35.6	9.7	23.4	1.4
cho2	choline	1.1	1.9	26.2	38.9	10.9	21.0	1.5
cho2	A.P.	1.8	2.3	28.2	34.4	12.8	20.3	1.2
cho2	D.E.	1.6	1.7	32.1	29.7	13.6	21.3	1.1
cho2	P.C.	1.3	2.1	26.7	34.9	11.3	23.7	1.4
cho2	S.C.	1.9	2.6	27.9	31.2	11.3	25.1	1.3

* Ratio of unsaturated to saturated fatty acid.

mitochondrial fatty acid composition of the various preparations of the mutant strain had changed. The levels of C14:0, C16:1 and C18:0 increased while those of C16:0 and C18:1 decreased. The redistribution of these membrane fatty acids are probably required to compensate for the incorporation of the charged and bulkier headgroups into mitochondrial phospholipids. Since no significant change was observed in the fatty acid ratio, it follows that changes in mitochondrial function are probably related to the alteration in phospholipid headgroups because these complex, bulky polar head groups may alter the geometry of the lipid bilayer and thus influence lipid-protein interactions. Experiments were undertaken to assess the effect of these membrane alterations on protein interactions, specifically cytochrome *c* oxidase. When studied under optimal conditions, steady-state biphasic kinetics for this enzyme have been shown to be first order and have been explained by two simultaneous reactions proceeding between cytochrome *c* and its oxidase, i.e., high and low affinity reaction sites (Trivedi *et al.*, 1986). It can be seen from the data in Table III that marked changes in the kinetics of this enzyme have been imposed by the choline analogue supplementation. The apparent Km value for the low affinity binding site in *cho2* unsupplemented cells increased when contrasted to the parental strain (12.6 μM vs. 4.9 μM). After choline supplemention the K_m for the low affinity site dropped to 5.7 μM. Since choline-supplemented mutant cells attempt to maintain their mitochondrial phospholipid composition in line with their

Table III. Cytochrome _c_ Oxidase Kinetics*

Strain	Medium addition (1mM)	Apparent "K_m"[a] Low Affinity	Apparent "K_m"[a] High Affinity	TN_{max}[b] Low Affinity	TN_{max}[b] High Affinity
ade5a	none	4.9	0.36	75	36
cho2	none	12.6	0.41	63	23
cho2	choline	5.7	0.37	82	37
cho2	A.P.	33.9	0.61	66	27
cho2	D.E.	22.3	0.67	67	33
cho2	P.C.	11.3	1.04	204	106
cho2	S.C.	12.4	1.63	182	135

* The kinetic data tabulated in this table were corrected for cryptic oxidase activity as previously described (Trivedi _et al._, 1986). The oxidase activity was measured by using 4 nM heme _aa_$_3$ in each assay.
a) μM; b) electron.s^{-1}.

parental strain (Table I), the changes brought about in the low affinity binding site can easily be related to the altered phospholipid head-groups.

The incorporation of either diethylethanolamine or 3-amino-1-propanol into membrane phospholipids increased the K_m of the low affinity reaction of the oxidase approximately six-fold over that of the parental or the mutant strain grown on its required substrate choline. The substrate requirements for the low affinity site for cells grown on phosphoniumcholine and on sulfoniumcholine were also affected but to a much lower degree (ca. two-fold increase). Alterations in the kinetics of the high affinity binding site were also evident in these mutant membrane preparations. The different analogue-enriched mitochondria demonstrated higher requirements for cytochrome _c_ for the oxidase's high affinity site than either the parental strain, the unsupplemented or the choline-grown mutant cells (Table III). Of these, mitochondria isolated from sulfoniumcholine and phosphoniumcholine grown cells demonstrated the greatest requirement for substrate, approximately two times more than the other two analogue mitochondrial preparations. It would appear from these results that the configuration of the phospholipid head group may play an important regulatory role in the enzyme's activity. The modulating effects are elicited not only by the nature of the electrical charge but are also conditioned by the size or bulk of the phospholipid's polar headgroup. The charge on the phospholipid headgroup as illustrated by both phosphonium- and sulfoniumcholine may have a discriminatory role which primarily affects the catalytic high affinity binding site of the oxidase while the bulkiness of the headgroup appears to influence the enzyme's low affinity binding site. It therefore, follows that the configuration of the phospholipid's head group is important in maintaining the oxidase activity on the basis of their physical chemical interactions.

Robinson (1982) studying the relationship between the phospholipid cardiolipin and cytochrome c oxidase, proposed that this phospholipid is necessary to maintain the enzyme complex in a proper orientation by clustering its negatively charged headgroups around the interface of the protein. In order to demonstrate the specificity between phospholipid headgroups and oxidase function, it is necessary to show how the electron transfer capability of this enzyme is affected by incorporating the different analogue head-groups into mitochondria. Surprisingly, the capability to transfer electrons at both binding sites was dramatically enhanced in those mitochondria whose membrane phospholipids were enriched with charged headgroups (Table III). The enzyme's maximal turn over rate (TN_{max}) in these membranes increased from 75 to 204 electron.s^{-1} at its low affinity binding site and from 36 to 135 electron.s^{-1} at its high affinity site. The enrichment of mitochondria with other analogue headgroups, however, did not elicit any change in their TN_{max} values. This would indicate that it is not the chemical interactions of phospholipid which maximize the molecular activity of the oxidase but probably their physical characteristics, i.e., the charge on the lipid may control the flow of electrons from cytochrome c to the enzyme and then to oxygen.

The differences observed in oxidase kinetics in the various membrane preparations may be attributed to modifications in the structure or expression of the enzyme during its synthesis. It creating a new membrane lipid environment, the translocation and processing of the various cytoplasmic oxidase polypeptides may have been interfered with. It is possible that the cho2 mutant may be pleiotropic and the observed changes in enzyme kinetics are due to altered gene expression of the enzyme's subunits. To examine if the levels of the oxidase had been affected, absolute low temperature cytochrome spectra were carried out on both parental and cho2 cells (Fig. 2). In contrast to the ade5a cells, the levels of cytochrome c were depressed and there were concomitant increases in the concentrations of cytochrome c_1 and b

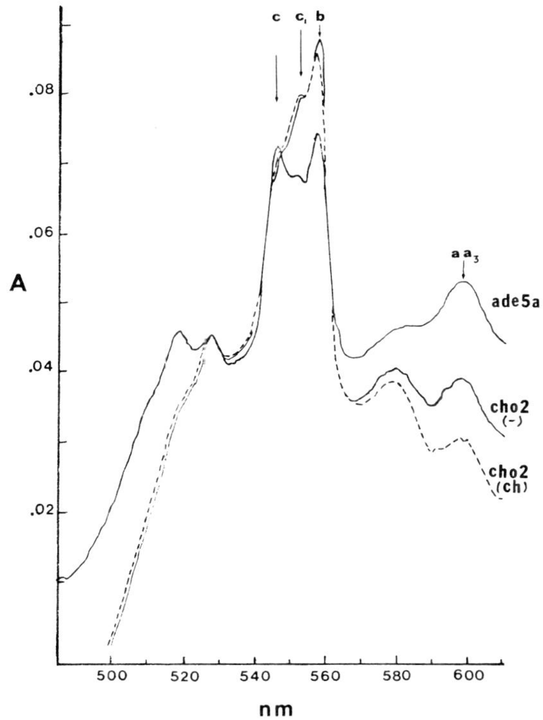

Fig. 2: Low Temperature Difference Spectra of cho2 and Wild-type Cells. Cell suspensions (5 mg protein/ ml 0.6 M sucrose) were reduced and oxidized respectively and then frozen in liquid N_2 prior to recording their spectra.

in the mutant cells. Cytochrome aa_3 levels in the mutant were markedly depressed, particularly in the presence of choline supplementation. The peak at 580 nm which was quite prominent in the mutant, was not identified but could be a precursor form of cytochrome aa_3. If this is the case, it is possible that the expression or incorporation of various mitochondrial proteins is indeed affected in the mutant strain when grown in the presence of the different organic bases. This would suggest that simultaneous synthesis of phospholipids and membrane proteins is necessary for mitochondrial membrane biogenesis. However, the kinetic data presented here can not be explained on the basis of these altered cytochrome ratios since the oxidase kinetics were carried out in the presence of 4nM heme aa_3.

From the results presented in this report, it would appear that the rate of oxidation of ferrocytochrome c by the oxidase is dependent on the nature of mitochondrial membrane phospholipid headgroups. Using a series of different physico-chemical approaches, it has been demonstrated by various investigators that cardiolipin, a constituent of the inner mitochondrial membrane is tightly bound or found to be the residual phospholipid in membrane and delipidated cytochrome c oxidase preparations (Dennis, 1986). Of all the mitochondrial phospholipids, cardiolipin is well suited for its role since it carries not only negative charges on its bulky headgroup but also contributes to the fluidity to the membrane by its high content of unsaturated fatty acid constituents. Therefore, any alterations in the size and nature of the other phospholipids may aid in the binding of cytochrome c to the oxidase and thus illustrates the importance of phospholipid headgroup in the kinetics of membrane-bound enzymes.

ACKNOWLEDGEMENT:

This work was funded through a grant-in-aid to ERT by the Medical Research Council of Canada (MT-1460).

REFERENCES:

Atkinson K, Fogel S and Henry S (1980) Yeast mutants defective in phosphatidylserine synthesis. J. Biol. Chem. 255: 6653-6661

Dennis M (1986) Structure and function of cytochrome c oxidase. Biochimie 68: 459-470

Errede B, Haight GP and Kamen MD (1976) Oxidation of ferrocytochrome c by mitochondrial cytochrome c oxidase. Proc. Natl. Acad. Sci. 73: 113-117

Fantin DJ (1987) The role of lipids in the structure and function of yeast membranes. Ph.D. Thesis, University of Western Ontario, London, Canada.

Hirsch JP and Henry SA (1986) Expression of the Saccharomyces cerevisiae inositol-1-phosphate synthase (inol) gene is regulated by factors that affect phospholipid synthesis. Mol. Cell. Biol. 6: 3320-3328

Lands W.E.M. and Graff G (1981) Effect of fatty acid structure on yeast cell physiology. In Current Developments in Yeast Research (Stewart GG and Russell I, eds.) pp 331-341, Pergamon Press, Toronto

Park LW (1978) Metabolism of sterols in yeast. CRC Crit. Rev. Microbiol. 6: 301-341

Rietveld A and de Kruijff B (1986) Phospholipids as a possible instruments for translocation of nascent proteins across biological membranes. Biosci. Rep. 6: 775-782

Robinson NC (1982) Specificity and binding affinity of phospholipids to the high-affinity cardiolipin sites of beef heart cytochrome c oxidase. Biochemistry 21: 184-188

Trivedi A, Fantin DJ and Tustanoff ER (1986) Role of phospholipid fatty acids on the kinetics of high and low affinity sites of cytochrome c oxidase. Biochem. Cell Biol. 64: 1195-1210

Trivedi A, Schwab M and Tustanoff ER (1987a) Significance of cardiolipin headgroups on yeast cytochrome c oxidase kinetics. In Membrane Receptors, Dynamics and Energetics (Wirtz KWA, ed.) pp 319-28, Plenum Press, New York

Van Gelder (1966) On cytochrome c oxidase. I. The extinction coefficient of cytochrome a and a_3. Biocim. Biophys. Acta 118: 36-46

THE REGULATION OF BILAYER STABILITY IN BACTERIA*

Howard Goldfine
Department of Microbiology
University of Pennsylvania
School of Medicine
Philadelphia, Pennsylvania 19104-6076
U.S.A.

The membranes of prokaryotic and eukaryotic cells contain mixtures of amphipathic lipids. When isolated, the major lipids can form different phase structures upon hydration at physiological temperatures. As discussed in other lectures in this volume (de Kruijff, Op den Kamp), the structures of lipid aggregates are related to the generalized shapes of the molecules. Bilayer-forming lipids (type II) have roughly cylindrical shapes whereas H_{II} phase lipids (type III) have a cone shape in which the area swept out by the motion of the hydrocarbon chains is larger than that occupied by the polar head groups. Hydration, charge repulsion, and hydrogen bonding, can affect the cross-sectional areas of the polar regions, and temperature and chain unsaturation influence the volumes occupied by the hydrocarbon chains (Rilfors et al., 1984; Cullis & Hope, 1985). Among the type II lipids found in prokaryotes are phosphatidylcholine (PC), phosphatidylglycerol (PG), phosphatidylserine (PS), cardiolipin (CL), and diglycosyldiacylglycerols. Among type III lipids are unsaturated species of phosphatidylethanolamine (PE), plasmenylethanolamine (PlaE) and monoglycosyldiacylglycerols (Goldfine, 1982). Lipids that aggregate to form micelles (Type I), for example lyso-phosphoglycerides, are rarely found in large amounts. The addition of Ca^{++} to CL or

*Research in the author's laboratory was supported by a research grant from the U.S. Public Health Service, National Institute of Allergy and Infectious Disease, #AI-08903.

NATO ASI Series, Vol. H16
Membrane Biogenesis. Edited by J. A. F. Op den Kamp

lowering the pH of PS* can convert them to type III lipids (Cullis and de Kruijff, 1979; Israelachvili et al., 1980; Gruner et al., 1985). A nearly continuous bilayer structure is probably essential for membrane function and the mixtures of lipids in biological membranes appear to be carefully balanced to include appropriate proportions of type II and type III lipids. The metabolic regulatory mechanisms that function to achieve the correct proportions of these lipids in a given membrane are little understood at the present time.

Regulation of Lipid Composition in Clostridia

We have studied this problem in two species of saccharolytic clostridia, *Clostridium butyricum* and *Clostridium beijerinckii*. These organisms are natural biotin auxotrophs. Since biotin is required in the formation of fatty acids at the step catalyzed by acetyl-CoA carboxylase, addition of long-chain unsaturated fatty acids can relieve the biotin requirement (Broquist and Snell, 1951). Under these conditions, the fed fatty acids are extensively incorporated into the membrane lipids. For example, when grown on oleic acid or its trans isomer, elaidic acid, in the complete absence of biotin, the cellular membrane lipids contain over 95% unsaturated chains or a combination of unsaturated and cyclopropane chains, which are derived from oleate (Khuller and Goldfine, 1975; Goldfine et al., 1977; Goldfine et al., 1981).

The phospholipids of these organisms normally contain mixtures of saturated, unsaturated and cyclopropane aliphatic chains when the cells are grown with biotin. Under these conditions the predominant amino lipids in *C. beijerinckii* are phosphatidyl N-methylethanolamine (PME), its plasmalogen (PlaME), and the glycerol acetal of plasmenyl-N-methylethanolamine. In *C. butyricum* the predominant amino lipids are PE,

---

*Abbreviations used: DGDG, diglucosyldiacylglycerol; CL, cardiolipin; GAPlaE, glycerol acetal of plasmenylethanolamine; MGDG, monoglucosyldiacylglycerol; PC, phosphatidylcholine; PG, phosphatidylglycerol; PME, phosphatidyl-N-monomethylethanolamine; PS, phosphatidylserine; PlaE, plasmenylethanolamine; PlaME, plasmenyl-N-monomethylethanolamine.

PlaE and the glycerol acetal of PlaE (GAPlaE)(Fig. 1). The other major polar lipids in these organism are PG and CL, which are approximately

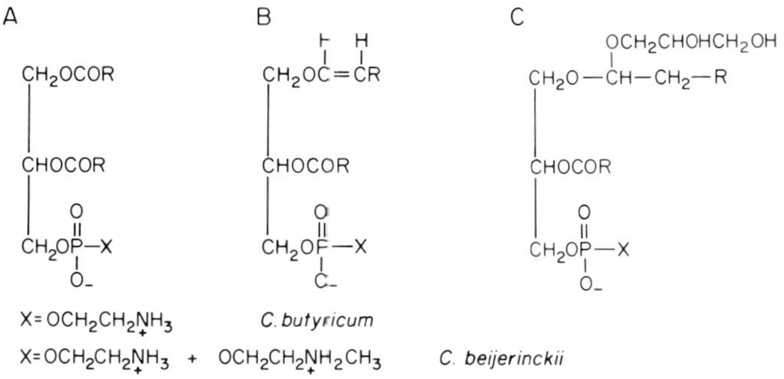

Figure 1. Major amino phospholipids of C. butyricum and C. beijerinckii. A. Diacyl; B. Plasmalogen; C. Glycerol acetal of a plasmalogen.

one-third of the plasmalogen type. Phospholipase C hydrolysis and trinitrobenzenesulfonate derivatization were used to probe the topology of these lipids in the cell membrane of C. butyricum. PE, PlaE, and GAPlaE appeared to be predominantly localized in the outer monolayer of the cell membrane in both biotin- and fatty acid-grown cells (Goldfine, et al., 1982; Johnston & Goldfine, 1985). When C. butyricum was grown on oleic acid in the absence of biotin, the ratio of GAPlaE to total PE increased from 0.86 found in biotin grown cells to 1.6. Growth of this organism on elaidic acid, the trans isomer of oleic acid, produced a ratio of 0.77. A similar increase in the glycerol acetal of plasmenyl-N-methylethanolamine content was observed in C. beijerinckii grown on oleic acid, but not on elaidic acid (Goldfine et al., 1977; Goldfine, 1985).

In order to modulate the content of unsaturated aliphatic chains, experiments were done in which C. butyricum was grown on mixtures of saturated and unsaturated fatty acids in the absence of biotin. Under these conditions the alkenyl chains of the plasmalogens and the alkyl chains of the glycerol acetal lipids were largely either unsaturated or cyclopropane even when the cells were grown on palmitate/oleate, 80:20 (wt/wt). However, as the proportion of oleate in the medium was increased, the content of unsaturated and cyclopropane acyl chains increased dramatically. Analysis of the phospholipid composition of these cells showed a concomitant increase in the GAPlaE/PE + PlaE ratio (Fig. 2).

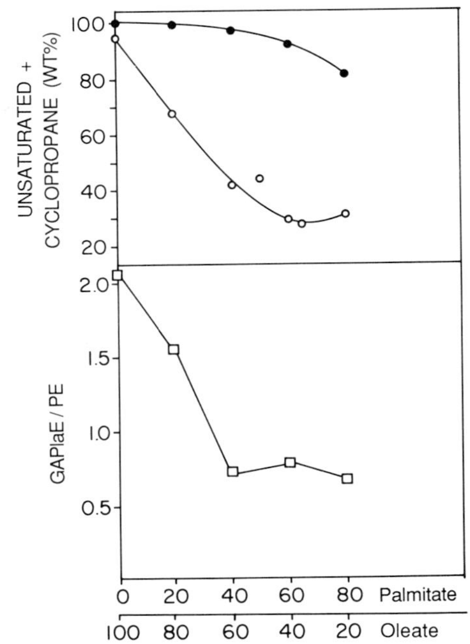

Figure 2. Upper panel; phospholipid acyl chain (o) and alkenyl plus alkyl chain (●) composition of C. butyricum grown on mixtures of palmitate plus oleate. Lower panal; The ratio of GAPlaE to total PE in these cells (□).

From earlier physical studies on PE it was evident that the lamellar to H_{II} phase transition temperature (T_{BH}) of this lipid decreases as the content of unsaturated fatty acids is increased (Cullis & de Kruijff, 1979). Recent work with PlaE containing saturated alkenyl chains and unsaturated acyl chains has demonstrated an even greater tendency to form the H_{II} phase (Lohner et al., 1984). Thus it was clear that an increase in the degree of unsaturation of PE and PlaE in C. butyricum would tend to destabilize the lamellar arrangement of the membrane lipids. We then asked whether the observed increase in GAPlaE serves to stabilize the bilayer arrangement. Does this lipid form a stable lamellar phase at and above the growth temperature of the organism when it is highly unsaturated, and would it stabilize the bilayer arrangement when mixed with unsaturated forms of PE and PlaE? Examination of space-filling models suggested that the glycerol moiety in acetal linkage, which appears to be highly hydrated (Goldfine et al., 1981), would increase the effective size of the polar head group leading to an increase in T_{BH}. Studies on the polymorphic behavior of these lipids were undertaken in collaboration with the laboratory of Graham Shipley.

Upon differential scanning calorimetry, the mixed PE/PlaE from C. butyricum grown on oleate produced an endothermic gel to liquid crystalline phase transition at -1.9 °C with a shoulder at 0.2 °C. ^{31}P NMR studies of this fraction gave a signal characteristic of an H_{II} phase at 0 °C. and the results of X-ray diffraction studies confirmed the low T_{BH} of this mixed lipid fraction. As predicted, the highly unsaturated glycerol acetal of PlaE isolated from the same cells behaved very differently. Its gel to liquid crystalline transition temperature was significantly higher (18 °C) with a large hysteresis in the reverse transition, which occurred at -2.2 °C, as observed previously (Goldfine et al., 1981). ^{31}P NMR and X-ray diffraction, indicated a transition from a lamellar phase to an isotropic phase between 50 and 60 °C (Goldfine et al., 1987). Thus the addition of a glycerol moiety on the sn-1 chain of PlaE resulted in an increase in the lamellar to non-lamellar transition temperature of over 45°.

Since cells grown on oleic acid produce relatively little of the PE fraction, we performed experiments in which oleate-enriched GAPlaE was mixed with natural and synthetic unsaturated species of PE. In addition to studies with egg and red blood cell PE, which provided evidence that oleate-enriched GAPlaE stabilized the bilayer arrangement, we undertook an in-depth study of its affect on dioleoyl PE (DOPE). This lipid undergoes a lamellar to H_{II} transition between 5 and 10 °C (Gruner, 1985). Admixture of varying amounts of oleate-enriched GAPlaE produced the following results obtained by ^{31}P NMR. Mixing 2 parts of DOPE with 1 of GAPlaE (wt/wt) resulted in a lamellar spectrum up to 25 °C with the appearance of isotropic phase signals at higher temperatures. A 1:1 mixture of the same two lipids produced a lamellar phase spectrum at 35 °C with little evidence of nonlamellar lipid at 46 °C. Low angle X-ray diffraction patterns of these mixtures produced a similar picture. At 46 °C an ~2:1 mixture showed two coexisting phases. At ~1:1, the lamellar reflections were dominant with only a very weak set of reflections corresponding to the H_{II} phase. Further addition of GAPlaE to 75.3 mol%

essentially abolished the hexagonal reflections (Goldfine et al., 1987). Similar experiments were carried out with the PE/PlaE mixture from cells grown on oleic acid. ^{31}P NMR again showed that an approximately 1:1 mixture of the two lipids, but not a 2:1 mixture, produced a predominantly lamellar spectrum at the growth temperature (37 °C). Again, X-ray diffraction produced similar findings. Thus it was clear that with the lipids isolated from oleate-grown cells, the total PE / GAPlaE ratio found in cells grown with biotin or on palmitic acid/oleic acid 40:60, would not produce a lamellar phase at the growth temperature, while the mixture found in oleic acid grown cells, ~1:2, would clearly do so.

These studies suggested that the lipid compositions of C. butyricum and C. beijerinckii were affected by lipid shape. Additional studies were carried out to support this conclusion. C. butyricum was grown on a variety of pure fatty acids in the absence of biotin. Again in each case, there was extensive incorporation of the fed fatty acids into the phospholipid aliphatic chains. The ratio of GAPlaE to total PE (note, the inverse of the ratio discussed above) was shown to decrease in the following order; cis-vaccenic acid $\geq$ oleic acid = C_{19}-cyclopropane acid > linoleic acid > petroselinic acid > elaidic acid > 14-methylhexadecanoic acid (anteiso-C_{17}) > 12-methyltridecanoic acid (iso-C_{14}). The last compound was extensively elongated by the cells to iso-C_{16}. These results are, with the possible exception of linoleic acid, entirely consistent with the hypothesis that the membrane lipid composition is strongly influenced by lipid shape. With a given polar head group, the branched chain fatty acids produce lipids which occupy smaller areas in monolayers than cis-unsaturated fatty acids, and lipids with iso acids occupy smaller areas than lipids with anteiso acids (Goldfine et al., to be published). Another way of looking at these results is illustrated in Figure 3 in which the GAPlaE/total PE ratio of C. butyricum grown on different fatty acids is compared with the T_{BH} for PEs containing these fatty acids. It is clear that fatty acids leading to PEs with low T_{BH} resulted in high GAPlaE/total PE ratios, while fatty acids leading to T_{BH} at or above the growth temperature, resulted in lower GAPlaE/total PE ratios.

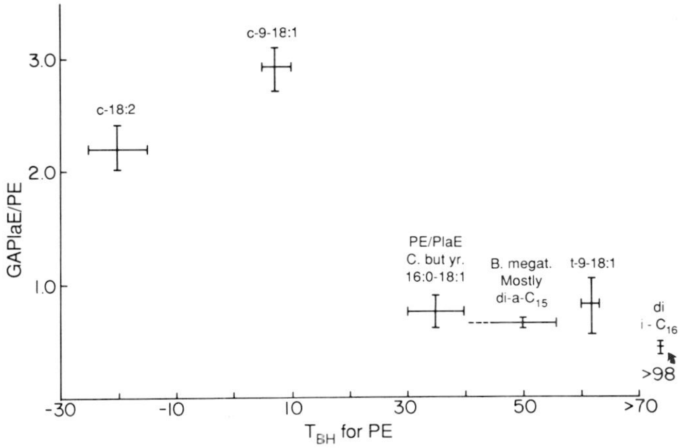

Figure 3. Relationship between the amino phospholipid composition of C. butyricum, expressed as the ratio of GAPlaE to total PE (diacyl plus plasmalogen), and the known T_{BH} (°C) of PEs enriched with specific fatty acids. Data for the PE/PlaE isolated from C. butyricum grown on a mixture of palmitic acid (16:0) and oleic acid (18:1) are from Malthaner et al. (1987). Data for anteiso-C_{15}-enriched PE and for di-iso-C_{16} PE are from Rilfors et al. (1982) and Silvius et al. (1985), respectively. Other values are from Cullis & Hope (1985).

Regulation of Bilayer Stability in Acholeplasma laidlawii

A. laidlawii, a natural fatty acid auxotroph, adjusts the ratio of two major polar glycolipids, monoglucosyldiacylglycerol (MGDG) and diglucosyldiacylglycerol (DGDG) in response to changes in growth temperature, degree of unsaturation of the acyl chains and the cholesterol content of the membranes (Reviewed by Rilfors et al., 1984). As the growth temperature or degree of unsaturation are increased, the ratio of MGDG to DGDG decreases. These workers have shown that unsaturated species of MGDG are unable to form a lamellar phase at physiological temperatures and are stabilized by addition of unsaturated DGDG, which is a type II lipid. Again the mixtures of these lipids produced by the cells grown on either mixtures of palmitic and oleic acid, or on pure oleic acid, appear to be supportive of the lamellar arrangement in the cell membrane. The results obtained with C. butyricum grown on these fatty acids are remarkably similar to those obtained with A. laidlawii when the PE/GAPlaE and MGDG/DGDG ratios are compared and suggest a similar regulatory mechanism (Goldfine, 1985). A. laidlawii responds to growth on iso- and anteiso-branched fatty acids by producing membranes with higher MGDG/DGDG ratios

(Rilfors, 1985). At present it is not known if the control of membrane lipid synthesis in these organisms is at the level of transcription or translation of the genes for the controlling enzymes, or at the level of the activities of the enzymes themselves. Although it is attractive to consider mechanisms in which the controlling enzymes are influenced by the incipient formation of microdomains of non-lamellar lipids as previously proposed (Goldfine, 1985), regulation at the level of the gene has not been ruled out.

Regulation of Bilayer Stability in Other Prokaryotes

We have recently studied an alternative regulatory system, in which a type II lipid is converted to a type III lipid. In _Megasphaera_ _elsdenii_ the two major polar lipids are PE and PS, both of which are predominantly present in their plasmalogen forms. An unidentified polar lipid represents less than 8% of lipid phosphorus. We had previously noted that although the degree of unsaturation of the acyl and alkenyl chains was little affected by the temperature of growth in the range 27 to 42.5 °C, there was a significant change in the PE to PS ratio, which decreased from 2.5 in cells grown at 27° to 1.8 in cells grown at 42.5 °C (Johnston & Goldfine, 1982). These results could not be easily interpreted in terms of the regulation of membrane fluidity since PE species have slightly higher gel to liquid crystalline phase transition temperatures than PS species (Cullis & Hope, 1985). In recent work with this organism we have isolated a plasmalogen-deficient variant by serial passage of the wild type strain. The ratio of PE to PS, 1.5, is somewhat lower in the plasmalogen deficient strain, and the acyl chains, which replace the alkenyl chains of the plasmalogens, are significantly more unsaturated. Physical studies showed relatively small differences in membrane fluidity (diphenylhexatriene polarization) or bilayer stability (^{31}P NMR and X-ray diffraction) (Kaufman, 1986). Recent preliminary reconstitution studies using ^{31}P NMR, have indicated that the mixtures of PE/PS produced by both the wild-type and plasmalogen-deficient strains produce bilayers when extracted from the cell membranes, separated and reconstituted. The ability of PS to stabilize the bilayer arrangement when mixed with PE has been well-studied with diacyl forms of these lipids (Tilcock et al., 1984).

Relatively few other prokaryotes have been studied with respect to the stabilization of the lamellar phase. As noted by Rilfors et al. (1984), the alteration of fatty acid composition with respect to growth temperature in bacteria has often been interpreted in terms of the regulation of membrane fluidity. However, in some cases, for example changes in the iso and anteiso fatty acid content of <u>Bacillus megaterium</u> grown at different temperatures, the observed variations can also be interpreted in terms of bilayer stabilization. In an early study with <u>Pseudomonas fluorescens</u> grown at different temperatures the acyl chains of the phospholipids were found to be more unsaturated at lower growth temperatures, and there was little variation in the lipid class composition. X-ray diffraction studies showed that the phase equilibria with respect to lamellar and H_{II} phases of PE isolated from cells grown at 5 and 22 °C were identical when compared at the respective growth temperatures (Cullen et al., 1971).

Other prokaryotes have not been the subject of experimentation with regard to the stabilization of the membrane bilayer; however, in a previous review article (Goldfine, 1984), I have pointed out that while PC is relatively rare among prokaryotes, it is often found in gram negative species in which the proportion of unsaturated fatty acyl chains is relatively high (≥70%). Since the predominant lipid of most gram negative species is PE, I have speculated that the formation of PC by the successive methylation of PE in those organisms with highly unsaturated membrane lipids serves to stabilize the bilayer arrangement. The various modes of bilayer stabilization reviewed in this talk are diagrammed in Figure 4.

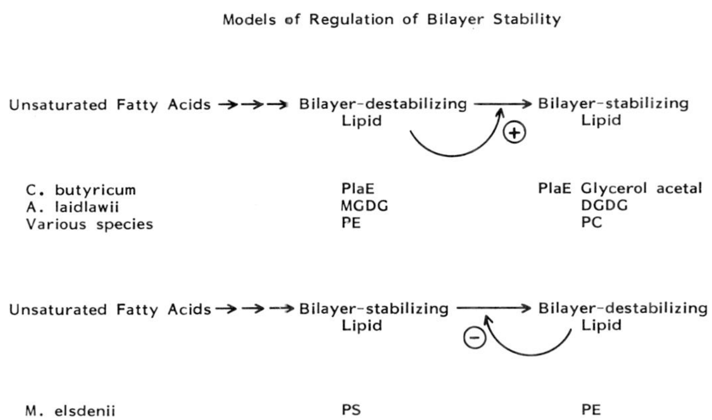

Figure 4. Models of regulation of bilayer stability proposed for prokaryotic organisms. Although arrows suggest direct affects on the enzymes involved in these reactions, other models have not been eliminated (see text).

Although these organisms have not been subjected to experimental manipulation of acyl chain composition, the hypothesis that polar lipid class compositional changes serve to stabilize the bilayer arrangement should be amenable to several experimental approaches including the isolation of appropriate fatty acid auxotrophs or phospholipid biosynthetic mutants. In the next several years, the hypothesis that cells can carry out this newly described regulation will undoubtedly be subjected to critical experimental tests.

References

Broquist HP, Snell EE (1951) Biotin and bacterial growth 1. Relation to aspartate, oleate, and carbon dioxide. J Biol Chem 188:431-444

Cullen J, Phillips MC, Shipley GG (1971) The effects of temperature on the composition and physical properties of the lipids of Pseudomonas fluorescens. Biochem J 125:733-742

Cullis PR, de Kruijff B (1979) Lipid polymorphism and the functional roles of lipids in biological membranes. Biochim Biophys Acta 559:399-420

Cullis PR, Hope MJ (1985) Physical properties and functional roles of lipids in membranes. In: Vance DE, Vance, JE (eds) Biochemistry of lipids and membranes. Benjamin/Cummings, Menlo Park, California, p 25

Goldfine H (1982) Lipids of prokaryotes-structure and distribution. Curr Top Membr Transp 17:1-43

Goldfine H (1984) Bacterial membranes and lipid packing theory. J Lipid Res 25:1501-1507

Goldfine H (1985) Modulation of polar lipid composition by aliphatic chain unsaturation in bacteria. Curr Top Cell Regul 26:163-174

Goldfine H, Khuller GK, Borie RP, Silverman B, Selick H, Johnston NC, Vanderkooi JM, Horwitz AF (1977) Effects of growth temperature and supplementation with exogenous fatty acids on some physical properties of Clostridium butyricum phospholipids. Biochim Biophys Acta 488:341-352

Goldfine H, Johnston NC, Phillips MC (1981) Phase behavior of ether lipids from Clostridium butyricum. Biochemistry 20:2908-2916

Goldfine H, Johnston NC, Bishop DG (1982) Ether phospholipid asymmetry in Clostridium butyricum. Biochem Biophys Res Commun 108:1502-1507

Goldfine H, Johnston NC, Mattai G, Shipley GG (1987) Biochemistry 26:2814-2822

Goldfine H, Rosenthal JJC, Johnston NC (to be published) Lipid shape as a determinant of lipid composition in Clostridium butyricum. The effects of incorporation of various fatty acids on the ratios of the major ether lipids. Biochim Biophys Acta

Gruner S (1985) Intrinsic curvature hypothesis for biomembrane lipid composition: A role for nonbilayer lipids. Proc Natl Acad Sci USA 82:3665-3669

Gruner S, Cullis PR, Hope MJ, Tilcock, CPS (1985) Lipid polymorphism: The molecular basis of nonbilayer phases. Ann Rev Biophys Biophys Chem 14:211-238

Israelachvili JN, Marcelja S, Horn RG (1980) Physical principles of membrane organization. Q Rev Biophys 13:121-200

Johnston NC, Goldfine H (1985) Phospholipid aliphatic chain composition modulates lipid class composition, but not lipid asymmetry in Clostridium butyricum. Biochim Biophys Acta 813:10-18

Kaufman A (1986) Studies of plasmalogen containing phospholipids in Clostridium pasteurianum and Megasphaera elsdenii. Ph.D. Thesis, University of Pennsylvania, Philadelphia.

Khuller GK, Goldfine H (1975) Replacement of acyl and alk-1-enyl groups in Clostridium butyricum phospholipids by exogenous fatty acids. Biochemistry 14:3642-3647

Lohner K, Hermetter A, Paltauf F (1984) Phase behavior of ethanolamine plasmalogen. Chem Phys Lipids 34:163-170

Malthaner M, Seelig J, Johnston NC, Goldfine (1987) Deuterium NMR studies on the plasmalogens and the glycerol acetals of plasmalogens of Clostridium butyricum and Clostridium beijerinckii. Biochemistry 26:5826-5833

Rilfors L (1985) Difference in packing properties between iso and anteiso methyl-branched fatty acids as revealed by incorporation into the membranes of Acholeplasma laidlawii strain A. Biochim Biophys Acta 813:151-160

Rilfors L, Khan A, Brentel I, Wieslander Å, Lindblom G (1982) Cubic liquid crystalline phase with phosphatidyl-ethanolamine from Bacillus megaterium containing branched acyl chains. FEBS Lett 149:293-298

Rilfors L, Lindblom G, Wieslander Å, Christiansson A (1984) Lipid bilayer stability in biological membranes. Biomembranes 12:205-245

Silvius JR, Lyons M, Yeagle PL, O'Leary TJ (1985) Thermotropic properties of bilayers containing branched-chain phospholipids. Calorimetric Raman and ^{31}P NMR studies. Biochemistry 24:5388-5395

Tilcock CPS, Bally MB, Farren SB, Cullis PR, Gruner S (1984) Cation-dependent segregation phenomena and phase behavior in model membrane systems containing phosphatidylserine: Influence of cholesterol and acyl chain composition. Biochemistry 23:2696-2703

MEMBRANE STRUCTURE AND DYNAMICS BY NMR. PART 1: EFFECT OF CYCLOPROPANE RINGS, DOUBLE BONDS AND STEROLS ON THE STRUCTURE AND DYNAMICS OF PHOSPHOLIPID MEMBRANES.

Erick J. Dufourc
C.R. Paul Pascal/CNRS
Domaine Universitaire
33405 TALENCE Cedex
FRANCE

INTRODUCTION

It is nowadays well recognized that many important biophysical events occur at the biological membrane level. The membrane can regulate, for instance, the vectorial transport of nutrients, waste, and exogenous components such as drugs. It can also provide a site of cell-specific antigens and hormone receptors. Unfortunately, the molecular mehanisms giving rise to such events are not always well characterized and often not known at all. This poor level of knowledge essentially resides in the complexity of the biological membrane. Lipids and proteins which compose the membrane can indeed have a wide variety of chemical structures. Moreover, it appears that membrane biological activity is closely linked to the relative amount of each of these compounds as well as to their spatial and/or temporal distribution in the membrane. The study of membrane structure and dynamics appears therefore as a potential tool in the understanding of biological mechanisms occurring at or in the membrane.

In view of the above, it is clear that many physical methods can be used to monitor structural and dynamical membrane properties. Although the situation is complicated by the multiplicity of components, significant advances have been nonetheless made, and nuclear magnetic resonance (NMR) is in large part responsible for this progress.

NATO ASI Series, Vol. H16
Membrane Biogenesis. Edited by J. A. F. Op den Kamp

SOME ASPECTS OF SOLID STATE NMR

Nuclear magnetic resonance requires the use of a nucleus with a non zero spin (proton (^1H), carbon-13 (^{13}C), deuterium (^2H), phosphorus-31 (^{31}P), ...). The total energy of such a nucleus in a static magnetic field may be decomposed into a sum of several terms, each of them corresponding to a given interaction between a nuclear property and the surroundings:

$$\begin{array}{cccccccccccc} H & = & H_Z & + & H_Q & + & H_{CSA} & + & H_D & + & H_J & (1) \\ & & 10^8 & & 10^6 & & 10^4\text{-}10^3 & & 10^4\text{-}10^3 & & 10. & \end{array}$$

H represents the total nuclear Hamiltonian, i.e. the quantum mechanical description of the total nuclear magnetic energy of interactions. H_Z describes the coupling between the nuclear magnetic moment with the external magnetic field, B_o. H_Q arises from an electronic interaction of the nuclear quadrupole moment of the nucleus with electric field gradients surrounding the nucleus (sensitive to phase symmetry and local dynamics). H_{CSA} represents the contribution relative to the anisotropy of the chemical shift (sensitive to chemical structure, phase symmetry and local dynamics), H_D the dipolar interaction (coupling through space with neighbouring spins) and H_J the scalar interaction (coupling via chemical bonding to neighbouring spin). The numbers below each term of the total Hamiltonian represent the orders of magnitude of static interactions (in Hertz).

From the above it is clear that NMR is able to probe modifications in the structure and/or dynamics of the surroundings of the nucleus under study. Changes will be reported by a variation of a given term in the interaction energy, certain terms will be more sensitive to structural modifications, while others will report on dynamics.

In liquids, where isotropic fast motions (random molecular tumbling) occur, H_Q, H_D and the anisotropic part of the H_{CSA} are averaged to zero. Spectra will only report on scalar and (isotropic) chemical shift interactions. These two couplings have been for many years the chemist's favourite tools to determine the molecular structures of their compounds. More recently, scalar and chemical shift couplings have been used in conjunction with sophisticated NMR methods (2D-NMR) to obtain structures in solution of large molecules such as proteins.

Biological membranes are by definition liquid crystals: in other words, they are anisotropic fluid media. The amphipathic molecules which comprise model membranes (phospholipids, sterols, ...) do not orient randomly in water, they usually align locally in the same direction - the normal to the bimolecular leaflet. Although well organized, the membrane components are nonetheless highly dynamic. Under such conditions, the quadrupolar, dipolar and chemical shift interactions no longer cancel. However, their magnitudes may be reduced by the anisotropic molecular motions, if the rates of motions have the appropriate frequency relative to that of the interaction energy.

NMR offers the possibility of selecting the interaction of interest. In what follows, two nuclei, i.e. to types of interactions, will be utilized, namely deuterium (spin I=1, quadrupolar interaction) and phosphorus-31 (spin I=½, chemical shift anisotropy interaction). In the case of ^{2}H-NMR, where the quadrupolar interaction dominates the CSA, dipolar and scalar couplings, calculation of energy levels and absorption lines gives (Seelig, 1977):

$$\Delta\nu(\beta) = \frac{3}{2} A_Q S_{C-^2H} \left(\frac{3\cos^2\beta - 1}{2}\right). \qquad (2)$$

In the case of ^{31}P-NMR, the quadrupolar interaction does not exist, the dipolar proton-phosphorus coupling is usually suppressed by proton decoupling and the scalar interaction can be neglected. The resonance lines are therefore governed by the anisotropy of the chemical shift (Seelig, 1978):

$$\sigma(\beta) = \sigma_{iso} + \frac{2\Delta\sigma}{3} \left(\frac{3\cos^2\beta - 1}{2}\right). \qquad (3)$$

In eq. (2), the frequency separation between the two absorption lines, called the quadrupolar splitting $\Delta\sigma(\beta)$, is a function of the static quadrupolar coupling constant, A_Q (A_Q = 170 kHz for methylene C-^{2}H bonds), the orientational order parameter, S_{C-^2H} and the angle β. In eq. (3) the position of the absorption line, called the chemical shift, $\sigma(\beta)$, is dependent on σ_{iso}, the isotropic chemical shift, on $\Delta\sigma$, the residual chemical shift anisotropy and on β. In the usual analysis of membrane spectra of both ^{2}H and ^{31}P, it is assumed that the system is axially symmetric and that the axis of symmetry is the bilayer normal, n. The angle β represents the orientation of that normal relative to the direction of the external magnetic field, B_o. A typical variation of the resonance lines (eqs. (2) and (3)) as a function of β is shown in Figure 1.

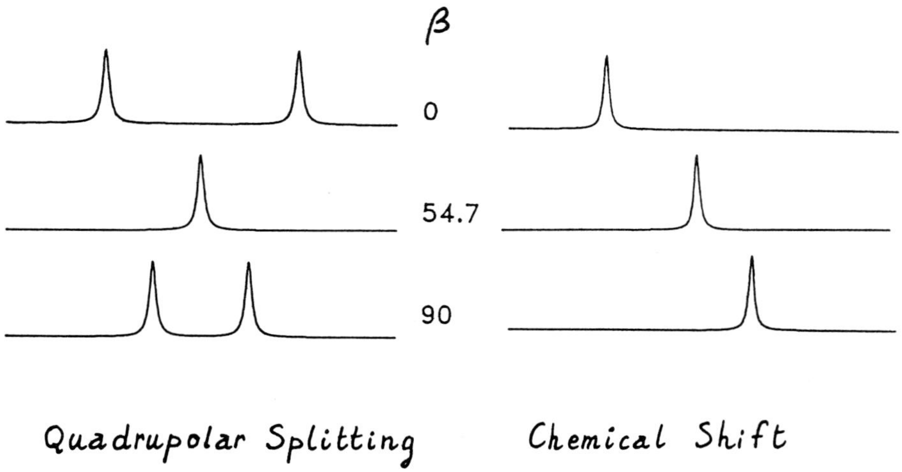

Figure 1. Variation of spectral frequencies, as a function of β, for nuclei subject to quadrupolar and chemical shift anisotropy interactions. The angle β is defined in the text.

The orientational order parameter S_{C-^2H} gives a direct measure of the time dependent angular excursions of a given $C-^2H$ bond with respect to the principal axis of motion, usually taken to be the bilayer normal. $\Delta\sigma$ leads in principle to a similar information for phosphate-containing lipid head groups; however, the order parameter cannot be extracted straightforwardly from measurement of the residual chemical shift anisotropy. The concept of order parameter will be further developed in the next section.

Biological membranes do not orient themselves in the magnetic field. In other words, in a typical dispersion of biomembranes in water, one will obtain a random distribution of local bilayer normals, with respect to B_o, i.e. the angle β will vary from 0° to 90° and spectra will be composed of a superimposition of all resonance lines originating from all bilayer normal orientations with respect to B_o. The contribution of each line to the total spectrum is described by a sin β function. This situation leads to typical non-oriented spectra as shown in Fig. 2, for nuclei subject to quadrupolar and chemical shift anisotropy interactions. Such composite spectra are called "powder patterns" and may appear rather complex. However, one can still measure the quadrupolar splitting for β=90°, called $\Delta\nu_Q$, and hence obtain S_{C-^2H}. The chemical shift anisotropy, $\Delta\sigma$, is measured directly from the chemical shifts due to β=0° and β=90° orientations.

The presence of powder patterns is often a good indication of non-oriented large bilayer structures. However, other physical techniques are often necessary to infer this structural information.

Detection of NMR signals is performed by means of radiofrequency (rF) pulse techniques. When a sample of resonant spins resides within a static magnetic field, the total magnetic interaction, described above, promotes a separation of the magnetic energy levels. At the equilibrium, the nuclear spins populate the energy levels according to Boltzmann statistics. The net effect is a resultant macroscopic magnetization along the external magnetic field direction. In a typical NMR

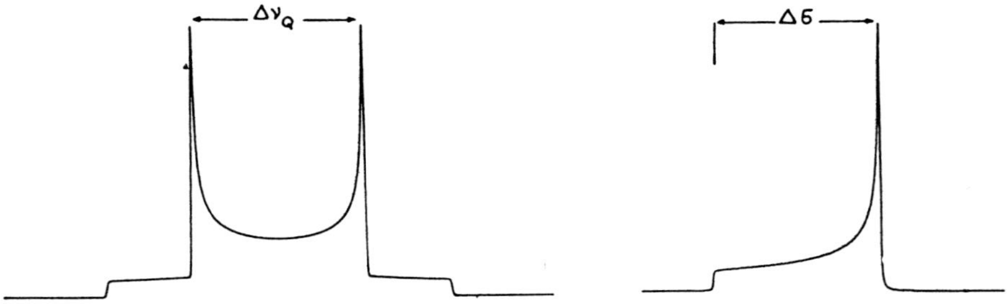

Figure 2. Axially symmetric powder patterns for nuclei subject to quadrupolar (left) and chemical shift anisotropy (right) interactions.

experiment, this magnetization is projected out of equilibrium, by application of a strong rF impulse. Monitoring return to equilibrium as a function of time allows the recording of time dependent NMR signals. Subsequent Fourier transformation leads to NMR signals in the frequency domain, i.e. NMR spectra. From the time dependent NMR signals, one can also extract time constants for the exponential processes (usually assumed to be so) by which the nuclear spins exchange energy and entropy with their surroundings. These time constants, called nuclear relaxation times, can bring additional information on dynamics (*vide infra*).

The magnitude of the various interactions (eq. (1)) indicates that we may be dealing with very wide spectra. This often implies the use of echo pulse sequences (Davis, 1979;

Rance and Byrd, 1983) to overcome technical problems which may arise in signal detection.

MEMBRANE DYNAMICS OR MEMBRANE FLUIDITY?

For many years the term fluidity has been utilized to characterize the dynamical state of the membrane. However, this attempt to define membrane motional behaviour by a single macroscopic parameter has not been very successful (fluidity is *stricto sensu* the reciprocal of viscosity, a parameter obtained via osmometric measurements). Figure 3 is an attempt to represent the dynamical character of the membrane at time t. If motions occur in a time interval dt, at t+dt, the position of the molecular units will have changed.

Figure 3 suggests that the dynamics of such a membrane, at the molecular level, cannot be described by a single parameter. Alternatively, the motions taking place in the membrane may be characterized, in a still simplistic way, by their nature (rotation, "wobbling", *gauche*-*trans* isomerizations, ...), rate (reciprocal of correlation time), amplitude (order parameters) and activation energy. Moreover, one may classify the motions according to the time domain in which they occur (Paddy et al., 1981). This classification is useful in choosing the physical technique to detect them. Motions in the picosecond time scale will affect the observables of vibrational spectroscopy (IR, Raman ...) whereas motions of correlation time of the order of microseconds will not.

In the case of ^{2}H-NMR, all motions smaller than microseconds will affect spectra. In principle, NMR can probe more motions than do fluorescence, IR or Raman spectroscopy. But, of course, the more motions affect spectra, the more observables are needed to identify them; this is often a limiting factor to achieve this sorting. Another important parameter in the definition of membrane dynamics is the amplitude of the motion. This parameter can be perceived as time dependent angular fluctuations of the molecule itself, if it is rigid

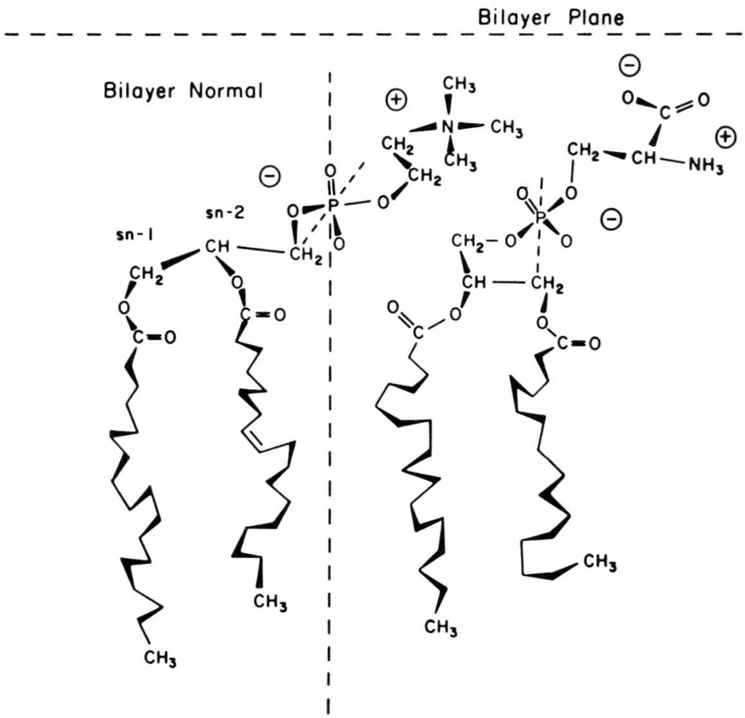

Figure 3. "Snap shot" representation of the various conformational states membrane lipids can adopt. Left: phosphatidylcholine; right: phosphatidylserine. Figure courtesy of Dr. I.C.P. Smith.

(e.g. the four rings of cholesterol) or of a molecular fragment, if the molecule is flexible (e.g. the methylene units of a phospholipid acyl chain). The net effect of this time and ensemble average is to reduce the static NMR interaction (eq. (1)). This is reflected on NMR spectra by, for instance, a decrease of the quadrupolar splitting. One has therefore a powerful means of probing the amplitude of motions, at a specific site in a membrane, by simply looking at spectra. If several motions are present, what is observed is the result of all averages produced by each motion. In ^{2}H-NMR, the orientational order parameter, S_{C-^2H}, may be considered as the product of all order parameters of motions having correlation times in the appropriate time scale (i.e. less than 10^{-6} s) (Dufourc and Smith, 1986).

$$S_{C-{}^2H} = \prod_{i}^{N} S_i \qquad (4)$$

In equation (4) S_i represents the ith motion of a hierarchy in time of N motions. If the motions are axially symmetric (which is usually assumed) $S_i = (\frac{3\cos^2\beta_i - 1}{2})$, where β_i is the angle between the principal axis of the coordinate system attached to motion i and the principal axis of the coordinate system attached to motion i-1. These order parameters are an easy way to characterize membrane states. A state in which there are no fluctuations of angles β_i will be called totally ordered and characterized by $S_i = 1$. Conversely, a totally desordered state, i.e. random fluctuations of angle β_i, will have $S_i = 0$. Unfortunately, the decomposition according to eq. (4) is not always possible and often one reports only the variation of $S_{C-{}^2H}$ under certain constraints. In this situation, $S_{C-{}^2H} = (\frac{3\cos^2\theta_{C-{}^2H} - 1}{2})$ is a measure of the fluctuations, promoted by all effective motions, of the angle $\theta_{C-{}^2H}$ between the $C-{}^2H$ bond and the bilayer normal. Although less precise, this information is nonetheless of first importance. For instance, $S_{C-{}^2H}$ has been successfully utilized to show the variation of the amplitude of motions as a function of the bilayer depth, for model membranes with fully saturated acyl chains (Seelig and Seelig, 1974). Figure 4 is an example of such a situation. The top of this figure shows the ^2H-NMR spectrum of dipalmitoylphosphatidylcholine (DPPC) model membranes where the sn-2 fatty acyl chain is perdeuterated. In fact what is displayed is the "de-Paked" spectrum, i.e. the result of applying a spectral deconvolution (Bloom et al., 198') to powder spectra (Fig. 2) in order to obtain sample oriented-like spectra (Fig. 1) in which the quadrupolar splittings are easily measurable. The bottom of this figure shows the variation of the quadrupolar splitting ($\Delta\nu_Q \alpha\ S_{C-{}^2H}$) as a function of the labelled position along the sn-2 acyl chain. This is a representation of the ordering as a function of bilayer depth. It can be noticed that the order parameter is constant from positions 3'-9', i.e. from the top of the

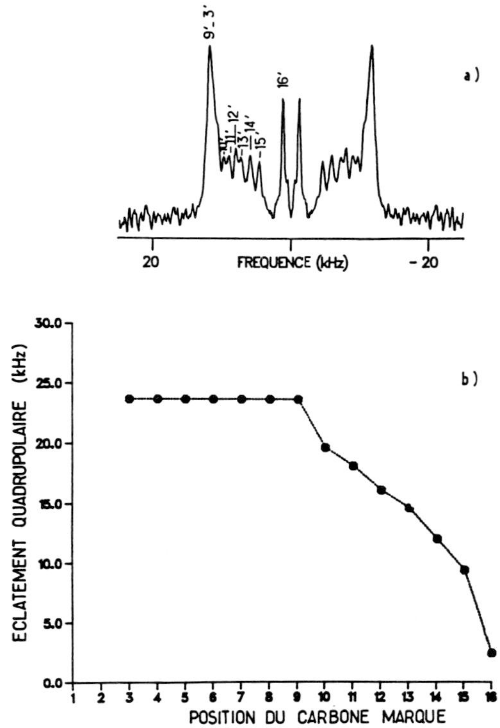

Figure 4. Order parameter profile for DPPC bilayers at 50°C. a) ^{2}H-NMR (de-Paked) spectrum of [*sn*-2-$^2H_{31}$] DPPC. Primed numbers indicate the position of labelling on the *sn*-2 chain. b) Variation of quadrupolar splittings ($\Delta\nu_Q \alpha\ S_{C-^2H}$) as function of labelled carbon position. From Dufourc, 1986.

bilayer to approximately half of the chain length. This is known as the "plateau" of order parameters (Seelig and Seelig, 1974). From position 9', the ordering rapidly decreases when going towards position 16' indicating the character highly dynamic (large amplitude of motions) of the center of the bilayer. This very minute description of the orientational order parameter as a function of bilayer depth is only possible in the so-called lamellar fluid phase of these model membranes. Below T_c, the temperature of the gel-to-fluid phase transition (T_c = 41°C for DPPC), spectra are broader (Fig. 5) and may not exhibit axially symmetric lineshapes like in the fluid phase (Davis, 1983). In these conditions, one cannot extract order parameters as a function of labelled positions. However, one may still characterize membrane ordering by estimating spectral moments (Davis, 1979). Figure 5 shows the temperature variation of the first moment, M_1, of [sn-2-$^2H_{31}$] DPPC. The first moment of such lipid spectra represents the overall acyl chain order parameter. This information does not bring as much detail as does the order parameter profile (Fig. 4) but nonetheless allows us to characterize spectra for all temperatures.

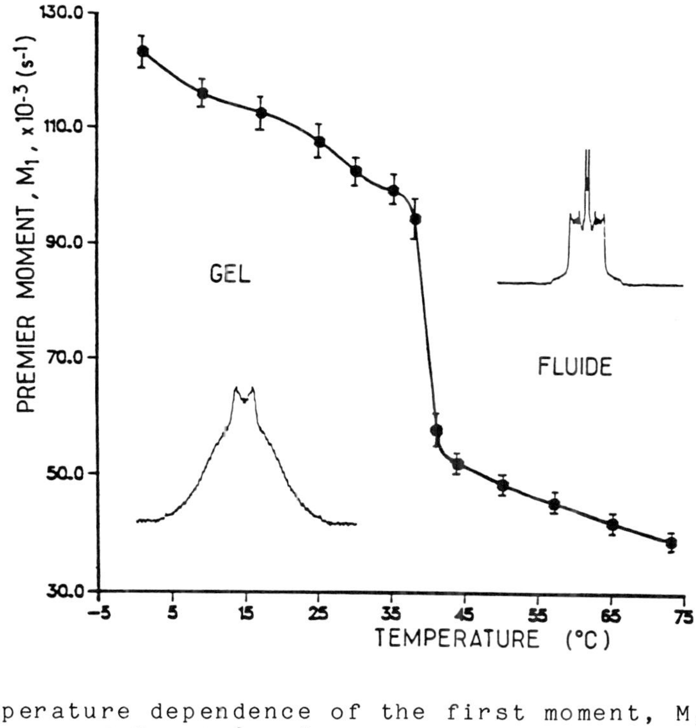

Figure 5. Temperature dependence of the first moment, M_1, of spectra of [sn-2-$^2H_{31}$] DPPC. Spectra inserts characterize the gel and fluid lamellar phases. From Dufourc, 1986.

The elevated M_1 values for temperatures below 41°C indicate a considerable reduction in the amplitude of acyl chain motions. Figure 5 shows that the well known order-disorder transition is well determined from the marked discontinuity in the M_1 temperature profile.

The above has shown that order parameter can bring very valuable information on the amplitude of motions. However, exhaustive determination of membrane dynamics requires the estimate of the rates of motions (from relaxation measurements) as well as their nature and activation energy.

CHANGES IN MEMBRANE DYNAMICS PROMOTED BY THE PRESENCE OF A CYCLOPROPANE GROUP IN LIPID ACYL CHAINS

^{2}H-NMR is a powerful tool for examining the dynamics of lipid membranes and has been extensively used to monitor the behaviour of lipids containing straight chain saturated and/or unsaturated fatty acids (Seelig, 1977; Davis, 1983; Smith, 1984). Little attention has been focused on systems in which fatty acids have branched chains or cyclopropane rings, despite their widespread occurrence in nature. Fatty acids containing cyclopropane groups are found in a variety of organisms and are common components of the membrane lipids of microorganisms (Christie, 1970). These acyl chains are formed from monoenoic fatty acids through the energy-expensive addition of a CH_2 group, in the late log phase of growth in some bacteria. Although occurrence of such cyclopropane containing fatty acids is not well understood, it has been postulated that their presence is associated in some way with membrane stability (Souzu, 1982).

In order to have a better understanding of the role of cyclopropane moieties in the lipid properties of biological membranes ^{2}H-NMR studies were undertaken on specifically ^{2}H-enriched 1-palmitoyl-2-dihydrosterculoyl-_sn_-glycero-3-phosphocholine (DPSPC). The cyclopropane group is at position 9-10 in the _cis_-9,10-methyleneoctadecanoic acid or "dihydrosterculoic" (DS) acid (Fig. 6). Both the _sn_-1 and _sn_-2 chains were separately labelled in order to monitor possible influences of the cyclopropane-containing chain on the fully saturated palmitoyl chain (Dufourc 1983; Perly et al., 1984). Above -10°C, i.e. in the PDSPC fluid phase, all spectra exhibit axially symmetric lineshapes similar to that of Figure 2. Spectra of

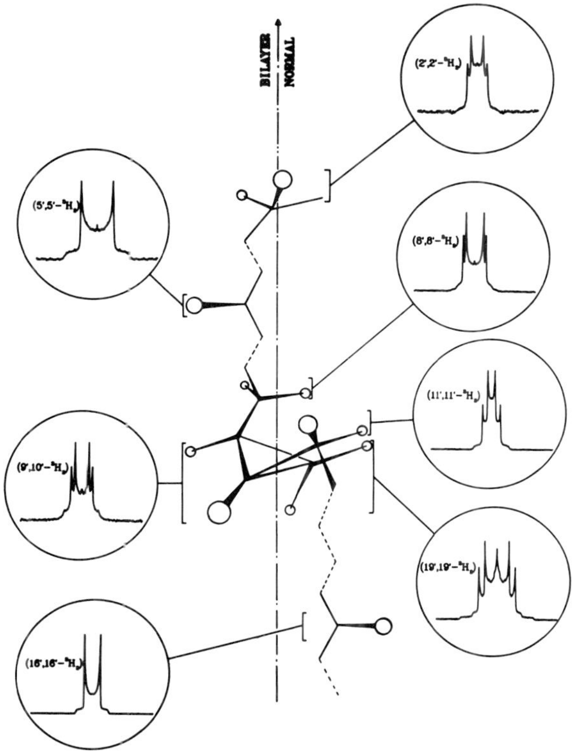

Figure 6. ^{2}H-NMR powder spectra (circles) at indicated carbon positions and average configuration of the sn-2 chain of PDSPC molecules embedded in a membrane (see text).

Figure 6 shows that one observes one or two quadrupolar splittings, depending upon the position of labelling. Spectra from sn-1 labelled positions exhibit only one splitting, for each deuterated position (Dufourc et al., 1984a). This behaviour is similar to what has been reported for DPPC (Davis, 1983; Seelig and Seelig, 1974).

The presence of one or several quadrupolar splittings for a given rigid subunit (methylene segment, cyclopropane ring) may be understood on the basis of eq. (4) which can be rewritten, in the case of 2 order parameters as (Seelig, 1977; Dufourc et al., 1983):

$$S_{C-^2H} = S_\alpha S_\gamma = \left(\frac{3\cos^2\alpha-1}{2}\right)\left(\frac{3\cos^2\gamma-1}{2}\right). \qquad (5)$$

S_γ stands for the change in axis of symmetry, from the C-^{2}H bond axis to the instantaneous axis of motional averaging of a rigid subunit, whereas S_α reports on the fluctuations of this instantaneous axis with respect to the average director of motion, the bilayer normal. Let us take a fully deuterated chain as an example. If there is only one motion, fast rotation, about the chain long axis, the methylene segments remaining in _all-trans_ conformation, S_α, will be equal to 1 and $S_\gamma = -\frac{1}{2}$ ($\gamma = 90°$ for both deuterons on a given carbon). If there are additional motions (segment wobbling, _gauche-trans_ isomerization, ...) which preserve _in the average_ the _all-trans_ conformation, the amplitude of motions at the level of the segment will be reported by a variation of the angle α so a decrease of S_α. The separation of S_{C-^2H} into 2 order parameters may appear subtle but allows us to express the ordering of any methylene unit with respect to the same axis of reference, the normal to the bilayer membrane. In this sense, S_γ may be viewed as a geometric transformation whereas S_α carries the information about the amplitude of motions of a given rigid subunit. In other words, S_α must be the same for all deuterons of a rigid unit (e.g. methyl segment, cyclopropane ring). Observing different quadrupolar splittings for deuterons attached to the same rigid segment indicates that the carbon-deuterium bonds have different average orientations, with respect to the axis of motional averaging.

The observed quadrupolar splittings for both chains of the PDSPC molecule are reported on Figure 7 (top). Dramatic differences are apparent. The _sn_-1 palmitoyl chain has a smooth order-position profile whereas that of the _sn_-2 DS chain shows the perturbation induced by the cyclopropane ring. As mentioned above, the appearance of several quadrupolar splittings, i.e. several S_{C-^2H} for a same rigid unit, indicates that the deuterons have different average orientations with respect to the bilayer normal. If one wishes to describe the amplitude of motions, S_α (also called S_{mol}), must be calculated. In other words, one must calculate the value of S_γ for each C-^{2}H bond of a rigid segment by making use of geo-

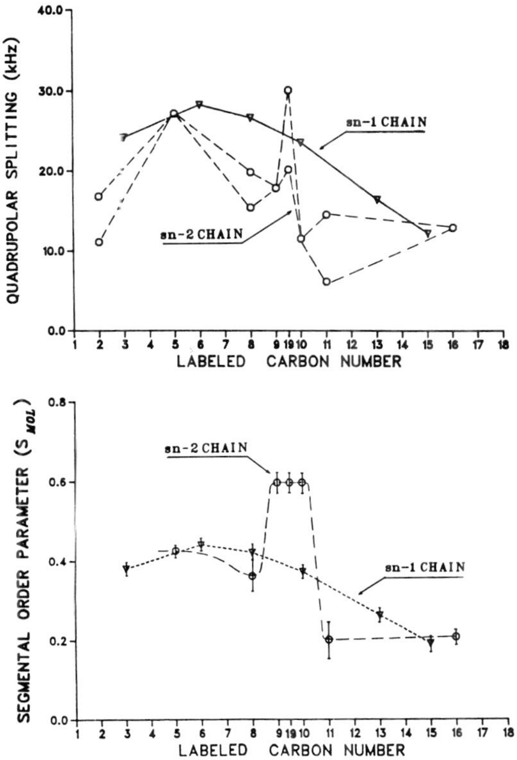

Figure 7. (Top panel) Variation of the quadrupolar splitting, $\Delta\nu_Q$, with the position of labelling for PDSPC at 25°C. (o) Dihydrosterculoyl sn-2 chain; (∇) palmitoyl sn-1 chain. (Bottom panel) Variation of the segmental order parameter, S_{mol}, with the position of labelling for PDSPC at 25°C. Same notations as for the top panel. The bars give an estimate of the error. From Dufourc et al., 1984a.

metrical relationships (angles, distances) between bonds (Dufourc et al., 1983). The lower panel of Figure 7 shows this "geometry-corrected" segmental order parameter profile, i.e. it reports on pure angular fluctuations of each rigid segment (methylene, cyclopropane) for both chains. The most dramatic effect of the cyclopropane ring is a large perturbation at its own level, in the order parameter profile: the amplitude of the motions of the ring is some 40% lower than that of the plateau positions (from position C-9 to the top of the bilayer). Interestingly, the effect of the cyclopropane is not significantly perceived by the sn-1 chain. This indicates that the effect of the cyclopropane ring is mainly localized on the sn-2 chain.

When calculating S_α values for each rigid segment one effectively varies the average orientation of the rigid unit to calculate S_γ values for each $C-^2H$ bond and try to obtain the best matching between $S_\alpha S_\gamma$ and the observed S_{C-^2H}. This means that in addition to the segmental order parameter one determines also the average orientation of each chain subunit. Figure 6 displays the average configuration of the sn-2 chain of PDSPC. One notices that the 9'-10' carbon-carbon bond is almost perpendicular to the axis of motional averaging, i.e. the bilayer normal. It is clearly understandable that such a configuration prevents an easy packing of the acyl chains, when lowering the temperature. This is reflected by a low transition temperature, i.e. betwen -10°C and -15°C.

Amplitude of motions as well as average configurations have been obtained from spectra. These informations can be correlated with measurement of nuclear relaxation times. The spin-lattice or longitudinal nuclear relaxation time, T_{1z}, is associated with the mechanism of energy exchange of the nuclear spins with their surroundings referred to as the lattice. This relaxation time contains information about the rates of the motions as well as about their amplitude. Of course, not all the motions will be effective in the mechanism of energy exchange. Only those with correlation times of approximately the inverse of the Larmor frequency will contribute to the spin-lattice relaxation. In other words, deuterium T_{1z} is sensitive to motions having correlation times in the nanosecond time scale.

The positional dependence of T_{1z} relaxation rate (inverse of T_{1z}) of both sn-1 and sn-2 chain labels is reported in Figure 8. This is the first case where the relaxation rate profiles of both chains of a lipid in a model membrane can be compared (Dufourc et al., 1984a). The comparison is dramatic. The sn-2 chain shows a considerable increase in the relaxation rate at the level of the cyclopropane ring whereas the sn-1 chain exhibits a plateau at the corresponding positions. The similarity between Figure 8 and Figure 7 (bottom) is striking. One can therefore state that the spin-lattice relaxation is strongly dependent on the segmental order parameter S_α (or S_{mol}), and hence dominated by motions associated with S_α. Unfortunately, the nature, correlation time and the number of effective motions contributing to the spin-lattice relaxation mechanism of both _sn_-1 and _sn_-2 chains cannot be obtained from this type of data, unless making use of speculative models. More observables are required to improve the level of description of the chain dynamics. An example of such a detailed analysis will be presented in the case of cholesterol (_vide infra_). However, information brought by Figure 8 is nonetheless of first importance since it directly reflects the ampli-

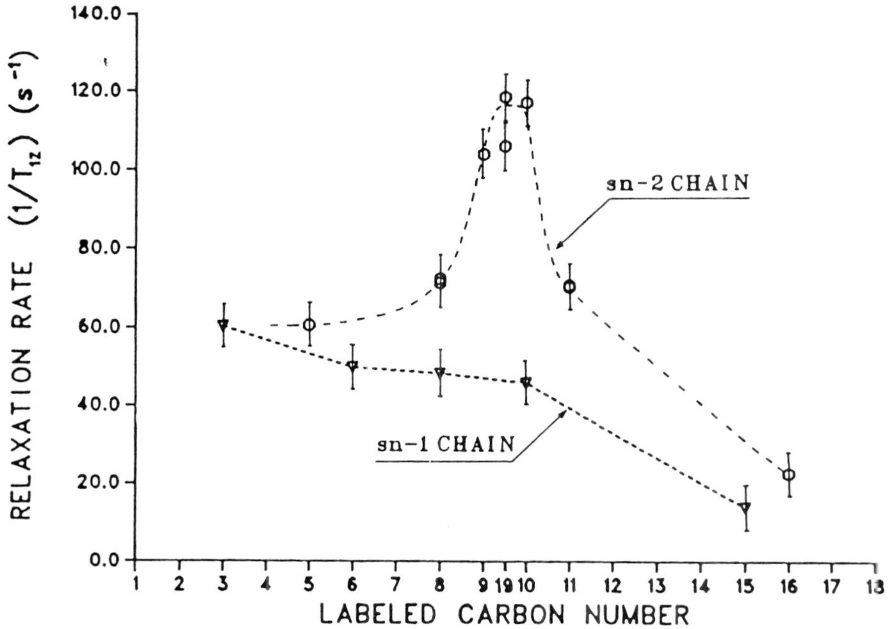

Figure 8. Comparison of the spin-lattice relaxation rates of both _sn_-1 and _sn_-2 chains of PDSPC as a function of labelled positions at 25°C. From Dufourc et al., 1984a.

tude of segmental motions. It must be reminded that such information had been obtained from quadrupolar splittings after correction for differences in average orientations of C-^{2}H bonds with respect to the bilayer normal (vide supra). This type of calculation is only possible when one has enough observables, i.e. enough S_{C-^2H}, as in the case of the cyclopropane ring. Moreover, the spin-lattice relaxation times were measured as a function of temperature and showed a marked increase when raising the temperature. This indicates that the motions of interest are in the so-called fast correlation time regime, i.e. such that $\omega_o^2\tau_i^2 << 1$ (ω_o is the Larmor angular frequency and is equal to $2\pi \times 46.1$ MHz in the experiment, τ_i which represents the correlation time of motion(s) i is therefore shorter than 10^{-9} s). Assuming that correlation times follow an Arrhenius dependence on temperature, one can determine from the temperature dependence of Ln (T_{1z}) the magnitude of E_a, the activation energy for molecular motions. The activation energy estimated for deuterons on the cyclopropane ring is 12.3 ± 1.0 kJ.mol^{-1} which is slightly less than the 14.6 ± 1.3 kJ.mol^{-1} found for bilayers of DPPC (Brown et al., 1979).

To summarize the findings about the presence of a cyclopropane group in lipid fatty acyl chains, it can be said that the ring in the sn-2 chain leads to a significant increase of the order parameter at the position of the cyclopropane ring. Such discontinuity is not seen in the order parameter profile of the sn-1 chain indicating that the perturbation induced by the ring is mainly localized to the sn-2 chain. The motions leading to such a behaviour are segmental, have correlation times shorter than the nanosecond and an activation energy smaller than that found in lipid bilayers with fully saturated acyl chains. The average orientation of the ring, in the sn-2 chain, leads to steric constraints which prevent easy packing of the chains, resulting in very low transition temperature. The implications of these findings will be discussed later.

COMPARISON OF MEMBRANE DYNAMICS FOR LIPIDS CONTAINING CYCLOPROPANE GROUPS, DOUBLE BONDS OR FULLY SATURATED CHAINS

In order to follow the changes in membrane dynamics upon replacement of a double bond by a cyclopropane group in the lipid acyl chains, the dynamical behaviour of PDSPC, POPC (palmitoyl oleoyl phosphatidylcholine) and DPPC was monitored by means of relaxation time, T_{1z}, measurements. These model membrane systems must be compared under conditions were they are in the same physical state, that is, subjected to the same average molecular forces. In theories of phase transitions, this is often assumed to occur when systems are at the same temperature relative to their individual phase transition temperatures (T_c); temperature on this scale is known as reduced temperature. Consequently, the relaxation rates as a function of labelled carbon position of _sn_-2 PDSPC at 25°C, _sn_-2 POPC at 36°C (Perly et al., 1985) and DPPC at 80°C (Brown et al., 1979) (Figure 9). All relaxation rates were therefore monitored at 35-40°C above the gel-to-fluid phase transition temperature of each of the three systems. Again, the behaviour of these model membranes is significantly different. The most obvious general comment is that the rates of relaxation increase from the fully saturated to the cyclopropane-containing lipids. Since it has been demonstrated that relaxation rate profiles reflect the amplitude of segmental motions, one can classify the three systems in a scale of increasing ordering as follows: DPPC < POPC < PDSPC, for the same reduced temperature. Figure 9 also shows that at the level of the functionality, i.e. the double bond or the cyclopropane ring, there is an increase in the rate of relaxation with respect to the positions preceding the functional group.

The above findings have revealed many important consequences of the presence of cyclopropane groups in membrane lipids. To put this in perspective, let us compare the behavior of a fully saatura̲ted lipid, DPPC, with that of PDSPC. A necessary property for lipid in membranes is that it be in a highly dynamic or "fluid" state (low order parameters, fast

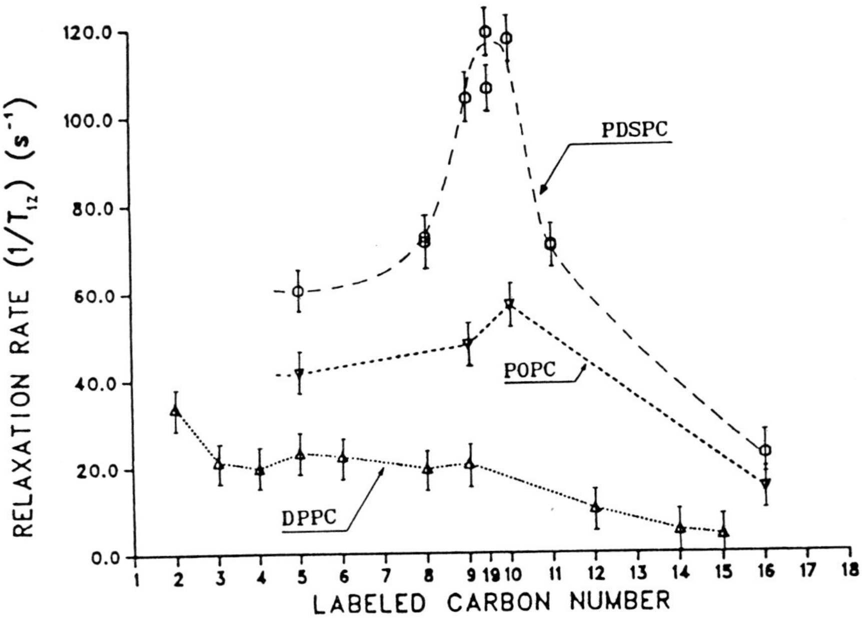

Figure 9. Comparison of the relaxation rate (T_{1z}^{-1}) profiles for several model membrane systems: sn-2 PDSPC at 25°C and 46.1 MHz (Dufourc et al., 1984a) (o); sn-2 POPC at 36°C and 46.1 MHz (Perly et al., 1985) (∇); DPPC at 80°C and 54.4 MHz (Brown et al., 1979) (Δ).

correlation times) in the biologically significant temperature range, T_L (35-45°C). The fully saturated lipids can pack so easily that their T_c is high and increases with chain length. Palmitoyl chains induce a T_c at 41°C, i.e. in the T_L range. These lipids are not good candidates for a "fluid" membrane around T_L. In contrast, the cyclopropane moiety, whose average orientation and bulkiness prevent an easy packing of the chains, induces T_c to be far from T_L. One might think that if the cyclopropane ring perturbs membrane packing, it would lead to a facile disruption of the membrane at high temperatures, relative to T_c. This is not the case; indeed, it has been shown herein that the ring possesses a high order parameter which does not change appreciably over a wide range of temperatures (Dufourc et al., 1984a). One can thus imagine that changing the temperature by 10 or 20°C about T_L will not affect the structural and dynamical properties of cyclopropane-containing membranes. The cyclopropyl moiety has therefore a certain inertia to motional variations and might play a regulatory role for motions involved in membrane stability.

With regard to the biological synthesis of cyclopropyl moiety from a *cis* double bond, one can imagine, from the profiles of Figure 9, that the unsaturated lipids are precursors of membrane stabilization by the above process. The unsaturated systems possess the properties of their cyclopropane analogues, but to a lesser extent. The final developmental stage in membrane growth in some bacteria would occur when the double bonds have been converted to cyclopropyl groups, the membrane thus reaching its highest stability with respect to large external perturbations.

Up to now, it appears that the above findings are general to various cyclopropane-containing membranes. Similar results have been reported for the phosphatidyl ethanolamine analog of PDSPC, PDSPE (Perly et al., 1985), as well as for the membranes of *A. laidlawii* in which the cyclopropane- or double bond-containing fatty acids have been incorporated biosynthetically (Jarrell et al., 1983).

THE EFFECT OF CHOLESTEROL ON MEMBRANE DYNAMICS

Cholesterol is a major component of plasma membranes, where it is often equimolar in concentration with the phospholipids. It has amphipathic properties due to the presence of a hydroxyl group in C3 and a hydrophobic body (the steroid skeleton and the aliphatic tail attached at C17). The dynamics and conformational properties of cholesterol-containing membranes have been investigated by a wide variety of techniques, and several modes of action have been postulated. One of them uses the ordering-disordering properties to suggest that the sterol molecule controls the membrane "fluidity" (Madden et al., 1979) whereas another alternative invokes a direct cholesterol-protein or cholesterol-lipid interaction without modification of the "viscosity" of membranes (Dahl et al., 1981). The following section reports on cholesterol-lipid interactions, at the molecular level. In order to provide a direct insight in the structure and dynamics of cholesterol-containing membranes, ^{2}H-NMR measurements have been performed with specifically deterium-enriched dimyristoyl phosphatidylcholine (DMPC) and cholesterol (Dufourc et al., 1984b). Both molecules were labelled at different positions in order to study the ^{2}H-NMR responses from the viewpoints of both lipid and cholesterol at various depths in the bilayer.

Figure 10 shows the powder and de-Paked spectra of β-[2,2,3,4,4,6-$^2H_6$]cholesterol in DMPC (3:7 molar ratio), at 25°C. The top of this figure is a good example of the implementation brought by the spectral deconvolution (Bloom et al., 1981); it is far easier to measure quadrupolar splittings on the de-Paked spectrum than on the powder spectrum.

The C-^{2}H bonds of the rigid sterol nucleus give very different quadrupolar splittings despite their undergoing same angular fluctuations, i.e. they have the same segmental order parameter S_α (or S_{mol}). As already discussed in the preceding sections, these different splittings are due to the different orientations of the C-^{2}H bonds (different S_γ) with respect to the axis of motional averaging. As shown earlier, this geometrical dependence can be accounted for by making use of geometrical relationships between bonds (obtained from X-ray data) and the pure angular fluctuations of the four rings of cholesterol can be determined. As a consequence of this calculation, the average position of the axis of motional averaging within the molecule may be found; hence the orientation of the steroid skeleton with respect to the bilayer normal. At 25°C the final result is $S_\alpha = 0.80 \pm 0.03$ and the orientation of the cholesterol molecule within the bilayer such that the C3-^{2}H bond vector makes an average angle of 84° ± 2° with respect to the axis of segmental motion. This result agrees

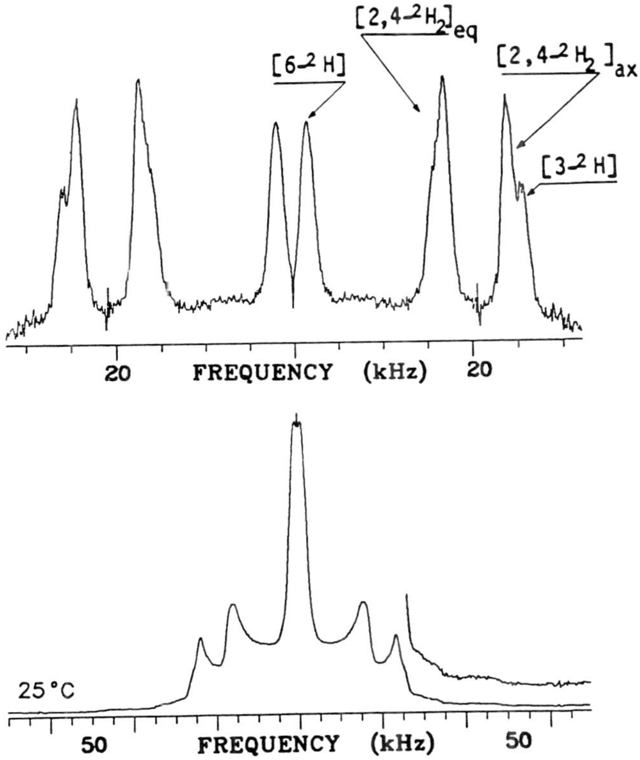

Figure 10. ^{2}H-NMR powder spectrum (bottom) and de-Paked spectrum (top) of β-[2,2,3,4,4,6-$^2H_6$]cholesterol in DMPC (3:7 molar ratio), at 25°C. From Dufourc et al., 1984b.

well with a value of S_α = 0.78 estimated by Oldfield et al. (1978) at 23°C for β-cholesterol in a similar system mixture. However, these authors did not calculate the position of the axis of motional averaging of the sterol; they assumed that the angle between the C3-^{2}H bond and the director of the motion was 90°. Our findings for the β-cholesterol-DMPC (3:7) system also correlate well with those obtained by Taylor et al. (1981) for a higher concentration of cholesterol in the membrane (β-cholesterol was in 1:1 molar ratio with egg phosphatidylcholine). The values are S_α = 0.87 and the angle between the C3-^{2}H bond and the axis of motional averaging equal to 79° ± 2°. These authors found a higher value of S_α; this is related to the property that the order parameter increases when the concentration of cholesterol increases in the bilayer "fluid" phase (Oldfield et al., 1978; Taylor et al., 1981). Once the average orientation of some C-^{2}H bonds is determined, the average orientation of the rigid body to which they are attached can be calculated. The result of such calculations can be represented using computer-assisted spatial drawing such as PLUTO 78 routines. The average orientation of the steroid skeleton is shown in Figure 13; the cholesterol molecule is quasi-perpendicular to the membrane surface.

All spectra obtained between 10 and 65°C exhibit axially symmetric lineshapes. Spectral deconvolutions and calculations as described above were performed. The values of the segmental order parameter S_α (or S_{mol}) are plotted in Figure 11.

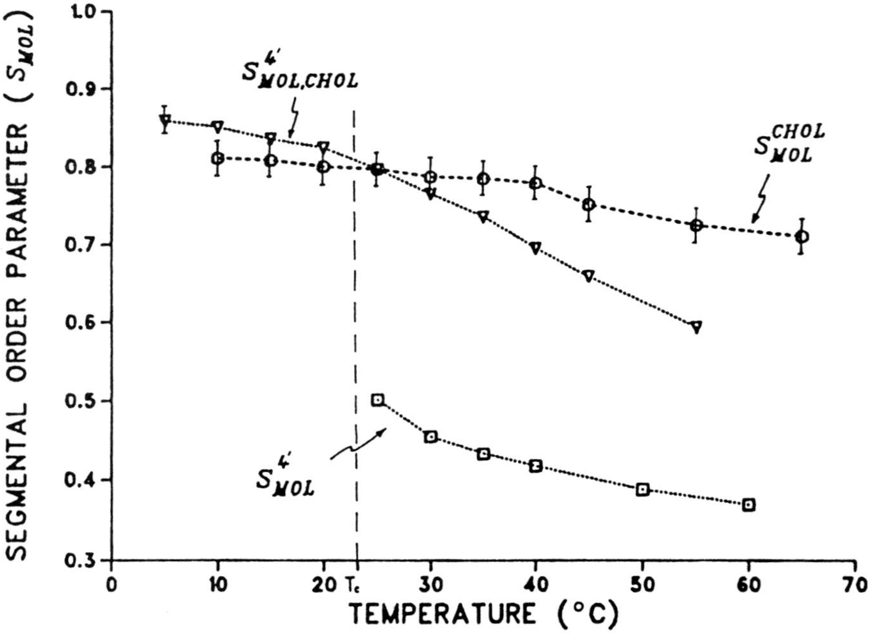

Figure 11. Temperature variation of the segmental order parameter, S_α, of the β-cholesterol-DMPC (3:7) system. $S^{4'}_{mol,chol}$ is S_α of DMPC at C4', $S^{4'}_{mol}$, but in the presence of sterol, and S^{chol}_{mol} is S_α of the four rings of cholesterol in DMPC. From Dufourc et al., 1984b.

In this figure is also reported S_α for the lipid labelled at C4', in the presence and absence of 30 mol % β-cholesterol. The average orientation of a chain methylene $C-^2H$ bond with respect to the axis of segmental order is taken to be 90°, i.e. the $C-^2H$ bond angular reorientations are equiprobable around $\gamma = 90°$, so $S_\alpha = |2S_{C-^2H}|$ for the lipid at C4' (Seelig and Niederberger, 1974).

The first noticeable feature on Figure 11 is the virtual temperature independence of the β-cholesterol segmental order parameter, S_{mol}^{chol}, between 10 and 40°C; even at 65°C, the S_{mol}^{chol} value differs by less than 10% from its value at 10°C. The segmental order parameter of the lipid at C4' in the presence of β-cholesterol, $S_{mol,chol}^{4'}$, changes rather drastically over the same temperature range. Below T_c = 23°C, the temperature of the gel-to-fluid phase transition of the pure lipid system, the change of $S_{mol,chol}^{4'}$ with temperature, is subtle, whereas above 23°C, the drop in $S_{mol,chol}^{4'}$ with increasing temperature is rather marked. It is interesting to notice that below T_c, the amplitudes of the motions of the $C-^2H$ segments at C4' are almost the same as those of the cholesterol rigid body. It should be recalled that below T_c, the lipid chains would not show any $C-^2H$ angular excursions at all if β-cholesterol were not present. One may therefore interpret this "fluidizing" effect as a disordering effect: β-cholesterol through its virtually temperature independent angular fluctuations forces the lipid chains to fluctuate even below 23°C. Since between 10 and 35°C S_α is almost the same for both the lipid at C4' and the steroid skeleton, it appears, therefore, that the cholesterol controls, through its own motions, the motions of the entire system, in this temperature range. In other words, the sterol increases the amplitude of lipid chain motions below T_c and decreases them above T_c (on comparing with systems lacking cholesterol); that is, it levels the ordering of the system to its own ordering level.

It is interesting to notice that at high temperatures, S_{mol}^{chol} is markedly greater than $S_{mol,chol}^{4'}$. The system is thus characterized by two kinds of angular fluctuations, i.e. those of the lipid and those of β-cholesterol. This point is rather important and emphasizes that measuring the segmental order parameter of β-cholesterol at 65°C would not give information about the amplitude of motions of the lipid chains. One would measure the property of the probe rather than that of the system itself.

The above has reported upon the amplitudes of the motions of both lipid and cholesterol, in the upper half of the membrane layer. The segmental order parameter was also determined for labelled positions near the center of the bilayer. The deuterium was on either the 12', 13', 14'-positions of the DMPC sn-2 chain or the 22-, 24-, 26- and 27-positions of the β-cholesterol alkyl tail. Carrying the same analysis as for the top of the bilayer, one finds that the cholesterol tail exhibits approximately the same amplitude of motions as does the steroid skeleton (Dufourc et al., 1984b). In addition, the segmental order parameter of the cholesterol tail shows almost the same temperature insensitivity as does that of the four rings. DMPC labelled near the end of the acyl chain shows a very similar behaviour, upon cholesterol addition, to what has been reported with labelled positions near the top of the bilayer. The regulatory role of cholesterol on the amplitude of acyl chain motions, as discussed for the plateau positions, can therefore be generalized throughout the membrane bilayer.

The motional behaviour of cholesterol in model membranes may be further detailed by measuring the spin-lattice relaxation time, T_{1z}, of [2,2,3,4,4,6-$^2H_6$]cholesterol in DMPC. Figure 12 reports the thermal variation of T_{1z} cholesterol labelled at position C4-equatorial and measured at two different Larmor frequencies.

The presence of T_{1z} minima at ν_o = 46.1 MHz and 30.7 MHz indicates that at the corresponding temperatures, the spin-lattice relaxation time is dominated by only one motion of correlation time $\tau_i = 0.62\ (2\pi\nu_o)^{-1}$ (Abragam, 1961; Davis, 1983). This leads to $\tau_i = 3.2\ 10^{-9}$ s at 25°C and $\tau_i = 2.1\ 10^{-9}$ s at 35°C. Assuming that this motion follows an

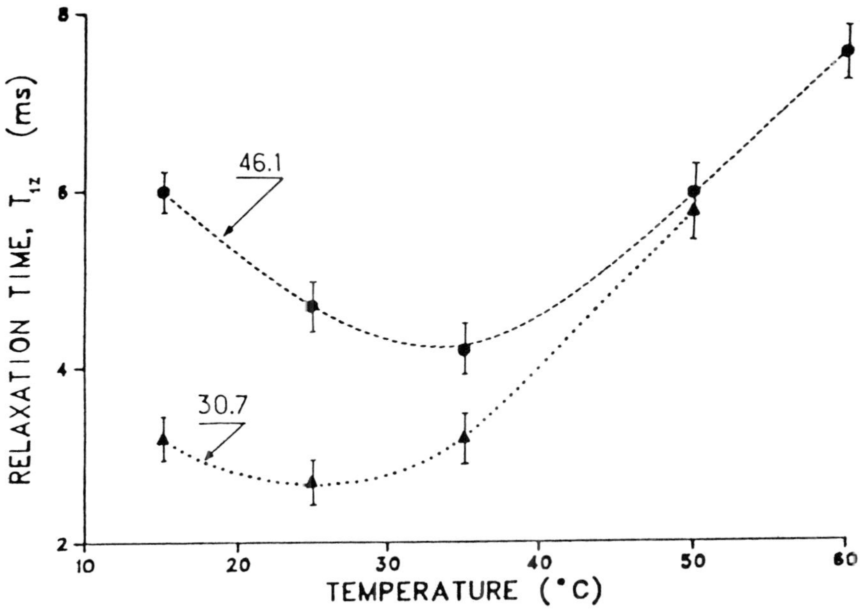

Figure 12. Temperature variation of the relaxation time (T_{1z}) of $[4\text{-}^2H]_{equat}$ cholesterol in DMPC. Figures on curves represent Larmor resonance frequencies in MHz. From Dufourc and Smith, 1986.

Arrhenius-type law of the form $\tau_i = \tau_{o,i} \exp\{E_a/RT\}$ one can calculate, <u>analytically</u>, the activation energy, $E_a = 32 \pm 5$ kJ mol^{-1} and $\tau_{o,i} = 8.0 \pm 0.5\ 10^{-15}$ s (Dufourc and Smith, 1986). It is worthwhile mentioning that these results are obtained without any need of motional model. It should also be emphasized that cholesterol may undergo several motions in DMPC; the above indicates that, at the temperature of the minimum, only one is efficient for T_{1z} relaxation at the Larmor frequencies considered.

Minima are also found for the other cholesterol labelled positions. The T_{1z} values at the minimum scale according to the orientation of the corresponding $C\text{-}^2H$ bond with respect to the axis of motional averaging of the steroid (Dufourc and Smith, 1986). This suggests that the axial rotation of cholesterol is a good candidate to induce such a peculiar variation of T_{1z} as a function of labelled position in the sterol rigid skeleton. This observation can be quantitated by modelling the relaxation with a hierarchy in time of motions, as described earlier (see Membrane Dynamics ...) (Paddy et al., 1981; Dufourc and Smith, 1986). According to this model two motions are sufficient to describe the observed deuterium T_{1z} relaxation for cholesterol. The fastest motion is not completely determined, it is possibly librational (wobbling of cholesterol), has a correlation time τ_1 in the fast correlation time regime ($\omega_0{}^2\tau_1{}^2 << 1$), does not contribute significantly to the T_{1z} relaxation but does reduce the ordering of the cholesterol molecule in DMPC from 1 to $S_1 = 0.79$ at 25°C. The second fastest motion is the sterol axial rotation characterized by a correlation time τ_2 as determined above (i.e. $\tau_2 = 8.0 \times 10^{-15}$ s $\times \exp\{32 \times 10^3$ J/RT$\}$) and a S_2 order parameter depending upon the orientation of the individual C-^{2}H bonds with respect to the sterol axis of rotation. S_1 and S_2 correspond to S_α and S_γ of eq. (5), respectively. It is noteworthy that S_1 and S_α have been determined with two different approaches: S_1 is obtained from relaxation measurements whereas S_α is obtained from quadrupolar splittings. The agreement between their values is remarkably good.

The above has shown that ^{2}H-NMR can bring very valuable information on structure and dynamics of cholesterol-containing membranes. Unfortunately, a very detailed analysis of cholesterol dynamics cannot always be performed with other biologically interesting molecules. The amplitude of motions is often easy to determine and can be utilized to account for some biological effects of a given species on a membrane. The rate of motions may lead to the time scale at which these effects may occur (see part II). However, it is often very difficult and time-consuming to obtain this information.

The next step in the analysis of membrane-cholesterol interactions is to monitor the dynamics of cholesterol in real membranes. This has been achieved by incorporation of deuterium-labelled cholesterol into human erythrocyte membranes

(Kelusky et al., 1983). Performing the analysis described above for model membranes, one obtains, from the deuterium quadrupolar splittings of cholesterol in this natural membrane, the average orientation and the segmental order parameter of cholesterol in a red blood cell membrane. The average orientation is such that the C3-^{2}H bond makes an angle of 81° ± 2 with the axis of rotation of cholesterol. This value is very close to that found for cholesterol in DMPC, i.e. 84 ± 2° (Dufourc et al., 1984b), and in egg PC, i.e. 79 ± 2° (Taylor et al., 1981). The orientation of cholesterol in the natural membrane is therefore that of Figure 13 (left). A value of S_α = 0.76 ± 0.05 is calculated for cholesterol in ghosts which, within the experimental error, is the same as that found for cholesterol in DMPC (S_α = 0.80 ± 0.03 at similar cholesterol to DMPC ratios). This is particularly interesting and indicates that the anisotropic motion of the cholesterol molecule is the same in natural and model membranes. Because erythrocyte membranes contain a large number of lipid classes and a high proportion of protein, the similarity in S_α suggests that the whole body motions of cholesterol are essentially independent of the composition of the matrix.

Unfortunately, the rates of motions of cholesterol in erythrocyte membranes could not be determined due to the lack of sensitivity of the technique on systems in which the incorporation of labelled material is very low.

CHANGES IN MEMBRANE DYNAMICS UPON MODIFICATION OF THE CHOLESTEROL CHEMICAL STRUCTURE

Is the cholesterol structure a necessary need to exert the previously discussed effects on membranes? In attempts to elucidate this question, studies have been carried out with the α-isomer of cholesterol, epicholesterol. The only structural change resides therefore in the position of the hydroxyl group in α or β with respect to the plane of the fused ring system. Experiments were performed with [2,2,3,4,4,6-$^2H_6$]α-cholesterol in DMPC. Segmental order parameters and average orientations have been calculated as previously described. A summary of the results is presented in Table I.

Table I. Temperature dependence of average orientation and S_α order parameter of α-, β-cholesterol in DMPC[a] and DPPC[b] systems

System	20°C	25°C	40°C	55°C	60°C	65°C
α-chol	0.92	0.89	0.87	0.77		0.72
DMPC	0.80	0.76		0.55		
$\gamma(C3\text{-}c,\vec{n})$	93.5	91.5	92	91.5		91
α-chol		0.99			0.72	
DPPC						
β-chol	0.80	0.80	0.78	0.72		0.71
DMPC	0.82	0.80	0.69	0.59		
$\gamma(C3\text{-}c,\vec{n})$	84	84	84.5	85		84
β-chol		0.77			0.75	
DPPC						

Cholesterol:Lipid ratio is 3:7 for all systems.

[a] From Dufourc et al., 1984b, S_α for sterols is obtained from $[2,2,3,4,4,6\text{-}^2H_6]$α- and β-cholesterol whereas S_α for DMPC is obtained from $[4'\text{-}^2H_2]$DMPC. $\gamma(C3\text{-}c,\vec{n})$ is the angle between the C3-c bond vector and the axis of motion $\vec{n}$, where c is axial.

[b] From Murari et al., 1986, S_α for sterols is obtained from $[3\text{-}^2H]$α- and β-cholesterol.

Several differences are noticed for α- and β-cholesterol in DMPC.

First, the average orientation of α-cholesterol with respect to the bilayer normal changes with temperature. This is noticed in Table I by a significant variation of the angle $\gamma(C3\text{-}c,\vec{n})$ of a particular C-^{2}H bond (c represents ^{2}H or OH in α- or β-cholesterol, respectively) with respect to the axis, $\vec{n}$, about which the rigid region of the sterol rotates as a function of temperature. No such variation is observed for β-cholesterol. Angles γ are different for α- and β-isomers, indicating that the α-isomer is tilted on comparing to the average orientation of the β-isomer (Figure 13, left).

Second, S_α values for epicholesterol are higher than those of cholesterol, especially at low temperatures. Results are comparable for both DMPC and DPPC systems. However, one must mention that the DPPC α-cholesterol, S_α, has been calculated upon the hypothesis that the axis of motional averaging is the same for both isomers (Murari et al., 1986). Unfortunately, this condition is not fulfilled (Dufourc et al., 1984b) and the reported S_α values for α-cholesterol:DPPC must be taken as indications rather than precise numbers.

Third, the S_α value of the lipid in the presence of epicholesterol, at approximately the same bilayer depth, is lower than in the presence of cholesterol. This increases the differences in S_α for both epicholesterol and lipid indicating that α-cholesterol possesses some of the ordering-disordering properties of its β-isomer, but to a lesser extent.

The above analysis leads to two main conclusions. First, a small modification in the structure of cholesterol (OH α or β) leads to important modifications of membrane structure and dynamics. Second, the tilted position of α-cholesterol within the bilayer membrane could account, at least in part, for the non-occurrence of this isomer of cholesterol in natural membranes - it would disturb too greatly the parallel packing of the lipid chains.

A more severe modification of the sterol structure can be promoted by the attachment of a palmitic chain on the β-hydroxyl group. Experiments have been carried out on ^{2}H-labelled cholesteryl palmitate embedded in a DPPC model membrane in order to investigate in which way the order parameter (S_α) and average orientation of this sterol were modified by the presence of the palmitic chain. Results are reported in Table II and Figure 13.

Table II indicates that the orientation of 8 mol % cholesterol in DMPC (Dufourc et al., 1984b) is identical to that found for 30 mol % cholesterol in DMPC, at 45°C. This corroborates the finding by Kelusky et al. (1983) that the average orientation of cholesterol in a membrane does not depend strongly upon the nature of the membrane lipid. S_c

Table II. Ordering and orientation of β-cholesterol and cholesteryl palmitate in model membranes at 45°C[a]

	5 mol % cholesteryl palmitate in DPPC	8 mol % cholesterol in DPPC	30 mol % cholesterol in DMPC
S_{α} ±0.03	0.47	0.63	0.75
$\gamma(C_3\text{-}c,\vec{n})$[b] ± 2°	75	84	84

[a]From Chana et al., 1985.

[b]Angle between the C_3-c bond and the rotational axis of the fused ring system, where c is axial.

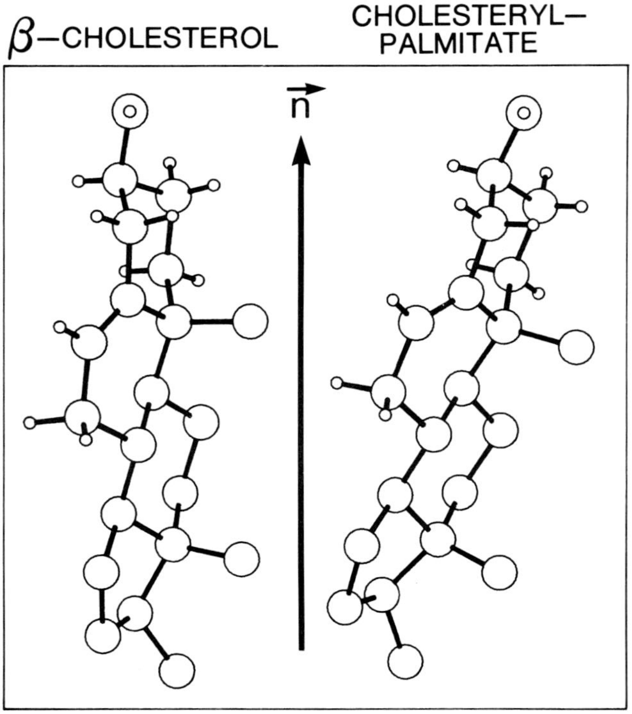

Figure 13. PLUTO 78 spatial representation of the fused ring system of β-cholesterol and cholesteryl palmitate; the arrow indicates the direction of the rotation axis, $\vec{n}$, which for S_{α} (or S_{mol}) = 1, corresponds to the director (bilayer normal) and lies in the plane of the page. The symbol o is the oxygen attached at C3. From Chana et al., 1985.

decreases on comparing cholesterol in DMPC and in DPPC, at the same temperature. This has already been accounted for by a concentration-dependent ordering effect of cholesterol in fluid phase membranes (Oldfield et al., 1978). S_α of cholesteryl palmitate is smaller than that of cholesterol in the same system, at the same temperature. This indicates that the amplitude of motions of the former is greater than that of the latter.

Figure 4 shows average orientations of both fused ring systems undergoing fast reorientations around their principal axis of rotational diffusion, $\vec{n}$. This representation does not account for the angular fluctuations ("wobbling") of the steroid moiety and can be considered as a "snap-shot" picture of molecules when $S_\alpha = 1$. One sees clearly that the steroid skeleton of cholesteryl palmitate is markedly tilted away from the position of that of cholesterol. This reflects the presence of the palmitoyl chain attached to the oxygen at C3. The palmitoyl chain is known to fold back into the membrane (Gorrissen et al., 1981) and therefore "pulls" the rigid cholesteryl skeleton away from the average orientation found for cholesterol.

CONCLUSIVE REMARKS

Solid state NMR can lead to very detailed analysis of the motional behaviour of membranes. Among the information which has been obtained from ^{2}H-NMR of rigid molecular units in membranes, one may retain that small structural modifications in a molecule may lead to considerable changes in the structure and dynamics of the membrane and that rigid moieties in bilayers (double bonds, cyclopropane groups, sterols), because of their relative inertia to motional changes, act as to increase membrane stability; they can be viewed as membrane motional regulators.

REFERENCES

Abragam A (1961) Principles of Nuclear Magnetism. Oxford University Press London

Bloom M, Davis JH, MacKay AL (1981) Direct determination of the oriented sample NMR spectrum from the powder spectrum for systems with local axial symmetry. Chem Phys Letters 80:198-202

Brown MF, Seelig J, Haeberlen U (1979) Structural dynamics in phospholipid bilayers from deuterium spin-lattice relaxation time measurements. J Chem Phys 70:5045-5053

Chana RS, Cushley RJ, Wassall SR, Smith ICP, Dufourc EJ (1985) Organization of cholesteryl esters in membranes. A deuterium nuclear magnetic resonance study. Chem Phys Lipids 37: 345-356

Christie WW (1970) Cyclopropane and cyclopropene fatty acids. In: Gunstone FD (ed) Topics in lipid chemistry, vol I. Lagos Press London p 1

Dahl JS, Dahl CE, Block K (1981) Effect of cholesterol on macromolecular synthesis and fatty-acid uptake by <u>mycoplasma-capricolum</u>. J Biol Chem 256:87-91

Davis JH (1979) Deuterium magnetic resonance study of the gel and liquid crystalline phases of dipalmitoyl phosphatidylcholine. Biophys J 27:339-358

Davis JH (1983) The description of membrane lipid conformation, order and dynamics by ^{2}H-NMR. Biochim Biophys Acta 737:117-171

Dufourc EJ (1983) The effects of cyclopropane rings, sterols and antibiotics on the structure and dynamics of phospholipid membranes. A deuterium solid state nuclear magnetic resonance approach. University of Ottawa Ottawa

Dufourc EJ (1986) La résonance magnétique nucléaire du déuterium et du phosphore dans les milieux organisés. Théorie et applications à l'étude de l'action de la mélittine et de la deltalysine sur les membranes biologiques. University of Bordeaux I Bordeaux

Dufourc EJ, Smith ICP (1986) A detailed analysis of the motions of cholesterol in biological membranes by ^{2}H-NMR relaxation. Chem Phys Lipids 41:123-135

Dufourc EJ, Smith ICP, Jarrell HC (1983) A ^{2}H-NMR analysis of dihydrosterculoyl-containing lipids in model membranes: structural effects of a cyclopropane ring. Chem Phys Lipids 33:153-177

Dufourc EJ, Smith ICP, Jarrell HC (1984a) Role of cyclopropane moieties in the lipid properties of biological membranes: a ^{2}H-NMR structural and dynamical approach. Biochemistry 23: 2300-2309

Dufourc EJ, Parish EJ, Chitrakorn S, Smith ICP (1984b) Structural and dynamical details of cholesterol-lipid interaction as revealed by deuterium NMR. Biochemistry 23:6062-6071

Gorrissen H, MacKay AL, Wassall SR, Valic MI, Tulloch AP, Cushley RJ (1981) Deuterium magnetic resonance of selectively deuterated cholesteryl esters in dipalmitoylphosphatidyl choline dispersions. Biochim Biophys Acta 644:266-272

Jarrell HC, Tulloch AP, Smith ICP (1983) Relative roles of cyclopropane-containing and cis-unsaturated fatty acids in determining membrane properties of Acholeplasma laidlawii: a deuterium nuclear magnetic resonance study. Biochemistry 22:5611-5619

Kelusky EC, Dufourc EJ, Smith ICP (1983) Direct observation of molecular ordering of cholesterol in human erythrocyte membranes. Biochim Biophys Acta 735:302-304

Madden TD, Chapman D, Quinn PJ (1979) Cholesterol modulates activity of calcium dependent ATPase of the sarcoplasmic reticulum. Nature (London) 279:538-541

Murari R, Murari MP, Baumann WJ (1986) Sterol orientation in phosphatidylcholine liposomes as determined by deuterium NMR. Biochemistry 25: 1062-1067

Oldfield E, Meadows M, Rice D, Jacobs R (1978) Spectroscopic studies of specifically deuterium labeled membrane systems. Nuclear magnetic resonance investigation of the effects of cholesterol in model systems. Biochemistry 17:2727-2740

Paddy MR, Dahlquist FW, Davis JH, Bloom M (1981) Dynamical and temperature dependent effects of lipid-protein interactions. Application of deuterium nuclear magnetic resonance and electron paramagnetic resonance spectroscopy to the same reconstitutions of Cytochrome c Oxidase. Biochemistry 20:3152-3162

Perly B, Dufourc EJ, Jarrell HC (1984) Facile and high yielding syntheses of phosphatidylcholines and phosphatidylethanolamines containing ^{2}H-labeled acyl chains. J Labelled Compd Radiopharm 21:1-13

Perly B, Smith ICP, Jarrell HC (1985) Acyl chain dynamics of phosphatidylethanolamines containing oleic acid and dihydrosterculic acid: ^{2}H-NMR relaxation studies. Biochemistry 24:4659-4665

Rance M, Byrd RA (1983) Obtaining high-fidelity spin 1/2 powder spectra in anisotropic media: phase-cycled Hahn echo spectroscopy. J Magn Res 52:221-240

Seelig A, Seelig J (1974) The dynamic structure of fatty acyl chains in a phospholipid bilayer measured by deuterium magnetic resonance. Biochemistry 13:4839-4845

Seelig J (1977) Deuterium magnetic resonance: theory and application to lipid membranes. Quart Rev Biophys 10:353-418

Seelig J (1978) Phosphorus-31 nuclear magnetic resonance and the head-group structure of phospholipids in membranes. Biochim Biophys Acta 515:105-141

Seelig J, Niederberger W (1974) Deuterium-labeled lipids as structural probes in liquid crystalline bilayers. A deuterium magnetic resonance study. J Am Chem Soc 96:2069-2072

Singer SJ, Nicolson GL (1972) The fluid mosaic model of the structure of cell membranes. Science 175:720-731

Souzu H (1982) *Escherichia coli* B membrane stability related to cell growth phase. Measurement of temperature dependent physical state change of the membrane over a wide range. Biochim Biophys Acta 691:161-170

Smith ICP (1984) Fluidity of membranes - an NMR approach. In: Burton RM, Guerra FC (eds) Biomembranes, dynamics and biology. Plenum Press New York p 81

Taylor MG, Akiyama T, Smith ICP (1981) The molecular dynamics of cholesterol in bilayer membranes: a deuterium NMR study. Chem Phys Lipids 29:327-339

MEMBRANE STRUCTURE AND DYNAMICS BY NMR. PART 2: THE ACTION OF POLYENE ANTIBIOTICS AND TOXINS ON BIOLOGICAL MEMBRANES.

Erick J. Dufourc
C.R. Paul Pascal/CNRS
Domaine Universitaire
33405 TALENCE Cedex
FRANCE

INTRODUCTION

Part 1 of this contribution dealt with properties of the membrane itself and has shown that relatively small modifications in the lipid structure or in the sterol structure can lead to very important changes in membrane dynamics.

In the following, I shall describe the effect of extraneous molecules, such as antibiotics or toxins, on model membranes. Here, we will look for modifications in the structural and dynamical properties of labelled membrane components (lipids, sterols) and attempt to relate, at the molecular level, the observed changes to well-known biochemical effects promoted by these molecules.

Polyene antibiotics are known to mediate changes in the membrane permeability of a number of organisms, thus inducing a leakage of important cellular constituents and ultimately lysis and death of the cell (Hazen and Brown, 1951; Kinsky, 1961a,b).

On the other hand, the bee venom toxin melittin is also known to lyse natural and artificial membranes (Haberman, 1972; Weissman et al., 1969; Olson et al., 1974; Hegner, 1968; Sessa et al., 1969).

As it will be shown below, considerable membrane modifications occur upon addition of these molecules. Deuterium (^{2}H) and Phosphorus-31 (^{31}P) NMR will be used to probe these changes in membrane structure and dynamics. Structural changes, when they happen, will also be monitored with other physical techniques such as Quasi-Elastic Light Scattering (QELS) and Freeze-fracture electron microscopy.

NATO ASI Series, Vol. H16
Membrane Biogenesis. Edited by J. A. F. Op den Kamp

POLYENE ANTIBIOTICS AND MEMBRANE STRUCTURE AND DYNAMICS

A. AMPHOTERICIN B

The principal characteristic of polyene antibiotics such as Amphotericin B and filipin is that they apparently require the presence of sterols in the cell membrane to promote changes in membrane permeability (Gottlieb et al., 1958). Permeability and calorimetry data on liposomes led to the hypothesis that polyene antibiotics and sterols formed molecular complexes to create channels or disrupt the membrane (Norman et al., 1976).

In order to obtain more insight into sterol-antibiotic interactions and to investigate the possible structure and dynamics of the proposed pores, ^2H-NMR studies were performed on model membranes, with and without cholesterol.

The effect of Amphotericin B on DMPC (dimyristoyl phosphatidylcholine) membranes lacking cholesterol was first investigated. Figure 1 shows deuterium powder spectra of [4'-$^2H_2$]DMPC in the absence and presence of Amphotericin B (30 mol %), as a function of temperature. At low temperatures (5°C), the spectra of lipids with and without antibiotic are almost identical. They exhibit the spectral shape characteristic of the so-called gel phase of pure lipids with a maximum spectral width of ca. 120 kHz. On increasing the temperature above 23°C (the gel-to-fluid phase transition temperature, T_c, of DMPC), the spectra of the pure lipid show axially symmetric shapes indicative of one environment. In contrast, those of the sample containing Amphotericin B exhibit, in addition to the usual quadrupolar Pake doublet, a broad spectral feature of maximum width ca. 120 kHz. The relative percentages of these two components can be estimated via spectral simulations or spectral moment calculations (Dufourc et al., 1984a). Above 25°C, the relative amount of the broad component is 30 ± 6°C, and appears to be invariant to increasing temperature.

The presence of this broad subspectrum at all temperatures above 25°C indicates that 30% of the lipids are in a motionally restricted environment. This state exhibits spec-

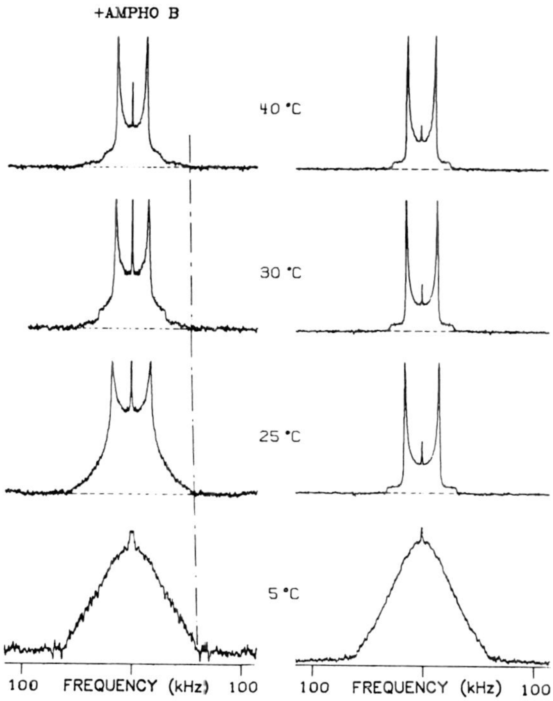

Figure 1. ^2H-NMR spectra of [4'-$^2H_2$]DMPC multilamellar membranes in the presence or absence of Amphotericin B (30 mol %), as a function of temperature. From Dufourc et al., 1984a.

troscopic features very similar to those of the gel phase, i.e. the lipid chains are "frozen" in the <u>all</u>-<u>trans</u> configuration, the only remaining motion being rotational diffusion around the long axis of the extended chain. This situation is achieved when $\Delta\nu_{max}$ = 127.5 kHz (the maximum spectral width is defined as the quadrupolar splitting produced by molecules whose axis of motion is oriented at 0° with respect to the magnetic field direction, i.e. the "shoulders" of powder spectra - see part 1 of this contribution), which is what is approximately observed on spectra. Furthermore, the lipids which are not strongly motionally restricted by the antibiotic manifest larger quadrupolar splittings than in the absence of Amphotericin B, at corresponding temperatures between 25°C and

40°C. This indicates that in this range of temperatures, the drug reduces the amplitude of motions of the lipid methylene segments, i.e. S_{C-^2H}, the orientational order parameter, increases. However, this ordering action is no longer perceived at high temperatures; above 50°C the lipids of the Amphotericin B-containing membranes face two very different physical environments: in one of these domains, the lipids are strongly coupled with the antibiotic (broad spectral component) whereas in the other they are as free as in the pure lipid system (axially symmetric powder pattern with $\Delta\nu_Q$ equal to that of pure lipids).

Experiments carried out with DMPC whose sn-2 fatty acyl chain was perdeuterated indicated that the above conclusions were true for all labelled positions, i.e. throughout the entire bilayer thickness (Dufourc et al., 1984a).

It appears, therefore, that at temperatures higher than T_c, Amphotericin B is in a lipid-aggregated form, in which the lipid is all-trans, and for which the only motion observable in the time scale of ^{2}H-NMR spectra (ca. 10^{-6} s) is possibly rotation about the long molecular axis. The drug:lipid ratio in the aggregate has to be approximately 1:1. At high temperatures, these aggregates are embedded in a lipid matrix whose properties are similar to those of the pure lipid system.

Effects of Amphotericin B on cholesterol-containing model membranes were monitored from the point of view of both the lipid and cholesterol. In order to compare the results with those of previous studies on cholesterol-containing systems (Dufourc et al., 1984b), the samples always contained 30 mol % sterol with respect to the lipid; the antibiotic, when present, was in equimolar amounts with the sterol.

Figures 2 and 3 show ^{2}H-NMR spectra of lipids in a cholesterol-containing system, with and without Amphotericin B. Figures 2 and 3 report on events occurring near the top of the bilayer (labelled position C4') and at the center of the bilayer (labelled position C14'), respectively. Upon addition of the antibiotic, no change is detected at C4' whereas a broad spectral component is observed at C14'. The relative area of this subspectrum varies from 28 ± 8% at 5°C to 10 ± 8% at 55°C. Despite the inaccuracy involved in estimating areas, it is clear that the broad component is favoured at low temperatures.

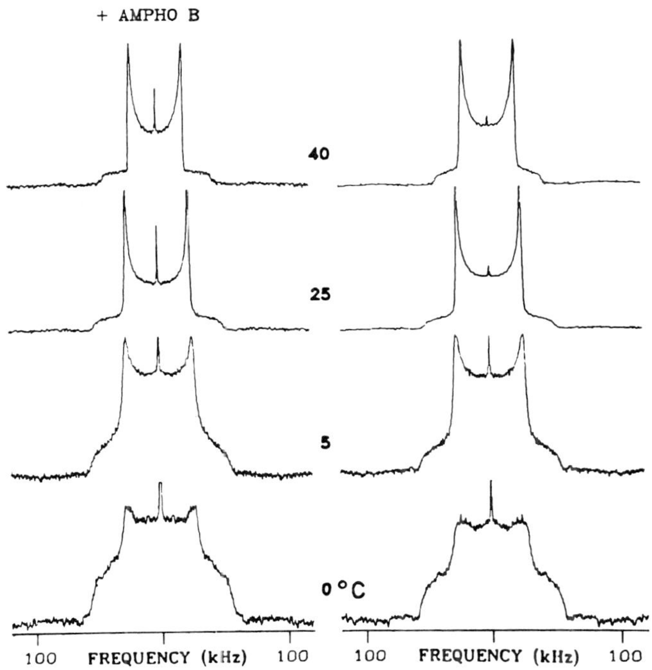

Figure 2. ^2H-NMR spectra of the β-cholesterol:[4'-$^2H_2$]DMPC system (3:7) in the absence or presence of Amphotericin B (equimolar to cholesterol). From Dufourc et al., 1984a.

Experiments with DMPC perdeuterated in the acyl chain at the sn-2 position indicate that the effect observed at C14' is only observed up to C13'. This suggests that the broad spectral component, which represents motionally restricted lipids, is characteristic of the bilayer center.

The effect of Amphotericin B on the amplitude of motions of membrane components may be followed by estimating the order parameter of the 4 rings of cholesterol (S_α or S_{mol}) (see part 1). At temperatures above 10°C, the cholesterol spectra with

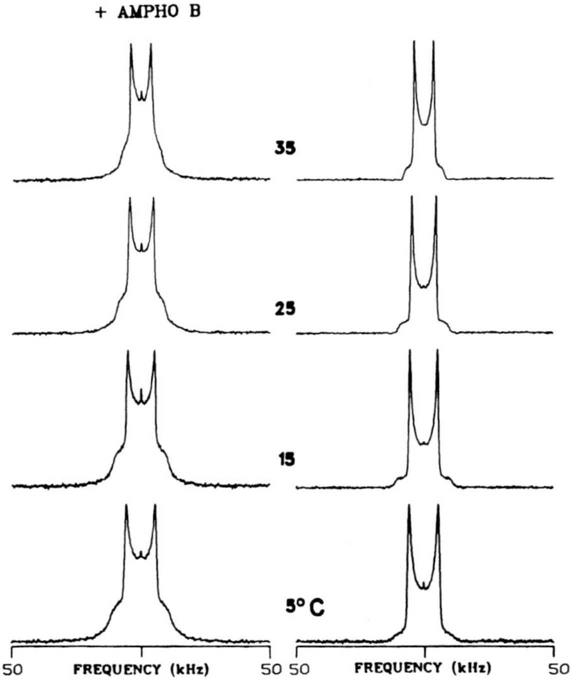

Figure 3. ^{2}H-NMR spectra of the β-cholesterol:[14'-$^2H_3$]DMPC system (3:7) in the absence or presence of Amphotericin B (equimolar to cholesterol). From Dufourc et al., 1984c.

and without the drug do not exhibit marked differences. This is reflected in Figure 4 by the same S_α for cholesterol in the presence and absence of Amphotericin B, between 10 and 40°C. The two profiles slightly diverge at high temperatures showing that the antibiotic extends the temperature-independent "wobbling" of cholesterol by some 20°C. The motional axis of the sterol, and thereby, the orientation of the cholesterol molecule with respect to the bilayer normal, was calculated as described in part 1. The orientation of cholesterol within the bilayer membrane was found to be identical for systems with and without the antibiotic, that is, if the "complex" formed between Amphotericin B and cholesterol exists, it must be such that the sterol long axis is essentially normal to the membrane surface.

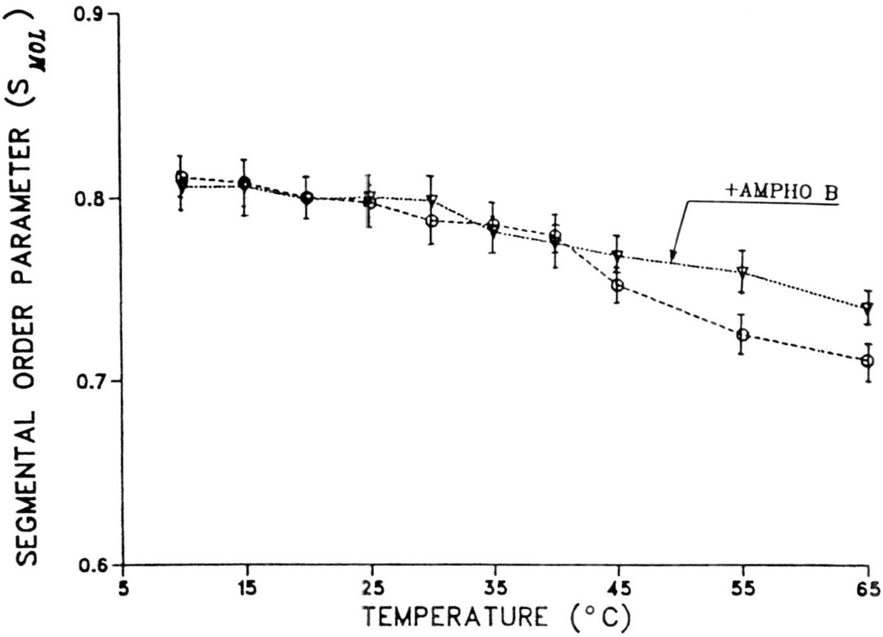

Figure 4. Temperature dependence of the order parameter, S_α (or S_{mol}) of the four rings of cholesterol, in DMPC, in the presence and absence of Amphotericin B. From Dufourc et al., 1984c.

[24-$^2H_2$]Cholesterol and [26,27-$^2H_6$]cholesterol were also used to monitor the influence of Amphotericin B near the bilayer center, from the sterol viewpoint. The presence of Amphotericin B causes very small changes on spectra, indicating that the antibiotic has little effect on the ordering properties of the cholesterol tail (Dufourc et al., 1984c).

Amphotericin B appears, therefore, not to modify the amplitude of motions of both the lipid and cholesterol, except at the center of the bilayer where the lipid senses two distinct environments.

The interpretation of these results in terms of the existence of a cholesterol-polyene antibiotic complex is not straightforward. The authors postulating such a complex indicated that the antibiotic is segregating the sterol (De

Kruijff et al., 1976a,b). If such a situation occurred in this model membrane system, two physical events would be detectable by ^{2}H-NMR: the first would be a decrease in the local order of the bulk lipid and the second would be the restoration of the phase transition. It is easily understandable that if the antibiotic complexed the sterol tightly, the ordering effect of the latter on the lipid would be modified or removed. The lipids would be free from the sterol interaction and thus capable of undergoing their cooperative gel- to-fluid phase transition. ^{2}H-NMR results show clearly that none of these events occur, as detected by ^{2}H-NMR spectra. Therefore, if the complex exists, it has to be dynamic with respect to the time scale of ^{2}H-NMR spectra (~10^{-6} s), that is, cholesterol has to be in fast exchange between a site in which it is associated with the antibiotic and a site in which it is not and therefore able to order the lipid.

However, it appears that some form of Amphotericin-cholesterol "complex" is present, since the antibiotic-induced motionally restricted lipid domain is no longer detected, except at the center of the bilayer, when cholesterol is present in the system. Cholesterol has been shown to possess an elevated order parameter up to carbons 22-24 (S_α or S_{mol} = 0.8-0.7), which indicates that little motional averaging of the quadrupolar interaction is taking place (see part 1). Bearing this in mind one can think of cholesterol as a "rigid cylinder" very insensitive to external ordering such as that possibly induced by Amphotericin B. With this peculiar property, the cholesterol molecule could act as a "screen", preventing the Amphotericin B from direct interaction with the lipid. The observation of a broad lipid component for positions near the bilayer center could be accounted for by a mismatching of the length of Amphotericin B by cholesterol, thus allowing the antibiotic to exert its ordering action on lipids at C14' and C13'.

The dynamical nature of the sterol-antibiotic complex may be probed by estimation of spin lattice relaxation times of deuterium-labelled cholesterol. Figure 5 shows such data in the presence and absence of Amphotericin B.

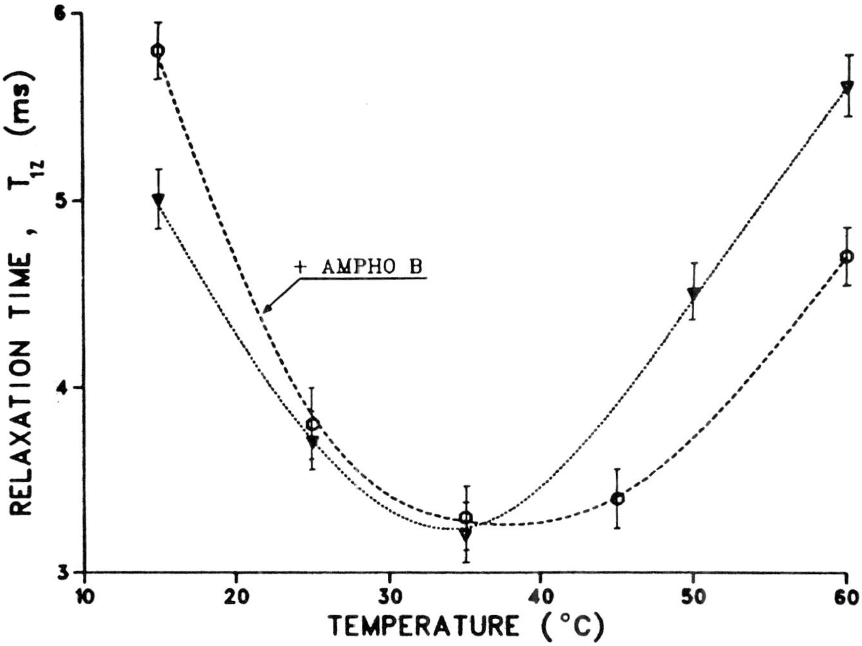

Figure 5. Temperature dependence of the relaxation time T_{1z} of [6-^{2}H]cholesterol in the presence and absence of Amphotericin B. From Dufourc et al., 1984c.

Both data sets yield a minimum in T_{1z} as a function of temperature. In the case of cholesterol in DMPC, this minimum has been shown to be promoted by the sterol axial rotation of correlation time 2.1 10^{-9} s at 35°C (Dufourc and Smith, 1986, and part 1). The observed minimum in the presence of Amphotericin B shows that the same correlation time is found at 38-40°C. This indicates that the system containing the antibiotic has to be warmed up by approximately 5°C in order to relax with the same efficiency as the system without the drug. This can be interpreted as a slowing down of the sterol axial rotation when the Amphotericin B is present in the membrane.

Permeability studies have indicated that Amphotericin B

induces greater leakage when cholesterol rather than its α-isomer, epicholesterol, is present in the membrane (Norman et al., 1976). This differential efficiency was attributed to the ability of the sterol OH group to form more hydrogen bonds with Amphotericin B when located in the β- rather than the α-orientation with respect to the steroid skeleton. In order to verify this hypothesis, some of the experiments described with cholesterol were repeated with the α-isomer. Spectral features obtained with either the α- or the β-isomer of cholesterol and the antibiotic were very similar (Dufourc et al., 1984c). The noticeable difference between systems containing epicholesterol or cholesterol resides in a ca. 10% increase in local order induced by the drug in the model membrane containing epicholesterol, especially for labelled positions near the top of the bilayer. The interaction of Amphotericin B with α-cholesterol might appear to contradict what has been found in the permeability studies. However, the structural nature of the sterol-antibiotic "complex" may be different for the two sterols. This structural difference may be due to the tilted configuration that epicholesterol exhibits in the membrane. Studies with labelled α-cholesterol indicate that the antibiotic does not change the average orientation of the sterol (Dufourc et al., 1984c). It is thus understandable that, although interacting, Amphotericin B and α-cholesterol in the "complex" may not have the right orientation with respect to the membrane surface to induce permeability.

B. FILIPIN

The polyene antibiotic filipin is known to produce rather dramatic effects on biological membranes. Its action on cholesterol-containing model membranes has been monitored from the viewpoints of both the lipid and cholesterol. In order to compare with results on cholesterol-containing DMPC (Dufourc et al., 1984b) and on cholesterol:Amphotericin B:DMPC (Dufourc et al., 1984c), cholesterol was at 30 mol % with respect to lipids and the antibiotic equimolar to the sterol.

The bottom spectrum of Figure 6 shows the characteristic powder pattern of [2,2,3,4,4,6-$^2H_6$]cholesterol in DMPC, whose features led to the calculation of a segmental order parameter, for the rigid fused ring system, of 0.8 under identical conditions of temperature and concentration (Dufourc et al., 1984b). The central resonance, which is due to the C6-^{2}H bond oriented close to the magic angle (Taylor et al., 1981; Dufourc et al., 1984b), has been deliberately truncated in order to enhance details in the shoulders of the powder spectra.

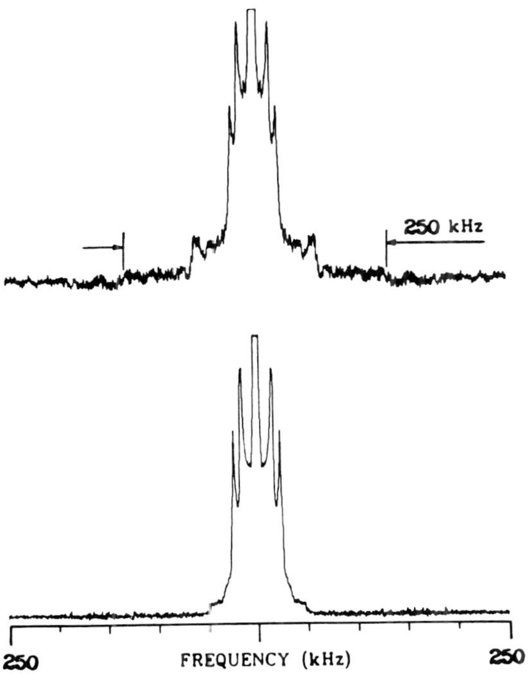

Figure 6. ^{2}H-NMR spectra of [2,2,3,4,4,6-$^2H_6$]cholesterol in DMPC (3:7 molar ratio) in the presence (top) and absence (bottom) of filipin at 25°C. The antibiotic, when present, is equimolar to cholesterol. From Dufourc and Smith, 1985.

When filipin is added to the sample containing labelled cholesterol, one obtains the top spectrum of Figure 6. In comparing top and bottom spectra, one notices that a new quadrupolar powder pattern of maximum width ca. 250 kHz appears in the spectrum arising from the sample containing the antibiotic. This new spectral component is superimposed on a spectrum similar to that of cholesterol in DMPC and exhibits an axially symmetric shape; the peaks of quadrupolar splitting $\Delta\nu_Q$, ca. 125 kHz, are clearly distinguished.

The interpretation of the subspectrum of total width 250 kHz is straightforward; cholesterol molecules giving rise to this NMR pattern no longer undergo motions that average the static quadrupolar interaction, i.e. they behave like a powder of monocrystals and, therefore, yield a spectrum of the maximum possible width (Seelig, 1977). Thus, in the presence of filipin, at 25°C, cholesterol senses two very distinct environments. In one of these it is immobilized, whereas in the other it undergoes axially symmetric motions similar to those in the membrane lacking the drug.

Figure 7 shows spectra resulting from a temperature variation performed on the sample giving rise to the top spectrum of Figure 6. At 0°C the spectrum is a single powder pattern, indicating that all cholesterol molecules are immobilized. As the temperature of the sample is increased, one notices that the spectrum characteristic of axially mobile cholesterol "grows" on the powder pattern of immobilized cholesterol. At 0°C the immobile site is stabilized whereas at 60°C cholesterol is almost entirely in the labile site. At intermediate temperatures, e.g. 25°C, one can distinguish cholesterol in the two sites, which indicates that the exchange rate is less than the frequency difference (ca. 10^5 s^{-1}) between the powder patterns associated with each domain.

Experiments carried out on the same system, using labelled lipids, exhibit axially symmetric powder patterns down to 5°C, that is, there is no trace of gel-type spectra down this temperature. The lipid spectra in the presence of cholesterol and filipin are very similar to those observed in the absence of the antibiotic. It is noteworthy that no rigid spectrum such as that observed with labelled cholesterol is detected for temperatures ranging between 5 and 65°C. This indicates

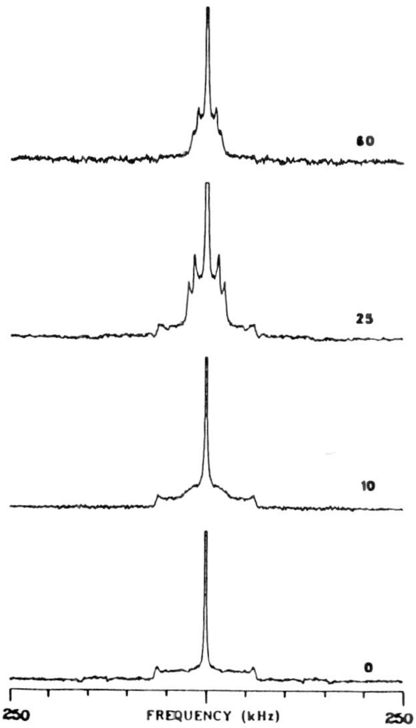

Figure 7. Temperature dependence of the ^{2}H-NMR spectra of cholesterol in the DMPC:cholesterol:filipin model membrane (7:3:3 molar ratio). Temperatures are indicated on each spectrum. From Dufourc and Smith, 1985.

that cholesterol acts as a screen, preventing the lipids from direct interaction with filipin. There is a slight decrease in the lipid quadrupolar splittings of the system containing the drug versus the system lacking it, for temperatures ranging from 5°C to 35°C. Above 35°C the lipid quadrupolar splittings markedly tend to be those of the pure lipid, i.e. the ordering action of cholesterol above T_c is partially removed by the presence of filipin.

On comparing the values of the S_α order parameter, at 60-65°C, as probed by the lipid at position C4' or by the four rings of the sterol, in the presence of the antibiotic, one finds different answers although the two probes monitor events occurring at approximately the same bilayer depth. This is due to the fact that one measures the ordering of a given molecular species in the system, rather than an average value for the entire system. The very important difference in local ordering of the lipid (S_α = 0.42) compared to that of cholesterol (S_α = 0.60) indicates that the filipin-sterol-containing model membrane may be pictured as an assembly of mobile domains (lipids) and rigid domains (macromolecular aggregates of filipin-cholesterol that lose their sterol molecules as the temperature is increased).

At 25°C, the spectrum of cholesterol in the DMPC (Figure 6, top) is identical to that of cholesterol in DMPC (Figure 6, bottom). This implies that labile cholesterol has the same ordering and average orientation in this lipid membrane, in the presence or absence of filipin. This implies that the steroid skeleton, in the labile site, is perpendicular to the membrane surface (Dufourc et al., 1984b). The orientation of cholesterol bound to the antibiotic cannot be calculated as for the labile site; however, it seems reasonable to imagine that the sterol has the same average orientation with respect to the membrane surface in labile and immobilized sites in order to facilitate the exchange. A configuration of the complex in which filipin and cholesterol are parallel to the membrane surface, as postulated by de Kruijff and Demel (1974), seems an unlikely way to favor this exchange.

The above has shown that the interaction between polyene antibiotics and membrane cholesterol is clearly monitored by ^{2}H-NMR. Cholesterol appears to prevent direct interaction between both antibiotics and lipids. The orientations of the complexes seem to adopt the average orientations that have the sterols in the membrane, i.e. perpendicular and tilted to the membrane surface with cholesterol and epicholesterol, respectively. The complex between Amphotericin B and cholesterol is dynamic in the ^{2}H-NMR time scale (the interaction is perceived on the time scale of 10^{-9} s at physiological temperatures) whereas that between filipin and cholesterol is static (the interaction occurs on a time scale greater than 10^{-5} s). This provides a means of differentiation between the action of both antibiotics on membranes.

THE ACTION OF THE BEE VENOM TOXIN MELITTIN ON MODEL MEMBRANES

Melittin is a 26 amino acid polypeptide that represents ca. 50% of bee venom (dry weight). This toxin is known to have two major actions: i) the lysis of natural and artificial lipid membranes (Haberman, 1972) and the potentiation of the activity of phospholipases A_2 found in bee venom (Mollay et al., 1976).

In order to obtain more information on the lysis mechanisms, ^2H- and ^{31}P-NMR experiments have been performed as a function of temperature and lipid-to-protein incubation molar ratio (R_i).

Figure 8 shows deuterium powder spectra of [sn-2-$^2H_{31}$] DPPC dispersions in the presence of melittin (R_i = 20) as a function of temperature; the arrows indicate the way the temperature was varied [i.e. a spectrum was taken at 25°C (after melittin addition), the temperature was brought to 65°C, and a spectrum was taken at each temperature step]. The spectrum at 25°C is typical of a pure lipid lamellar gel phase in which the acyl chains are almost in all-trans conformation (Davis, 1983); i.e. there is no effect of melittin detectable. Above T_c = 41°C, the spectra exhibit axially symmetric line shapes characteristic of lipids undergoing fast reorientations around their long molecular axes in a uniaxial mesophase (Seelig, 1977). When the temperature is lowered below T_c (e.g. spectra at 38, 30 and 0°C), isotropic central lines superimposed on gel-type powder patterns are observed. The observation of two subspectra indicates that the lipid molecules are in two distinct environments and that their rate of exchange from one region to the other is slow on the time scale of solid-state deuterium NMR (ca. 10^{-6} s). When the temperature is raised again above T_c, the two region spectra disappear and one observes axially symmetric powder patterns identical with those obtained in the first runs above T_c, at corresponding temperatures. Reproducible spectral shapes with isotropic lines, at corresponding temperatures, were also obtained on going below T_c again. The transition from an axially symmetric fluid

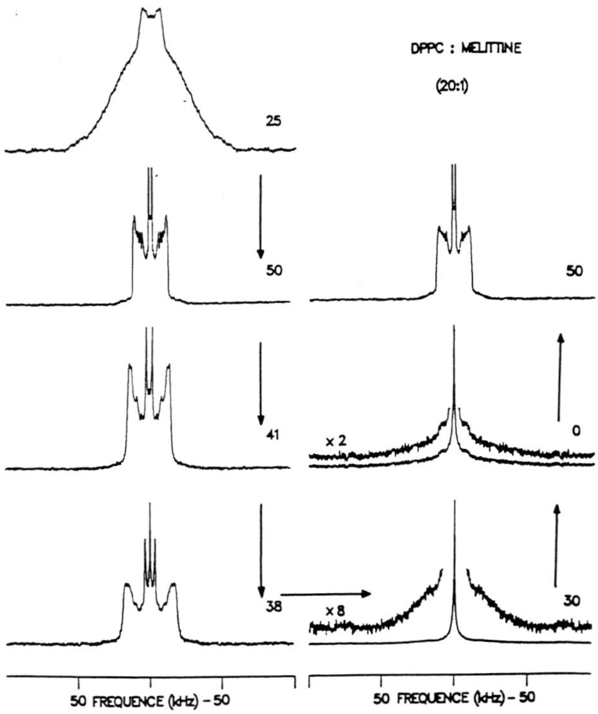

Figure 8. Temperature variation of ^{2}H-NMR powder spectra of aqueous dispersions of [sn-2-$^2H_{31}$]DPPC in the presence of melittin (R_i = 20). Temperatures (°C) are indicated on the right hand side of each spectrum. From Dufourc et al., 1986a.

phase to gel plus isotropic phases is thus reversible (in the time scale of the thermal variation, i.e. 1-2 days) when the system melittin-lipid dispersions, mixed below T_c, has been brought once above T_c.

Experiments carried out with R_i = 50 (lower amount of melittin *versus* lipids) gave essentially the same results as for R_i = 20, except that the proportion of isotropic sharp lines was smaller for R_i = 50 than for R_i = 20, at corresponding temperatures in the gel phase.

When more melittin is added to the system (R_i = 4) very dramatic changes happen. One notices in Figure 9 a gel-type powder pattern at 25°C, after melittin addition, when the system has never been brought above T_c. On increasing the temperature above T_c, a single isotropic line is observed. This line is also observed on lowering the temperature again below T_c. It is noteworthy that powder patterns are no longer observed once the system is brought in the fluid phase.

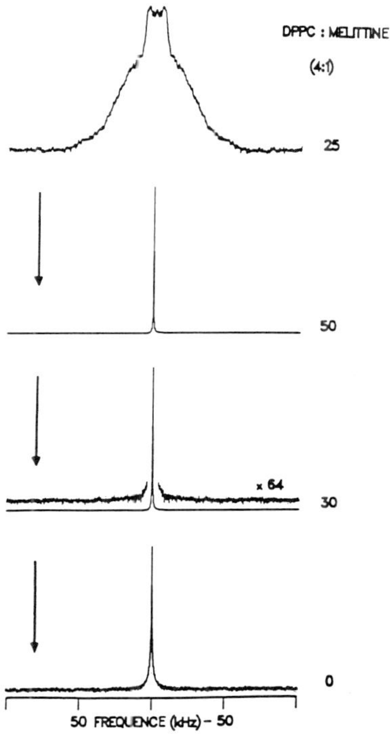

Figure 9. As in Figure 8, except $R_i = 4$. From Dufourc et al., 1986a.

Similar isotropic narrow lines are also observed with ^{31}P-NMR, in the same experimental conditions (Dufourc et al., 1986a,b). These narrow resonances can be accounted for by fast isotropic tumbling of a molecule (case of pure liquid) or by tumbling of supramolecular aggregates of melittin and lipids (bilayer vesicles, mixed micelles, ...). Release of a fatty acid chain into the water medium due to possible phospholipid degradation could also lead to single ^{2}H-NMR resonances; this possibility is discounted on the basis of TLC control experiments (Dufourc et al., 1986a). Formation of mesophases with spherical symmetry (cubic phases, ...) would also give an isotropic NMR line. If the observed narrow lines were reflecting total isotropic molecular tumbling (as is liquids) the deuterium T_{1z} and T_2 relaxation times would be equal (Abragam, 1961). In contrast, it is found that T_{1z} is 20 times greater than T_2 at 65°C and $R_i = 4$. This indicates that the lipid chains possess local ordering, due to molecular packing, as in bilayer vesicles.

The nature of the phase giving rise to narrow NMR lines must be determined by other physical techniques. Figure 10 shows how the action of melittin can be characterized by freeze-fracture electron microscopy, quasi-elastic light scattering (QELS) and ^{31}P-NMR (Dufourc et al., 1986b).

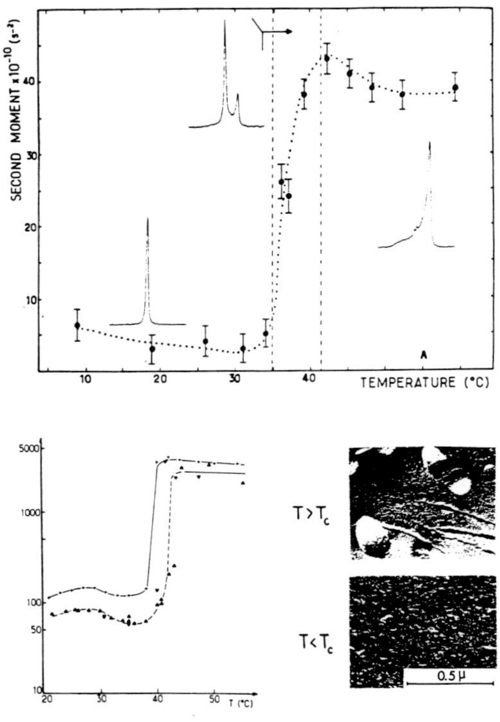

Figure 10. Evidence of a temperature-induced size transition of the DPPC:melittin system, as detected by three physical techniques. A) Second spectral moment by ^{31}P-NMR, $R_i = 15$; hydrodynamic radius, R_H = (in Å), by QELS, (+) $R_i = 20$, (Δ) $R_i = 10$; C) freeze-fracture electron microscopy, $R_i = 15$. From Dufourc, 1986.

The three physical techniques show that melittin reorganizes the lipid assembly. Above T_c, and for low to intermediate amounts of protein <u>versus</u> lipids ($R_i \geqq 10$), the DPPC multilayers are transformed into very large unilamellar vesicles (Figure 10B,C). Lowering the temperature below T_c leads to the formation of very small discoidal structures (Figure 10B, C). As observed by QELS and ^{31}P-NMR this effect is reversible and the transition occurs at around T_c, i.e. when the acyl chains melt. The changes in the physical state of lipids (gel-to-fluid) therefore appears to trigger this morphological transition.

Above T_c, at low lipid-to-protein ratio (e.g. $R_i = 4$) the large vesicles are no longer detected; instead, small objects of diameter ~120 Å are observed. This correlates well with the appearance of isotropic NMR lines above and below T_c, at the same R_i.

The lipid-to-protein ratio, R_c, in the very small discoidal structures can be estimated from R_i and the integrated areas of subspectra, for cases where one observed an isotropic NMR line superimposed on a powder pattern, one obtains R_c ~ 10-13. However, this value is calculated under the assumption that all melittin molecules are bound to lipids. If there is still melittin in solution the R_c value, as determined by NMR must be taken as a lower limit. Gel-filtration studies of the complexes indeed indicate that R_c for the same system is about 18 ± 2 (Dufourcq et al., 1986).

For low to intermediate amounts of melittin <u>versus</u> lipids ($R_i \geqq 10$) one can calculate that several hundred discs merge, when lipid chains melt, into a single vesicle (Figure 11). Conversely, the freezing of lipid chains induces the fragmentation of a large vesicle into a plethora of small discs. It is worth mentioning that vesicle-to-disc changes on varying the amount of amphiphiles to lipids have been reported (Mazer and Carey, 1983), but this is the first time, to date, that a temperature-triggered reversible disc-to-vesicle transition has been demonstrated.

The mechanism by which melittin promotes such morphologi-

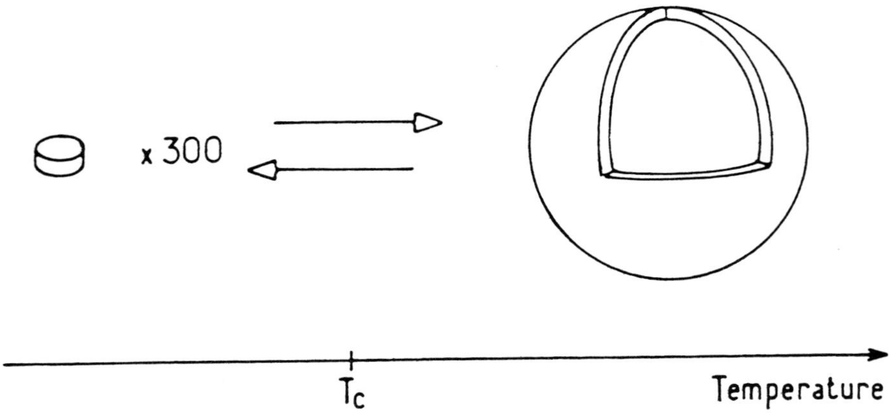

Figure 11. Schematic representation of the reversible disc-to-vesicle transition occurring at about T_c, for DPPC-melittin complexes. From Dufourc et al., 1986b).

cal changes may be investigated by looking at order parameter changes in the lipid fluid phase, when powder patterns are observed, i.e. for $R_i \geqq 10$.

Figure 12 shows the quadrupolar splitting, $\Delta\nu_Q$ (directly proportional to S_{C-^2H}, see part 1), as a function of the labelled carbon position on the <u>sn</u>-2 chain of DPPC, in the presence (R_i = 20) and absence of melittin. This pictures the ordering as a function of bilayer depth. One notices in this figure that melittin induces a general decrease in $\Delta\nu_Q$ of the DPPC acyl chain, at 65°C; the magnitude of this effect appears to be greater at the "plateau" level (labelled positions 3-9) than near the center of the bilayer. At 41°C, melittin leads to practically no change of the "plateau" quadrupolar splittings concomitantly with a marked increase of those for positions near the end of the acyl chains. Melittin thus induces different ordering effects depending upon temperature and bilayer depth.

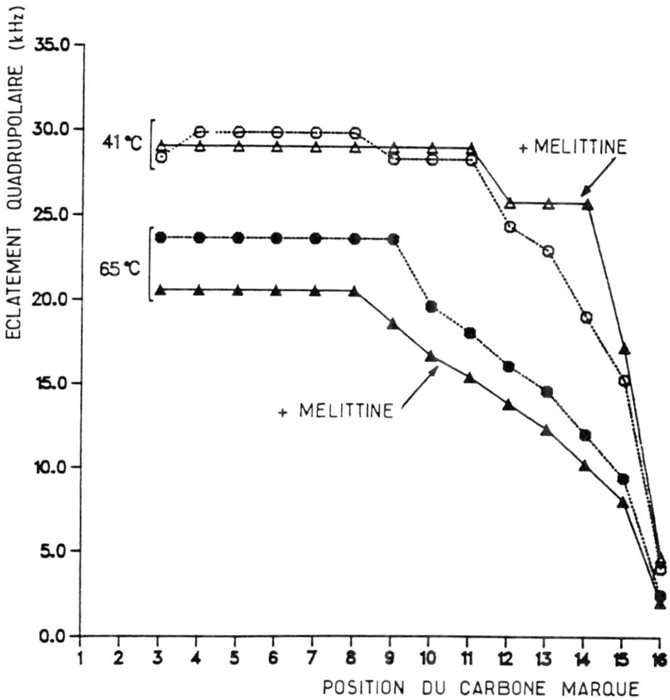

Figure 12. Quadrupolar splittings of [sn-2-$^2H_{31}$]DPPC in the presence (R_i = 20) and absence of melittin, at 41°C and 65°C, as a function of labelled position on the sn-2 chain. From Dufourc et al., 1986a.

Keeping in mind that the ordering is related to the amplitude of fluctuations of a given C-^{2}H, one may rationalize the observed differential ordering upon consideration of melittin location with regard to the lipid bilayer. Disordering for all labelled positions may be accounted for by a position in which melittin is inserted into the top of the bilayer, i.e. acting as a spacer of lipid head groups and glycerol backbone regions. This situation provides more conformational space for methylene and methyl motions which leads to a larger amplitude of $C-^2H$ bond fluctuations, i.e. more disorder. On the other hand, a preferential ordering at the center of the bilayer may be accounted for by a situation in which melittin is in contact with the acyl chains, down to the core of the bilayer. This position may restrict segmental motions by reducing the conformational space of CH_2 and CH_3 groups, i.e. increase in ordering. To summarize, melittin may occupy several positions in the bilayer depending on the temperature with respect to T_c. Disordering at high temperatures and differential ordering near T_c may reflect the progressive penetration of melittin into the bilayer when the temperature is lowered toward T_c. At T_c, when the dynamics of the system is dramatically reduced, the melittin molecules may aggregate laterally in the bilayer and lead to membrane fragmentation when the temperature is lowered below T_c. Conversely, warming up the small discoidal particles by entering into the lipid fluid phase represents a large increase in acyl chain dynamics. This could allow the fusion of the small structures, hence the formation of the very large vesicles.

The formation of very small structures (melittin-lipid complexes) upon addition of melittin to very large lipid bilayers can be proposed as a mechanism for membrane lysis. This <u>supralysis</u> (or direct-lysis) mechanism may be viewed as a large change in the morphology of the biological membrane, which leads to membrane disruption and consequently to leakage of the cell contents. The above mechanism must be differentiated from the indirect-lysis or molecular lysis, which resides in the destruction, by chemical hydrolysis, of a molecular unit (the phospholipid).

REFERENCES

Abragam A (1961) Principles of nuclear magnetism. Oxford University Press London

Davis JH (1983) The description of membrane lipid conformation order and dynamics by ^{2}H-NMR. Biochim Biophys Acta 737:117-171

De Kruijff B, Demel RA (1974) Polyene antibiotic sterol interactions in membranes of _Acholeplasma laidlawii_ cells and lecithin liposomes. Part 3: molecular structure of the polyene antibiotic cholesterol complexes. Biochim Biophys Acta 339:57-70

De Kruijff B, Gerritsen WJ, Oerlemans A, Demel R, Van Deenen LLM (1974a) Polyene antibiotic sterol interactions in membranes of _Acholeplasma laidlawii_ cells and lecithin liposomes. Part 1: specificity of the membrane permeability changes induced by the polyene antibiotics. Biochim Biophys Acta 339:30-43

De Kruijff B, Gerritsen WJ, Oerlemans A, Demel R, Van Deenen LLM (1974b) Polyene antibiotic sterol interactions in membranes of _Acholeplasma laidlawii_ cells and lecithin liposomes. Part 2: temperature dependence of the polyene antibiotic sterol complex formation. Biochim Biophys Acta 339: 44-56

Dufourc EJ (1986) La résonance magnétique nucléaire du deutérium et du phosphore dans les milieux organisés. Théorie et applications à l'étude de l'action de la melittine et de la delta-lysine sur les membranes biologiques. University of Bordeaux I Bordeaux

Dufourc EJ, Smith ICP (1985) ^{2}H-NMR evidence for antibiotic-induced cholesterol immobilization in biological model membranes. Biochemistry 24:2420-2424

Dufourc EJ, Smith ICP (1986) A detailed analysis of the motions of cholesterol in biological membranes by ^{2}H-NMR relaxation. Chem Phys Lipids 41:123-135

Dufourc EJ, Smith ICP, Jarrell HC (1984a) Interaction of Amphotericin B with membrane lipids as viewed by ^{2}H-NMR. Biochim Biophys Acta 778:435-442

Dufourc EJ, Parish EJ, Chitrakorn S, Smith ICP (1984b) Structural and dynamical details of cholesterol-lipid interaction as revealed by deuterium NMR. Biochemistry 23:6062-6071

Dufourc EJ, Smith ICP, Jarrell HC (1984c) Amphotericin B and model membranes: the effect of Amphotericin B on cholesterol-containing systems as viewed by ^{2}H-NMR. Biochim Biophys Acta 776:317-329

Dufourc EJ, Smith ICP, Dufourcq J (1986a) Molecular details of melittin-induced lysis of phospholipid membranes as revealed by deuterium and phosphorus NMR. Biochemistry 25:6448-6455

Dufourc EJ, Faucon JF, Fourche G, Dufourcq J, Gulik-Krzywicki T, Le Maire M (1986b) Reversible disc-to-vesicle transition of melittin-DPPC complexes triggered by the phospholipid acyl chain melting. FEBS Letters 201:205-209

Dufourcq J, Faucon JF, Fourche G, Dasseux JL, Le Maire M, Gulik-Krzywicki T (1986) Morphological changes of phosphatidylcholine bilayers induced by melittin:vesicularization, fusion, discoidal particles. Biochim Biophys Acta 859:33-48

Gottlieb D, Carter HE, Sloneker JH, Ammon A (1958) Protection of fungi against polyene antibiotic by sterols. Science 128:361

Haberman E (1972) Bee and wasp venoms. Science (Washington DC) 177:314-222

Hazen EL, Brown R (1951) Fungicidin, an antibiotic produced by soil actinomycete. Proc Soc Exp Biol Med 76:93-97

Hegner D (1968) The effect of melittin hematal on isolated lysosomal granules and polymorphonuclear leukocytes in vitro bovine. Naunyn-Schmiedebergs Arch Pharmakol Exp Pathol 261:118-127

Kinsky SC (1961a) Effect of polyene antibiotics on permeability in _Neurospora crassa_. Biochem Biophys Res Commun 4:353-357

Kinsky SC (1961b) Alterations in the permeability of _Neurospora crassa_ due to polyene antibiotics. J Bacteriol 82: 889-897

Mollay C, Kreil G, Berger H (1976) Action of phospholipase, on the cytoplasmic membrane of _Escherichia coli_ stimulation by melittin. Biochim Biophys Acta 426:317-324

Norman AW, Spielvogel AM, Wong RG (1976) Polyene antibiotic sterol interaction. Adv Lipid Res 14:127-170

Olson FC, Munjal D, Malriya AN (1974) Structural and respiratory effects of melittin apis-mellifera on rat liver mitochondria. Toxicon 12:419-425

Seelig J (1977) Deuterium magnetic resonance: theory and application to lipid membranes. Quart Rev Biophys 10:353-418

Sessa G, Freer JH, Colacicco G, Weismann G (1969) Interaction of a lytic polypeptide melittin with lipid membrane systems. J Biol Chem 244:3575-3582

Taylor MG, Akiyama T, Smith ICP (1981) The molecular dynamics of cholesterol in bilayer membranes: a deuterium NMR study. Chem Phys Lipids 29:327-339

Weissmann G, Hirschlorn R, Kreakaver K (1969) Effect of melittin upon cellular and lysosomal membranes. Biochem Pharmacol 18:1771-1775

INTERACTION OF THE BIBENZIMIDAZOLE DERIVATIVE HOECHST 33258 WITH LIPID BILYERS - A FLUORESCENCE STUDY

M.R.RAJESWARI and G.S. SINGHAL
Lab of Photobiology and Membrane Biophysics,
School of Life Sciences,
Jawaharlal Nehru University,
New Delhi-110067.
INDIA.

ABSTRACT

Interaction of Hoechst 33258 (2-[2-(4-hydroxy Phenyl)-6-Benzimidalzolyl]-6-(1-methyl-4-piperazyl)-Benzimidazole) with model membranes of EPC, DPPC, DMPC and PS was investigated using fluorescence spectroscopy. Hoechst 33258 alone is a weak fluorescent while bound to lipid bilayer gives high fluorescence at 460 nm. The fluidity changes in the membranes were followed by steady state fluorescence polarization of Hoechst 33258 as it is embedded in the lipid bilayer. We propose that the Hoechst 33258 could be used as a probe to study the membrane dynamics.

INTRODUCTION

Hoechst 33258 (HOE), 2-[2-(4-hydroxy Phenyl)-6-Benzimidalzolyl]-6-(1-methyl-4-piperazyl)-benzimadazole - trihydrochloride, a dye which is widely used for staining chromosomes. (Comings, 1975; Hilwig and Gropp 1972; Weisblum and Hanseller (1974).

STRUCTURE OF HOECHST 33258

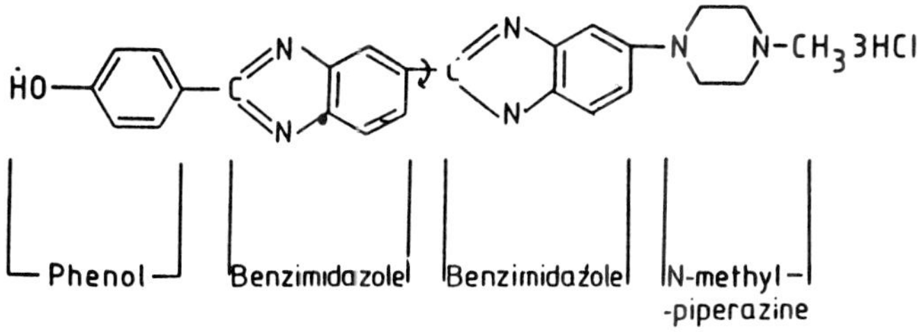

The results of its action on living cells in vitro were reported by Hilwig (1970). But, there are no reports on the interaction of HOE with the cells and parti-

NATO ASI Series, Vol. H16
Membrane Biogenesis. Edited by J.A.F. Op den Kamp

cularly cell membrane. In the present paper we present some spectroscopic investigations on the interaction of HOE with the model membranes. We propose that HOE could be used as a probe to study the membrane dynamics.

MATERIALS AND METHODS

The dye Hoechst 33258 (Loewe and Urbanietz, 1974) was purchased from Aldrich Chem. Co., The lipids, Egg Phosphatidylcholine (EPC) Dipalmitoyl phosphatidylcholine (DPPC), Dimerisytyl phosphatidyl choline (DMPC) and phosphatidyl serine (PS) were Sigma products. All the other chemicals used for making buffers were guarenteed reagents. Small unilamellar vesicles were prepared by the method described earlier (Le Doan etal, 1986). The pH of the system was always maintained at 7.0 using Tris ($1x10^{-3}$M) buffer.

The M.Wt of HOE used was 623.97 and a $\epsilon^{M}335$ of 40,700 was used (Comings 1975). The stock solutions of HOE were stored in a cool and dark place as it is light sensitive.

Absorption spectra were recorded on a Schimadzu 260 spectrophotometer and fluorescence spectra on a Schimadzu FP 450 Spectrofluorometer using 1cm pathlength cuvettes.

RESULTS AND DISCUSSION

(The emission spectrum of HOE at pH 7.0 has a very weak fluorescent band which is broad and is situated around 480 nm) The fluorescence spectra of HOE in presence of EPC visicles at various mole ratios are shown in Fig.1 (right half). Two major features may be noted from Fig.1 i.e.) The emission maximum is blue shifted from 480 nm to 460 nm and the intensity is enhanced by 10 folds (at mole ratio of 0.1) compared to the intensity of HOE in absence of EPC. The excitation spectra recorded using this emission maximum (460 nm) show that the excitation maximum is red shifted to 350 nm (from 335 nm). These results suggest that HOE when added to the lipid vesicles, is situated in apolar medium, while free dye is in fairly polar environment. This means that HOE is transferred from the aqeous medium to the lipid bilayer which is hydrophobic in nature.

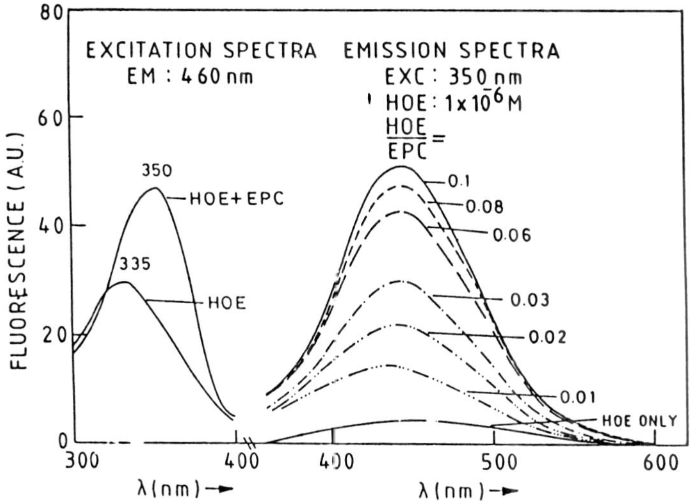

Fig. 1 Excitation spectra (with em:460 nm) and emission spectra (with λex:350 nm) of HOE of 1×10^{-6} M in presence and absence of EPC vesicles at pH 7.0 and 25°C.

The binding of HOE with other lipid vesicles, DPPC, DMPC, and PS were also studied. The fluorescence intensity changes (at 460 nm) are shown in Fig. 2. It can be seen that the maximum changes in intensity are obtained with DPPC. So the binding affinity to different lipid vesicles is in the order: DPPC $>$ DMPC $>$ EPC $>$ PS. However, the changes in the emission wavelength is nearly the same ($\Delta\lambda$ = 20 nm) in all the cases. This can be considered as an advantage as the dye, HOE can be used as a probe to study with any given model membrane. Looking at the structure of HOE, it appears that the bibenzimidazole part of the molecule must be interacting with the lipid bilayer, exposing the relatively polar phenolic and piperazine groups to the surface. Then one generally expects a maximum interaction of HOE with PS Vesicles as it is -vely charged. Surprisingly the dye did not show preference to PS compared to the other lipids studied here. It may be noted that under the experimental conditions used in the present stucy (pH 7.0) HOE exists as a hydrochloride salt and at alkaline pHs, intensity of HOE decreases due to the deprotonation of phenolic hydroxyl group.

After these binding experiments, we were interested to know whether HOE is sensitive to the fluidity of the bilayer. This was studied by determining the fluorescence polarisation. The two lipid systems chosen are : EPC (T_C = 4°C) which exists in liquid crystalline phase and the other DPPC (T_C=41°C). Which exists in gel phase at room temp (25°C).

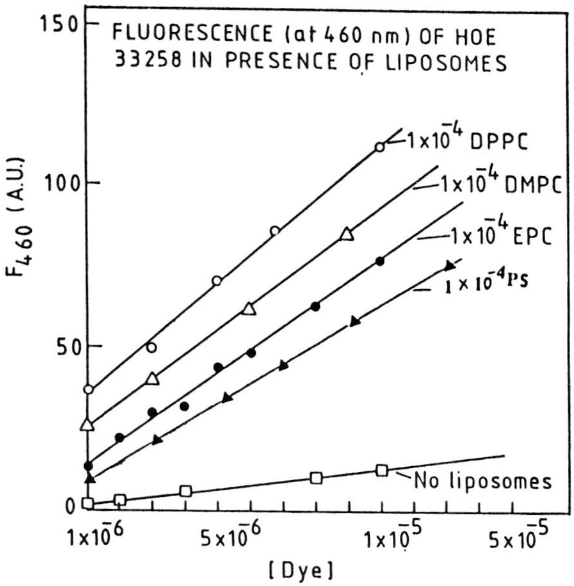

Fig. 2 : Fluorescence intensity (at 460 nm) of HOE in absence (□-□) and in presence of lipid vesicles DPPC (-0-0), DMPC (-△-△-), EPC (-●-●) and PS (-▲-▲) as a function of concentration of HOE.

It can be seen from Fig.3 that the changes in P are marginal in presence of EPC while the change is significant (0.30 to 0.36) in presence of DPPC.

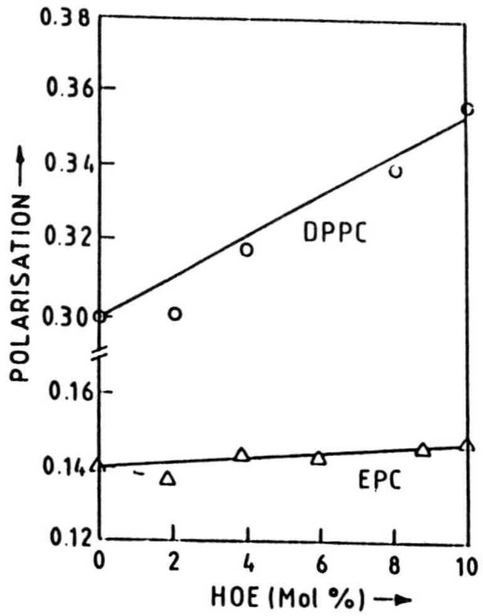

Fig. 3 : Fluorescence polarisation of HOE in presence of EPC (-△-) and DPPC (-0-0) vesicles vs conc. of HOE at 25°C.

degree of polarisation of the dye reflect the packing of lipids influencing the motional freedom of lipid molecules in the plane of the membrane.

We have also studied the spectral characteristics of HOE in water:dioxane mixtures to know the effect of dielectric constant on HOE and inturn to exploit this data to estimate the dielectric of constant of the given bilayer (results not given here).

Finally we conclude that the dye, Hoechst is a 33258 could be used as a fluorescent probe in the studies where lipid vesicles are used as model membrances and perhaps also in the studies with natural cells but in the latter case one has to check whether there are side effects before using.

ACKNOWLEDGEMENTS

MRR is thankful to the Council for Scientific and Industrial Research for Financial assistance.

REFERENCES

Comings DE (1975) Mechanisms of chromosome binding, Chromosoma 47:229-245.

Hilwig I(1970) Verhalten tierscher Zellen Und der gewebekutter gegenuber einem basisch substituerten Benzimidazole derivative mit flurochrome eigenshaften, Z.Zelforsch 104:127-137.

Hilwig I and Gropp A (1972) staining of constitutive hetero chromatin in mammalian chromosomes with a new fluorochrome, Exp. Cell Res 75:122-126.

Le Doan T, Hajji M EL, Rebuffat S, Rajeswari MR and Bodo S (1986) Fluorescence studies of the enteraction of trichorzianine AIIIC with model membranes, Biochim biophys Acta 858:1-5.

Loewe H and Urbanietz J (1974) Basisch substitutiere 2, 6-Bis-Benzimicazole derivative eine neue chemotherapeutisch aktive Korperklasse, Arzeneimittel forsch 24: 1927-1933.

Weisblum B and Hanseller E (1974) Fluorometric Properties of the Bibenzimidazole Derivative Hoechst 33258, a fluorescent probe specific for AT concentration in chromosomal DNA, Chromosoma 46:225-260.

"PHYSICAL" MEMBRANES : A SHORT REVIEW

Didier SORNETTE
Laboratoire de Physique de la Matière Condensée
CNRS UA190, Faculté des Sciences
Parc Valrose, 06034 NICE Cedex, FRANCE

Abstrast: This paper reviews some recent developments in the theoretical modeling of membranes. Results on 1)the structure and phase transitions in single membranes, 2)the interactions between two approaching membranes and 3)the coagulation and fusion of quenched vesicles are presented.

"Physicist always have the habit of taking the simplest example of any phenomenon and calling it "physics", leaving the more complicated examples to become the concern of other fields."

R.P.Feynman

1. INTRODUCTION

These last few years have seen a proliferation of works on models of biological membranes carried on in the physics community. From the physicist point of view, the spectacular property of membranes, compared to other two-dimensional systems such as fluid interfaces, is the vanishing value of their surface tension. Therefore, notwithstanding the similarity between the problems they pose and those from surface physics, they offer qualitatively new phenomena and call for the introductions of new concepts.

I will consider essentially idealized membranes made of pure amphiphilic molecules and will neglect complications such as most internal degrees of freedom, the influence of heterogeneities in their composition, small intercalated molecules such as cholesterol and the intramembrane proteins. This is a natural model to start with before facing the full complexity of real membranes. The physical idealization of real membranes also illustrates the specificity of physical approach which is characterized by a quest of "universality" transcending the specificity and complexity of "real life". However, I have the feeling that we have reached a state where the most interesting problems that are posed to the physics community come from a confrontation with the richness and "complexity" of "real" membranes of biological cells.

The paper briefly summarizes the main concepts which have emerged recently. Emphasis is put on the works carried on in our laboratory.

NATO ASI Series, Vol. H16
Membrane Biogenesis. Edited by J. A. F. Op den Kamp

2.STRUCTURE AND PHASE TRANSITIONS

2.1 The membrane model

Bilayer membranes are constituted of amphiphilic molecules presenting an affinity both with the polar solvent, such as water, and apolar organic compounds. This remarquable structure of the building blocks of membranes is at the basis of the extraordinary properties of membranes. The structure, stability and the transport properties of membranes are conditioned by the fine physical-chemistry which is acting at the microscopic level at the interfaces. The general approach in physics has been to try to relate the macroscopic or cooperative behaviour of these systems to the microscopic structure.

In this respect, the most fruitful approach from the physicist point of view has consisted in describing the membranes as two-dimensional continuous objects and constructing a statistical mechanics of such "surfaces". This approach is valid for properties which do not involve internal degrees of freedom. Notwithstanding the roughness of the approximation, a number of results have been obtained which are succintely summarized.

The 2D-membrane model amounts to assuming that the deformation from planarity is controlled by a free energy density f which reads

$$f= (\kappa/2)\ (H-H_0)^2 + \underline{\kappa}\, K \qquad (1)$$

where κ and $\underline{\kappa}$ are rigidity constants, H and K are the mean and Gaussian local curvatures and H_0 is the spontaneous curvature of the membrane. The statistical properties of the equilibrium structure of the membrane is known from a statistical sum performed over all allowed configurations of the membranes weighted by the Boltzmann factor $\exp\{-\ (k_BT)^{-1} \int H\, dS\ \}$ where T is the temperature and the integral is performed over the area of the membrane.

Note that expression (1) allows the prediction of the equilibrium shape of closed vesicles (Deuling and Helfrich,1976; Sackmann,1987); the typical biconcave shape of red-blood cells can be obtained by minimizing the curvature (bending) energy (1) with appropriate constraints (constant enclosed volume and imposed spontaneous curvature H_0).

2.2 Renormalization of the bending rigidity

Due to thermal excitations, membranes will present spontaneous curvature fluctuations. The spectrum of such fluctuations (known in red-blood cells as the "flicker phenomenon" (Brochard and Lennon,1975)) can be obtained from a simple hydrodynamical theory which includes the energy (1) as the dominant contribution.

In exploring the statistical theory based on (1), it has been recently discovered that thermal induced curvature fluctuations change ("renormalize") the value of the bending rigidity κ (Helfrich 1985). This result has been derived within the framework of the renormalization group which exhibits the strong links between curvature and rigidity by introducing the concept of a scale-dependent rigidity (Peliti and Leibler 1985; Foerster 1987 ; Sornette 1985). This notion has proven essential to understanding the structure of microemulsions (oil-water-amphiphilic compounds-cosurfactants ternary and quaternary systems) (Safran et al 1986) and the size distribution of vesicles in equilibrium (Helfrich 1986). The possible existence of a "crumpling" transition has been announced between a state in which the membrane is flexible and crumpled at large scales from another in which it is rigid and flat. This transition appears when the membrane is in the solid (gel) state as verified by a recent Monte-Carlo simulation (Kantor and Nelson 1987) or maybe even in the so-called hexatic state (Nelson and Peliti 1987). Existence of long range in-plane membrane interactions (for example mediated by proteins) could also trigger this transition. It is at present not clear if it has any biological relevance.

It is interesting to mention the connection between the renormalization properties of the 2D-membranes and of the string models of elementary particles. In this last context, the cosmological constant plays the role of a surface tension and must be faced (Polyakov 1986).

2.3 Liquid-gel phase transition

Model membranes made of a pure single amphiphilic species exhibit several phases and in particular a phase transition between a low temperature state in which the membrane is in a "solid" or gel state and a high temperature state in which it forms a quasi two dimensional liquid similar to the fluid model of biological membranes (Singer and Nicolson 1972). Several physical properties of real membranes have been tentatively proposed to being connected to the existence of such a neighbouring phase transition. One major consequence is the increased thermal fluctuations of the concentration in the vicinity of the transition point which could correlate with anomalies in "passive" transport properties of molecules and ions across the membrane (Georgallas et al 1987).

From the theoretical side, a solid-liquid phase transition in a quasi two dimensional system poses the fascinating problem of its relation with the Kosterlitz-thouless dislocation unbinding mechanism of 2D-melting (Kosterlitz and Thouless 1973). However, the hydrocarbon chains degrees of freedom make the membrane more similar to a finite three dimensional object. I have analyzed some effects of this reduced dimensionality and of the 2D-3D cross over in the first order liquid-gel transition. The rounding of the transition has been proposed to being related to a finite size cross over (Sornette and Ostrowsky 1984). The experimental observation of finite clusters of the gel phase floating in the liquid phase at the transition point is also derived from this idea of a dimensionality cross over. A perfect first order transition as would occur for infinite chains has its transition temperature decreased in inverse proportion to its thickness (Sornette 1985), a fact well verified experimentally. Below a thickness threshold, the first order transition vanishes and possibly let place to a Kosterlitz-Thouless melting mechanism (Holz et al 1985).

These ideas shed new light on the causes of rounding in the liquid-gel phase transition. This problem also illustrates the importance of the "internal" degrees of freedom, here the hydrocarbon chains, for determining the structural properties of the membrane.

3. INTERACTIONS BETWEEN MEMBRANES

3.1 "Microscopic" forces

The main "microscopic" forces between neighbouring membranes, a distance z from each other, are:

i) The Van der Waals attraction is relatively short range since it goes as z^{-2} for z less than the lamellar thickness to z^{-4} for larger spacings (Nir 1976)(unretarded case):

$V_W = -(W/12\pi) \cdot \{1/z^2 - 2/(z+h)^2 + 1/(z+2h)^2\}$, where W is the Hamaker constant (#10^{-22} -10^{-21}J) and h#40Å denotes the lamella thickness.

ii) The electrostatic forces can be very short range when screened by added salt (Kjellander and Marcelja 1985). For charged membranes, V_E # $A_E \exp\{-z/\lambda_E\}$ valid for z greater than the Debye length λ_E#$1/\sqrt{\chi}$ where χ is the ionic concentration in the aqueous solution. A_E can be estimated from the Debye-Huckel approximation or from more elaborate theories.

iii) The "hydration" or "structural" forces are extremely short-ranged but dominate in "the last thirty angtroems" (Evans and Ninham 1986). It has the form

V_H # $A_H \exp\{-z/\lambda_H\}$ with A_H#0.2J/m^2 and λ_H#3Å.

iv)Membranes carrying macromolecules and proteins will interact via the overlap of the polymer chains. This general problem of polymers at interfaces and the corresponding forces is just beginning to be studied experimentally (Klein and Luckham, Klein 1983) and theoretically (de Gennes 1987). Long range forces can appear as a result of reduction of the entropy of conformation of the macromolecules. Metastable states are also very important due to the slow dynamics of adhesion and desorption of polymer at interfaces and can lead to both attractive and repulsive forces.

v) Lastly, membranes cannot (or can) overlap due to the existence (absence) of a steric hindrance. This reduces the entropy of the membranes which develops a part dependent on the spacing z (note that the electrostatic and hydration interactions also play the role of a steric hindrance, preventing membranes overlaps). This leads to the so-called repulsive steric interaction introduced by Helfrich (Helfrich 1978; Fisher and Fisher 1982; Sornette and Ostrowsky 1984; Sornette 1986) . It is long ranged and decreases as $\{(kT)^2/\kappa\}\, z^{-2}$ for infinite membranes with vanishing surface tension and in absence of other long range forces.

3.2 The hydration or structural force

A still very controversial question concerns the nature, strength and sign of the so-called hydration or structural forces. It is generally accepted that these forces appear via a distortion of the local structure of the solvent induced by the presence of a membrane which overlaps with the structure distortion created by the other approaching membrane. In apolar solvents, the situation is clear and can be analyzed within the formalism of liquid state theory. For very associated liquids as water, the complex structure of the hydrogen bonds network makes difficult a precise description of the effect at a microscopic level. In particular, the dynamics of the water network is poorly understood. It has recently been recognized that it should play a role in the transfer and stockage of protons across membranes (Chen and Teixeira 1985). Biological processes and more generally life seem to be very dependent upon the dynamics of the water structure. It is striking to observe that life occurs in water but is not possible in such a comparable solvent as deuterated water!

A wealth of works are concerned with the understanding of hydration forces. It has been shown to exist an interdependence between forces responsible for the self-association of amphiphiles in membranes and the forces occuring between such structures when they come close together (Israelachvili and Sornette 1985). This is particularly apparent in the quantitative correlation between the measured head-groups areas of isolated bilayers and the repulsive pressure between bilayers.

More fondamentally, I have proposed recently to view the hydration forces as resulting not only from the discrete structure of the aqueous solvent but also from the fine structure of the membranes (Sornette 1985; Sornette 1987). The competition and mismatch between the two structures lead to the appearence of an additional energy cost in a way similar to the problem of commensurate-incommensurate phase transition analyzed in surface physics. The shift from "commensurability" of the water molecules near the membrane is a function of the distance between the two approaching interfaces. This leads to an interaction steming from the overlap of the two "incommensurabilities" induced by each membrane. This picture allows therefore to quantify somewhat the intuitive idea behind the hydration forces.

3.3 Steric interaction and renormalization of microscopic forces

The important new fact to recognize is that , due to the low dimensionality of the systems (2D-interfaces in 3D-bulk) and the vanishing value of the surface tension, thermal fluctuations are very important and therefore change drastically the "bare" interactions described in §3.1. The complex interplay between the thermal fluctuations and the interactions must be fully addressed to be able to understand the physics of systems made of membranes.

We have recently addressed this problem and introduced the essential idea that <u>curvature fluctuations renormalize and enhance the bare microscopic interactions</u> (Ostrowsky and Sornette 1985). This stems simply from the fact that fluctuations carry some portions of the membranes to within a very small distance to each other and therefore a short range interaction which would otherwise be outside of range is felt in presence of the membranes fluctuations. This theory allows to interpret the increase of the short range repulsion between bilayers as the temperature is decreased below the liquid-gel transition temperature (Lis et al 1982) as resulting from the increased freedom of their curvature fluctuations as the membranes become fluid (Sornette and Ostrowsky 1986).

3.4 The unbinding transitions

In presence of both attractive and repulsive forces, the full determination of the structure of interacting membranes involves taking into account the coupling between the fluctuations and both the repulsive and attractive forces. The careful description of these non-linear coupling has been done within various version of the functional renormalization group (Lipowsky and Leibler 1986; Sornette 1987). If the technical details are rather involved, the idea behind these treatments is the same as the one initially introduced in (Ostrowsky and Sornette 1985). In (Sornette 1987), a renormalized effective interaction potential between two membranes is constructed which takes into account the influence of curvature fluctuations. It consists in a steric entropic force, as discussed in point v) of §3.1 and in §3.3, balanced by the attractive renormalized Van der Waals force. A continuous "unbinding" transition is predicted which separates a phase in which the membranes are close together from a phase where the membranes are completely separated (unbound). As the driving parameter decreases (which can be the Hamaker constant, the curvature rigidity κ, or the thickness of the membranes), the distance of separation of the membranes increases continuously and diverges at a certain threshold with a critical powerlaw. A tricritical behaviour is also predicted in presence of very short range strong attractive forces. It also corresponds to an unbinding transition but with a discontinuous jump in the distance of separation of the membranes.

Open problems in these directions are

-the constraint of self-avoiding crossing for distant portions of the same membrane and its role on interactions and structure,

-the intra membrane motion of molecules and its coupling with out of plane fluctuations (Foerster 1987),

-the structural and dynamical role of other molecules imbedded into the membranes such as cholesterol and proteins...

4. COAGULATION AND GROWTH

Here, I briefly summarize an experimental and theoretical work carried on in our laboratory on the formation and stability of phospholipidic vesicles studied around the gel-liquid phase transition.

4.1 Fusion kinetics on quenched destabilized vesicles

A large number of experiments have studied the triggering of the coagulation kinetics of non-ionic vesicles which can be followed by a membrane fusion leading to the formation of a larger vesicle. This problem is of great interest since it can be viewed as a paradigm of coalescence and fusion processes occuring in cells. This kinetics is very dependent upon the nature and range of interactions between membranes which are coupled to the physical state and chemical composition of the membrane.

Using quasi-elastic light scattering, small unilamellar sonicated vesicles prepared in the "liquid" state above the temperature T_m of the gel-liquid phase transition have been cooled down below T_m. This quenching triggers a growth process of the vesicles which become unstable and spontaneously fuse to form larger vesicles (Sornette et al 1981). This kinetics has been analyzed (Ostrowsky and Sornette 1984) using a generalized Smoluchowski coagulation model assuming that collisions between vesicles occur as a result of their random Brownian motion. At each collision, a partial reversible aggregation occurs which may be followed by a fusion. We have introduced a characteristic time for fusion which decreases as the quenched temperature is decreased away from the gel-liquid transition temperature. The good quantitative agreement with the experimental results suggests that fusion could proceed similarly to a nucleation process with possible metastability.

4.2 Membranes fusion mechanisms

Endocytosis and exocytosis are crucial action by which a cell rejects or collects biochemical substances essential to its functioning. In these processes, vesicles see their membrane fusioning with the cell membrane. This phenomenon is triggered by active chemical agents and the precise mechanism by which the membrane fusion takes place is not known. This phenomenon can be mimicked by model vesicles made of a single closed bilayer constituted of a single type of amphiphilic molecule. The triggering factor which induces fusion can be the temperature as described above in §4.1, an osmotic shock or addition of ionic species. Here, I would like to speculate on some mechanisms which can be proposed to explain the fusion process at the microscopic level.

Fusion has been studied by different teams using several techniques. Let us summarize a few facts:

-i)addition of Ca^{++} favors fusion of vesicles with ionic heads (role of the electronic screening) (Fraley et al 1980; Wilschut et al 1981)

-ii)the presence of molecules added to the bilayers, which spontaneously aggregate as an inverse hexagonal structure when pure, favors the fusion, (Borovjagin et al 1982; Verkleij et al 1979; Verkleij et al 1980)

-iii)our results show that as one decreases the temperature, the fusion kinetics is accelerated in small vesicles (Ostrowsky and Sornette 1984). In concentrated solutions, fusion is increased as T approaches the melting temperature (Prestegard and Fellmeth 1974).

-iv)the characteristic time of fusion increases with the length of the chains of the amphiphilic molecules ($\tau_f \approx$ 1 hour for C_{16} and $\tau_f \approx$ 1 minute for C_{14}) (Prestegard and Fellmeth 1974).

Several mechanism can therefore be proposed:

α) fusion as a kind of nucleation process which needs defects to "nucleate". This leads to the intringuing question of the structure and dynamics of defects in membranes. One can distinguish between essentially

*structural defects at the molecular scale (such as holes or pores...)

*topological defects which control the large scale structures (dislocations, disclinations...)

See (Helfrich 1980) for a review of the different types of defects in membranes. The proposed relationship between fusion and defects is underlined by numerous unknown : coupling between defects in adjacent membranes, role of the structural interactions on the nucleation and stability of defects... (see §3 for a discussion on structural forces).

β) A second idea relies on the fact that fusion is preceded by the existence of very large local curvatures which must lead to an important restructuration of the local molecules concentration. The role of thermal fluctuations described in §2.2 can be important. In particular, one must describe the role of the coupling between neighbouring membranes on the renormalization of the curvature rigidity κ. A partial answer to the renormalization of κ by the interactions between membranes is presented in (Leibler and Lipowsky 1987). A local decrease of κ induced by the interaction would be favorable and allow for the high curvature necessary for the fusion to take place.

γ)The Brussel school has developped a formalism describing linear dynamical instabilities of hydrodynamical modes which appear when two membranes approach to within a small distance (Prevost et al 1983; Gallez et al 1983). It is tempting to associate an instability as the triggering factor of the fusion .

δ)The fact ii) suggests that fusion could be related to the lamellar-inverse hexagonal phase transition. This could explain the appearance of inverse micelles at the point of contact of two membranes.

ε)Recent experiments (Alperine et al 1985) suggest to take into account the coupling between the local concentrations of lipids and the local membrane curvature. Within a model using as the essential ingredient a deformation energy of the type (1), (Leibler 1985) has shown that such a coupling could lead to a curvature instability, which is a possible explanation of the echinocytosis transformation of the red blood cell. Is this curvature instability also relevant in fusion?

φ)One should also consider the possible local dissolution of the lipids in the solvant due to a shift in the critical micelle concentration induced by the membrane interactions.

It is possible that different mechanism take place in different systems. I feel that this problem is very much worth studying and calls for further investigations.

5. FINAL REMARKS

Future investigations could address, among others, the following topics:

-influence of the hydrocarbon chains which lie inside the membranes. It has been proposed that this leads to an additional *streching energy* term (Sadoc and Charvolin 1986) and could describe the cubic phases (Leibler et al 1987),

-the bioenergetics of biomembranes including the intricate "integrated" electrical circuitry in the membrane plane, electronic transport and related oxydo-reduction reactions...

-the excitations of electric signals, their sources and propagations mechanisms...

Let us hope that progress, interesting from the physical but also from the biological point of view, will result from a strengthened confrontation between biochemists, biophysicists and physicists.

Many of the early ideas which have been presented have been shared with N.Ostrowsky.

REFERENCES

Alperine S., Hendrikx Y. and Charvolin J.(1985) J.Phys.Lett.**46**:L27

Borovjagin V.L., Vergara J.A. and Mc Intosh T.J. (1982) J.Membr.Biol.**69**:199

Brochard F. and Lennon J.-F. (1975) J.Physique **36**: 1035

Chen S.H. and Teixeira J. (1985) Adv.Chem.Phys.**64**: 1

Deuling H.J. and Helfrich W.(1976) Biophys.J.**16**: 861

Evans D.F. and Ninham B.W. (1986) J. Phys. Chem.**90** :226

Fisher M.E. and Fisher D.S. (1982) Phys. Rev. **B25**:3192

Foerster D. (1987) Euro.Phys.Lett.**4** :65

Fraley R., Wilschut J., Duzgunes N., Smith G. and Papahadjopoulos D. (1980) Biochemistry **19** :6021

Gallez D.,Bisch P.M. and Wendel H. (1983) J.Coll.Int.Sci. **92**:121

de Gennes P.G.(1987) Model polymers at interfaces in Physical basis of cell/cell adhesion CRC press, P.Bongrand ed.

Georgallas A., MacArthru J.D., Ma X.-P., Nguyen C.V. and Palmer G.R.(1987) J.Chem.Phys.**86**: 7218

Helfrich W. (1978) Z.Naturforsh **33a**:305

Helfrich W. (1980) Les Houches : Session XXXV Eds R.Balian et al North Holland :716

Helfrich W.(1985) J.Physique **46**: 1263

Helfrich W.(1986) J.Physique **47**: 321

Holz A., Vigren D.T. and Zuckermann M. (1985): Phys.Rev.**B31**: 420

Israelachvili J.N. and Sornette D. (1985) The interdependence of intra-aggregate and inter-aggregate forces. J.Physique **46**: 2125-2136

Kantor Y. and Nelson D.R.(1987) Phys.Rev.Lett.**58**:2774

Kjellander R. and Marcelja S. (1985) Chemica Scripta **25** : 1

Klein J. and Luckham P.F. (1987) Long range attractive forces between two mica surfaces immersed in a good aqueous polymer solution. Preprint

Klein J. (1983) J.Chem.Soc.Farad.Trans.I **79**:99

Kosterlitz J.M. and Thouless D.J.(1973) J.Phys.**C6**: 1181

Leibler S. (1985) J.Physique**47** : 507

Leibler S. and Lipowsky R. (1987) Phys.Rev.**B35**: 7004

Leibler S.,Lipowsky R. and Peliti L. (1987) Curvature and fluctuations of amphiphilic membranes in Les Houches Physics of amphiphilic layers. eds D.Langevin and J.Meunier, Springer Verlag

Lipowsky R. and Leibler S. (1986) Phys.Rev.Lett.**56**:2541

Lis L.J., McAllister M., Fuller N., Rand R.P. and Parsegian V.A. (1982) Biophys.J. **37**:657

Nelson D.R. and Peliti L.(1987) Fluctuations in membranes with crystalline and hexatic order. J.Physique **48**: 1085

Nir S. (1976) Progress.Surf.Sci.**8**:1

Ostrowsky N. and Sornette D.(1984) Interactions between non-ionic vesicles. Importance of the aliphatic chain phase transition. in Proc. of the international school E.Fermi on Physics of amphiphiles: micelles, vesicles and microemulsions. eds. by M.Corti and V.Digiorgio, Plenum Press

Ostrowsky N. and Sornette D.(1985) Role of membrane fluidity on bilayers; Short-range interactions. Chemica Scripta **25**:108

Peliti L. and Leibler S.(1985) Phys.Rev.Lett.**54**:1960

Polyakov A.M. (1986) Nucl.Phys.**B268**: 406

Prestegard J.H. and Fellmeth B. (1974) Biochemistry **13**:1122

Prevost M., Gallez D. and Sanfeld A. (1983) J.Chem.Soc.Farad.Trans.II **79**:961

Sackmann E. (1987) in Physics of amphiphilic layers. Les Houches ed. by Langevin D. and Meunier J. SPRINGER-VERLAG

Sadoc J.-F and Charvolin J. (1986) J. Physique **47**:683

Safran S.A., Roux D., Cates M.E. and Andelman D.(1986) Phys.Rev.Lett.**56**:491

Singer S.J. and Nicolson G.L. (1972) The fluid mosaic model of the structure of cell membranes. Science VolCLXXV:720

Sornette D., Bezot C. and Ostrowsky N. (1981) Fusion kinetics of dimyristoyl phosphatidylcholine vesicles around the phase transition of the aliphatic chains. Biochimie **63**:955

Sornette D. and N.Ostrowsky (1984) Lipid phase transition and interdimensionality. in Proc.

of the international school E.Fermi on Physics of amphiphiles: micelles, vesicles and microemulsions. eds. by M.Corti and V.Digiorgio. Plenum Press

Sornette D. and Ostrowsky N. (1984) Repulsive steric interaction between membranes of finite size. J. Physique **45**:265

Sornette D.(1985) Thèse d'état : Comportements statiques et dynamiques d'interfaces libres ou contraintes. Chapter 1. University of Nice

Sornette D.(1985) Thèse d'état : Comportements statiques et dynamiques d'interfaces libres ou contraintes. Chapter 2. University of Nice :189

Sornette D.(1985) First order transition in systems of finite thickness. Phys.Rev.**B31**:4672

Sornette D. and Ostrowsky N. (1986) Importance of membrane fluidity on bilayers interactions. J. Chem. Phys. **84**:4062

Sornette D.(1986) Steric interaction between wandering walls. Study of the strong deviation from mean-field theory. Euro. Phys. Lett.**2**:715

Sornette D. (1987) Lock-in transition and structural forces. unpublished

Sornette D. (1987) Tricritical behaviour in lamellar phases. J.Phys.**C20**

Sornette D. (1987) Fluctuations and interactions between membranes. in Les Houches Physics of amphiphilic layers. eds D.Langevin and J.Meunier. Springer Verlag

Verkleij A.J., Mombers C., Gerritsen W.J., Leunissen-Bijvelt L. and Cullis P.R. (1979) Biochim.Biophys.Acta **555**: 358

Verkleij A.J., Van Echteld C.J.A., Gerritsen W.J., Cullis P.R. and de Kruijff B.(1980) Biochim.Biophys.Acta **555**: 358

Wilschut J., Duzgunes N., and Papahadjcpoulos D. (1981) Biochemistry **20**: 3126

ANCHORING OF MEMBRANE PROTEINS BY GLYCOSYL-PHOSPHATIDYLINOSITOL

Martin G. Low
Department of Physiology and Cellular Biophysics
College of Physicians and Surgeons
Columbia University
630 W. 168th Street
New York, NY 10032

In 1985 a novel post-translational modification found in certain membrane proteins was described (see Low et al., 1986). This involves a complex lipid molecule, glycosyl-phosphatidylinositol (GPI), which is covalently attached to the C-terminus of the protein and serves as its membrane anchoring domain. Since that time a considerable amount of information has emerged on the structure of this modification although its functional significance still remains a matter of conjecture. In this chapter a brief review of the discovery, distribution, structure and biosynthesis of these lipid anchors will be given followed by a more detailed description of recent work from this laboratory on their metabolism. Several reviews on GPI-anchoring are available and the reader is referred to those for topics not covered in depth here (Cross, 1987;Low, 1987;Low, Saltiel, 1988;Ferguson, Williams,1988).

Discovery of a role for phosphatidylinositol in membrane protein anchoring

The first indications of a specific role for phosphatidylinositol (PI) in anchoring of a membrane protein was suggested by the work of Slein and Logan during the early 1960's. Their studies, originally concerned with the pathogenesis of anthrax, showed that alkaline phosphatase could be released from tissues by the action of a soluble factor secreted by *Bacillus anthracis* and several other Bacillus species (Slein,Logan, 1962). This soluble factor (isolated from *B.cereus*) was subsequently suggested to be due to a phospholipase C with specificity for phosphatidylinositol (Slein, Logan, 1965). However since the enzyme was not fully purified the significance of these observations was not appreciated. The central role of phosphatidylinositol in these observations was not established untill, over a decade later when the phosphatidylinositol-specific phospholipase C (PI-PLC) was purified from *B.cereus* and several other bacteria (Ikezawa et al.,1976;Low, Finean,1977a;Taguchi, Ikezawa, 1978;Taguchi et al. 1980).

NATO ASI Series, Vol. H16
Membrane Biogenesis. Edited by J. A. F. Op den Kamp

It was also observed that two additional membrane enzymes, 5'-nucleotidase and acetylcholinesterase could be released by PI-PLC but several other membrane components were not (Low, Finean 1977b, 1978). The specificity of this process and the molecular characteristics of the released alkaline phosphatase indicated a unique and specific role for the polar headgroup of phosphatidylinositol in its attachment to membranes. It was therefore proposed in 1980 that a covalently attached phosphatidylinositol molecule was responsible for anchoring of this protein (Low, Zilversmit,

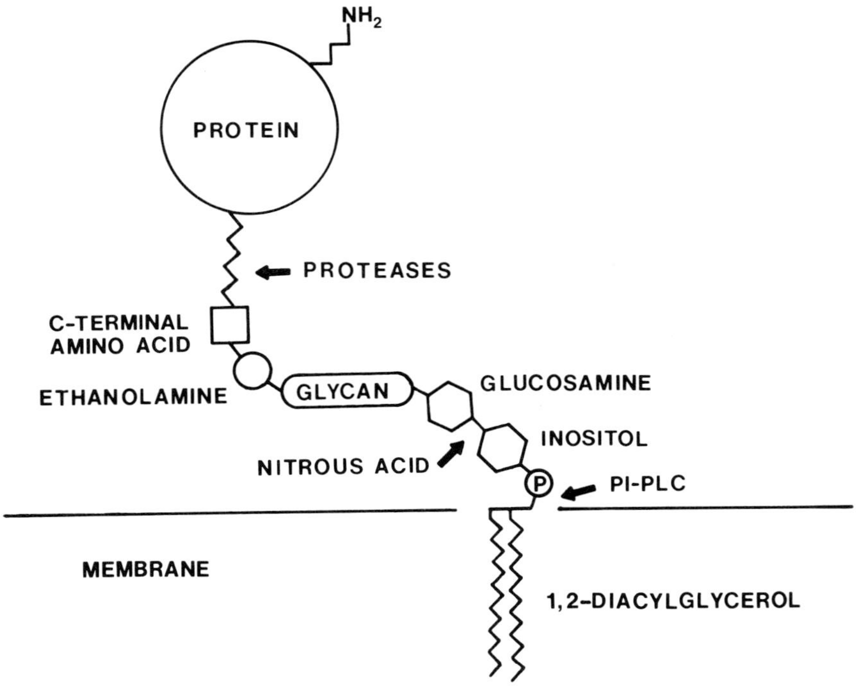

Figure 1. Structure of the glycosyl-phosphatidylinositol anchor (from Low et al., 1986)

1980). However this novel concept was not widely accepted untill 1985 when detailed chemical and enzymatic degradation studies of several proteins confirmed the existence of such anchoring structures (Futerman et al., 1985;Ferguson et al., 1985;Tse et al., 1985;Low, Kincade, 1985;for additional references see below). These studies also revealed that the anchoring structure was considerably more complex than was originally suggested on the basis of the release studies using PI-PLC (Fig.1). Since the structure included a number of sugars the term glycosyl-phosphatidylinositol (GPI) has been adopted. In spite of this unexpected structural complexity these findings established PI-PLC as a convenient tool for identifying and characterizing proteins anchored by this mechanism and soon led to the identification of many more proteins that utilise these anchoring structures (currently 25-30;see Table 1 for some examples). The identification of the GPI anchoring structure on such a large group of proteins in a relatively short time suggests that many more proteins may use this anchoring mechanism than are presently supposed. Although the chemical and PI-PLC release studies on these proteins reveal that this novel and rather complex anchoring structure is conserved, this group of proteins is also distinguished by extreme diversity of function and evolutionary origin. One common feature of the proteins in Table 1 does exist; they all are located at the cell surface. However since the anchoring of intracellular membrane proteins has received little attention it cannot be assumed that the glycosyl-phosphatidylinositol anchors are restricted to cell surface proteins.

Table 1. Cell Surface Proteins with a Glycosyl-phosphatidylinositol Membrane Anchor

<u>Hydrolytic Enzymes</u>	
Alkaline Phosphatase (APase)	5'-Nucleotidase
Acetylcholinesterase (AChE)	Trehalase
Alkaline phosphodiesterase I	p63 protease
Renal dipeptidase	
<u>Antigens</u>	
Thy-1	RT-6
T-cell activating protein	Qa
(and other Ly-6 antigens)	Carcinoembryonic antigen
<u>Cell Adhesion</u>	
LFA-3	Heparan Sulfate Proteoglycan
Neural Cell Adhesion Molecule (N-CAM)	
<u>Coat Proteins</u>	
Variant Surface Glycoprotein (VSG)	Surface proteins (<u>Paramecium</u>)
<u>Miscellaneous</u>	
Decay accelerating factor (DAF)	130 kDa Hepatoma glycoprotein
34 kDa Placental growth factor	Scrapie prion protein (PrP)

Structural analysis of the glycosyl-phosphatidylinositol anchors

The complete structure of the glycosyl-phosphatidylinositol anchor of the variant surface glycoprotein (VSG) of *Trypanosoma brucei*, has been determined (see Ferguson, Williams, 1988) and similar studies on other GPI-anchored proteins, although less well advanced, provide strong support (Table 2) for the general model illustrated in Figure 1 (Low et al., 1986). The protein is attached by an amide linkage between its C-terminal amino acid and an ethanolamine residue. The ethanolamine is phosphodiester linked to a glycan that contains three mannose residues. This in turn is linked to a glucosamine which is unusual in that its amino group is not acylated or sulfated and allows specific cleavage of the structure at this point by nitrous acid deamination. The glucosamine is glycosidically linked to the inositol ring of a phosphatidylinositol molecule (Futerman et al.,1985;Ferguson, Cross, 1984;Ferguson et al., 1985ab;Tse et al., 1985;Medof et al., 1986;Roberts, Rosenberry, 1985,1986;Haas et al.,1986;Low et al., 1987;Roberts et al., 1987;Ferguson, Williams, 1988). As indicated in Table 3 several major structural variations have been observed or can be inferred from the limited data available; (i) additional ethanolamine residues (with free amino group); (ii) variations in the structure and composition of the glycan; although it is probable that the mannose 'core'

Table 2. Common structural features of GPI-anchors

Structural feature	Proteins analysed
<u>myo-inositol</u>	
detected by GC/MS	APase, AChE, VSG, Thy-1, DAF, PrP
nitrous acid sensitive linkage	APase, AChE, VSG
<u>Ethanolamine</u>	
biosynthetic labelling	APase, DAF, VSG, Thy-1
chemical analysis	VSG, AChE, Thy-1, PrP
<u>Fatty acids</u>	
biosynthetic labelling	APase, VSG, Thy-1
chemical analysis	AChE, VSG, Thy-1, DAF, PrP
<u>Phospholipase sensitivity</u>	
PI-PLC/GPI-PLC	APase, AChE, VSG, Thy-1, DAF, PrP
GPI-PLD	APase, AChE, VSG, DAF
<u>Cross-Reacting Determinant</u>	APase, AChE, VSG, DAF, PrP

Table 3. Structural Variations in GPI-anchors

Structural feature	Proteins analysed
Inositol	
variable amounts of *chiro* isomer	AChE, APase, DAF
Ethanolamine	
1 residue	VSG
2-3 residues	Thy-1, AChE, DAF, PrP
Glycan	
mannose	VSG, Thy-1
galactose	VSG
galactosamine	Thy-1
Fatty acids	
myristate only	VSG
stearate only	PrP
mixture	Thy-1, AChE
Fatty acid linkages	
1,2-diacylglycerol	VSG
1,2-alkylacylglycerol	AChE
C-terminal attachment site	
Asp or Ser	VSG
Cys	Thy-1
Gly	AChE

connecting the protein and the lipid moiety is relatively conserved with the variable sugars being attached to it; (iii) variations in the lipid moiety both at the level of fatty acid composition and nature of the fatty acid linkages. In this context it should be emphasized that the only protein for which 1,2-diacylglycerol has been demonstrated directly is VSG and it is not known whether, in higher eukaryotes, the 1,2-alkylacylglycerol structure will predominate. Neither the distribution of these variations between the different proteins nor their functional significance are understood at present.

Biosynthesis of the glycosyl-phosphatidylinositol anchor

Current models of the assembly of the anchored protein indicate at least three important stages in the process. In the first stage it is proposed that a GPI-anchor precursor lipid is biosynthesized and a candidate lipid with appropriate structural features has been identified in *T.brucei* (Krakow et al., 1986). This lipid is GPI-PLC/PI-PLC and nitrous acid sensitive and is specifically labelled with [^{3}H] myristate. Although its complete structure is not yet known identification of this lipid has given preliminary support to the hypothesis that the protein is attached to a preassembled lipid rather than assembly of lipid on the protein. In the next stage a newly-synthesized polypeptide is recognized as being destined for post-translational modification of the lipid. The mechanism of this recognition process is unclear but is has been proposed that the polypeptide may contain a 'signal' sequence which directs the attachment of lipid (Low, 1987;Williams, Ferguson, 1988). This is based on the following rather indirect evidence: (i) All the cDNA and genomic sequences currently available predict the existence of a short (approx. 10-20 residues) hydrophobic segment at the C-terminus (for references see Low, 1987;Williams, Ferguson, 1988), (ii) C-terminal sequences of this sort are rather unusual and (iii) construction of hybrid proteins by DNA cloning techniques indicates that the information necessary for attachment is located in the C-terminal region (Stroynowski et al., 1987;Waneck et al., 1987). The third stage in the biosynthesis is the attachment of the protein to the lipid. This is believed to involve simultaneous removal of a C-terminal peptide and formation of an amide bond between the GPI-anchor lipid, through the amino group on the ethanolamine, and the newly exposed C-terminal carboxyl group on the protein. Removal of the C-terminal peptide can be inferred from a comparison of the amino acid sequence for the mature polypeptide and cDNA-predicted sequences for some of these proteins (Cross, 1985;Seki et al., 1985). Although the possibility of separate proteolytic and attachment steps cannot be ruled out at present such a process would seem to be less efficient. Furthermore the attachment process appears to be rapid (within one minute for the VSG of *T.brucei*) and probably occurs immediately after translation in the lumen of the rough endoplasmic reticulum. Clearly most of the steps in this biosynthetic process remain to be established and it is likely that such studies will reveal the existence of novel enzymes and metabolic pathways.

Function of the glycosyl-phosphatidylinositol anchors

One function of the covalently-attached GPI is undoubtedly to anchor the protein to the cell surface. However in view of the relative complexity of this anchoring structure, compared to that utilised by most other cell surface proteins, two additional functions have been proposed. First, utilization of a lipid anchor which is located only in one leaflet of the lipid bilayer will enable the protein to move relatively rapidly in the plane of the membrane (Ishihara et al., 1987;Noda et al., 1987;Thomas et al., 1987). Why these particular proteins require the ability to move rapidly in the membrane in order to carry out their function is not obvious except in the case of decay accelerating factor (Davitz et al. 1986). Second, the glycosyl-phosphatidylinositol anchor may be the target for specific enzymes. Cleavage by these enzymes would have two consequences (i) the biologically-active lipids, 1,2-diacylglycerol and phosphatidic acid which form part of the anchoring structure would be released and might affect intracellular metabolism. (ii) the protein would also be released from the cell surface with consequent loss or alteration of its function. Although this would provide the cell with a convenient mechanism for coordinately regulating cell surface protein release and changes in intracellular metabolism it should be emphasized that there is no direct evidence for this. Several GPI-anchored proteins are involved in receptor/transduction of mitogenic signals and it is possible that released lipids are involved in these events (Rock et al., 1986;Kroczek et al., 1986;Gunter et al., 1987;Le et al., 1987). There are also indications that release of GPI-anchored proteins does occur *in vivo* but whether this involves the specific phospholipase-mediated cleavage described above or what its physiological function might be is unknown.

In spite of all this uncertainty regarding the physiological relevance of a phospholipase-mediated release mechanism there is now subtantial evidence for the existence of the requisite GPI-anchor specific phospholipases. The original identification of eukaryotic enzyme activities which could cleave the GPI-anchor (i.e. of alkaline phosphatase in pig kidney and of VSG in *T.brucei*;Low, Zilversmit, 1980;Cardoso de Almeida, Turner, 1983) actually preceded the structural work on the anchors of these proteins by several years. Consequently, it is only recently that these enzymes have been characterized and their potential functional significance fully appreciated. Both activities were discovered as a result of their activation by procedures (butanol or detergent extraction or osmotic lysis) used in the purification of the proteins; under these conditions the proteins were converted from the hydrophobic, membrane form to a hydrophilic form (Low, Zilversmit, 1980;Cardoso de Almeida, Turner, 1983). However more recent studies have indicated that these two anchor-degrading activities are in fact due to two quite distinct phospholipases as described in detail below.

The glycosyl-phosphatidylinositol specific phospholipase C

The anchor-degrading enzyme in trypanosomes has been purified and shown to be a phospholipase C but several properties distinguish it from the other phospholipases C that hydrolyse inositol phospholipids (Bulow, Overath, 1986;Hereld et al., 1987;Fox et al., 1987). Thus it is insensitive to divalent cation chelators and relatively inactive against free phosphatidylinositol (Table 4). Although the VSG coat of the trypanosome is changed during antigenic variation in the mammalian host or lost during transformation in the insect host there is no evidence that rapid cell surface release of VSG by the GPI-specific phospholipase C (GPI-PLC) is involved. On the contrary recent evidence suggests that the enzyme is located on intracellular membranes and may be involved in normal metabolic turnover of the surface coat by an endocytic mechanism (reviewed in Ferguson, Williams, 1988). A phospholipase C with similar substrate specificity and chelator insensitivity to the trypanosome enzyme has also been purified from rat liver plasma membranes (Fox et al., 1987). However this enzyme has a higher molecular weight. It

Table 4. Properties of the phospholipases C that are specific for phosphatidylinositol and its derivatives

Type I:

1. Soluble or membrane-associated enzymes in many mammalian tissues; molecular wt. 70-140 kDa.
2. Inhibited by thiol blockers and EDTA or EGTA; Ca^{2+}-dependent.
3. Can hydrolyse PI and PI-phosphates.

Type II:

1. Soluble enzymes secreted by Bacillus cereus, B. thuringiensis, Clostridium novyi and Staphylococcus aureus; molecular wt. approx 31-34 kDa.
2 Inhibited by thiol blockers but not by EDTA or EGTA.
3. Can hydrolyse PI and GPI but not PI-phosphates.

Type III:

1. Membrane-bound enzyme purified from Trypanosoma brucei (37-40 kDa) and rat liver plasma membranes (52 kDa).
2. Inhibited by thiol blockers but not by EDTA or EGTA.
3. Can hydrolyse GPI but not PI or PI-phosphates.

has been proposed that the rat liver GPI-PLC is responsible for the insulin-sensitive degradation of free GPI inside the cell (reviewed in Low, Saltiel, 1988); a function which appears to be incompatible with its involvement in protein release from the cell surface.

The glycosyl-phosphatidylinositol specific phospholipase D

The observation that the butanol-activated anchor-degrading activity in pig kidney was sensitive to inhibition by EDTA originally suggested that it was due to the Ca^{2+}-dependent phosphoinositide-specific phospholipase C (i.e. Type I in Table 4) that is widely distributed in mammalian tissues (Low, Zilversmit, 1980). This conclusion was strengthened when it was also shown that the activity shared other properties with this phospholipase, i.e. acid pH optimum and stimulation by deoxycholate or Ca^{2+} ions (Low, Weglicki, 1981;Miki et al., 1985;Kominami et al., 1985;Balasubramanian, 1985;Ikehara et al., 1987). However it was subsequently shown that the phospholipase C activity could be inhibited approx. 95% by heating at 60^{0}C without significant effects on anchor-degrading activity (Malik, Low, 1986). The two activities were distinguished even further when it was found that the product of hydrolysis of [^{3}H] myristate-labelled VSG was [^{3}H] phosphatidic acid instead of the expected product of a phospholipase C mediated hydrolysis, [^{3}H] 1,2-diacylglycerol. This suggested that the anchor-degrading activity in mammalian tissues is a phospholipase D which is relatively specific for glycosyl-phosphatidylinositol structures (i.e. a GPI-PLD).

Recently additional evidence has accumulated supporting the concept that this activity is due to a distinct enzyme (Low, Prasad, 1987,1988). Thus it is readily separated from the phospholipase C by gel filtration on Sephacryl S-300, exhibiting an apparent molecular weight of approx. 500 kDa. Furthermore it is sensitive to inhibition by 1,10-phenanthroline, which does not chelate Ca^{2+}ions, whereas the phospholipase C activity is unaffected by this inhibitor (Low, Prasad, 1988). Attempts have also been made to show that the site of cleavage of the alkaline phosphatase by the anchor-degrading enzyme is consistent with a phospholipase D. Due to a variety of technical limitations it has proven impracticable to prepare sufficient [^{3}H] fatty acid-labelled alkaline phosphatase to permit identification of the lipid product of the reaction. Consequently two less direct techniques, which attempt to characterize the protein product, had to be used. First, treatment of alkaline phosphatase with the anchor degrading enzyme from rat heart did not expose the cross-reacting determinant (CRD) whereas parallel treatments with a bacterial PI-PLC did (Table 5). This suggested that the phosphate ester on the inositol ring (which is believed to be a necessary component of the CRD) had been removed by the rat heart enzyme. The second technique utilized

Table 5. The anchor degrading enzyme from rat heart does not expose the Cross Reacting Determinant (CRD) on human placental alkaline phophatase.

	CRD Activity (absorbance units)			
Incubation conditions	mAb/E5	mAb/B2	mAb/B10	no mAb
Buffer	0.034	0.020	0.013	0.005
PI-PLC 0.25 ng/ml	0.054	0.030	0.016	0.003
PI-PLC 2.5 ng/ml	0.571	0.339	0.134	0.003
PI-PLC 25 ng/ml	1.094	0.958	0.628	0.005
PI-PLC 250 ng/ml	1.017	0.828	0.469	0.005
Rat heart enzyme I	0.040	0.029	0.015	0.001
Rat heart enzyme II	0.051	0.034	0.017	0.000

polyacrylamide gel electrophoresis at alkaline pH (in the absence of SDS) to analyse the alkaline phosphatase produced after removal of the anchoring domain by enzymic hydrolysis. The mobility of the product formed when the mammalian anchor-degrading enzyme was used to remove the anchor was substantially lower than that observed using the bacterial PI-PLC. This is consistent with a lower negative charge on the protein due to removal of the phosphate from the inositol ring by a phospholipase D. Furthermore sequential treatments with phospholipase C followed by the phospholipase D and *vice versa* indicate that their effects are mutually exclusive. By contrast treatment with a protease which cleaves near the C-terminus produces the same result regardless of prior treatment with the phospholipases. These results provide indirect support for the conclusion that the mammalian anchor-degrading enzyme acts on alkaline phosphatase by a phospholipase D mechanism.

One puzzling feature of the GPI-PLD has been its apparent abundance in the cytosolic fraction of several tissues (compared to membranes) where it would normally have little or no access to its substrates on the cell surface (Low, Weglicki, 1981;Majumdar, Balasubramanian, 1985;Low, Prasad, 1987). Furthermore, different tissues revealed markedly different activities in the supernatant fraction (compare liver, brain and heart in Expt.I, Table 6). This paradox was resolved when it was found that an apparently similar enzyme activity was also present in the plasma (Low, Prasad, 1987,1988). In fact the plasma level of this enzyme is sufficiently that it probably accounts for most if not all of the activity observed in the cytosolic fraction of unperfused tissues (Expt.I, Table 6).

The availability of a rich source of the enzyme has also permitted partial purification and characterization of the phospholipase D. Thus it has now been shown that the plasma GPI-PLD can act on two genetically distinct types of alkaline phosphatase, VSG and 5′-nucleotidase (Low, Prasad,1987,1988). A similar activity in human serum has also been reported to remove the anchor from decay accelerating factor (Davitz et al., 1987). In contrast to its wide specificity for different GPI-anchored protein this enzyme is unable to act on phosphatidylinositol (PI), phosphatidylcholine (PC) or phosphatidylethanolamine

Table 6 Specificity and distribution of phospholipase C and anchor-degrading activities

Source of enzyme		* Relative activity		
	Substrate:	PC	PI	APase
Expt I				
rat liver supt.		-	7	10
rat heart supt.		-	8	10
rat brain supt.		-	10	1.5
rat plasma		-	-	60
Expt II				
purified rat plasma		<0.01	<0.01	200
purified rabbit plasma		<0.01	<0.01	500

*Expts. I and II done under different assay conditions

(Expt.II, Table 6;Davitz et al., 1987;Low, Prasad,1988).

In summary, recent studies have indicated the presence of a novel enzyme which is specific for the GPI-anchors of cell surface proteins. The function of this and other GPI anchor-specific enzymes remains to be determined but it is not unreasonable to suppose that they are involved in the degradation of the GPI anchors. Whether they also play some additional role in releasing proteins from the cell surface and/or generating phosphatidic acid and 1,2-diacylglycerol molecules that might themselves have biological activities (Moolenaar et al., 1986;Murayami, Ui, 1987) is an exciting area for futher investigation.

References and Notes

Bangs JD, Hereld D, Krakow JL, Hart GW, Englund PT (1985) Rapid processing of the carboxyl terminus of a trypanosome variant surface glycoprotein.Proc. Natl. Acad. Sci. USA 82:3207-3211

Bangs JD, Andrews NW, Hart GW, Englund PT (1986) Posttranslational modification and intracellular transport of a trypanosome varaiant surface glycoprotein. J. Cell Biol. 103:255-263

Bulow R, Overath P (1986) Purification and characterization of the membrane-form variant surface glycoprotein hydrolase of Trypanosoma brucei. J. Biol. Chem. 261:11918-11923

Cardoso de Almeida ML, Turner MJ (1983) The membrane form of variant suface glycoproteins of _Trypanosoma brucei_. Nature (London) 302:349-352

Cross GAM (1984) Structure of the variant glycoproteins and surface coat of _Trypanosoma brucei_. Philos. Trans. R. Soc. London Ser. B 307:3-12

Cross GAM (1987) Eukaryotic protein modification and membrane attachment via phosphatidylinositol. Cell 48:179-181

Davitz MA, Low MG, Nussenzweig V (1986) Release of decay accelerating factor (DAF) from the cell membrane by phosphatidylinositol-specific phospholipase C (PI-PLC) J. ExP. Med. 163:1150-1161

Davitz MA, Hereld D, Shak S, Krakow J, Englund PT, Nussenzweig V (1987) A glycan-phosphatidylinositol-specific phospholpase D in human serum. Science

Ferguson MAJ, Cross GAM (1984) Myristylation of the membrane form of a _Trypanosoma brucei_ variant surface glycoprotein. J. Biol. Chem. 259:3011-3015

Ferguson MAJ, Haldar K, Cross GAM (1985) _Trypanosoma brucei_ variant surface glycoprotein has a _sn_-1,2-dimyristylglycerol membrane anchor at its COOH-terminus. J. Biol. Chem. 260:4963-4968

Ferguson MAJ, Low MG, Cross GAM (1985) Glycosyl-_sn_-1,2-dimyristylphosphatidylinositol is covalently linked to _Trypanosoma brucei_ variant surface glycoprotein. J. Biol. Chem. 260:14547-14555

Ferguson MAJ, Duszenko M, Lamont GS, Overath P, Cross GAM (1986) Biosynthesis of _Trypanosoma brucei_ variant surface glycoproteins J. Biol. Chem. 261:356-362

Ferguson MAJ, Williams AF (1988) Cell surface anchoring of proteins via glycosyl-phosphatidylinositol structures. Ann. Rev. Biochem.

Fox JA, Duszenko M, Ferguson MAJ, Low MG, Cross GAM (1986) Purification and characterization of a novel glycan-phosphatidylinositol-specific phospholipase C from _Trypanosoma brucei_ J. Biol. Chem. 261:15767-15771

Fox JA, Soliz NM, Saltiel AR (1987) Purification of a phosphatidylinositol-glycan-specific phospholipase C from liver plasma membranes; a possible target of insulin action. Proc. Natl. Acad. Sci. USA 84:2663-2667

Futerman AH, Fiorini RM, Roth E, Low MG, Silman I (1985) Physico-chemical behaviour and structural characteristics of membrane-bound acetylcholinsterase from _Torpedo_ electric organ: effect of phosphatidylinositol-specific phospholipase C. Biochem. J. 226:369-377

Futerman AH, Low MG, Ackermann KE, Sherman WR, Silman I (1985) Identification of covalently bound inositol in the hydrophobic membrane anchoring domain of _Torpedo_ acetylcholinesterase. Biochem. Biophys. Res. Commun. 129:312-317

Gunter KC, Germain RN, Kroczek RA, Saito T, Yokoyama WM, Chan C, Weiss A, Shevach EM (1987) Thy-1-mediated T-cell activation requires co-expression of CD3/Ti complex. 326:505-507

Low MG, (1987) Biochemistry of the glycosyl-phosphatidylinositol membrane protein anchors. Biochem. J. 244:1-13
Low MG, Prasad ARS (1987) Characterization of an enzyme activity in rat tissues which degrades the phosphatidylinositol anchor of human placental alkaline phosphatase. Fed. Proc. 46:2110-2110
Low MG, Prasad ARS (1988) A phospholipase D specific for the phosphatidylinositol anchor of cell surface proteins is abundant in plasma. Proc. Natl. Acad. Sci. USA in press
Low MG, Saltiel AR (1988) Structural and functional roles of glycosyl-phosphatidylinositol in biological membranes. Science in press
Majumdar R, Balasubramanian AS, (1985) The solubilization of platelet membrane-bound acetylcholinesterase and arylacylamidase by exogenous or endogenous phosphatidylinositol-specific phospholipase C. Biochem. Pharmacol. 34:4109-4115
Malik AS, Low MG, (1986) Conversion of human placental alkaline phosphatase from a high M_r form to a low M_r form during butanol extraction: an investigation of the role of endogenous phosphoinositide-specific phospholipases. Biochem. J. 240:519-527
Medof ME, Walter EI, Roberts WL, Haas R, Rosenberry TL (1986) Decay accelerating factor of complement id anchored to cells by a C-terminal glycolipid. Biochemistry 25:6740-6747
Miki A, Kominami T, Ikehara Y (1985) pH-dependent conversion of liver-membranous alkaline phosphatase to a serum-soluble form by n-butanol extraction. Biochem. Biophys. Res. Commun. 126:89-95
Moolenaar WH,Kruijer W, Tilly BC, Verlaan I, Bierman AJ, de Laat SW (1986) Growth factor-like action of phosphatidic acid.Nature 323:171-173
Murayama T, Ui M, (1987) Phosphatidic acid may stimulate membrane receptors mediating adenylate cyclase inhibition and phospholipid breakdown in 3T3 fibroblasts. J. Biol. Chem. 262:5522-5529
Noda N, Yoon K, Low MG, Rodan GA, Koppel DE (1987) High lateral mobility of endogenous and transfected alkaline phosphatase, a phosphatidylinositol-anchored protein. Fed. Proc. 46, 2200-2200
Roberts WL, Rosenberry TL, (1985) Identification of covalently attached fatty acids in the hydrophobic membrane-binding domain of human erythrocyte acetylcholinesterase. Biochem. Biophys. Res. Commun. 133:621-627
Roberts WL, Rosenberry TL (1986) Selective radiolabeling and isolation of the hydrophobic membrane-binding domain of human erythrocyte acetylcholinesterase. Biochemistry 25, 3091-3098
Roberts WL, Kim BH, Rosenberry TL (1988) Differences in the glycolipid membrane anchors of bovine and human erythrocyte acetylcholinesterases. Proc. Natl. Acad. Sci. USA in press
Rock KL, Yeh TH, Gramm CF, Haber SI, Reiser H, Benacerraf B (1986) TAP, a novel T cell-activating protein involved in the stimulation of MHC-restricted T lymphocytes. J. Exp. Med. 163:315-333
Seki T, Chang HC, Moriuchi T, Denome R, Ploegh H, Silver J (1985) A hydrophobic transmembrane segment at the carboxyl terminus of Thy-1. Science 227:649-651
Slein MW, Logan GF (1962) Mechanism of action of the toxin of Bacillus anthracis: alkaline phosphatasemia produced by culture filtrates of various bacilli. J. Bactericl. 83, 359-369
Slein MW, Logan GF (1965) Characterization of the phospholipase of Bacillus cereus and their effects on erythrocytes, bone and kidney cells. J. Bacteriol. 90:69-81

Haas R, Brandt PT, Knight J, Rosenberry TL (1986) Identification of amine components in a glycolipid membrane-binding domain at the C-terminus human erythrocyte acetylcholinesterase. Biochemistry 25:3098-3105

Hereld D, Krakow JL, Bangs JD, Hart GW, Englund PT (1986) A phospholipase C from Trypanosoma brucei which selectively cleaves the glycolipid on the varriant surface glycoprotein. 261:13813-13819

Ikehara Y, Hayashi Y, Ogata S, Miki A, Kominami T (1987) Purification and characterization of a major glycoprotein in rat hepatoma plasma membranes: one of the membrane proteins released by phosphatidylinositol-specific phospholipase C Biochem. J. 241:63-70

Ikezawa H, Yamanegi M, Taguchi R, Miyashita T, Ohyabu T (1976) Studies on phosphatidylinositol phosphodiesterase (phospholipase C type) of Bacillus cereus. Biochim. Biophys. Acta 450:154-164

Ishihara A, Hou Y, Jacobson K (1987) The Thy-1 antigen exhibits rapid lateral diffusion in the plasma membrane of rodent lymphoid cells and fibroblasts. Proc. Natl Acad. Sci. USA 84:1290-1293

Kominami T, Miki A, Ikehara Y (1985) Electrophoretic characterization of hepatic alkaline phosphatase released by phosphatidylinositol-specific phospholipase C: a comparison with liver membrane and serum-soluble forms. Biochem. J. 227:183-189

Krakow JL, Hereld D, Bangs JD, Hart GW, Englund PT (1986) Identification of a glycolipid precursor of the Trypanosoma brucei variant surface glycoprotein. J. Biol. Chem. 261:12147-12153

Kroczek RA, Gunter KC, Germain RN, Shevach EM (1986) Thy-1 functions as a signal transduction molecule in T lymphocytes and transfected B lymphocytes. Nature (London) 322:181-184

Le P, Denning S, Springer TA, Haynes B, Singer K (1987) Anti-LFA-3 monoclonal antibody induces interleukin 1 (IL 1) release by thymic epithelial (TE) cells and monocytes. Fed. Proc. 46:447-447

Low MG, Finean JB (1977) Release of alkaline phosphatase from membranes by a phosphatidylinositol-specific phospholipase C. Biochem. J. 167:281-284

Low MG, Finean JB (1977) Non-lytic release of acetylcholinesterase from erythrocytes by a phosphatidylinositol-specific phospholipase C. FEBS Lett. 82:143-146

Low MG, Finean JB (1978) Specific release of plasmamembrane enzymes by a phosphatidylinositol-specific phospholipase C. Biochim. Biophys. Acta 508:565-570

Low MG, Zilversmit DB (1980) Role of phosphatidylinositol in attachment of alkaline phosphatase to membranes. Biochemistry 19:3913-3918

Low MG, Weglicki WB (1981) Release of membrane-bound alkaline phosphatase by activation of cardiac or renal phosphatidyinositol-specific phospholipase C. Fed. Proc. 40:826-826

Low MG, Kincade PW (1985) Phosphatidylinositol is the membrane anchoring domain of the Thy-1 glycoprotein. Nature (London) 318:62-64

Low MG, Futerman AH, Ferguson MAJ,Silman I (1986) Covalently attached phosphatidylinositol as a hydrophobic anchor for membrane proteins.Trends Biochem. Sci. 11:212-215

Low MG, Futerman AH, Ackermann KE, Sherman WR, Silman I (1987) Removal of covalently bound inositol from Torpedo acetylcholinesterase and mammalian alkaline phosphatases by deamination with nitrous acid: evidence for a common membrane-anchoring structure. Biochem. J. 241:615-619

Stahl N, Borchelt DB, Hsiao K, Prusiner SB (1987) Scrapie prion protein contains a phosphatidylinositol glycolipid. Cell 51: in press

Stroynowski I, Soloski M, Low MG, Hood L (1987) A single gene encodes soluble and membrane-bound forms of the major histocompatibility Qa-2 antigen: anchoring of the product by a phospholipid tail. Cell 50:759-768

Taguchi R, Ikezawa H (1978) Phosphatidylinositol-specific phospholipase C from _Clostridium novyi_ Type A. Arch. Biochem. Biophys. 186:196-201

Taguchi R, Asahi Y, Ikezawa H (1980) Purification and properties of phosphatidylinositol-specific phospholipase C of _Bacillus thuringiensis_

Thomas J, Webb W, Davitz MA, Nussenzweig V (1987) Decay accelerating factor diffuses rapidly on HeLaAE cell surfaces. Biophys. J. 51, 522-522

Tse AGD, Barclay AN, Watts AN, Williams AF (1985) A glycophospholipid tail at the carboxyl terminus of the Thy-1 glycoprotein of neurons and thymocytes. Science 230:1003-1008

Waneck GL, Sherman DH, Kincade PW, Low MG, Flavell RA (1987) Molecular mapping of signals in Qa-2 required for attachment of the phosphatidylinositol membrane anchor. Proc. Natl. Acad. Sci. USA im press

Wonigeit K, Schwinzer R (1987) Polyclonal activation of rat T lymphocytes by RT6 alloantisera. Transpl. Proc. 19:296-299

FATTY ACYLATION OF PROTEINS

Michael F.G. Schmidt

Faculty of Medicine-Department of Biochemistry
Kuwait University
P.O. Box 24923 / Code 13110 - SAFAT
Kuwait - Arabian Gulf

and

Marion Schmidt

Institut für Virologie
Justus-Liebig-Universität Giessen
Frankfurter Strasse 107
D-6300 Giessen, F.R.G.

1. Introduction

The various biosynthetic modifications of proteins, as for instance the addition of oligosaccharides, phosphate groups and fatty acid chains have become an important area of study in the biomedical sciences. Especially in a time where cloned eukaryotic genes are expressed in bacterial (or other alien) systems yielding non-modified protein products for medical application it would be of interest to know what the biological implications are of using such "unnatural" proteins. Accordingly, three lines of investigations are in progress which aim at the structural analysis of a given modification, elucidating its various biological functions and at an understanding of the biosynthetic event of the modification.

NATO ASI Series, Vol. H16
Membrane Biogenesis. Edited by J. A. F. Op den Kamp

Here we would like to focus on the hydrophobic modification through covalent linkage of fatty acyl chains, mainly palmitic acid, to a polypeptide backbone. In nature this seems to be verified by three biosynthetic mechanisms, palmitoylation, myristilation and through covalent linkage of phosphotidylinositol to an extended carboxyterminus (for reviews see Schmidt, 1983; Sefton and Buss, 1987; Low et al., 1986).

Following the first reports of fatty acids in polypeptides (Folch-Pi, and Lees, 1951; MacLennan et al., 1972) it was generally believed that this modification would be restricted to a highly specialized class of proteins soluble in organic solvents. When fatty acylation was observed also in typical membrane glycoproteins (Schmidt et al., 1979; Schmidt and Schlesinger, 1979) a subsequent quest for such acylation revealed its wide distribution in nature across all levels of development. Myristic acid in a protein was first detected with cAMP-dependent protein kinase (Carr et al., 1982) and later shown to occur also with various tumor virus related proteins (Schultz et al., 1985). It was then also established that the palmitoylation mentioned above differs principally from this latter myristilation since different sets of proteins were acylated with palmitic acid (Pal) or myristic acid (Myr) in a given cell species. Likewise, the linkage properties differed in these two types of acylproteins (Magee and Courtneidge, 1985; McIlhinney et al., 1985; see below). In order to give an idea of the wide distribution of hydrophobic modifications of proteins in nature, such proteins are listed in tables 1 - 5. Especially with the cellular acylproteins the great functional variety of the respective protein species is striking, whereas with the virus specific acylproteins some common features have become apparent (see below).

TABLE 1

CELLULAR PROTEINS ACYLATED WITH PALMITATE

ACYLPROTEIN SPECIES	REFERENCE
VINCULIN	Burn and Burger, 1987
MHC, HLA B7 HEAVY CHAIN	Kaufman et al., 1984
MHC, HLA DR HEAVY CHAIN	Kaufman et al., 1984
Ia INVARIANT CHAIN	Koch and Hammerling, 1985
GOLGI MANNOSIDASE II	Moreman and Touster, 1985
CONTACT SITE A GP, ACTIN *(D.discoideum)*	Stadler et al., 1984
ANKYRIN	Staufenbiel and Lazarides, 1985
APOLIPOPROTEIN I	Hoeg et al., 1986
INTERPHOTORECEPTOR RETINOID-BINDING PROTEIN	Bazan et al., 1985
MURINE ERYTHROCYTE SIALOGLYCOPROTEINS	Dolci and Palade, 1985
IgE RECEPTOR	Kinet et al., 1985
TRANSFERRIN RECEPTOR	Omary and Trowbridge, 1981
GOLGI GALACTOSYLTRANSFERASE	Strous et al., 1983
SALIVARY MUCOUS GP's	Slomiany et al., 1983
INSULIN RECEPTOR	Magee and Siddle, 1986
$p21^{ras}$	Sefton et al., 1982
PROTEOLIPOPROTEIN (MYELIN)	Agrawal et al, 1982
RHODOPSIN	O'Brien and Zatz, 1984
LIPOPHILIN	Folch-Pi and Lees, 1951
α-FACTOR (YEAST)	Betz et al., 1987
TGF-α PRECURSOR	Bringman et al., 1987
BUTYROPHILIN	Keenan et al., 1982
XANTHINOXIDASE	Keenan et al., 1982
RAS1, RAS2, (YEAST)	Fujiyama and Tamanoi, 1986
Ca^{2+}-ATPáse (SARCOPLASMIC)	Mac Lennan et al., 1982
DM20 (BRAIN MYELIN)	Agrawal et al., 1982
PO (SCIATIC NERVE MYELIN)	Agrawal et al., 1983
ACETYLCHOLINE RECEPTOR	Olson et al., 1984
DEVELOPMENTAL GP's (SEA URCHIN)	Bolanowski et al., 1984

TABLE 2

VIRAL PROTEINS ACYLATED WITH PALMITATE

VIRUS SPECIES	ACYLPROTEIN SPECIES	REFERENCE
SEMLIKI FOREST VIRUS	E1, E2	Schmidt, 1982
SINDBIS VIRUS	E1, E2	Schmidt et al., 1979
VESICULAR STOMATITIS VIRUS	G	Schmidt et al., 1979
INFLUENZA VIRUSES	HA_2	Schmidt, 1982
SIMIAN VIRUS 40	TAg	Klockman and Deppert, 1983
CORONAVIRUS (MURINE, BOVINE)	E2	Schmidt, 1982
ADENOVIRUS 12 (Ad 12)	Eb 18K	Grand et al., 1985
HERPES SIMPLEX VIRUS	gE	Johnson and Spear, 1983
LA CROSSE VIRUS	G1, G2	Madoff and Lenard, 1982
SPLEEN FOCUS FORMING VIRUS	gp 65	Srinivas and Compans,1983
MOUSE MAMMARY TUMOR VIRUS	gp 78, gp 70, 38	Firestone et al., 1982
VACCINIA VIRUS	p37	Hiller and Weber, 1985
ADENOVIRUS 2 (Ad 2)	72K	Deppert et al., 1985
NEWCASTLE DISEASE VIRUS	F	Schmidt, 1982
MUMPS VIRUS	F (HN)	Merz and Wolinsky, 1983
PARAMYXO VIRUS YUCAIPA	F	Schmidt, unpublished
ROUS SARCOMA VIRUS	gP35	Gebhardt etal., 1984
FRIEND LEUKEMIA VIRUS	p15E	Srinivas and Compans, 1983

TABLE 3

CELLULAR PROTEINS ACYLATED WITH MYRISTATE

$pp60^{c-SRC}$	Buss and Sefton, 1985
p36	Soric and Gordon, 1985
$p56^{tlk}$ (LSTRA MURINE THYOMA)	Voronova et al., 1984
CALCINEURIN B	Aitken et al, 1982
cAMP-DEPENDENT P.K. (c SUBUNIT)	Carr et al., 1982
NADH-CYT B5 REDUCTASE	Ozols et al., 1984
NICOTINIC AChR (α and β SUBUNITS)	Olson et al., 1984
FIBRONECTIN	Magee, unpublished
VINCULIN	Kellie and Wrigglesworth, 1987

TABLE 4

VIRAL PROTEINS ACYLATED WITH MYRISTATE

$pp60^{v-src}$	Sefton et al., 1982
$p120^{gag-abl}$	Sefton et al., 1982
GAG PROTEINS OF MAMMALIAN RETROVIRUSES (TYPES B, C, D)	Schultz and Oroszlan, 1983
GAG-ONC FUSION PROTEINS	Schultz and Oroszlan, 1984
$p75^{gag-raf}$	Schultz et al., 1985
$p85^{gag-fes}$	Schultz and Oroszlan, 1984
$p110^{gag-fes}$	Schultz and Oroszlan, 1984
$p29^{gag-ras}$	Schultz and Oroszlan, 1984
POLIO VIRUS, VP4	Chow et al., 1987
FOOT AND MOUTH DISEASE, VP4	Chow et al., 1987
POLYOMA VIRUS, VP2	Streuli and Griffin, 1987
SIMIAN VIRUS 40, VP2	Streuli and Griffin, 1987

TABLE 5

PROTEINS WITH ACYL CHAINS IN THEIR MEMBRANE ANCHOR (PI-TAIL)

DELAY ACCELERATING FACTOR	Medof et al., 1986
P.falciparum TFN-R	Haldar et al., 1985
TRYPANOSOME VSG	Ferguson et al., 1985
N-CAM 120	He et al., 1986
Leishmania gp63	Bordier et al., 1986
ALKALINE PHOSPHATASE (RAT KIDNEY)	Low and Zilversmit, 1980
ALKALINE PHOSPHODIESTERASE (RAT KIDNEY)	Nakabayashi and Ikezawa, 1984
5'NUCLEOTIDASE	Nakabayashi and Ikezawa, 1984
TREHALASE	Takesue et al., 1986
ACh 'ASE	Roberts and Rosenberry, 1985
MYELIN BASIC PROTEIN	Yang et al., 1986
Thy-1	Tse et al., 1985
ALKALINE PHOSPHATASE (BRUSHBORDER MEMBRANE)	Nayudu et al., unpublished

2. Structure and Function of Protein Bound Fatty Acyl Chains

Understanding the function of protein linked acylchains in molecular terms requires some knowledge of the overall structure of the acylprotein and of the linkage site in particular. Since the small amounts of fatty acid (usually 1 - 2 chains per polypeptide; Schmidt, 1986) are not easily detected in low abundance proteins, radiolabeling with ^{3}H-myristic acid or ^{3}H-palmitic acid and subsequent isolation of the protein with various established methods is applied. Figure 1 gives an example for the visualization of the acylated protein of influenza viruses, the small subunit HA_2 of the viral hemagglutinin. Neither the large glycoprotein fragment of HA, HA_1, nor the membrane linked neuraminidase (NA) or the hydrophobic M-protein contain fatty acids.

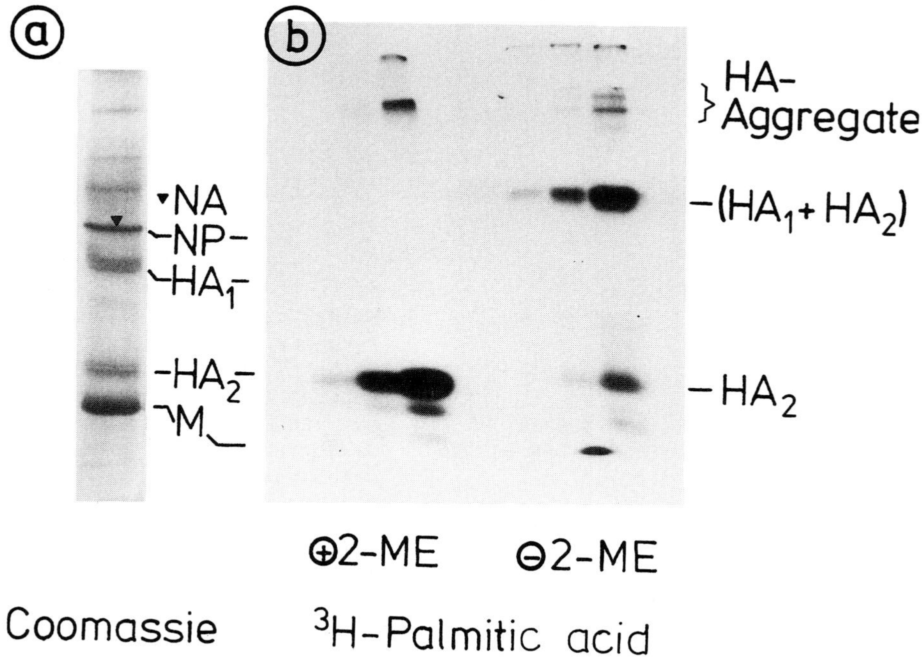

Figure 1: Acylproteins of Influenza Virus

PAGE-analysis of the proteins of ^{3}H-Pal labeled influenza A virus (fowl plague virus) on 12% SDS-polyacrylamide gels with 6 M urea. Panel a shows the Coomassie staining pattern and panel b a fluorogram of viral proteins. In panel b 50 µg, 300 µg or 1 mg viral protein were applied and the gel run under reducing (left side) or non-reducing conditions (right side).

One step towards the topographical identification of acyl linkage sites in acylproteins has been the method of limited proteolysis. Since most of the virus particles with fatty acylated proteins are enveloped by a lipid bilayer, proteases can only digest the exposed peripheral portions of the relevant spike proteins. Thus, if ^{3}H-Pal labeled virus is digested and subsequently centrifuged, supernatants with peripheral spike-peptides and sedimented "naked" virus particles can be analysed separately. Typical results obtained from such experiments are documented in table 6 and Fig. 2. After digesting ^{3}H-acylated and ^{3}H-glycosylated virus particles, the distribution of radioactivity in supernatants and sediments is opposite for ^{3}H-glucosamine labeled and ^{3}H-palmitoylated peptides.

TABLE 6

LIMITED DIGESTION OF ^{3}H-PAL LABELED INFLUENZA VIRUS

		Radioactivity (cpm) in fractions		
Label	Treatment	Sediment	Supernatant	Total
^{3}H-Palmitic acid	Control	263300	20600	283900
	Bromelain	268500	6600	275100
^{3}H-Glucosamine	Control	55100	8000	63100
	Bromelain	15500	48200	63700

About 300 µg of labeled fowl plague virus was digested for 3 min at 37°C in 100 µl phosphate buffered saline (PBS) pH 7.4 in the presence of 5 mM dithiothreitol. After dilution with ice cold PBS the samples were centrifuged for 30 min at 45000 rpm in a Beckman SW 50 Ti-rotor and radioactivity measured in aliquots of supernatants and pellets.

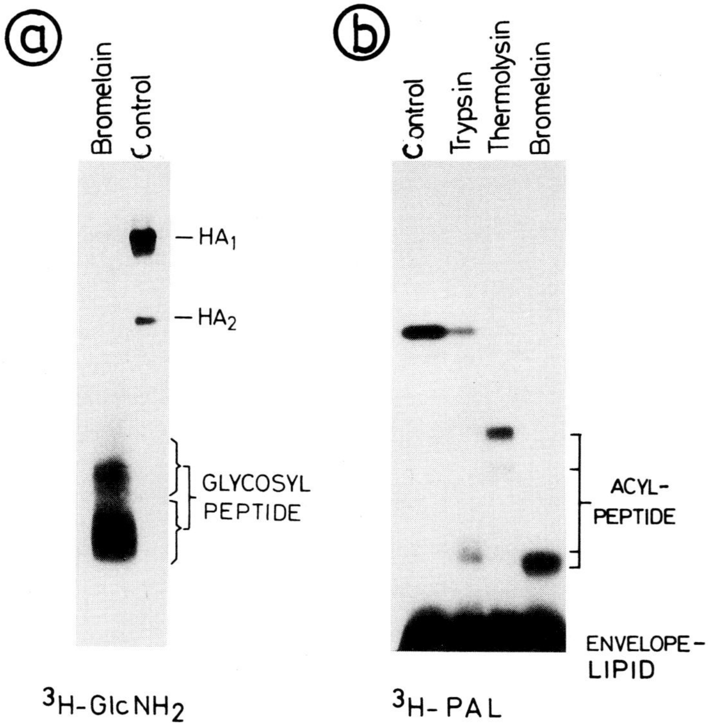

Figure 2: Fatty Acylated Peptides of Influenza HA_2

^3H-Glucosamine ($GlcNH_2$) or ^3H-palmitic acid (Pal) labeled influenza A virus (fowl plague) was subjected to limited proteolysis as described in the legend of table 6. ^3H-$GlcNH_2$ labeled peptides in the supernatant after bromelain digestion (panel a) and ^3H-palmitoylated material in the sediments after trypsin, thermolysin and bromelain digestion were analysed on 15% SDS-gels. Fluorograms of the respective gels are shown.

In order to identify the nature of the radioactive material, the relevant fractions were subjected to PAGE-analysis on 15% SDS-polyacrylamide gels. The fluorograms of such gels shown in Fig. 2 reveal, that the ^3H-Pal material in the sediments contains indeed defined ^3H-acylated peptides beside the ^3H-labeled bulk lipid of the viral envelope still present in "naked" particles.

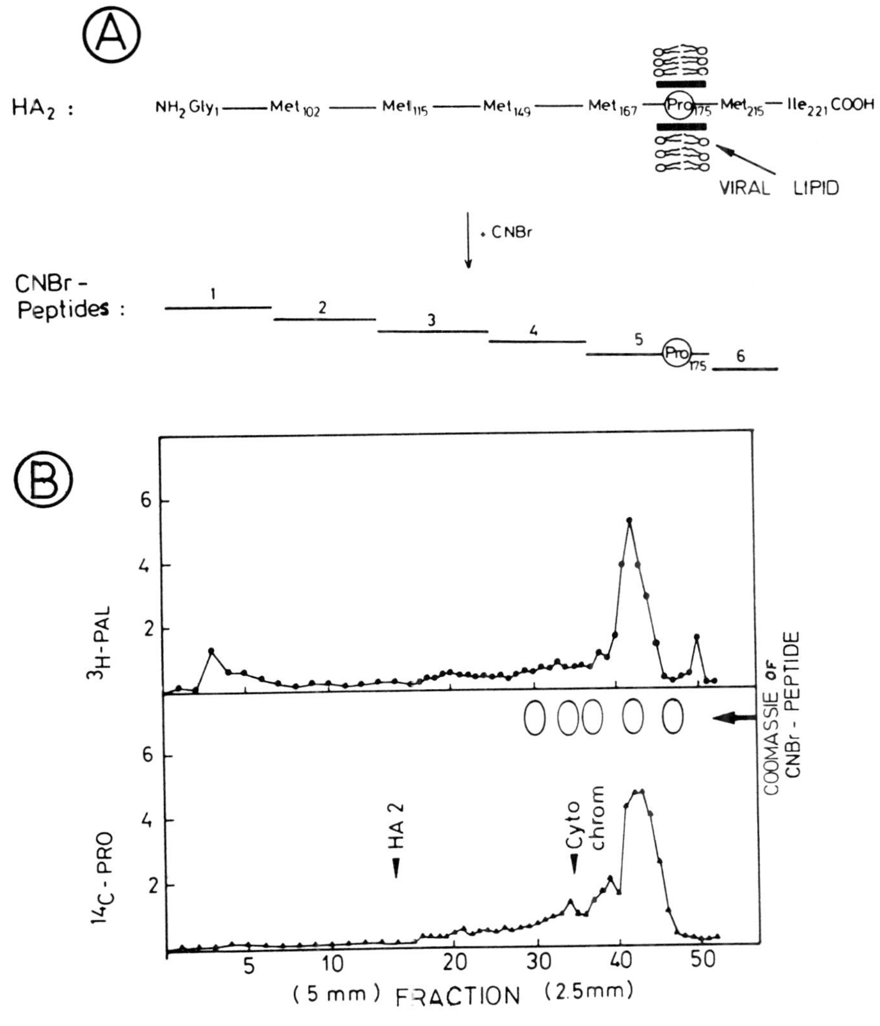

Figure 3: Cyanogen Bromide Cleavage of HA_2

A, Schematic of the expected cleavage pattern of fowl plague virus HA at methionine residues.

B, PAGE-analysis of CNBr-peptides on 15% cylindrical SDS-polyacrylamide gels. The distribution of ^{3}H-palmitic acid (PAL) and ^{14}C-proline (PRO) in gel slices is shown. Prior to fractionation gels were stained with Coomassie and staining pattern is indicated by ovals in lower panel. The positions of non-treated HA_2 and cytochrome C (Cyto) run as references are indicated by arrows.

This can be taken as an indication that the linkage site is close to the region of HA_2 which interacts with the viral lipid envelope.However more is required to prove this. One approach is to start with purified acylproteins, which seems most promising for proteins of which the aminoacid sequence is known. This is the case for the hemagglutinin (HA) of fowl plague virus (Porter et al., 1979) as well as for E1 and E2 of Semliki Forest virus (SFV; Garoff et al., 1980) which are discussed below.

As indicated in panel A of figure 3 fowl plague HA offers a promising fragment pattern if cyanogen bromide (CNBr) is used as the cleavage reagent. If fatty acids are located in the membrane segment, then this same fragment should also contain the only proline residue present in this HA_2. PAGE-analysis of CNBr-fragments generated in such an experiment show that ^{3}H-Pal and ^{14}C-Pro labeled peptides run to the same position, which substantiates the above hypothesis that the acylation site is located near to the membrane. Furthermore, the realistic possibility that more peripheral regions of HA_2 are acylated which potentially could loop back into the viral bilayer (compare ^{3}H-GlcNH$_2$ radioactivity in sediment in table 6) has been excluded by this latter finding (Figure 3, panel B).

All our attempts to cleave the acylated CNBR-peptide of HA_2 further in order to isolate short peptides for sequence studies failed because of apparently adverse properties of such material. For this reason we chose a viral acylprotein, which can be purified to higher yields and the primary structure of which is known, the SFV-E1 (Garoff et al., 1980). E1-protein from ^{3}H-palmitoylated SFV digested with chymotrypsin yielded only one ^{3}H-acylated peptide of about 6000 dalton (Figure 4, lanes 1 and 2). Likewise, digestion of the more heavily acylated SFV-E2 with Staph. aureus protease V8 yielded several defined ^{3}H-acylpeptides (Figure 4, lanes 3 and 4).

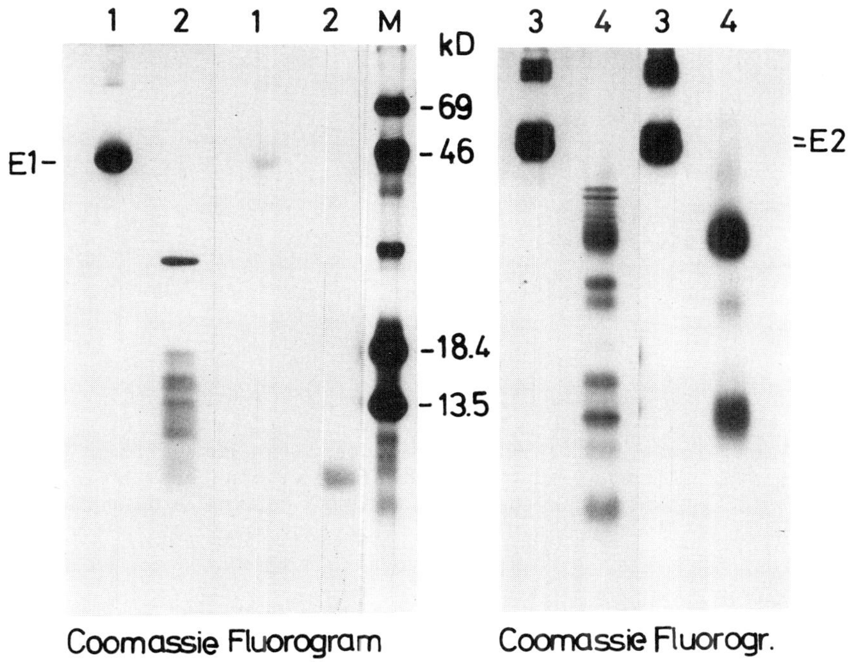

Figure 4: Proteolytic Digestion of ^{3}H-Acyl E1 and -E2 from Semliki Forest Virus

Acylproteins were isolated from purified ^{3}H-palmitic acid labeled SFV and digested with chymotrypsin (E1-protein, lanes 2) or with V8-protease (E2-protein, lanes 4) and run on 15% SDS-polyacrylamide gels. Coomassie staining patterns and fluorograms of the same gels are shown. Nontreated E1 and E2 are shown in lanes 1 and lanes 3, respectively.

Since it was found unsuitable to isolate acylpeptides from polyacrylamide gels (unacceptable losses), separation was attempted by utilizing reversed phase materials. The best results have been achieved with chymotryptic peptides of E1. The digest was loaded onto Sep-Pak cartridges in 60% formic acid and then eluted stepwise with increasing concentrations of n-propanol. The distribution of ^{3}H-Pal in the various fractions shown in table 7 reveals the expected high solubility of ^{3}H-acylated peptide in fractions with high n-propanol concentration.

TABLE 7

ELUTION OF CHYMOTRYPTIC PEPTIDES FROM ^{3}H-PALMITOYLATED SFV-E1

Fraction	Eluent	^{3}H-Pal (cpm)	Percent of Total
1	SDS (0.1%)	63875	30.3
2	0% n-Propanol	9430	4.5
3	3% "	4980	2.4
4	8% "	3910	1.8
5	25% "	16890	8
6	40% "	55480	26.3
7	80% "	56200	26.7

Table 7: The digest in 0.1% SDS was loaded onto C18 reversed phase Sep-Pak cartridges. After washing several times with water step elution was started with 60% formic acid and increasing concentrations of n-propanol. Radioactivity was measured by scintillation counting of aliquots.

Amino acid analysis of fractions revealed that after elution with 25% n-propanol more than 80% of all peptides had been eluted. Fraction 6 (40% n-propanol in 60% formic acid) was chosen for further purification by HPLC on a C4-wide pore column. As shown in figure 5 the ^{3}H-acylpeptide elutes at a concentration of about 65% 2-propanol in 0.1% trifluoroacetic acid (TFA).

Our attempts to sequence the fraction containing radiolabeled peptides failed because of the extremely hydrophobic properties of this material. To identify the peptide we utilized amino acid analysis by the Pico-Tag method (Waters) by which the phenylthiocarbamyl (PTC) amino

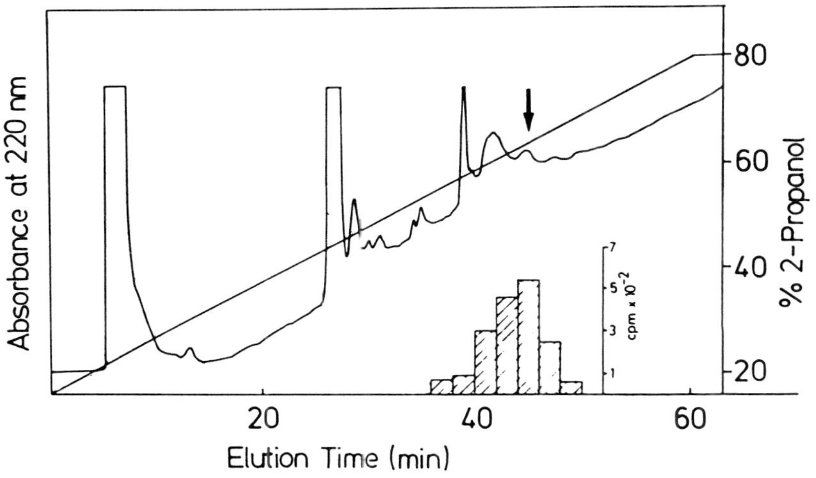

Figure 5: HPLC of Chymotryptic Peptides of ^{3}H-Pal labeled SFV-E1

Prepurified ^{3}H-Acylpeptides from E1 (fraction 6 from elution shown in table 7) were applied to a wide pore reversed phase column (Organogen, Heidelberg). Elution was with a gradient from 20-80% 2-propanol in TFA at a flow rate of 0.6 ml/min. Protein was monitored at 220 nm and radioactivity was measured by scintillation counting.

acids are separated. The results shown in table 8 strongly suggest that, as with the influenza HA, also the ^{3}H-acylated peptide isolated from SFV-E1 originates from the carboxyterminnus of E1 (compare figure 6).

Recently experiments are underway to sequence this material on solid phases. Preliminary results indicated alanin residues both in the first and second step (compare Fig. 5). Thus, we have purified fatty acylated peptide material which upon sequence analysis will eventually yield the precise linkage site of palmitic acid. This type of analysis, the protein chemical study of the "real thing" is tidious and hitherto only in one case has an acyl linkage site for palmitic acid been determined, a threonine residue in lipophilin from bovine brain (Stoffel, 1983).

TABLE 8

AMINO ACID ANALYSIS OF ^{3}H-ACYLPEPTIDE FROM E1

Amino Acid	^{3}H-Acylpeptide	Theoretical Peptide (Pos. 400-438 in Fig. 6)
Asp	1	1
Glu	1 - 2	1
Ser	4	3
Gly	(10?)	6
His	-	-
Arg	1 - 2	2
Thr	2	2
Ala	4	4
Pro	1	1
Tyr	-	-
Val	2 - 3	5
Met	0 - 1	1
Cys	-	1
Ile		4
Leu	4	5
Phe	1	1
Lys	1 - 2	1

Table 8: HPLC-fractions containing the ^{3}H-Pal labeled peptides (arrow in Fig. 5) were hydrolysed for 24 h at 110°C in 6N HCl. After derivation with phenylisothiocyanate according to the PICO-TAG procedure (Waters) the PTC-derivatives were separated by HPLC with a gradient from 6 to 46% acetonitril in 0.14 M sodium acetate with 0.05% (v/v) triethylamin. PTC-amino acids were detected at 254 nm.

NH_2-Tyr-Glu-. . .-Leu-Gys-Ser-Ala-Arg-Ala-Thr-Cys-Ser-Ala-
1 378
Ser-Cys-Glu-Pro-Pro-Lys-Asp-His-Ile-Val-Pro-Tyr-Ala-
391
Ala-Ser-His-Ser-Asn-Val-Val-Phe-Pro-Asp-Met-Ser-Gly-
404
Thr-Ala-Leu-Ser-Trp-Val-Gln-Lys-Ile-Ser-Gly-Gly-Leu-
417
Gly-Ala-Phe-Ala-Ile-Gly-Ala-Ile-Leu-Val-Leu-Val-Val-
430
Val-Thr-Cys-Ile-Gly-Leu-Arg-Arg-COOH

Figure 6: Genome derived amino acid sequence of the carboxyterminal region of SFV-1

Only the aminoterminus and the last 70 carboxyterminal amino acids given as determined by Garoff et al. (1980). The potential chymotrypsin cleavage sites are indicated by arrows. The membrane spanning segment is underlined.

However, identification and sequencing of acylpeptides seems advantageous over the gene technological approach. With this latter technique acylation is studied by expresion of genes in which potential acylation sites (Cys, Ser, Thr) have been replaced. In a number of cases Cys has been suggested to be the linkage site, because its replacement by Ser or other amino acids led to a loss of acylation (Willumsen et al., 1984; Koch and Hämmerling, 1986; Rose et al., 1984; Jing and Trowbridge, 1987). Such interpretations are certainly valid but not compelling. Replacement of cystein residues may lead to an alteration of the structure of a protein such that other acylation sites may not be available for the biosynthetic acylation which seems to be a membrane located process involving specific enzymes (Berger and Schmidt, 1984; Schmidt, 1984; Mack et al., 1987). For this reason further results of palmitic acid linkage analyses on the protein level will have to be awaited before any general conclusions can be drawn on consensus sequences for palmitoylation. It should be noted that for myristilation much more structural information has become available due to the stability of the amide linkage between myristic acid and the aminoterminal glycine residue (see Sefton and Buss, 1987). On the basis of this information Glaser and coworkers recently succeeded in isolating a protein N-myristoyl-transferase from yeast (Towler et al., 1987).

Concerning the function of fatty acids in a protein more speculations are available in the literature than hard experimental results, particularly with palmitoylated proteins. The most intrinsing hypothesis for a general function of palmitoylation of cellular proteins was recently put forward by Glick and Rothman (1987). These authors reported that in an *in vitro* system protein transport through the Golgi-stack was stimulated by palmitoyl-CoA, wich is the known precursor for palmitoylation of proteins (Schmidt, 1984).

With viral structural acylproteins it was noted that many of these were fusogenic as their most prominent biological feature (Schmidt, 1982). In order to test the hypothesis that acyl chains are in some way involved in fusion induction, virus particles with acylated protein or detergent solubilized acylproteins were compared with their hydroxylamin deacylated counterparts in their fusogenic properties as measured by hemolysis and resonance energy transfer with N-NBD-PE and N-Rho-PE labeled liposomes (Struck et al., 1981). The results of such experiments utilizing influenza virus show clearly, that no fusion occurs between virus particles and cell membranes if the viral hemagglutinin HA_2 is devoid of fatty acids. Similar results have been obtained with Semliki Forest- and Newcastle disease virus (Schmidt and Lambrecht, 1985; Lambrecht and Schmidt, 1986). It is therefore likely that with enveloped viruses protein bound fatty acids may be essential to initiate an infection by triggering the fusion between viral and cellular membranes.

In the myristilated transforming proteins (table 4) fatty acids are essential for anchoring the protein to the cytoplasmic face of the plasma membrane. If no myristic acid is present at their aminoterminus these proteins are found in the cytoplasm. As a consequence, the cells cannot be transformed and retain their normal phenotype (Kamps et al., 1986; Pellman et al., 1985).

3. Conclusion

Although many acylproteins have been identified and many more with as yet unknown identities have been reported, quite little is known on the structure and function of protein bound acyl chains. Thus far, in large the viral acylproteins can be regarded as the "avantgarde class of molecules" in the study of the various aspects of fatty acylation. Hopefully, as perhaps indicated by the recent identification of a protein myristoyl transferase (Towler et al., 1987) more data will become available in the near future on the linkage, enzymology and functional role of acyl chains in the ever increasing number of cellular acylproteins.

Acknowledgement

We thank Rudi Rott for continuous support and encouragement in our studies on acylation. All effort by Eva Kroell and Margot Seitz with editorial work and typing of the manuscript are gratefully acknowledged. Work herein has been supported by Sonderforschungsbereich 47 of the Deutsche Forschungsgemeinschaft and research initiation grant MB 173 of Kuwait University. This work is presented in partial fulfillment of the requirements for Dr. rer. nat. of Marion Schmidt at the FB 15 of Justus-Liebig-Universität Giessen, FRG.

4. References

Agrawal HC, Randle CL, Agrawal D (1982) In vivo acylation of rat brain myelin proteolipid protein. J Biol Chem 257:4588-4592

Agrawal HC, Schmidt RE, Agrawal D (1983) In vivo incorporation of (^{3}H)palmitic acid into PO protein, the major intrinsic protein of rat sciatic nerve myelin. J Biol Chem 258:6556-6560

Aitken A, Cohen P, Santikarn S, Williams DN, Calder AB, Smith A, Klee CB (1982) Identificationof the NH_2-terminal blocking group of calcineurin B as myristic acid. FEBS-Lett 150:7-12

Aitken A, Cohen P (1984) Identification of N-terminal myristyl blocking groups in proteins. Methods Enzymol 106:205-210

Bazan NG, Reddy TS, Redmond TM, Wiggert B, Chader GJ (1985) Endogeneous fatty acids are covalently and noncovalently bound to interphotoreceptor retinoid-binding protein in the monkey retina. J Biol Chem 260:13677-13680

Berger M, Schmidt MFG (1984) Cell free fatty acid acylation of Semliki Forest viral polypeptides with microsomal membranes from eukaryotic cells. J Biol Chem 259:7245-7252

Berger M, Schmidt MFG, (1986) Characterization of a protein fatty acylesterase present in microsomal membranes of diverse origin. J Biol Chem 261:14912-14918

Betz R, Crabb JW, Meyer HE, Wittig R, Duntze W (1986) Amino acid sequences of α factor mating peptides from Saccharomyces cerevisiae. J Biol Chem 262:546-548

Bolanowski MA, Earles BJ, Lennarz WJ (1984) Fatty acylation of proteins during development of sea urchin embryos. J Biol Chem 259: 4934-4940

Bordier C, Etges RJ, Ward J, Turner MJ, Cardoso de Almeida ML (1986) Leishmania and trypanosoma surface glycoproteins have a common glycophospholipid membrane anchor. Proc Natl Acad Sci 83:5988-5991

Bringman TS, Lindquist PB, Derynck R (1987) Different transforming growth factor-α species are derived from a glycosylated and palmitoylated precursor. Cell 48:429-440

Burn P, Burger MM (1987) The cytoskeletal protein vinculin contains transformation sensitive, covalently bound lipid. Science 235:476-479

Buss, JE, Sefton BM (1985) Myristic acid, a rare fatty acid, is the lipid attached to the transforming protein of Rous sarcoma virus and its cellular homolog. J Virol 53:7-12

Carr SA, Biemann K, Shoji S, Parmelee DS, Titani K (1982) n-Tetradecanoyl is the NH_2-terminal blocking group of the catalytic subunit of the cyclic AMP-dependent protein kinase from bovine cardiac muscle. Proc Natl Acad Sci USA 79:6128-6131

Chow M, Newman JFE, Filman D, Hogle JM, Rawlands DJ, Brown F (1987) Myristilation of picornavirus capsid protein VP4 and its structural significance. Nature 327:482-486

Dolci ED, Palade GE (1985 The biosynthesis and fatty acid acylation of the murine erythrocyte sialoglycoproteins. J Biol Chem 260:10728-10735

Ferguson MAJ, Haldar K, Cross GAM (1985) Trypanosoma brucei Variant Surface Glycoprotein has a sn-1,2-dimyristyl glycerol membrane anchor at its COOH terminus. J Biol Chem 260:4963-4968

Firestone GL, Payvar F, Yamamoto KR (1982) Glucocorticoid regulation of protein processing and compartmentatization. Nature 300:221-225

Folch-Pi J, Lees M (1951) Proteolipids, a new type of tissue lipoproteins. Their isolation from brain. J Biol Chem 191:807-817

Fujiyama A, Tamanoi F (1986) Processing and fatty acylation of RAS1 and RAS2 proteins in Saccharomyces cerevisiae. Proc Natl Acad Sci USA 83:1266-1270

Garoff H, Frischauf AM, Simons K, Lehrach H, Delius H (1980) Nucleotide sequence of cDNA coding for Semliki Forest virus membrane glycoproteins. Nature 288:236-241

Gebhardt A, Bosch JV, Ziemiecki A, Friis RR (1984) Rous sarcoma virus p19 and gp35 can be chemically cross-linked to high molecular weight complexes - an insight into virus assembly. J Mol Biol 174:297-317

Grand RJA, Roberts C, Gallimore PH (1985) Acylation of Adenovirus type 12 early region 1b 18-kDa protein. FEBS-Lett 181:229-235

Haldar K, Ferguson MAJ, Cross GAM (1985) Acylation of Plasmodium falciparum merozite surface antigen via sn-1,2-diacyl glycerol. J Biol Chem 260:4969-4974

He H-T, Barbet J, Chaix J-C, Goridis C (1986) Phosphatidylinositol is involved in the membrane attachment of NCAM-120, the smallest component of the natural cell adhesion molecule. EMBO-J 2:2489-2494

Hiller G, Weber K (1985) Golgi-derived membranes that contain an acylated viral polypeptide are used for vaccinia virus envelopment. J Virol 55:651-659

Hoeg JM, Meng MS, Ronan R, Fairwell T, Brewer jr HB (1986) Human apoliprotein A-I. Post-translational modification by fatty acid acylation. J Biol Chem 261:3911-3914

Jing S, Trowbridge IS (1987) Identification of the intermolecular disulfide bonds of the human transferrin receptor and its lipid-attachment site. EMBO-J 6:327-331

Johnson DC, Spear PG (1983) O-linked oligosaccharides are acquired by Herpes simplex virus glycoproteins in the Golgi apparatus. Cell 32:987-997

Kamps MP, Buss JE, Sefton BM (1986) Rous sarcoma virus transforming proteins lacking myristic acid phosphorylates known polypeptide substrates without inducing formation. Cell 45:105-112

Kaufmann JF, Krangel MS, Strominger JL (1984) Cysteines in the transmembrane region of major histocompatibility complex antigens are fatty acylated via thioester bonds. J Biol Chem 259:1-9

Keenan TW, Heid HW, Stadler J, Jarasch ED, Frankke WW (1982) Tight attachment of fatty acids to proteins associated with milk lipid globule membrane. Eur J Cell Biol 26:270-276

Kellie AE, Wrigglesworth P (1987) Vinculin is myristilated. FEBS-Lett 213:428-432

Kinet JP, Quarto R, Perez-Monfort R, Metzger H (1985) Noncovalently and covalently bound lipid on the receptor for immunoglobulin E. Biochemistry 24:7342-7348

Klockmann U, Deppert W (1983) Acylated Simian virus 40 large T-antigen: a new subclass associated with a detergent-resistant lamina of the plasma membrane. EMBO J 2:1151-1157

Koch N, Hämmerling GJ (1985) The Ia associated invariant chain is fatty acylated before addition of sialic acid. Biochemistry 24:6185-6190

Koch N, Hämmerling GJ (1986) The HLA-D-associated invariant chain binds palmitic acid at the cysteine adjacent to the membrane segment. J Biol Chem 261:3434-3440

Lambrecht B, Schmidt MFG (1986) Membrane fusion induced by influenza virus hemagglutinin requires protein bound fatty acids. FEBS-Lett 202:127-132

Lange-Mutschler J (1986) Acylated fibronectin: a new type of post-translational modification of cellular fibronectin. FEBS-Lett 201:210-214

Low MG, Zilversmit DB (1980) Role of phosphatidylinositol in attachment of alkaline phosphatase to membranes. Biochemistry 19:3913-3918

Low MG, Ferguson MAJ, Futerman AH, Silman I (1986) Covalently attached phosphatidylinositolas a hydrophobic anchor for membrane proteins. Trends Biochem Sci 244.1-13

Mac Lennan DH, Yip CC, Iles GH, Seeman P (1972) Isolation of sarcoplasmic reticulum protein. Cold Spring Harbour Symp Quant Biol 27:469-477

Madoff DH, Lenard J (1982) A membrane glycoprotein that accumulates intracellularly: cellular processing of the large glycoprotein of La Crosse virus. Cell 28:821-829

Mack D, Berger M, Kruppa J, Schmidt MFG (1987) In vitro-acylation of the VSV G-protein with microsomal membranes from cultured cells. J Biol Chem 262:4297-4302

Magee AI, Courtneidge SA (1985) Two classes of fatty acid acylated proteins exist in eukaryotic cells. EMBO-J 4:1137-1144

Magee AI, Siddle K (1986) Human insulin rezeptor contains covalently bound palmitic acid. Biochem Soc Trans 14:1103-1104

McIlhinney RAJ, Pelly SJ, Chadwick JK, Cowley GP (1985) Studies on the attachment of myristic and palmitic acid to cell proteins in human squamous carcinoma cell lines: evidence for two pathways. EMBO-J 4:1145-1152

Medof ME, Walter EI, Roberts WL, Haas R, Rosenberry TL (1986) Decay accelerating factor of complement is anchored to cells by a C-terminal glycolipid. Biochemistry 25:6740-6747

Merz DC, Wolinski JS (1983) Conversion of nonfusing mumps virus infections by selective proteolysis of the HN glycoprotein. Virology 131:328-340

Moremen KW, Touster O (1985) Biosynthesis and modification of Golgi mannosidase II in HeLa and 3T3 cells. J Biol Chem 260:6654-6662

Nakabayashi T, Ikezawa H (1984) PI-Anchors of alkaline phosphodiesterase and 5'nucleotidase. Cell Struct Funct 9:247

O'Brien PJ, Zatz M (1984) Acylation of bovine rhodopsin by ^{3}H-palmitic acid. J Biol Chem 259:5054-5057

Olson EN, Glaser L, Merlie JP (1984) Alpha- and beta-subunits of the nicotinic acetylcholine receptor contain covalently bound lipid. J Biol Chem 259:5364-5367

Omary MB, Trowbridge JS (1981) Biosynthesis of the human transferrin receptor in cultured cells. J Biol Chem 256:12888-12892

Ozols J, Carr SA, Strittmatter P (1984) Identification of the NH_2-terminal blocking group of NADH-cytochrome b_5 reductase as myristic acid and the complete amino acid sequence of the membrane binding domain. J Biol Chem 259:13349-13354

Pellman D, Garber EA, Cross FR, Hanafusa H (1985) An N-terminal peptide from p60^{src} can direct myristilation and plasma membrane localization when fused to heterologous proteins. Nature 314:344-347

Porter AG, Barber C, Carey NH, Hallewell RA, Threlfall G, Emtage JS (1979) Complete nucleotide sequence of an influenza virus hemagglutinin gene from cloned DNA. Nature 282:471-477

Roberts WL, Rosenberry TL (1985) Identification of covalently attached fatty acids in the hydrophobic membrane-binding domain of human erythrocyte acetylcholin esterase. Biochem Biophys Res Com 133:621-627

Rose JK, Adams GA, Gallione CJ (1984) The presence of cysteine in the cytoplasmic domain of the Vesicular Stomatitis virus glycoprotein is required for palmitate addition. Proc Natl Acad Sci USA 81:2050-2054

Schmidt MFG, Bracha M, Schlesinger MJ (1979) Evidence for covalent attachment of fatty acids to Sindbis virus glycoproteins. Proc Natl Acad Sci USA 76:1687-1691

Schmidt MFG, Schlesinger MJ (1979) Fatty acid binding to Vesicular Stomatitis virus glycoprotein - a new type of posttranslational modification of the viral glycoprotein. Cell 17:813-819

Schmidt MFG (1982) Acylation of viral spike glycoproteins - a feature of enveloped RNA-viruses. Virology 116:327-338

Schmidt MFG (1983) Fatty acid binding: a new kind of posttranslational modification of membrane proteins. Curr Top Microbiol Immunol 102:101-129

Schmidt MFG (1984) The transfer of myristic- and other fatty acids on lipid- and viral protein acceptors in cultured cells infected with Semliki Forest- and Influenza virus. EMBO-J 3:2295-2300

Schmidt MFG (1986) Die Acylierung von Proteinen mit Fettsäuren, Habilitationsschrift, Justus-Liebig-Universität Giessen, FRG

Schmidt MFG, Lambrecht B (1985) On the structure of the acyl linkage and the function of fatty acyl chains in the influenza virus hemagglutinin and the glycoproteins of Semliki Forest virus. J gen Virol 66:2635-2647

Schultz AM, Oroszlan S (1983) In vivo modification of retroviral gag gene-encoded polyproteins by myristic acid. J Virol 46:355-361

Schultz AM, Oroszlan S (1984) Myristilation of gag-onc fusion proteins in mammalian transforming retroviruses. Virology 133:431-437

Schultz AM, Copeland TD, Mark GE, Rapp UR, Oroszlan S (1985) Detection of the myristylated gag-raf transforming proptein with raf-specific antipeptide sera. Virology 146:78-83

Schultz AM, Henderson LE, Oroszlan S, Garber EA, Hanafusa H (1985) Aminoterminal myristilation of protein kinase p60^{src}, a retroviral transforming protein. Science 227:427-429

Sefton BB, Trowbridge IS, Cooper JA, Scolnick EM (1982) The transforming proteins of Rous sarcoma virus and Abelson virus contain tightly bound lipid. Cell 31:465-474

Sefton BM, Buss JE (1987) The covalent modification of eukaryotic proteins with lipid. J Cell Biol 104:1449-1453

Slomiany A, Witas H, Aono M, Slomiany BL (1983) Covalently linked fatty acids in gastric mucus glycoproteins of cystic fibrosis patients. J Biol Chem 258:8535-8538

Soric J, Gordon JA (1985) The 36-kilodalton substrate of $pp60^{v-src}$ is myristilated in a transformation-sensitive manner. Science 230:563-566

Staufenbiel M, Lazarides E (1986) Ankyrin is fatty acid acylated in red blood cells. Jahrestagung d. Dt. Gesellsch. f. Zellbiol. Tübingen.

Stoffel W., Hillen H., Schröder W, Deutzmann W (1983) The primary structure of bovine brain myelin lipophilin (proteolipid apoprotein). Hoppe-Seyler's Z Physiol Chem 364:1455-1466

Streuli CH, Griffin BE (1987 Myristic acid is coupled to a structural protein of polyoma virus and SV40. Nature 326:619-622

Strous GJ, van Kerkhof P, WillemsenR, Berger EG (1983) Role of oligosaccharides in the metabolism of the Golgi enzyme galactosyltransferase. In: Proceedings of the 7th International Symposium on Glycoconjugates (Chester MA, Heinegard D, Lundblad A, Svenson S, Lund S, eds) pp 753-754

Struck DK, Hoekstra D, Pagano RE (1981) Use of resonance energy transfer to monitor membrane fusion. Biochemistry 20:4093-4099

Takesue Y, Yokota K, Nishi Y, Taguchi R, Ikezawa H (1986) Solubilization of trehalase from rabbit renal and intestinal brush-border membranes by a phosphatidylinositol-specific phospholipase G. FEBS-Lett 201:5-8

Towler DA, Adams SP, Eubanks SR, Towery DS, Jackson-Macheliski E, Glaser L, Gordon JI (1987) Purification and characterisation of yeast myristoyl CoA: protein N-myristoyltransferase. Proc Natl Acad Sci 84:2708-2712

Tse AGD, Barclay AN, Watts A, Williams AF (1985) A glycophospholipid tail at the carboxyl terminus of the Thy-1 glycoprotein of neurons and thymocytes. Science 230:1003-1008

Voronova AF, Buss JE, Patschinsky T, Hunter T, Sefton BM (1984) Characterization of the protein apparently responsible for the elevated tyrosine protein kinase activity in LSTRA cells. Mol Cell Biol 4:2705-2713

Willumsen BW, Norris K, Papageorge AG, Hubbert NL, Lowy D (1984) Harvey murine sarcoma virus p21 ras protein: biological and biochemical significance of cysteine nearest carboxyterminus. EMBO-J 3:2581-2585

Yang JC, Chang PC, Fujitaki JM, Chiu KC, Smith RA (1986) Covalent linkage of phospholipid to myelin basic protein: Identification of phosphatidylinositol bisphosphate as the attached phospholipid. Biochemistry 25:2677-2681

INTERACTION OF THE MITOCHONDRIAL PRECURSOR PROTEIN APOCYTOCHROME C WITH MODELMEMBRANES AND ITS IMPLICATIONS FOR PROTEIN TRANSLOCATION

B. de Kruijff[1], A. Rietveld[2], W. Jordi[2], T.A. Berkhout[2], R.A. Demel[2], H. Görrissen[3] and D. Marsh[3]

INTRODUCTION

The majority of mitochondrial proteins are synthesized in the cytosol on free ribosomes in the form of precursors which are subsequently imported into the organelle by a posttranslational transport step (Hay et al., 1984). Depending on the final destination of the protein, either insertion into or translocation across one or two membranes has to occur. During this process the precursor (apoprotein) is converted into the mature holoprotein.

Apocytochrome c (apo-c), the heme-free precursor of cytochrome c, has been extensively used to study the molecular details of posttranslational protein import. These studies have not only given insight into the import possible mechanism of protein translocation across the outer mitochondrial membrane but also might serve as a more general model for protein insertion and translocation. Apo-c and cytochrome c have identical polypeptide chains of 104 amino acids (Fig. 1), but they differ in secondary and tertiary structure. Cytochrome c is a highly structured, nearly spherical protein, whereas apo-c has virtually no structure (Fisher et al., 1987). After synthesis on free polysomes apo-c binds to the mitochondrial outer membrane possibly involving a specific receptor (Hennig et al., 1983). During transport across this membrane the heme group is attached (Hennig and Neupert, 1981), whereafter the cytochrome c reaches its final location at the outside of the inner mitochondrial membrane (Capaldi, 1982).

[1]Inst. Mol. Biol. and Medical Biotechnol., [2]Dept. Biochem., Univ. of Utrecht, Padualaan 8, 3584 CH Utrecht, NL, and [3]Max-Planck Inst. Biophysikalische Chemie, Abt. Spektroskopie, D-3400 Göttingen, FRG

NATO ASI Series, Vol. H16
Membrane Biogenesis. Edited by J. A. F. Op den Kamp

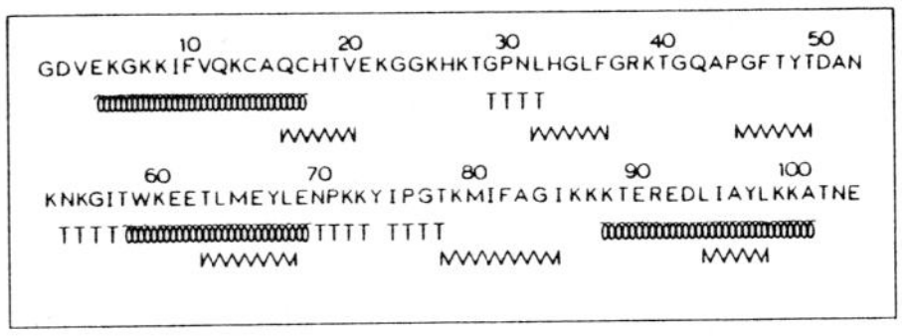

Fig. 1. Primary structure of apo-c and predicted secondary structure according to (Rietveld et al., 1986).

The molecular details of the actual insertion and translocation steps of apo-c and for precursor proteins in general are obscure. In principle, transport can occur either via protein tunnels, protein-lipid complexes or solely through the lipid part of membranes (Fig. 2). Protein tunnels are at yet fully hypothetical. However, there are several arguments (De Kruijff et al., 1985; Rietveld et al., 1986), which suggest that membrane lipids are involved in the insertion and translocation processes.

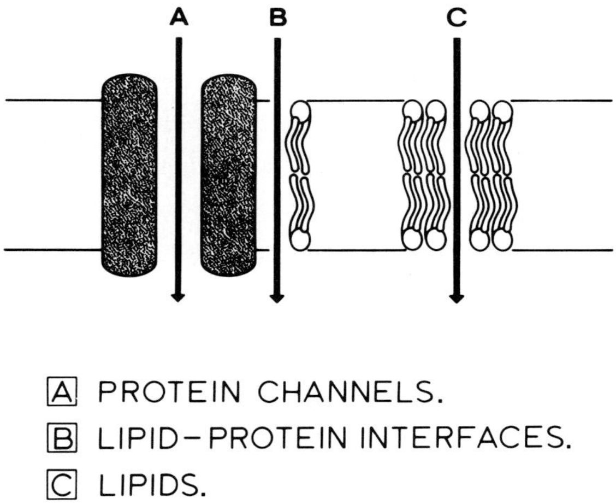

Fig. 2. Possible protein insertion/translocation pathways in membranes.

Potential sites for protein insertion and translocation involving membrane lipids in a bilayer organization are for instance (De Kruijff et al., 1985) the structural defects which can occur at the interface of integral membrane proteins and lipids or at the interface between lipids in the gel and liquid-crystalline state (Fig. 3). Also inverted non-lamellar lipid structures (De Kruijff et al., 1985) of the kind schematically illustrated in Fig. 4 for instance might provide transient sites especially for the translocation of the more polar parts of proteins.

Apo-c can be readily made in large quantities by the chemical removal of the heme group from the holoprotein (Fisher et al., 1973). This allows for detailed biophysical investigations of this precursor protein. It is the aim of this chapter to review our recent studies on interactions between apo-c and lipids and to construct from the results of these studies a model for the membrane insertion and translocation of this protein.

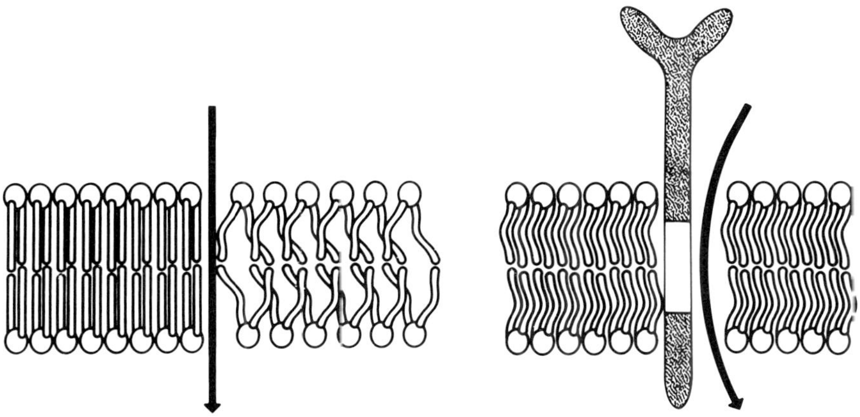

Fig. 3. Packing defects are potential sites for protein insertion and translocation in a bilayer.

APO-C INTERACTS WITH VESICLES CONTAINING NEGATIVELY CHARGED LIPIDS

Horse heart apo-c is a highly basic protein containing a net +9 charge at neutral pH. It therefore can be expected that the protein will strongly bind to negatively charged lipids of which phosphatidylserine (PS) is the most abundant species in the outer mitochondrial membrane. The results of binding experiments with different types of lipid vesicles indeed demonstrated (Rietveld et al., 1983; Rietveld et al., 1986; Berkhout et al, 1987) that the protein can bind with very high affinity to PS vesicles, but virtually not to vesicles of the neutral phosphatidylcholine (PC). In the physiological range of ionic strength and pH the binding is accompanied by vesicle aggregation but not by vesicle fusion (Rietveld et al., 1986). In

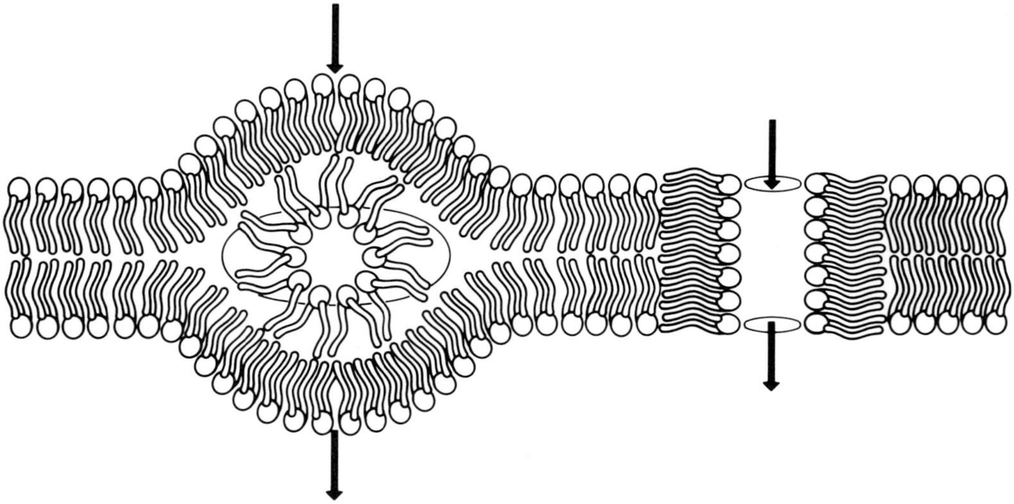

Fig. 4 Potential non-bilayer structures involved in protein insertion and translocation.

case of small unilamellar PS vesicles, a dissociation constant of 0.04 µM was determined by fluorescence measurements of the single Trp-59 (Berkhout et al., 1987). This value is very close to the value of binding to the mitochondrial outer membrane (Hennig et al., 1983). The affinity of the protein for the bilayer of the less curved large unilamellar vesicles (LUV) is less but still in the µM range (Table I).

Table I. Apo-c binding to PC/PS vesicles. For details see (Rietveld et al., 1986).

PC/PS (M/M)	Kd (µM)	n(PS/apo-c)
85/15	21 ± 4	10 ± 2
70/30	14 ± 6	11 ± 2
50/50	23 ± 5	8 ± 1
30/70	16 ± 5	10 ± 1
0/100	15 ± 4	9 ± 1

This binding is nearly charge stoichiometric and thus primarily electrostatic and interestingly nearly independent of the PS fraction suggesting that the protein creates its own micro domain which is enriched in PS.

Direct evidence for the creating of such micro domains came from fluorescence spectroscopy (Berkhout et al., 1987; Rietveld et al., 1986) and electron spin resonance (ESR) measurements (Rietveld et al., 1986). For instance, when apo-c is titrated with mixed PC/PS vesicles in which either the PC or PS component contained a parinaric acid chain at the 2-position (Rietveld et al., 1986) the fluorescence of the Trp-59 residue was quenched by an energy transfer mechanism, most when the parinaric acid was present in the PS component (Fig. 5). Thus, the immediate environment of the protein in the model membrane is enriched in the PS component.

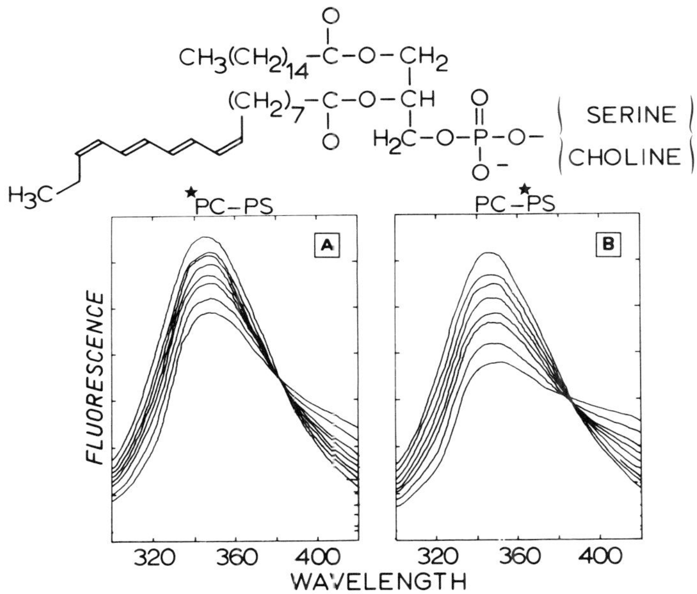

Fig. 5. Quenching of Trp-59 fluorescence of apo-c by 1-palmitoyl-2-parinaroyl-PC/PS in mixed PC/PS (1:1) LUV. In A and B the parinaric acid is in PC or PS, respectively. The top traces represent the fluorescence emission spectra of a 50 µg/ml apo-c solution. Subsequent traces are in the presence of increasing amounts of vesicles. For details see (Rietveld et al., 1986).

APO-C PENETRATES INTO THE HYDROPHOBIC PART OF A MODEL MEMBRANE

Injection of apo-c under a monolayer of either PS or PC results in a large increase in surface pressure only in case of negatively charged lipids (Rietveld et al., 1985), demonstrating that following the electrostatic interaction the protein penetrates into the hydrophobic part of the lipid monolayer (Fig. 6).

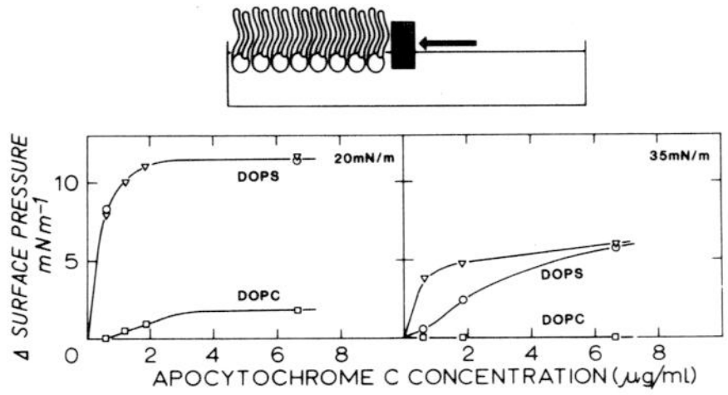

Fig. 6. Apo-c penetration in monolayers of cardiolipin (∇-∇), DOPS (o-o) and DOPC (□-□) at initial surface pressures of 20 and 35 mN/m. For details see (Rietveld et al., 1985).

The area of the protein occupied in the monolayer was determined by quantifying the surface expansion at constant surface pressure and by determining the amount of protein in the monolayer by using radiolabelled proteins and was found to be 700-900 Å^2/molecule (Pilon et al., 1987). The existence of extensive hydrophobic interactions between apo-c and the acyl chains of negatively charged lipids could also be demonstrated using differential scanning calorimetry (DSC). Figure 7 illustrates this for the

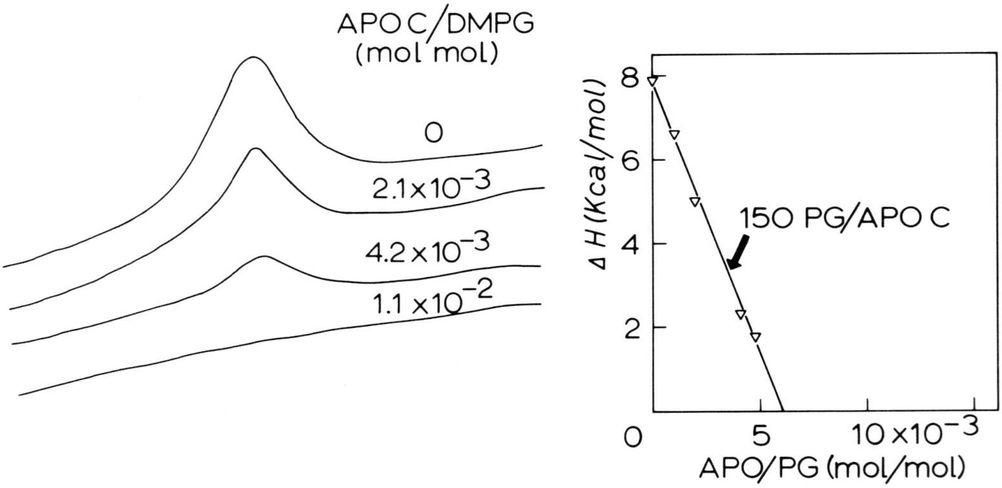

Fig. 7. Apo-c eliminates the cooperative gel -> liquid-crystalline transition ofDMPG (mid point transition temperature 23°C). For details see (Rietveld et al., 1985).

synthetic dimyristoylphosphatidyl glycerol (DMPG), a negatively charged lipid which resembles PS in its interactions with apo-c. Addition of relatively small amounts of protein causes a drastic reduction in the energy content of the gel to liquid-crystalline phase transition. From the proportional decrease in H as a function of the apo-c/PG ratio it can be concluded that one apo-c molecule can prevent 150 PG molecules from undergoing the cooperative phase transition. Such strong effects are typical for integral membrane proteins or very hydrophobic polypeptides and thus support the view that apo-c penetrates the hydrocarbon part of the bilayer.

The depth of penetration can be monitored with fluorescence spectroscopy (Berkhout et al., 1987) and ESR (Görrissen et al., 1986). Bromine atoms covalently attached to the acyl chain of phospholipids can quench the fluorescence of nearby polypeptide Trp residues by means of a collisional quenching process. When increasing amounts of mixed non-bromine PC/PS (1:1) vesicles are added to an apo-c solution the Trp fluorescence intensity increases (Berkhout et al., 1987; Rietveld et al., 1985) most likely due to a more hydrophobic environment of the indole ring system (Fig. 8). When in such mixed vesicles the 2-sn-acyl chain of PC contains bromine atoms at position 2, 6 and 7 or 11 and 12, the the fluorescence increase is less due to fluorescence quenching (Berkhout et al, 1987). The quenching efficiency is strongest when the bromines are present at position 6, 7, suggesting a penetration of the Trp-containing part of the protein some 7-8 Å deep in the bilayer (Berkhout et al., 1987).

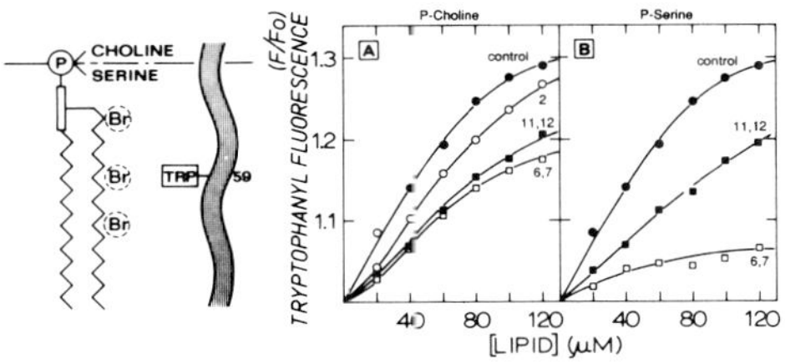

Fig. 8. Quenching of apo-c (4 μM) tryptophanyl fluorescence by PS/PC (1:1) vesicles containing non-brominated (control) or brominated acyl chains. The bromine atoms are located at the indicated positions in either PC (P-choline) or PS (P-serine). The left part gives the rationale of the experiment. For details see (Berkhout et al., 1987).

The quenching is much more obvious when the bromine atoms are present on the acyl chain of the PS component in line with the preferential interaction between apo-c and the negatively charged lipids.

ESR measurements of phospholipids spin-labelled at different positions on the sn-2 chain indicate a generalized decrease in mobility of the lipids while the characteristic flexibility gradient towards the terminal methyl end of the chain is maintained on binding of apo-c to PS dispersions (Görrissen et al., 1986). In addition, a second, more motionally restricted component is observed with lipids labelled close to the terminal methyl ends of the chains (Fig. 9). This second component is not observed on binding of the holocytochrome c and can be taken as direct evidence for deep penetration of apo-c into the lipid bilayer.

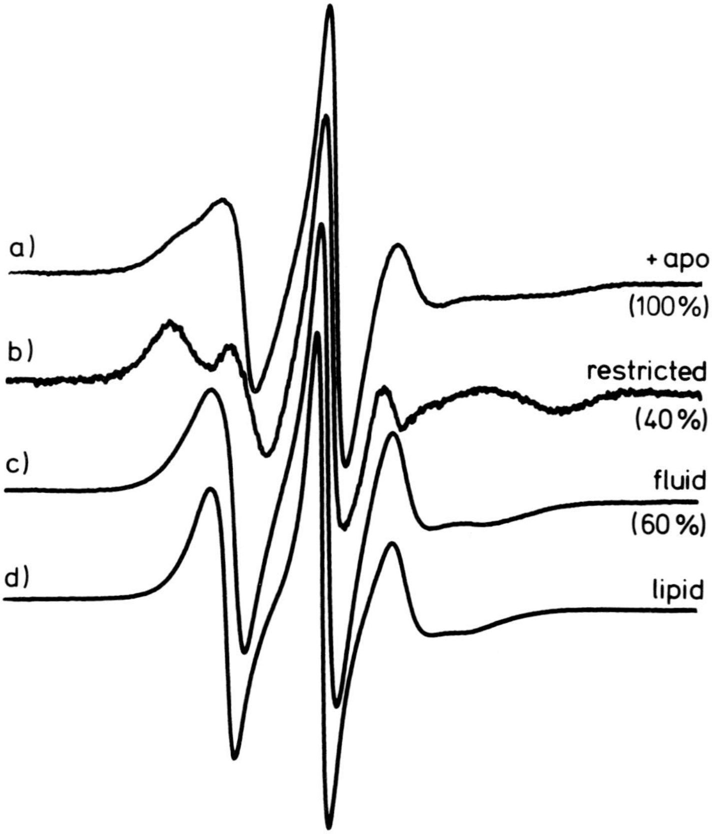

Fig. 9. ESR spectra of 12-PG spin label in bovine PS dispersions containing a saturating amount of apo-c (a), difference spectrum (b) obtained by subtracting (c) from (a), (d) experimental spectrum in the absence of apo-c. For details see (Görissen et al., 1986).

APO-C CAN (PARTIALLY) TRANSLOCATE ACROSS A PURE LIPID BILAYER

Figure 10 illustrates a translocation assay for apo-c in a lipid vesicle system (Rietveld et al., 1986; Dumont and Richards, 1984; Rietveld and De Kruijff, 1984). PC/PS (1:1) LUV are prepared containing enclosed trypsin (Rietveld et al., 1986). Outside trypsin is washed away by centrifugation and the residual outside trypsin activity is blocked by soy bean trypsin inhibitor (SBTI).

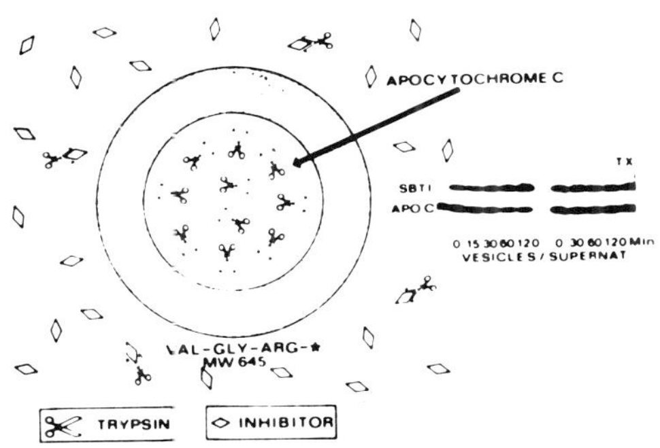

Fig. 10. Schematic representation of the apo-c translocation assay with LUV of PS/PC (1:1). For details see text and (Rietveld et al., 1986).

When apo-c is added to these vesicles a time-dependent degradation of apo-c in the vesicles could be demonstrated (Rietveld et al., 1986). This is for instance illustrated by a time-dependent decrease in intensity of the apo-c band after gel electrophoresis (compare with constant SBTI band). This is not due to apo-c degradation outside the vesicles as incubation of supernatant of such vesicles with a similar amount of apo-c does not result in a decrease in the apo-c band (Fig. 10). Apo-c degradation is also not due to transient rupture of the vesicle bilayer by the protein causing rapid release of trypsin because disrupting the bilayer with the detergent Triton X-100 (TX) immediately after the addition of the protein to the vesicle does not result in apo-c digestion due to the large excess of SBTI which immediately blocks the trypsin activity. Simultaneous addition of apo-c and a chromogenic tripeptide substrate of trypsin to the vesicles only results in apo-c digestion proving the specificity of the translocation process (Rietveld et al., 1986). These experiments thus demonstrate that a substantial amount of apo-c reaches the inner monolayer surface upon external addition to these pure lipid vesicles. The apo-c translocation is dependent on the lipid composition of the vesicles. As expected no translocation occurs for pure PC vesicles (Rietveld et al., 1986; Rietveld and De Kruijff, 1984). Inclusion of equimolar amounts of cholesterol or gel state lipids decreases acyl chain packing and causes a reduction in the rate of translocation (Fig. 11). Increasing acyl chain unsaturation by replacing beef brain PS by beef heart PS in the vesicles results in an increase in the rate of translocation. These differences are not due to differences in apo-c binding which was comparable for the different vesicles. Also for lipid vesicles made of the total mitochondrial lipid extract apo-c translocation could be demonstrated (Rietveld et al., 1986).

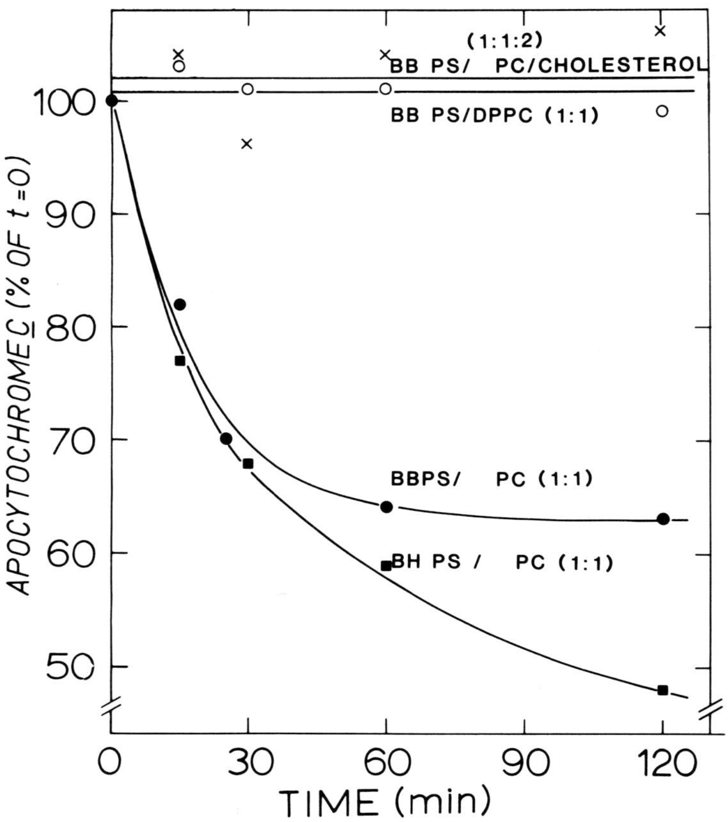

Fig. 11. Influence of lipid composition on apo-c translocation. For details see (Rietveld et al., 1986). BBPS: PS from beef brain, PC: PC from egg, DPPC: dipalmitoyl-PC.

APO-C CHANGES STRUCTURE UPON BINDING TO NEGATIVELY CHARGED LIPIDS

Apo-c is a highly disordered protein when compared to the holocytochrome c (Fischer et al., 1973). Figure 12 illustrates this with circular dichroism (CD) which reveal a typical random coil profile for apo-c which is unaltered by the presence of the neutral detergent lubrol (Rietveld et al., 1985). However, addition of negatively charged detergent micelles of SDS (Rietveld et al., 1985) or small unilamellar PS vesicles (Walter et al., 1986) results in the induction of extrema at 209 and 222 nm typical for α-helical structures (α-helix ± content 27%). For comparison, the CD spectrum of cytochrome c in buffer is included which is characterized by a similar content of α-helix. The α-helices of apo-c induced after interaction with negatively charged lipids are probably located at residues 4-17, 58-69 and 88-101 because their structure prediction methods (Fig. 1) indicate the highest probability for α-helix (ᘓᘓᘓ) formation (Rietveld et al., 1985). In addition, some small stretches of β-strand (ᐯᐯ) and β-turns (TT) are predicted. Nowhere in the primary structure continuous stretches of hydrophobic amino acid residues occur which are long enough to span either as α-helix or β-strand the hydrophobic part of a bilayer.

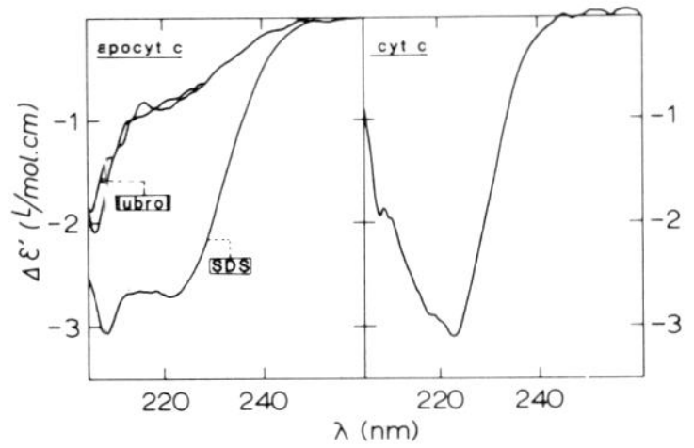

Fig. 12. CD spectra of cytochrome c and apo-c in the absence and presence of lubrol and SDS. For details see (Rietveld et al., 1985).

APO-C-LIPID VERSUS THE CYTOCHROME C-LIPID INTERACTION

The net charge of the polypeptide part of both proteins is similar therefore it is not unexpected that both proteins bind electrostatically to model membranes of negatively charged lipids (Rietveld et al., 1983). However, apo-c binding to PS is stronger. More interesting differences are found in the interaction of these proteins with the phospholipid acyl chains. DSC experiments demonstrate more significant acyl chain perturbation of the apo-protein (Rietveld et al. , 1983) ESR experiments revealed that only the apo-protein can deeply insert into the bilayer (Görrissen et al., 1986). Accordingly only the apo-c (Rietveld et al., 1983; Dumont and Richards, 1984; Rietveld and De Kruijff, 1984) and not cytochrome c (Geller and Wickner, 1985) can translocate across a vesicle bilayer. The covalent coupling of the heme group thus renders the cytochrome c translocation incompetent, probably because it fixes the protein into a specific folded conformation.

AN APO-C MEMBRANE TRANSLOCATION MODEL

Figure 13 is a pictorial description of the apo-c-lipid interaction in model membranes and schematically represents the insertion and translocation process (Rietveld and De Kruijff, 1986). The initially non-structured protein binds electrostatically to the negatively charged phospholipids causing them to cluster and to form a translocation site. Subsequently, the protein penetrates into the hydrophobic part of the bilayer which probably is accompanied by the formation of 3 α-helical segments. In order to accommodate the charged amino acid residues into the bilayer either some of the negatively charged phospholipids have to adopt temporarily a non-bilayer conformation (shown in this figure) or alternatively polypeptide-polypeptide interactions cause a favorable hydrophobic exterior of part of the polypeptide aggregate (not shown in Fig. 13).

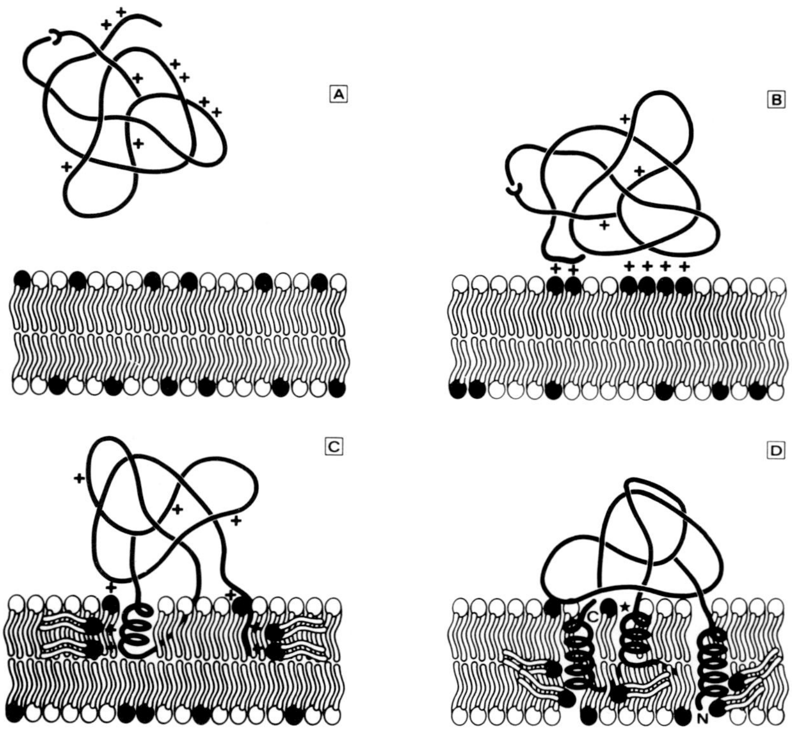

Fig. 13. Schematic representation of the interaction of apo-c with mixed PC (o)/PS (o) bilayers. For details see text and (Rietveld and De Kruijff, 1986).

In this respect the observation of PS induced oligomer formation of apo-c in model membranes could be of relevance. After the insertion process the C-terminus is assumed to be still located at the exterior surface of the vesicle. The N-terminus where at cysteine 14 and 17 the heme group has to be covalently coupled is assumed to cross the bilayer. In extrapolation this model to the apo-c import in mitochondria we suggested (Rietveld and De Kruijff, 1986) that the interaction of apo-c with the outer membrane proceeds similarly but that subsequently cardiolipin molecules of the outer layer of the inner membrane are involved in the charge neutralization of the penetrated positively charged apo-c segments, possibly in the form of (transient) contact sites between the outer and inner membrane. Subsequent enzymatic heme coupling and folding of the polypeptide chain around the heme moiety results in formation of cytochrome c which then is attached to the outside of the inner membrane by high affinity binding to the cytochrome c oxidase and cardiolipin. We furthermore proposed (Walter et al., 1986) that apo-c import and possibly import of other mitochondrial precursor protein with positively charged N-terminal presequences might (in part) be coupled to the inward flow of PS molecules from endoplasmic reticulum to inner mitochondrial membrane.

REFERENCES

Berkhout TA, Rietveld A, De Kruijff B (1987) Biochim Biophys Acta 897:1

Capaldi RA (1982) Biochim Biophys Acta 694: 291

De Kruijff B, Cullis PR, Verkleij AJ, Hope MJ, Van Echteld CJA, Taraschi TF, Van Hoogevest P, Killian JA, Rietveld A, Van der Steen ATM (1985) In: A. Watts and J.J.H.M. de Pont (eds) Progress in Protein-Lipid Interactions. Elsevier Publ Co., Amsterdam, Vol 1 pp 89-142

Dumont ME, Richards FM (1984) J Biol Chem 259: 4147

Fisher WR, Taniuchi H, Anfinsen CB (1973) J Biol Chem 248: 3188

Geller BL, Wickner W (1985) J. Biol. Chem. 260: 13281

Görrissen H, Marsh D, Rietveld A, De Kruijff B (1986) Biochemistry 25: 2904

Hay R, Böhni P, Gasser S (1984) Biochim Biophys Acta 779: 65

Hennig B, Neupert W (1981) Eur J Biochem 121: 203

Hennig B, Koehler H, Neupert W (1983) Proc Natl Acad Sci USA 80:4963

Pilon M, Jordi W, De Kruijff B, Demel RD (1987) Biochim Biophys Acta 902: 207-216

Rietveld A, Sijens P, Verkleij AJ, De Kruijff B (1983) EMBO J 2:907

Rietveld A, Ponjee GAE, Schiffers P, Jordi W, Van de Coolwijk PJFM, Demel RA, Marsh D, De Kruijff B (1985) Biochim Biophys Acta 818: 398

Rietveld A, De Kruijff B (1986) Biosci Reports 6: 775

Rietveld A, Jordi W, De Kruijff B (1986) J Biol Chem 261: 3846

Rietveld A, Berkhout TA, Roenhorst A, Marsh D, De Kruijff B (1986) Biochim Biophys. Acta 858:38

Rietveld A, De Kruijff B (1984) J Biol Chem 259: 6704

Walter E, Margolis D, Mohan R, Blumenthal R (1986) Membrane Biochem 6: 217

THE LDL RECEPTOR: STRUCTURAL INSIGHT FROM HUMAN MUTATIONS

Wolfgang J. Schneider
Department of Biochemistry
The University of Alberta
Edmonton, Alberta, Canada

INTRODUCTION

The cell surface receptor for low density lipoprotein (LDL) plays an important role in the regulation of cholesterol metabolism in humans and animals (Brown and Goldstein, 1986). This receptor binds LDL, the major plasma cholesterol transport protein and carries it into cells. Since its discovery 13 years ago, the LDL receptor has proven to be an extremely useful system for the study of receptor-mediated endocytosis, the process by which macromolecules enter cells after binding to receptors in coated pits. This receptor is also of interest because its expression is regulated at the transcriptional level by a feedback mechanism that is responsive to the level of cholesterol in the cell (Russell *et al.*, 1983). A third and distinctive feature of the LDL receptor is the frequent occurrence of mutations that disrupt its structure and function. These mutations occur in patients with the genetic disease, familial hypercholesterolemia (FH) (Goldstein *et al.*, 1985). When mutant receptor genes are present in the heterozygous or homozygous state, the plasma level of LDL rises and atherosclerosis ensues.

The LDL receptor is an integral membrane protein that is synthesized on the ribosomes of the rough endoplasmic reticulum as a 120-kilodalton (kDa) precursor that contains both N-linked and O-linked core sugars (Tolleshaug *et al.*, 1982). After the receptor is transported to the Golgi apparatus, the precursor sugars are processed to their mature form, which shifts the apparent molecular mass of the glycoprotein to 160 kDa as determined by migration in SDS polyacrylamide gels. The 160 kDa LDL receptor is then transported to the cell surface, where it clusters in coated pits and mediates the binding and internalization of cholesterol-rich LDL particles circulating in plasma. The receptor and its bound ligand enter the cell by receptor-mediated endocytosis, a process in which

NATO ASI Series, Vol. H16
Membrane Biogenesis. Edited by J. A. F. Op den Kamp

coated pits invaginate to form coated vesicles. After shedding their clathrin coats, these vesicles fuse with each other to form endosomes, within which the LDL receptor and its ligand are sorted away from each other. The LDL receptor is recycled back to the cell surface where it again clusters in coated pits and mediates the binding and internalization of additional LDL particles. After its liberation from the receptor in an endosomal compartment, the LDL is targeted to the lysosome where it is degraded. The released cholesterol is used by the cell for membrane synthesis and to coordinately regulate several genes involved in cholesterol metabolism, a series of events termed the LDL receptor pathway (Goldstein *et al.*, 1979).

The LDL Receptor Pathway

This pathway, depicted in Fig. 1, describes how cells, through receptor-mediated endocytosis of LDL, control their internal cholesterol concentration. The cholesterol derived from the lysosomal hydrolysis of LDL cholesteryl esters mediates a complex array of feedback control mechanisms that protects the cell from overaccumulation of cholesterol. First, the LDL-derived cholesterol suppresses the activity of the enzyme 3-hydroxy-3-methylglutaryl CoA reductase (HMG-CoA reductase), the rate controlling enzyme in cholesterol biosynthesis, thereby turning off cholesterol synthesis in the cell. Second, the cholesterol activates a cholesterol-esterifying enzyme called acyl-CoA:cholesterol acyltransferase (ACAT) so that excess cholesterol can be stored as droplets of cholesteryl esters. Third, the cholesterol suppresses the synthesis of new LDL receptors, preventing further cellular entry of LDL and thus overloading with cholesterol.

The overall effect of this regulatory system is to coordinate the intracellular and extracellular sources of cholesterol in order to maintain a constant level of cholesterol within the cell while the external supply in the form of lipoproteins undergoes fluctuation. Human fibroblasts and other mammalian cells are able to subsist in the absence of lipoproteins because they can synthesize cholesterol from acetyl-CoA. In contrast, when LDL is accessible, the cells primarily use the LDL receptor to take up LDL and keep their own cholesterol synthesis suppressed.

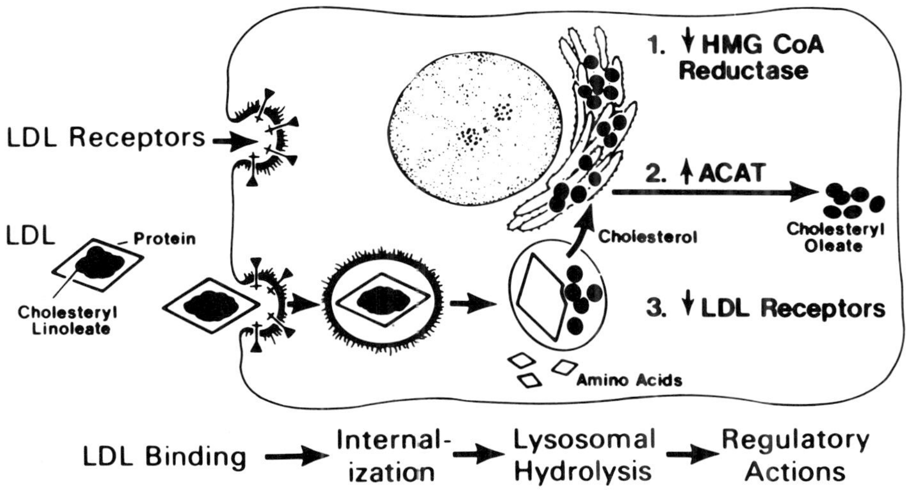

Fig. 1 Sequential steps in the LDL pathway in cultured mammalian cells. LDL denotes low density lipoprotein; HMG CoA reductase denotes 3-hydroxy-3-methylglutaryl CoA reductase; and ACAT denotes acyl-CoA: cholesterol acyltransferase.

Familial Hypercholesterolemia: Clinical Consequences of the Genetic Disruption of the LDL Receptor Pathway

The delineation of the LDL receptor pathway in fibroblasts was greatly facilitated by studies on fibroblasts from a large number of patients with FH. In 1974, it was recognized that mutations affecting the function of the LDL receptor are responsible for FH. Now, 13 years later, at least 13 different alleles at the LDL receptor locus have been identified in fibroblast lines from patients with the clinical features of FH homozygotes.

FH is characterized by three cardinal features: (i) a selective elevation in the plasma level of LDL; (ii) deposition of cholesterol in abnormal sites, in particular in tendons (formation of xanthomas) and in arteries (forming atheromas); and (iii) inheritance as an autosomal dominant trait with a gene dosage effect. This means that individuals who inherit two mutant alleles (FH homozygotes) are more severely affected than those with one mutant allele (FH heterozygotes). Heterozygotes occur at a frequency of about 1 in 500 persons among European and American populations while homozygotes occur at a frequency of about 1 in one million persons. Through the delineation of the molecular defects in certain patients (see below), it is now clear that the severely affected offspring of a marriage between two heterozygotes can be either true homozygous or actually a heteroallelic genetic compound (e.g., Tolleshaug *et al.*, 1982). Thus, the term homozygote in most cases is not strictly correct in a genetic sense, yet is a convenient one and is generally applied to patients with two mutant alleles at the LDL receptor locus.

Familial Hypercholesterolemia is the outstanding example of a single-gene mutation that regularly produces atherosclerosis. The deposition of LDL-cholesterol within the intima of major arteries in homozygotes becomes rapidly progressive, and leads to remarkably uniform clinical findings. Myocardial infarction, angina pectoris, and sudden death usually occur in homozygotes before the age of 15.

The clinical features of heterozygous FH are more variable and less severe than those of the homozygous form. Heterozygous men experience a myocardial infarction before 60 years of age with a 75% probability; in normal men, this risk is 15%. Despite the expression of the same genetic abnormality and similarly elevated plasma LDL levels, heterozygous women suffer from coronary heart disease less often and at a later age than do heterozygous men. The risk for female heterozygotes to develop coronary artery disease by age 60 is about 45% as compared with 10% in unaffected females.

The LDL Receptor - Molecular Aspects

Many of the steps in the complex itinerary of the LDL receptor have been elucidated through the biochemical analysis of mutations in subjects with FH. These mutations fall into several classes, namely those affecting 1) the synthesis of the receptor, 2) its transport from the endoplasmic reticulum to the cell surface, 3) its ability to bind LDL, and 4) its ability to cluster in clathrin-coated pits and therefore to be internalized. The molecular genetics of the LDL receptor in FH became amenable through purification of the LDL receptor protein (Schneider *et al.*, 1982), the isolation of full-length cDNA clones corresponding to the LDL receptor messenger RNA (Yamamoto *et al.*, 1984), and elucidation of the structure of the normal receptor gene (Südhof *et al.*, 1985). These cloned DNAs in turn have been used to determine the molecular basis of at least one example of each of the different classes of mutations discussed above.

Here, our knowledge of these FH mutations is summarized, and some studies that revealed interesting structure-function relationships in the LDL receptor protein are described.

The LDL Receptor Protein

The structure of the LDL receptor from four species (human, bovine, rabbit, and hamster) has been elucidated through the use of protein chemistry techniques, cDNA cloning, and DNA sequence analysis (Schneider *et al.*, 1982; Yamamoto *et al.*, 1984; Russell *et al.*, 1984; Yamamoto *et al.*, 1987). These studies have revealed an evolutionarily conserved integral membrane protein consisting of six domains (Fig. 2). In order of appearance from the NH_2-terminus these domains are: 1) a signal sequence; 2) the ligand binding domain; 3) a domain homologous to the epidermal growth factor (EGF) precursor; 4) a domain of clustered O-linked sugars; 5) a transmembrane domain; and 6) a cytoplasmic domain. Functional roles to several of these domains have been assigned by comparing the structure of the LDL receptor to other membrane proteins and by molecular analysis of mutant receptors. Some details are described below.

The <u>first domain</u> in the human receptor, a 21-amino acid signal sequence, is analogous to signal sequences in many membrane proteins. It is thought to mediate the insertion of the nascent LDL receptor polypeptide into the lumen of the endoplasmic reticulum, thus initiating the receptor's multistep intracellular itinerary.

The <u>second domain</u>, the ligand binding domain, mediates the interaction between the receptor and lipoproteins that contain apoprotein B and/or apoprotein E. It has been postulated that the binding function is localized to a 292-amino acid sequence at the NH_2-terminus of the mature protein. This domain is composed of seven repeats of approximately 40 amino acids each that are arranged in a head-to-tail array (Südhof *et. al.*, 1985). Each repeat contains six cysteine residues that appear

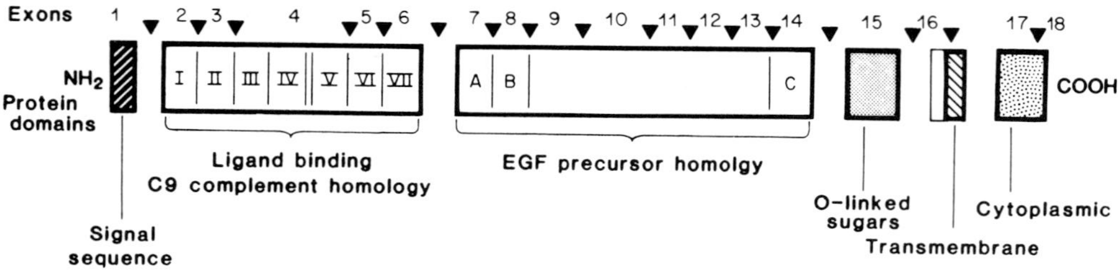

Fig. 2 Exon organization and protein domains in the human LDL receptor. The domains of the protein are delimited by thick black lines and are labeled in the lower portion of the figure. The 7 cysteine-rich, 40-amino acid repeats in the LDL binding domain are assigned roman numerals I-VII. Repeats IV and V are separated by 8 amino acids. The cysteine-rich repeats in the EGF precursor homology domain are lettered A-C. The positions at which introns interrupt the coding region are indicated by arrow heads. Exon numbers are shown between the arrow heads (reprinted with permission from Südhof *et al.*, (1985); Copyright 1985 by the AAAS).

to form three intra-repeat disulfide bonds and a cluster of negatively-charged amino acids located at the COOH-terminal end. Computer modeling indicates that this region contains alternating hydrophilic and hydrophobic regions. The center of each repeat is relatively hydrophobic, and it seems likely that many of the disulfide-bonded cysteines are buried in these hydrophobic pockets, which are flanked by polar residues. Such a conformation presumably allows the negatively charged residues to appear at the surface of the receptor where they bind to a cluster of positively-charged amino acids on apoprotein E (Innerarity *et al.*, 1984), or apoprotein b-100 (Knott *et al.*, 1985). For a comparison of amino acid sequences on apoproteins vs. the receptor involved in ligand binding, refer to Fig. 3.

Consensus Sequence of Receptor and Ligand

Protein	Sequence
Apo-E (Res. 140 - 150)	H - L - R - K - L - R - K - R - L - L - R
Apo-B (Res. 3147 - 3157)	K - A - Q - Y - K - K - M - K - H - R - H
(Res. 3357 - 3367)	T - T - R - L - T - R - K - R - G - L - K
Apo-B,E Receptor (Consensus region)	C - D - xxx - D - C - x - D - G - S - D - E

Fig. 3 Structural complementarity between the LDL (B,E)-receptor ligand binding domain and regions of apo-B and apo-E. The homologous positively charged residues are filled in the sequence of the apoproteins. The consensus sequence of the ligand binding site of the receptor, which is enriched in negatively charged residues (aspartate and glutamate, outlined), is shown for comparison.

The third domain of the LDL receptor consists of a stretch of approximately 400 amino acids whose most striking structural feature is a strong degree of sequence similarity to the EGF precursor (Russell *et al.*, 1984). In this homologous region, approximately 35% of the amino acids are held in common by the two proteins. The function of this domain is unknown. However, several possibilities are being considered: 1) it might function as part of the ligand binding domain; 2) it might serve to extend the ligand binding domain away from the cell surface, thus removing a steric hindrance to the interaction of the receptor and the large LDL particle (~ 220 Å diameter); 3) it might play a role in the binding of LDL receptors to each other or to other proteins in the plane of the membrane; or 4) it might play a role in the recycling of the LDL receptor from the endosome back to the cell surface.

The fourth domain of the LDL receptor consists of a sequence of 58 amino acids that are highly enriched in serine and threonine residues.

Biochemical and genetic studies have shown that this domain contains clustered O-linked sugar chains that retard the movement of the LDL receptor upon SDS polyacrylamide gel electrophoresis (Russell *et al.*, 1984). The function of these clustered O-linked sugars is at present not known. This region of the protein can be deleted without adverse effects on receptor function in cultured fibroblasts (Davis *et al.*, 1986a). It is conceivable that in other cell types these clustered O-linked sugars play a critical role in LDL receptor function or turnover. For instance, in polarized epithelial cells, they may serve as targeting signals for movement of the receptor to the basolateral or apical cell surface.

In addition to the O-linked sugars, the LDL receptor contains a few scattered O-linked sugar chains that are located in different parts of the molecule. Their precise location is as yet unknown (Davis *et al.*, 1986a). Through study of mutant Chinese hamster cells that are unable to synthesize O-linked sugars, Kingsley, *et al.*, (1986) found that these scattered O-linked sugars play a role in stabilizing the LDL receptor. When these O-linked sugars were not present, the LDL receptor reached the cell surface but was rapidly degraded. It is not yet known whether this effect is directly attributable to the loss of the scattered O-linked sugars on the receptor itself or whether it is attributable to the lack of O-linked sugars on some other membrane protein whose activity is necessary to prevent the receptor from being degraded.

The <u>fifth domain</u> of the LDL receptor consists of 22 to 25 hydrophobic amino acids that span the membrane, thus anchoring the protein. The deletion of this region in certain naturally-occurring mutations results in the secretion of LDL receptors from the cell (Lehrman *et al.*, 1985b). A comparison of membrane-spanning domains from LDL receptors of four different species reveals the maintenance of a hydrophobic character, but little actual sequence conservation. Thus, it is unlikely that the transmembrane segment has a specific function, such as transducing a signal across the plasma membrane (after ligand binding) or serving as an ion channel. These conclusions are consistent with the finding that the LDL receptor recycles constitutively in the absence of ligand and thus does not require that a signal to the cytoplasmic domain be transduced in order for internalization to occur.

The <u>last domain</u> of the LDL receptor is found at the extreme COOH-terminus of the molecule and consists of 50 amino acids that are located on the cytoplasmic side of the plasma membrane (Russell *et al.*, 1984).

The analysis of naturally-occurring and artificial mutations has revealed that this domain targets the LDL receptor to coated pits in the plasma membrane (Lehrman *et al.*, 1985a; Davis *et al.*, 1986b). The proteins with which the LDL receptor interacts in the coated pit are at present not known, nor is it known whether other receptors that cluster in coated pits bind to the same or different proteins.

Molecular Defects in LDL Receptors of Patients with Familial Hypercholesterolemia

At least 13 different mutant alleles at the LDL receptor locus have been described (Lehrman *et al.*, 1985a; Tolleshaug *et al.*, 1983; Schneider *et al.*, 1983). Fibroblasts from over 100 FH homozygotes were analyzed in terms of LDL receptor structure and function; these studies revealed that the mutations could be divided into four classes (for review, see Goldstein *et al.*, 1985). The mutant alleles are summarized in Table 1 and described in the following.

Class 1 mutations: no detectable precursor. These alleles, designated R-0, fail to express receptor proteins as measured by functional assays (binding of ^{125}-I-LDL) or attempts to immunologically identify proteins with a variety of monoclonal and polyclonal antibodies directed against the LDL receptor. R-0 alleles are the most frequent of the mutant alleles and probably account for about 40% of all mutations that introduce termination codons early in the protein coding region, but point mutations in the promoter that block transcription, mutations in intron/exon junctions that lead to abnormal splicing of the mRNA, and large deletions may also be included in this class of mutant alleles. Examples are described in Tolleshaug *et al.*, (1982) and Tolleshaug *et al.*, (1983).

Class 2 mutations: precursor not processed. These alleles specify receptor precursors that are produced in normal or reduced amounts, have normal (120 kDa) or abnormal (100 kDa, 135 kDa) molecular weight, but do not undergo any apparent increase in molecular weight after synthesis. Receptors encoded by these alleles never reach the cell surface (Tolleshaug *et al.*, 1983).

Most of the mutant receptor precursors in this class have an apparent M_r of 120,000 (allele designation, R-120) but precursors with apparent molecular weights of 100,000 (R-100 allele) and 135,000 (R-135 allele) have also been identified as members of this class. These M_r abnormalities are likely due to alterations in the length of the protein chain and do not result from changes in the carbohydrate moieties (Tolleshaug *et al.*, 1983).

Variants of the class 2 mutations were observed in a black American family, in Afrikaners, and in WHHL rabbits (Schneider *et al.*, 1983). In these variants the receptor is produced as a precursor of apparently normal M_r that is processed to the mature form, but much slower than the normal precursor. Eventually, about 5-10% of the receptors reach the cell surface as 160 kDa mature proteins. Once on the cell surface, these receptors display a reduced ability to bind LDL; that means that the mutation has disrupted both transport and binding capacity of the receptor (Schneider *et al.*, 1983). Very recently, a mutation in the WHHL rabbit receptor gene has been identified that causes the deletion of four amino acids in the LDL binding domain (Yamamoto *et al.*, 1987). While this might explain the inability of the mutant receptors to bind LDL normally, the reason for their delayed processing is less obvious. In any event, expression of the mutant gene in transformed cells will provide the answer as to whether this mutation is sufficient to bring about the WHHL phenotype.

The molecular basis of other defects in the class 2 mutations and their variants are not known. It seems likely that the failure of transport arises from some subtle changes in structure, possibly similar to the one occurring in the WHHL rabbit.

Class 3 mutations: precursor processed, abnormal binding. Receptors belonging to this class of mutants reach the cell surface at normal rates, but are able to bind only less than 15% of the normal amount of LDL. Most of class 3 receptors have normal M_r in SDS-PAGE (allele designation, R-160 b^-). Receptors with apparent M_r of 140,000 (R-140 b^-) and 210,000 (R-210 b^-) have been described as well. All mature receptor proteins in this class originate as precursors with apparent molecular weights that are 40,000 less, *i.e.*, 120,000, 100,000, and 170,000, respectively. It appears that the abnormal molecular weights are due to alterations in the amino acid sequence and not to changes in carbohydrate content.

The structure of the LDL binding domain - 7 repeats of a 40 amino acid sequence - suggests an explanation for the abnormally sized, binding deficient receptors in class 3. The DNA encoding the repeat structure might be susceptible to deletion or duplication following mispairing and recombination of homologous regions during meiosis. Such deletions or duplications would change size and binding capacity of the receptors.

Class 4 mutations: precursor processed, receptor binds LDL but does not cluster in coated pits. These mutations have been termed internalization-defective (Brown and Goldstein, 1976). From the description of the process of receptor-mediated endocytosis (see Introduction) it is apparent that clustering of receptors in coated pits is a prerequisite for subsequent internalization of ligand; it is the clustering of LDL receptors that is disrupted in class 4 mutations. To date, the molecular defects in five patients with this phenotype have been elucidated. The original example was patient J.D. (Brown and Goldstein, 1976). While he inherited a R-0 (class 1) allele from his mother, he inherited from his father a gene that produces a cell surface receptor of normal size and normal binding capacity, but which is unable to carry the bound LDL into the cell (R-160 i^- allele). Thus, J.D. is a compound heterozygote with regards to his genotype at the LDL receptor locus.

It was suspected early that the R-160 i^- allele is producing a receptor altered in its intracellular portion of 50 amino acid residues, because this region of the receptor is believed to mediate the clustering of receptors into coated pits through interaction with cytoplasmic, coated pit-specific proteins. The molecular defect, indeed, was found to reside in the intracellular domain (Davis *et al.*, 1986b). A point mutation has converted the codon for tyrosine 807 into one for cysteine. Subsequently to the discovery of J.D., four other patients with internalization defects have been identified and their genes analyzed. In one of them, the internalization-defective allele produces a receptor that lacks the intracellular domain and the membrane-spanning region, and the truncated protein is secreted from these cells (Lehrman *et al.*, 1985b). The secreted receptor has an apparent M_r of 150,000 (allele designation, R-150 i^-,sec). Another is a patient from Japan (Miyake *et al.*, 1981), the offspring of consanguineous parents, apparently homozygous for a R-150 i^-,sec allele. Two more internalization-defective alleles were identified and the mutant genes analyzed (Lehrman *et al.*, 1985a). In one (FH patients 682 and 683), the cytoplasmic domain consists of only two residues due to a point mutation that changes a tryptophan codon into a termination codon (Lehrman *et al.*, 1985a). The other mutant gene (found in patient GH 763) contains a four-base duplication, producing a frameshift that alters the reading frame. The cytoplasmic tail of this mutant receptor is made up of six of the normal plus eight additional amino acids instead of the normal 50 (Lehrman *et al.*, 1985a).

In conclusion, the analysis of class 4 mutations at the LDL receptor locus strongly suggests that the signal for movement of the receptor into coated pits resides in its cytoplasmic domain. In addition, the function of the cysteine-rich region at the aminoterminus as the ligand binding site is now well established. Future analysis of mutant receptor genes that affect posttranslational processing can be expected to shed light on the function of other LDL receptor domains.

Table 1. Mutations at the LDL receptor locus that produce familial hypercholesterolemia (FH)

Class of mutation	Allele designation[1]	Apparent receptor mass on SDS gels (kDa)		Receptor location: Plasma membrane				LDL binding to intact cells	Frequency in FH patients
		Precursor	Mature	Intra-cellular	Coated pits	Noncoated regions	Extra-cellular		
Class 1: No detectable precursor	R-0	None	None					None	Common
Class 2: Precursor not processed	R-100	100	100	+				None	Rare; found in Lebanese
	R-120	120	120	+				None	Common
	R-135	135	135	+				None	Rare
Class 2 variant: Precursor processed slowly, mature receptor binds LDL poorly	R-120 ^{slow} 160 b⁻	120	160	+	(+)[2]			Reduced	Rare; found in Afrikaners and in WHHL rabbits
Class 3: Precursor processed normally, mature receptor binds LDL poorly	R-140 b⁻	100	140	+				Reduced	Rare
	R-160 b⁻	120	160	+				Reduced	Common
	R-210 b⁻	170	210	+				Reduced	Rare
Class 4: Precursor processed normally, mature receptor binds LDL normally, but does not enter coated pits	R-150 i⁻, sec	110	150			(+)	+	Normal binding; defective internalization	Rare
	R-160 i⁻	120	160			+		Normal binding; defective internalization	Rare
	R-155 i⁻	115	155			+			

[1] Allele designations are based on the apparent molecular weight (in kilodaltons on SDS polyacrylamide gel electrophoresis of the mature forms of the receptor, i.e. the predominant form observed after a 2 hr pulse followed by a 2 hr chase. b⁻ denotes defective LDL binding; i⁻ denotes defective LDL binding; sec denotes secretion from the cell.

[2] Symbols in parentheses refer to minor populations.

REFERENCES

Beisiegel U, Schneider WJ, Goldstein JL, Anderson RGW, Brown MS (1981) Monoclonal antibodies to the LDL receptor as probes for study of receptor-mediated endocytosis and the genetics of Familial Hypercholesterolemia. J Biol Chem 256:11923-11931

Brown MS, Goldstein JL (1976) Analysis of a mutant strain of human fibroblasts with a defect in the internalization of receptor-bound low density lipoprotein. Cell 9:663-674.

Brown MS, Goldstein JL (1986) A receptor-mediated pathway for cholesterol homeostasis. Science 232:34-47

Davis CG, Elhammer A, Russell DW, Schneider WJ, Kornfeld S, Brown MS, Goldstein JL (1986a) Deletion of clustered O-linked carbohydrates does not impair function of low density lipoprotein receptor in transfected fibroblasts. J Biol Chem 261:2828-2838

Davis CG, Lehrman MA, Russell DW, Anderson RGW, Brown MS, Goldstein JL (1986b) The J.D. mutation in familial hypercholesterolemia: Substitution of cysteine for tyrosine in cytoplasmic domain impedes internalization of LDL receptors. Cell 45:15-24

Goldstein JL, Anderson RGW, Brown MS (1979) Coated pits, coated vesicles, and receptor-mediated endocytosis. Nature 279:679-685

Goldstein JL, Brown MS, Anderson RGW, Russell DW, Schneider WJ (1985) Receptor-mediated endocytosis: Concepts emerging from the LDL receptor system. Ann Rev Cell Biol 1:1-39

Innerarity TL, Weisgraber KH, Arnold KS, Rall SC Jr, Mahley RW (1984) Normalization of receptor binding of apolipoprotein E2: Evidence for modulation of the binding site conformation. J Biol Chem 259:7261-7267

Kingsley DM, Kozarsky KF, Hobbie L, Krieger M (1986) Reversible defects in O-linked glycosylation and LDL receptor expression in a UDP-Gal/UDP-GalNAc 4-epimerase deficient mutant. Cell 44:749-759

Knott TJ, Rall SC Jr, Innerarity TL, Jacobson SF, Urdea MS, Levy-Wilson B, Powell LM, Pease RJ, Eddy R, Nakai H, Byers M, Priestley LM, Robertson E, Rall LB, Betsholtz C, Shows TB, Mahley RW, Scott J (1985) Human apolipoprotein B: Structure of carboxyl-terminal domains, sites of gene expression, and chromosomal location. Science 230:37-43

Lehrman MA, Goldstein JL, Brown MS, Russell DW, Schneider WJ (1985a) Internalization-defective LDL receptors produced by genes with nonsense and frameshift mutations that truncate the cytoplasmic domain. Cell 41:735-753.

Lehrman MA, Schneider WJ, Südhof TC, Brown MS, Goldstein JL, Russell DW (1985b) Mutation in LDL receptor: Alu-Alu recombination deletes exons encoding transmembrane and cytoplasmic domains. Science 227:140-146.

Miyake Y, Tajima S, Yamamura T, Yamamoto A (1981) Homozygous familial hypercholesterolemia mutant with a defect in internalization of LDL. Proc Natl Acad Sci USA 78:5151-5155

Russell DW, Yamamoto T, Schneider WJ, Slaughter CJ, Brown MS, Goldstein JL (1983) cDNA cloning of the bovine low density lipoprotein receptor: Feedback regulation of a receptor mRNA. Proc Natl Acad Sci USA 80:7501-7505

Russell DW, Schneider WJ, Yamamoto T, Luskey KL, Brown MS, Goldstein JL (1984) Domain map of the LDL receptor: Sequence homology with the epidermal growth factor precursor. Cell 37:577-585

Schneider WJ, Beisiegel U, Goldstein JL, Brown MS (1982) Purification of the low density lipoprotein receptor, an acidic glycoprotein of 164,000 molecular weight. J Biol Chem 257:2664-2673

Schneider WJ, Brown MS, Goldstein JL (1983) Kinetic defects in the processing of the LDL receptor in fibroblasts from WHHL rabbits and a family with Familial Hypercholesterolemia. Mol Biol Med 1:355-367

Südhof TC, Goldstein JL, Brown, MS, Russell, DW (1985) The LDL receptor gene: A mosaic of exons shared with different proteins. Science 228:815-822

Tolleshaug H, Goldstein JL, Schneider WJ, Brown MS (1982) Posttranslational processing of the LDL receptor and its genetic disruption in familial hypercholesterolemia. Cell 30:715-724

Tolleshaug H, Hobgood KK, Brown MS, Goldstein JL (1983) The LDL receptor locus in familial hypercholesterolemia: Multiple mutations disrupting the transport and processing of a membrane receptor. Cell 32:941-951

Yamamoto T, Davis CG, Brown MS, Schneider WJ, Casey ML, Goldstein JL, Russell DW (1984) The human LDL receptor: A cysteine-rich protein with multiple Alu sequences in its mRNA. Cell 39:27-38

Yamamoto T, Bishop RW, Brown MS, Goldstein JL, Russell DW (1987) Deletion in cysteine-rich region of LDL receptor impedes transport to cell surface in WHHL rabbit. Science 232:1230-1237

THE ASSEMBLY AND TRANSFER OF OLIGOSACCHARIDE CHAINS TO PROTEINS

William J. Lennarz
Department of Biochemistry and Molecular Biology
The University of Texas System Cancer Center
M. D. Anderson Hospital and Tumor Institute
1515 Holcombe Boulevard
Houston, Texas 77030
USA

INTRODUCTION

Starting about two decades ago a series of developments provided us with a basic understanding of the mechanism of synthesis of membrane and secretory glycoproteins. These were the distinction between membrane-bound and free polysomes (Redman, 1969), the discovery of the signal peptide (Blobel & Sabatini, 1971; Milstein et al., 1972), the demonstration of the cotranslational glycosylation of ovalbumin (Kiely et al., 1976), the findings that, in accord with the signal hypothesis, newly translated and glycosylated proteins were sequestered within the lumen of the endoplasmic reticulum (Lingappa et al., 1978) and that viral coat proteins were inserted with their carbohydrate chains facing the lumen (Katz et al., 1977). All of these findings set the scene for in-depth biochemical studies on the mechanisms of assembly of the oligosaccharide chains of the glycoproteins. These studies (summarized in Parodi & Leloir, 1979; Struck & Lennarz, 1980; Hubbard & Ivatt, 1981; and Kornfeld & Kornfeld, 1985) have led to the following general conclusions:

1. All eukaryotic cells are capable of the synthesis of glycoproteins that are destined to become components of the plasma membrane. In addition, many cell types commit a significant portion of their protein biosynthetic activity to the synthesis of secreted and/or lysosome-packaged glycoproteins.

2. The synthesis of membrane, secretory, or lysosomal glycoproteins is a highly segregated process that occurs within an intracellular membrane system composed of the endoplasmic reticulum, transfer vesicles, Golgi apparatus, and secretory vesicles. During their translation, glycosylation and processing the glycoproteins are completely isolated from the cytoplasm and travel to the cell surface (in the case of membrane glycoproteins), to the extracellular environment (in the case of secretory glycoproteins), or to the lysosomes (in the case of lysosomal enzymes) as part of, or within, these membrane compartments.

3. The assembly of N-linked oligosaccharide chains occurs in the

NATO ASI Series, Vol. H16
Membrane Biogenesis. Edited by J. A. F. Op den Kamp

endoplasmic reticulum and involves the stepwise preassembly of the oligosaccharide chain on dolichyl phosphate followed by en bloc transfer of the oligosaccharyl unit to the growing polypeptide chain.

4. Subsequent modifications to the oligosaccharide chains of N-linked glycoproteins are initiated in the rough endoplasmic reticulum and completed in the Golgi apparatus.

As a result of the efforts of many laboratories, the individual steps in the assembly and transfer of the oligosaccharide chain are known to occur as shown in Figure 1.

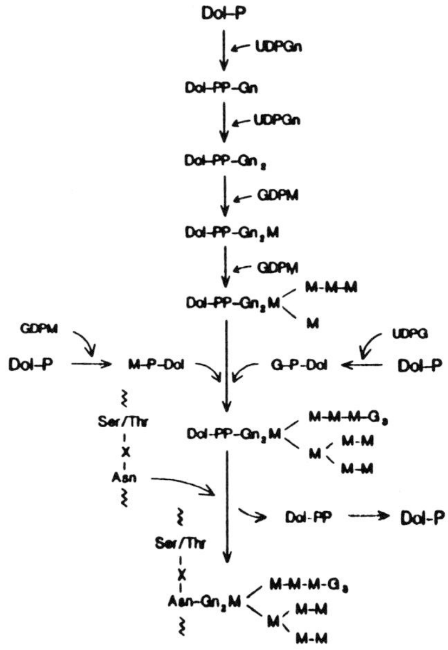

Figure 1 - Steps in the assembly of the oligosaccharide chain that becomes N-linked to proteins.

Recent studies on the Golgi apparatus with respect to protein glycosylation (Farquhar, 1985) and sugar nucleotide transport (Perez & Hirschberg, 1986a) have been reviewed. Therefore, this review will 1) take up questions of a topological nature that pertain to the assembly of the oligosaccharide chains and 2) discuss work on the process of transfer to Asn sites in polypeptides (cf. Kaplan et al., 1987).

KEY OBSERVATIONS RELATING TO TOPOLOGY

The oligosaccharide chain assembled on dolichyl phosphate and subsequently transferred to nascent polypeptide chains contains three sugars: N-acetylglucosamine, mannose, and glucose. All three are synthesized in their activated form as sugar nucleotides in the cytoplasm (Coates et al., 1980). A variety of studies (Rothman & Lodish, 1975; Bergman & Kuehl, 1977; Katz et al., 1977; Lingappa et al., 1978; Hanover & Lennarz, 1980) utilizing _in vivo_ pulse-chase techniques and _in vitro_ translation in the presence of microsomes revealed that these three sugars, in the form of an oligosaccharide chain, were linked to the polypeptide chains in the lumen of the rough endoplasmic reticulum. Consistent with these observations, it was determined that the Asn residue in ovalbumin destined to be glycosylated did not acquire its carbohydrate chain until at least another 32 amino acid residues beyond this Asn had been added to the growing polypeptide chain (Glabe et al., 1980). Because this number of residues was more than sufficient to span the lipid bilayer, it was concluded that the most likely site of glycosylation of the growing polypeptide chain was at the lumenal face of the rough endoplasmic reticulum.

Three observations indicated that the precursor to the protein-linked oligosaccharide chain, namely oligosaccharylpyrophosphoryl dolichol, was also lumenally oriented:

1. Studies discussed below using the impermeable lectin concanavalin A to study the topology of the completed oligosaccharylpyrophosphoryl dolichol in microsomes indicated that the oligosaccharide moiety was not exposed at the cytoplasmic face of the bilayer (Snider & Robbins, 1982).

2. Oligosaccharylpyrophosphoryl dolichol in "aged" preparations of oviduct microsomes was shown to undergo hydrolytic cleavage, yielding free oligosaccharide that remained entrapped in the lumen (Hanover & Lennarz, 1982). Studies by Anumula and Spiro (1983) subsequently revealed that an analogous process occurred in thyroid microsomes.

3. Experiments with acceptor peptide substrates for oligosaccharyl transferase indicated that these peptides freely entered and could be washed from the rough endoplasmic reticulum, but that after their glycosylation most were entrapped in the lumen (Welply et al., 1983).

BASIC QUESTIONS PERTAINING TO TOPOLOGY

These observations led to the general question of how sugar nucleotides in the cytoplasm are converted to an oligosaccharide chain facing the lumen of the rough endoplasmic reticulum. Perhaps it is more useful to rephrase this general question in terms of four more sharply defined questions:

1. Are sugar nucleotides transported into the lumen of the rough endoplasmic reticulum? If all three sugar nucleotides are transported into the lumen, the topological aspects of the enzymes and their products become simple: After their transport, the sugar nucleotides in the lumen interact with enzymes whose active sites are exposed at the lumenal face of the endoplasmic reticulum. Under these circumstances the dolichol-linked intermediates, the dolichol-linked end product oligosaccharylpyrophosphoryl dolichol, and the glycosylated polypeptide are all lumenally disposed.

However, it appears that this relatively simple situation does not obtain. Initial in vivo and in vitro evidence showed that oviduct microsomes lacked the ability to transport GDP-M (Hanover & Lennarz, 1982). Subsequent studies in liver microsomes supported the idea that GDP-M transport does not occur (Perez & Hirschberg, 1986b). However, positive evidence was reported for the transport of both UDP-Gn (Perez & Hirschberg, 1985) and UDP-G (Perez & Hirschberg, 1986b). In the case of UDP-Gn, the translocation process was temperature dependent, saturable, and inactivated by protease treatment of the endoplasmic reticulum (Perez & Hirschberg, 1985). Preliminary studies suggested that the import of UDP-Gn was coupled to the export of nucleotide monophosphate. Similar observations were made for UDP-G (Perez & Hirschberg, 1986b). Thus, on the basis of the available evidence, the simplest model, involving transport of all three sugar nucleotides into the lumen, followed by assembly and transfer of the oligosaccharide chain at the lumenal face, does not appear to be tenable. However, it is important to remember that the available evidence on transport of GDP-M is negative and that it is always possible that the kinetics of utilization subsequent to transport might be so rapid that it makes detection of this sugar nucleotide in the lumen impossible.

2. What is the topological distribution of Dol-P in the bilayer of the rough endoplasmic reticulum? As noted a number of years ago (Struck & Lennarz, 1980), the hydrocarbon chain of dolichol in its extended form is longer than the width of a typical biological membrane. Thus, the polyisoprenoid chain of dolichyl phosphate is almost certainly in some type of nonextended conformation. The precise nature of this conformation and the location of the polar phosphate head group in what has been called a "superlipid" (McCloskey & Troy, 1980) are important issues that remain to be resolved. Model studies using NMR to examine polyprenoids site specifically labeled with ^{2}H in phospholipid liposomes suggest that the acetyl esters of geraniol (C_{10}), farnesol (C_{15}) and solanesol (C_{45}) are not arranged in the lipid bilayer with their polar head groups exposed at the bilayer interface (deRopp & Troy, 1984; deRopp & Troy, 1985; deRopp et al., 1987). In contrast, the spin label studies of McCloskey and Troy (1980) and the studies of Valtersson et al. (1985) indicate that dolichyl phosphate does assume a conformation such that the polar portion of the molecule bearing the phosphoryl group is exposed at the bilayer interface. These studies also reveal that spin labeled analogues of phosphorylated dolichol remain in monomolecular form in phosphatidycholine vesicles.

With respect to the possible role of dolichyl phosphate in the translocation of saccharides across the membrane, knowledge of the dynamic properties of these molecules in bilayers is very important. However, the available evidence indicates that the $T_{1/2}$ for "flip-flop" of polyprenyl phosphates in artifical bilayers containing phosphatidylcholine is exceedingly long (>5 h at 25°C) (McCloskey & Troy, 1980; deRopp et al., 1987). The disparity between the observed ratio of $T_{1/2}$ > 5 hr and the rate calculated for dolichol transposition if it is involved in saccharide translocation ($T_{1/2}$ < 1 sec) led McCloskey and Troy (1980) to conclude that neither the unusual length nor poly-cis unsaturation of isoprenoid carriers affords them a uniquely rapid rate of unassisted flip-flop. On this basis, they proposed that incipient oligosaccharide chains are not transported unassisted across the bilayer while covalently bound to lipid; if they do transit the membrane, the process is probably mediated by a specific protein. There is also evidence that an early intermediate in the assembly pathway, Gn_2-PP-Dol, that is formed from dolichyl phosphate does not undergo "flip-flop" in phosphatidylcholine-containing liposomes (see below). If dolichyl phosphate and its glycosylated derivatives in the rough endoplasmic reticulum also do not undergo transverse movements, severe constraints are placed on models for the mechanism of assembly of oligosaccharylpyrophosphoryl dolichol (see below).

3. What is the topological distribution of the saccharide chains of the various intermediates involved in assembly of oligosaccharylpyrophosphoryl dolichol? Of the various questions pertaining to the overall assembly process, the most effort has been devoted to the issue of the sidedness of the saccharide chains. Initially it was established that galactosyl transferase could serve as a probe of the disaccharide chain of Gn_2-PP-Dol, because in the presence of detergent and UDP-Gal it catalyzed its quantitative conversion to Gal-Gn_2-PP-Dol (Hanover & Lennarz, 1979). Futhermore, when Gn_2-PP-Dol was incorporated into unilamellar liposomes, galactosyl transferase derivativized only approximately 50% of the disaccharide chains. This indicated that a) isolated Gn_2-PP-Dol was incorporated into the artificial bilayer in an essentially symmetric fashion and b) galactosyl transferase could act on this glycolipid in a phospholipid bilayer.

When this same probe was applied to hen oviduct rough endoplasmic reticulum that had been preincubated with labeled UDP-Gn in order to generate labeled Gn_2-PP-Dol, virtually none of the labeled disaccharyl lipid was found to be accessible to the enzyme unless detergent was added. This result suggested that the disaccharyl unit of endogenously synthesized Gn_2-PP-Dol was not exposed on the cytoplasmic face of the rough endoplasmic reticulum.

The above experiment determining the topology of labeled Gn_2-PP-Dol after it had been synthesized in a preincubation did not exclude the possibility that assembly occurred on the cytoplasmic face but a subsequent translocation process moved the Gn_2-containing molecule to the lumen where it was inaccessible to the probe. Therefore, an experiment was performed in which the assembly reaction with labeled UDP-Gn and the probing reaction with galactosyl transferase and UDP-Gal were carried out concurrently rather than sequentially (Hanover & Lennarz, 1981). Because even under these conditions no conversion was observed, it was concluded that synthesis of Gn_2-PP-Dol did not occur at the cytoplasmic face.

In an extension of the earlier mentioned studies on the topology of completed oligosaccharide chains, concanavalin A was used to probe the α-linked mannosyl units of the growing oligosaccharide chain (Snider & Rogers, 1984). Experiments with liver microsomes labeled with GDP-M and then probed with concanavalin A revealed that the short chain oligosaccharylpyrophosphoryl dolichol, containing from two to five mannosyl units, was accessible to the lectin. This indicated that these intermediates are located at the cytoplasmic face of the bilayer. The longer intermediates and the completed oligosaccharylpyrophosphoryl dolichol could not be detected with concanavalin A in intact microsomes, implying that unlike the shorter intermediates, they were not present at the cytoplasmic face of the membrane and therefore could not interact with the lectin.

A model that incorporates all of the experimental observations discussed in this section is shown in Figure 2. This model has the attractive feature of having individual sugar transfer reactions compartmentalized; such compartmentalization could provide multiple sites for control of assembly of oligosaccharide chains.

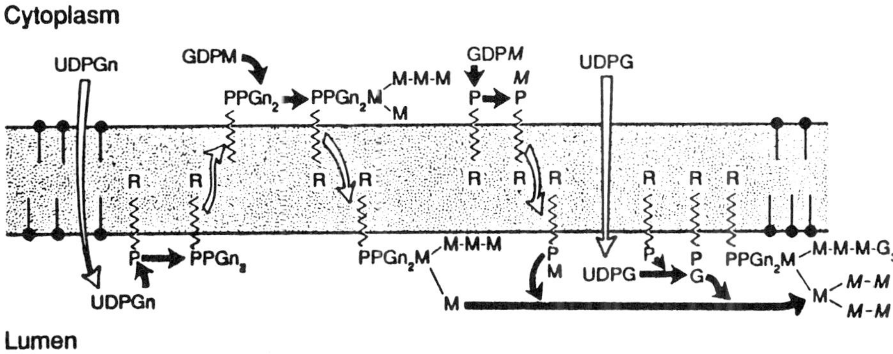

Figure 2 - "Consensus" model derived from the studies discussed in the text depicting assembly of the oligosaccharide chain of oligosaccharylpyrophosphoryl dolichol. R~~P represents dolichyl phosphate. The solid, bold arrows represent the progression of reaction sequences from left to right. The open, tapered arrows represent translocation events. The italic M indicates the mannose units that arise from M-P-Dol.

However, it must be emphasized that this represents an extreme interpretation of the data. Thus, the fact that Gn_2-PP-Dol is not accessible to an impermeable probe does not mean that it must be at the lumenal face; it merely means that it is not at the cytoplasmic face. Similarly, the finding that oligosaccharide-lipids containing chains longer than M_5Gn_2 are not accessible to concanavalin A does not mean that they must be oriented toward the lumen. Moreover, the model shown in Figure 2 has two very unattractive features: First, as noted above, although there is evidence for translocation of certain sugar nucleotides, there is no evidence for flip-flop of glycosyl-lipids. Haselbeck and Tanner (1982) inferred that flip-flop might occur based on the apparent ability of crude mannosylphosphoryl dolichol synthase incorporated into liposomes to translocate mannosyl units to entrapped GDP. However, aside from this observation there is no experimental support for such translocation. Second is the apparent energy requirement for translocation. F.A. Troy (personal communication) has estimated that the movement of one saccharide unit attached to dolichyl phosphate from an aqueous environment to the hydrophobic environment of the bilayer, with the concommitant disruption of four hydrogen bonds between the four hydroxyl groups and water, would require the expenditure of 25-50 kcal/mol or the equivalent of 2.5-5 ATP molecules per translocation. Clearly this expenditure would increase as the number of glycosyl units in the oligosaccharide chain increased; i.e., for M_5Gn_2 linked to dolichylpyrophosphate it could be as large as 130-260 kcal/mol.

Thus, if such translocations occur, either the saccharide chain is not directly exposed to the hydrophobic environment or a great deal of energy is expended in accomplishing the process. Based on all of these considerations, it seems reasonable to conclude that this "consensus" model, incorporating the results of all the published studies, may require considerable modification in the future as additional experimental information becomes available.

4. What is the topological distribution of the enzymes involved in the synthesis of oligosaccharylpyrophosphoryl dolichol? A question obviously related to the sidedness of the saccharide-lipids in the endoplasmic reticulum is the topological orientation of the enzymes that catalyze their synthesis. However, aside from oligosaccharyl transferase (cf. below), nothing is known about the structural features of the enzymes in this pathway. Consequently, approaches to defining the topological orientation of their active sites have been limited to studying the effect of proteolysis of sealed microsomes on the activity of these enzymes. This experimental approach makes the assumption that if inactivation of enzyme occurs, the active site of the enzyme faces the cytoplasm, and conversely, if no inactivation occurs it faces the lumen. However, there are several potential circumstances that could confound conclusions on the sidedness of the active site. For example, cleavage of the polypeptide chain of the enzyme at a site other than the active site could lead to inactivation, or cleavage might not occur if the enzyme is largely buried in the membrane. If either of these conditions obtain, it would be impossible to deduce the topological orientation of the active site. At best, one can conclude from the results of these early studies (Snider et al., 1980; Hanover & Lennarz, 1982) that some of the enzymes in the pathway may differ in the sidedness of their active sites. Recently, more detailed comparative studies of the properties of M-P-Dol and G-P-Dol synthases, including their sensitivity to trypsin, suggest that the former enzyme is exposed at the cytoplasmic face, whereas the latter is not (Koro et al., unpublished observations). This would be consistent with the earlier discussed difference in transport of the sugar nucleotides utilized by these two enzymes.

OLIGOSACCHARYL TRANSFERASE

Upon its completion the oligosaccharide chain of oligosaccharylpyrophosphoryl dolichol is transferred to the carboxyamido side chain of an asparagine residue in a polypeptide (cf. Figure 1). Processing of the oligosaccharide chain of the glycosylated polypeptide product of this reaction is initiated in the rough endoplasmic reticulum and continues during its transport through the Golgi complex. Based on this consideration alone, the enzyme that catalyzes this reaction is of interest, because it is situated in a central position in the overall pathway, after synthesis of the oligosaccharide chain but before its processing. This enzyme is particularly interesting for other reasons: First, in most cases the reaction it catalyzes is cotranslational. Second, the substrates are unusual, one being a molecule consisting of a polar oligosaccharide chain attached to an extremely hydrophobic lipid and the other being the side chain of an amino acid residue in a growing polypeptide chain. Third, this reaction apparently involves a carboxyamido group as a nucleophile and Dol-PP as a leaving group. Fourth, only specific Asn-X-Ser/Thr sites in a polypeptide are glycosylated.

Peptide Substrates for Oligosaccharyl Transferase. Marshall (1974) first proposed that the Asn must be part of the sequence -Asn-X-Ser- or -Asn-X-Thr-, where X was any one of the 20 amino acids. Moreover, the observation that many such tripeptide sequences in proteins are not glycosylated led to the conclusion that the primary structure of this tripeptide was necessary, but not sufficient, for glycosylation. Subsequent surveys of known protein sequences have fully confirmed these generalizations. Because of the great increase in the number of proteins that have been sequenced, an earlier survey (Struck & Lennarz, 1980) of the distribution of the 20 amino acids in the X position has been updated (Kaplan et al., 1987). Approximately 2000 -Asn-X-Ser/Thr- tripeptides were identified by examination of the protein sequence data base of the Protein Identification Resource, National Biomedical Research Foundation. About one half (1000) were found to be -Asn-X-Ser-. Of these, 4.2% (85 sites) were glycosylated. Approximately 1000 -Asn-X-Thr- were found; thus, both types of tripeptides occur with the same frequency. Of interest was the finding that -Asn-X-Thr- is almost three times more likely to be glycosylated than -Asn-X-Ser- because 11.3% (227 sites) of the former were glycosylated. Earlier possible exclusions of several amino acids from the X position (Struck & Lennarz, 1980) were not substantiated in the current survey, probably because more data were available. The results also revealed that Cys, Trp and Pro in the X position are very rare in glycosylated -Asn-X-Ser/Thr- sites. Overall, 7.5% of the -Asn-X-Ser/Thr- tripeptides in sequenced proteins are known with certainty to be glycosylated. It is important to emphasize, however, that the true percentage is much higher in membrane or secretory proteins because (a) in many other proteins that have been sequenced oligosaccharide chains are known to be present but have not yet been assigned to a specific site, and (b) it is unrealistic to include in this tally the cytoplasmic proteins, because they do not have an opportunity to interact with oligosaccharyl transferase during their synthesis and hence are unlikely to ever exist in a glycosylated form.

Further insight into the factors controlling glycosylation was initially provided by the finding that unfolded proteins (Pless & Lennarz, 1977; Kronquist & Lennarz, 1978), peptides generated from such proteins (Struck et al., 1978), or synthetic peptides (Hart et al., 1979) could serve as substrates for oligosaccharyl transferase. Several more recent studies have increased our understanding of the glycosylation process. Bause (1983), studying structural analogues of the tripeptide containing modifications in the hydroxyamino acid (Ser or Thr), demonstrated that a tripeptide containing as the third residue the epoxy analogue of Thr, epoxyethylglycine, was an irreversible inhibitor of the enzyme, probably because it became covalently linked to the protein in the vicinity of the active site. With respect to the other amino acid that obviously is crucial, studies on modification of the Asn residue by Welply _et al_. (1983) revealed that peptides containing altered Asn residues were neither substrates nor inhibitors. In addition, it was shown (Welply et al., 1983) that increased hydrophobicity of the Asn-x-Ser(Thr) tripeptide by use of hydrophobic blocking groups decreased the apparent K_m of the peptide. Finally, it was found that such acceptor tripeptides, when added to an _in vitro_ system for cotranslational glycosylation of the nascent chain of VSV G protein or ovalbumin, inhibited the glycosylation of these polypeptides, presumably by saturating the -Asn-x-Ser(Thr)- recognition site of oligosaccharyl transferase (Lau et al., 1983).

Saccharide-lipid Substrates for Oligosaccharyl Transferase. Much less information is available about the specificity of oligosaccharyl transferase for the other substrate, oligosaccharyl-PP-dolichol. The results of some earlier studies are difficult to interpret because of the possible presence of endogenous oligosaccharide-lipid that competes with exogenous substrate. Nevertheless, the overall conclusion of such studies (reviewed in Waechter & Lennarz (1976) and Hubbard & Ivatt (1981)) is that the complete oligosaccharide-lipid, containing three glucosyl units, is a more effective substrate than shorter oligosaccharides lacking these residues. However, shorter oligosaccharide-lipids, including N,N'-diacetylchitobiosyl-lipid, can be transferred to protein acceptors in oviduct (Chen & Lennarz, 1977) and yeast (Sharma et al., 1981) membrane preparations. In fact, in yeast membranes both the disaccharide-lipid, $GlcNAc_2$-PP-Dol, and the oligosaccharide-lipid, $Glc_3Man_9GlcNAc_2$-PP-Dol, exhibit the same K_m (Sharma et al., 1981). Furthermore, _in vivo_ studies (Kornfeld et al., 1979) using a mutant that lacks one of the enzymes in the assembly pathway, mannosylphosphoryldolichol synthase, indicate that transfer of a shorter oligosaccharide-lipid containing Glc_3 but lacking the Man residues that arise from Man-P-Dol does occur.

Genetic evidence concerning yeast mutants supports the idea that the glucosyl residues on oligosaccharyl-PP-Dol can enhance its efficiency as a donor (Huftaker & Robbins, 1983; Ballou et al., 1986). One such mutant (Ballou et al., 1986) known to lack the ability to synthesize glucosylphosphoryl dolichol, recently was shown to exhibit reduced efficiency of glycosylation of the major yeast glycoprotein, invertase, presumably as a result of the absence of glucose residues on the oligosaccharide chain. Thus, based on both the in vivo and in vitro studies, oligosaccharyl transferase seems to exhibit a rather broad specificity with respect to the saccharide-lipid. At present, however, little is known about how the presence of the "correct" oligosaccharide-lipid substrate affects recognition of the polypeptide substrate. Studies with oligosaccharyl transferase in yeast membranes by Sharma et al. (1981) indicate that the structure of the oligosaccharide-lipid affects the apparent K_m of the enzyme for the peptide substrate. Thus, the K_m for the peptide in the presence of the "natural" substrate, $Glc_3Man_9GlcNAc_2$-PP-Dol, is 10-fold lower than the K_m observed when the "unnatural" substrate, $GlcNAc_2$-PP-Dol, is added.

Other Considerations of Specificity. As is often the case, extrapolation from in vitro findings using a simplified system results in conclusions not supported by the results of in vivo studies. Presumably this apparent conflict arises because in vivo factors not present in the simplified in vitro system come into play. This clearly is the case with respect to peptide site recognition, because there are many examples of potential glycosylation sites in polypeptides that are not glycosylated in vivo. Studies with native and denatured proteins containing potential acceptor sites suggested that accessibility of the glycosylation site can determine whether or not that site is glycosylated (Pless & Lennarz, 1977; Kronquist & Lennarz, 1978). However, findings with ovalbumin suggested that other factors may also be involved. Ovalbumin isolated from egg white contains a carbohydrate chain at one Asn site but is not glycosylated at a second site 19 residues away (Catterall et al., 1978). This protein is only glycosylated in vitro at this second site when the molecule is first subjected to reduction and alkylation (Pless & Lennarz, 1977). Moreover, a tryptic peptide from ovalbumin containing both the glycosylated first site and the unglycosylated second site is glycosylated in vitro (Glabe et al., 1980). The simplest interpretation of these findings is that the second site is not glycosylated in vivo during translation because the conformation of the peptide in the vicinity of the second site prevents it from interacting with oligosaccharyl transferase. Consistent with this idea is the observation that even when ovalbumin is synthesized _in vivo_ in a heterologous system, e.g., in mouse cells infected with herpes simplex virus carrying a plasmid containing the ovalbumin gene, it is still only glycosylated at the first site (Sheares & Robbins, 1986). However, the idea that the conformation is the sole determining factor does not explain the finding that both the first and the second site are glycosylated when ovalbumin is synthesized in an _in vitro_ translation/glycosylation system (Palmiter et al., 1980). One explanation for this puzzling observation (which has been confirmed in our laboratory) is that translation _in vitro_ occurs at a much lower rate than _in vivo_. The _in vitro_ system may, therefore, offer a much greater opportunity than the _in vivo_ translation system for oligosaccharyl transferase to interact with the second glycosylation site in the growing polypeptide chain.

Another example of the intervention of other factors is the relative rate of glycosylation of Asn residues in -Asn-X-Ser- versus -Asn-X-Thr- glycosylation sites. A comparative study by Bause (1984) in which these two types of peptides served as acceptors _in vitro_, indicated that the Thr-containing peptide is glycosylated 40-fold more rapidly than the Ser-containing peptide. _In vivo_, however, other factors must play an important role because the frequency of glycoproteins containing glycosylated -Asn-X-Thr- is 3-fold rather than 40-fold greater than those containing glycosylated -Asn-X-Ser-.

Yet another example of additional factors comes from studies on the conversion of RNase A (unglycosylated) to RNase B (glycosylated) (Carson et al., 1981). _In vitro_, denatured but not native RNase A can be converted to RNase B (Pless & Lennarz, 1977). However, in pancreatic tissue slices, in which only a small fraction of the newly synthesized RNase (15%) is glycosylated under normal experimental conditions, addition of exogenous dolichyl phosphate results in glycosylation of more than 90% of the protein (Carson et al., 1981). In this case, the controlling factor seems to be the availability of oligosaccharyl-PP-Dol, which presumably is synthesized as a result of the availability of dolichyl phosphate.

Finally, the specificity of oligosaccharyl transferase raises the question of post-translational glycosylation. Studies by Pless & Lennarz (1977) and Kronquist and Lennarz (1978) using reduced and alkylated proteins containing -Asn-X-Ser/Thr- sites revealed that oligosaccharyl transferase can recognize and catalyze the glycosylation of completed polypeptide chains. This finding suggested that translation and passage through the membrane (i.e., translocation) are not inherent mechanistic prerequisites for glycosylation. Although a number of studies (cited in Bergman & Kuehl, 1978) provided limited evidence for the post-translational glycosylation of proteins, the clearest demonstration that this process can occur was provided by in vitro (Tucker & Pestka, 1977) and in vivo (Bergman & Kuehl, 1978) studies on the synthesis of an immunoglobulin kappa light chain in the MOPC-46B cell system. Subsequent studies of the insulin receptor and the effect of tunicamycin on its synthesis, glycosylation and receptor activity provided indirect evidence that, after removal of tunicamycin, this protein could undergo post-translational glycosylation (Ronnett & Lane, 1981). Definitive evidence that translocation and glycosylation can occur post-translationally was obtained recently in an in vitro yeast translation/translocation/ glycosylation system programmed to synthesize prepro-α-factor (Rothblatt & Meyer, 1986; Hansen et al., 1986; Waters & Blobel, 1986).

Thus, a variety of studies have shown that in most instances glycosylation occurs cotranslationally but may occur post-translationally in certain cases. This raises the interesting question of protein folding and accessibility of the glycosylation site. In cases in which the protein can be post-translationally glycosylated, perhaps the glycosylation site is located in an exposed region of the completed, folded protein or the rate of folding is very slow. Under these circumstances, it is easy to understand how such proteins, while they remain in the lumen of the rough endoplasmic reticulum, could interact with oligosaccharyl transferase if the active site of this enzyme is exposed at the membrane's lumenal face (cf. below).

Topological Arrangement of Oligosaccharyl Transferase. As discussed above, all the available evidence indicates that oligosaccharide transfer to the growing nascent polypeptide chain occurs at the lumenal rather than the cytoplasmic face. Based on these findings and the observation that oligosaccharyl transferase seems to be tightly associated with the membrane of the rough endoplasmic reticulum (see below), the model shown in Figure 3 may be considered. This model contains several components that may participate in mediating association of the growing polypeptide chain with the rough endoplasmic reticulum and its processing. The mechanism of translocation of the polypeptide through the membrane remains unclear, and other than limited evidence suggesting that the route through the membrane is aqueous (Gilmore & Blobel, 1985) nothing is known. Similarly, the possible interaction of oligosaccharyl transferase with potential pore components involved in translocation of nascent chains is not known. As noted earlier, what is clear is that small acceptor peptides do freely permeate the membrane and are accessible to the peptide-binding site of oligosaccharyl transferase, since they inhibit cotranslational glycosylation (Lau et al., 1983). One interesting possibility is that oligosaccharyl transferase serves as a component of the nascent chain translocation system to ensure that all polypeptides moving through the membrane are accessible to the glycosylating enzyme. This might occur with the enzyme complexed to the signal recognition particle, signal recognition particle receptor, signal peptidase or ribophorin. Because signal recognition particle and its receptor are present in levels substoichiometric to that of membrane-bound ribosomes (cf. Evans et al., 1986) more reasonable candidates for formation of such a complex with oligosaccharyl transferase might be the recently purified signal peptidase complex (Evans et al., 1986) or the ribophorins (Kreibich et al., 1978).

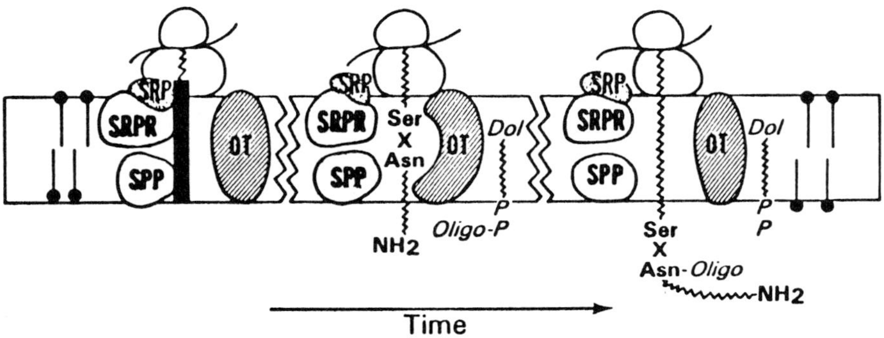

Figure 3 - Model of oligosaccharyltransferase in the membrane of the rough endoplasmic reticulum. The solid bar shown on the left side of the model represents the signal peptide. SRP, signal recognition particle; SRPR, signal recognition particle receptor; SPP, signal peptide peptidase; OT, oligosaccharyl transferase.

Properties of oligosaccharyl transferase. Because oligosaccharyl transferase has proved to be an extremely labile enzyme, a novel approach was taken to identify the protein. Studies with a variety of -Asn-X-Thr- tripeptide derivatives established the structural modifications that would permit the tripeptide to retain recognition by oligosaccharyl transferase (Welply et al., 1983). A photoaffinity reagent, N^{α}-[^{3}H]Ac-Asn-Lys(N^{ε}-p-azidobenzoyl)-Thr-NH_2, was developed to label the active site of oligosaccharyl transferase from hen oviduct rough endoplasmic reticulum (Welply et al., 1985). This probe revealed that the apparent molecular mass of ^{3}H-labeled oligosaccharyl transferase polypeptide was 60 kDa when it was analyzed by sodium dodecyl sulfate-polyacrylamide gel electrophoresis.

Some of the physical and topological properties of soluble and membrane-associated oligosaccharyl transferase were examined using a new active-site directed reagent, ^{125}I-labeled N^{α}-3-(4-hydroxyphenylpropionyl)-Asn-Lys(N^{ε}-p-azidobenzoyl)-Thr-NH_2 (Welply et al., 1986). Trypsin was used as a probe to study the topological orientation of oligosaccharyl transferase in the membrane. Oligosaccharyl transferase in sealed microsomes was photoaffinity-labeled either before or after trypsin treatment; the effect of such trypsin treatment on both the enzymatic activity of oligosaccharyl transferase and the amount of membrane-associated, ^{125}I-labeled enzyme was then determined. Trypsin treatment, regardless of whether it was done before or after photoaffinity labeling, caused a time-dependent fall of enzymatic activity to 30-40% of initial values after 30 min of protease treatment, as assessed by measuring glycopeptide formation from a ^{3}H-labeled peptide acceptor. However, the loss of catalytic activity was not accompanied by either a reduction in the amount of membrane-associated ^{125}I-labeled oligosaccharyl transferase or a detectable change in the molecular weight of the enzyme, as determined by sodium dodecyl sulfate-polyacrylamide gel electrophoresis. Indeed, degradation of the ^{125}I-labeled polypeptide by trypsin was observed only when detergent was present to disrupt membrane integrity. One possible explanation for this result is that oligosaccharyl transferase contains a second subunit required for catalysis that is partially exposed on the cytoplasmic side of the membrane, thus rendering it trypsin-sensitive. A second possibility is that cleavage of only a very small portion of the polypeptide, undetectable by gel electrophoresis, occurred at the cytoplasmic face and this cleavage affected enzymatic activity. A third possibility is that, although trypsinization did not proteolyze the enzyme, it affected the membrane sufficiently to cause enzyme inactivation. Although this study could not distinguish between these possibilities, the results clearly indicated that the oligosaccharyl transferase domain to which the photoreactive tripeptide acceptor covalently bound was either lumenally oriented or buried within the lipid bilayer rather than exposed on the cytoplasmic face.

Phylogenetic Distribution of Oligosaccharyl Transferase. N-Linked glycoproteins are widespread in nature, being found in single-cell eukaryotes, such as yeast, and in infinitely more complex multicellular eukaryotes, including higher plants and animals. In addition, although N-linked glycoproteins are absent from most prokaryotes, they are found in halobacteria (Wieland et al., 1983). Thus, the ability to modify proteins by addition of N-linked oligosaccharide chains seems to be of major consequence to the survival of a very broad spectrum of organisms. Studies of diverse biological systems have confirmed the universal features of the substrates for the N-glycosylation reaction; the donor substrate consi ts of lipid-linked oligosaccharide and the acceptor substrate consists of an asparagine residue within a defined sequence of a polypeptide chain. Based on this information, it seems reasonable to assume that the N-glycosylation reaction is catalyzed by an enzyme that has not undergone radical changes in primary structure during evolution. To test this hypothesis the active site-directed photoreactive probe for oligosaccharyl transferase was used to identify the enzyme in a number of diverse biological organisms and tissues (Kaplan et al., 1987). As shown in Figure 4, except for yeast, a 60 kDa photolabeled oligosaccharyl transferase was detected in tissues derived from a number of species known to synthesize N-linked glycoproteins. Using a competing peptide acceptor (Welply et al., 1983; Welply et al., 1985) it was established that all radiolabeled bands associated with the samples of radish roots and Escherichia coli (Figure 4, panels g and h) were the result of nonspecific interaction with the ^{125}I-labeled photoreactive probe. This finding was not unexpected because these two organisms are not known to synthesize N-linked glycoproteins.

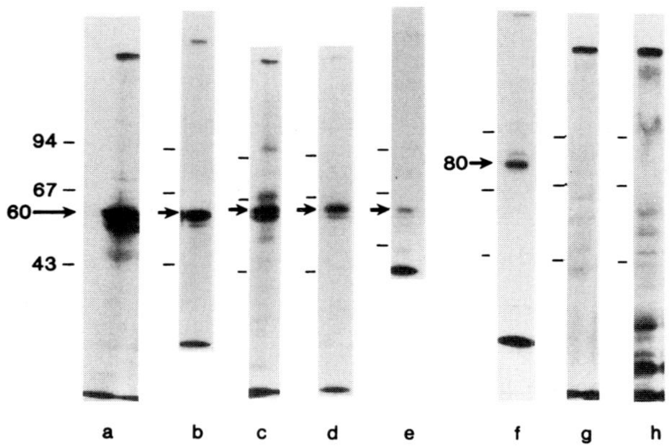

Figure 4. Photoaffinity labeling of oligosaccharyl transferase in different tissues or organisms. The apparent molecular weight of the enzyme was determined by gel electrophoresis in membranes prepared from: a, mouse liver; b, rat hepatoma cell line (FAZA); c, rabbit cervix; d, dog pancreas; e, Drosophila melanogaster; f, yeast (Saccharomyces cerevisiae); g, radish roots; h, Escherichia coli. The gels were run at different times under different conditions. Molecular mass markers and oligosaccharyl transferase at 60 kDa (80 kDa in yeast) are shown.

The finding of an apparent molecular mass of 80 kDa for yeast oligosaccharyl transferase is of interest in view of the earlier observation that the enzyme from this source retains catalytic activity after detergent solubilization (Sharma et al.,

1981). Because of the lability of solubilized oligosaccharyl transferase observed in other biological systems, one interesting question related to the possible evolutionary divergence of the enzyme is: Does the apparently more stable yeast oligosaccharyl transferase consist of a single polypeptide, whereas the more labile enzyme found in higher organisms consist of multiple subunits that dissociate upon solubilization? Clearly, a definitive understanding of possible differences between oligosaccharyl transferase from yeast and higher organisms awaits purification and subsequent reconstitution of active enzyme from these sources.

PROSPECTUS

The uncertainties and equivocations in the answers to topological questions about the assembly process in the endoplasmic reticulum reflect the difficulties in dealing with the structural organization of lipids and proteins in membranes. Clearly, "solid state" biochemistry has a number of inherent limitations that are not present when one is dealing with enzymes and substrates that are soluble in an aqueous environment. Before a clear picture of the topological features of the glycosylation process emerges it will be necessary to known more about three aspects: 1) The physical organization of dolichyl phosphate in the membrane: at present the best way to study this appears to be by use of physical chemical techniques in model membrane systems, but ultimately one will have to devise ways to study this in biological membranes. 2) The sidedness of the saccharide-lipid intermediates: new impermeable probes must be developed that exhibit high specificity for the molecule of interest. 3) The enzymes in the pathway: these proteins must be isolated and sequenced, using either state-of-the-art biochemical techniques, the methods of recombinant DNA, or molecular genetics. Such sequence information, coupled with proteolysis and inhibitor studies on these enzymes in the endoplasmic reticulum, should allow one to deduce their arrangement in the bilayer. Only by combining the information obtained by all three approaches will we be able to understand fully the topological features of this complex assembly process.

ACKNOWLEDGMENTS

This work was supported by NIH grants (GM 33184 and GM 33185) to WJL. Dr. Lennarz, a Robert A. Welch Professor of Chemistry, gratefully acknowledges the Robert A. Welch Foundation.

REFERENCES

Anumula KR, Spiro RG (1983) Release of glucose-containing polymannose oligosaccharides during glycoprotein biosynthesis. J Biol Chem 258:15274-15282

Ballou L, Gopal P, Krummel B, Tammi M, Ballou C (1986) A mutation that prevents glycosylation of the lipid-linked oligosaccharide precursor leads to undergly cosylation of secreted yeast invertase. Proc Natl Acad Sci USA 83: 3081-3085

Bause E (1983) Active-site-directed inhibition of asparagine N-glycosyltransferases with epoxy-peptide derivatives. Biochem J 209:323-330

Bause E (1984) Model studies on N-glycosylation of proteins. Biochem Soc Trans 12:514-517

Bergman LW, Kuehl NM (1977) Addition of glucosamine and mannose to nascent immunoglobulin heavy chains. Biochemistry 16:4490-4497

Bergman LW, Kuehl WM (1978) Temporal relationship of translation and glycosylation of immunoglobulin heavy and light chains. Biochem 17:5174-5180

Blobel G, Sabatini DD (1971) Ribosome-membrane interaction in eukaryotic cells. In: Mason, LA (ed) Biomembranes, Plenum Publishing Co., New York, p. 193

Carson DD, Earles BJ, Lennarz WJ (1981) Enhancement of protein glycosylation in tissue slices by dolichylphosphate. J Biol Chem 256:11552-11557

Catterall JB, O'Malley BW, Robertson MA, Staden R, Tanaka Y, Brownlee GG (1978) Nucleotide sequence homology at 12 intron-exon junctions in the chick ovalbumin gene. Nature (London) 275: 510-513

Chen WW, Lennarz WJ (1977) Metabolism of lipid-linked N-acetylglucosamine intermediates. J Biol Chem 252:3473-3479

Coates SW, Gurney T, Sommers LW, Yeh M, Hirschberg CB (1980) Subcellular localization of sugar nucleotide synthetases. J Biol Chem 255:9225-9229

deRopp JS, Troy FA (1984) Chemical synthesis and ^{2}H NMR investigations of polyisoprenols: Dynamics in model membranes. Biochemistry 23:2691-2695

deRopp JS, Troy FA (1985) ^{2}H NMR investigation of the organization and dynamics of polyisoprenols in membranes. J Biol Chem 260:15669-15674

deRopp JS, Knudsen MJ, Troy FA (1987) ^{2}H NMR investigation of the dynamics and conformation of polyisoprenols in model membranes. Chemica Scripta 27:101-108

Evans EA, Gilmore R, Blobel G (1986) Purification of microsomal signal peptidase as a complex. Proc Natl Acad Sci USA 83: 581-585

Farquhar MG (1985) Progress in unraveling pathways of Golgi traffic. Annu Rev Cell Biol 1:447-488

Gilmore R, Blobel G (1985) Translocation of secretory proteins across the microsomal membrane occurs through an environment accessible to aqueous perturbants. Cell 42:497-505

Glabe CG, Hannover JA, Lennarz WJ (1980) Glycosylation of ovalbumin nascent chains. The spatial relationship between translation and glycosylation. J Biol Chem 255:9236-9242

Hanover JA, Lennarz WJ (1979) The topoloical orientation of N,N^{1}-diacetylchito biosylpyrophosphoryldolichol in artificial and natural membranes. J Biol Chem 254:9237-9246

Hanover JA, Lennarz WJ (1980) N-Linked glycoprotein assembly: evidence that oligosaccharide attachment occurs within the lumen of the endoplasmic reticulum. J Biol Chem 255:3600-3604

Hanover JA, Lennarz WJ (1981) Transmembrane assembly of membrane and secretory glycoproteins. Arch Biochem Biophys 211:1-19

Hanover JA, Lennarz WJ (1982) Transmembrane assembly of N-linked glycopro teins: Studies on the topology of saccharide-lipid synthesis. J Biol Chem 257:2787-2794

Hansen W, Garcia DP, Walter, P (1986) In vitro protein translocation across the yeast endoplasmic reticulum: ATP-dependent post-translational

translocation of the prepro-α-factor. Cell 45:397-406
Hart GW, Brew K, Grant GA, Bradshaw RA, Lennarz WJ (1979) Primary structural requirements for the enzymatic formation of the N-glycosidic bond in glycoproteins. J Biol Chem 254:9747-9753
Haselbeck A, Tanner W (1982) Dolichol phosphate-mediated mannosyl transfer through liposomal membranes. Proc Natl Acad Sci USA 79:1520-1524
Hubbard SC, Ivatt RJ (1981) Synthesis and processing of asparagine-linked oligosaccharides. Annu Rev Biochem 50:555-583
Huftaker TC, Robbins PW (1983) Yeast mutants deficient in protein glycosylation. Proc Natl Acad Sci USA 80:7466-7470
Kaplan HA, Welply JK, Lennarz WJ (1987) Oligosaccharyl transferase: the central enzyme in the pathway of glycoprotein assembly. BBA - Reviews on Biomembranes, 906:161-173
Katz FN, Rothman JE, Lingappa UR, Blobel G, Lodish HR (1977) Membrane assembly in vitro: synthesis, glycosylation, and asymmetric insertion of a transmembrane protein. Proc Natl Acad Sci USA 74:3278-3282
Kiely ML, McKnight GS, Schimke RT (1976) Studies on the attachment of carbohydrate to ovalbumin nascent chains in hen oviduct. J Biol Chem 251:5490-5495
Kornfeld R, Kornfeld S (1985) Assembly of asparagine-linked oligosaccharides. Annu Rev Biochem 54:631-664
Kornfeld S, Gregory W, Chapman A (1979) Class E thy-1 negative mouse lymphoma cells utilize an alternate pathway of oligosaccharide processing to synthesize complex-type oligosaccharides. J Biol Chem 254:11649-11654
Kreibich G, Ulrich BC, Sabatini DD (1978) Proteins of rough microsomal membranes related to ribosome binding. I. Identification of ribophorins I and II membrane proteins characteristic of rough microsomes. J Cell Biol 77:464-487
Kronquist KE, Lennarz WJ (1978) Enzymatic conversion of proteins to glycoproteins by lipid-linked saccharides. A study of potential exogenous acceptor proteins. J Supramol Struct, 8:51-65
Lau JTY, Welply JK, Shenbagamurthi P, Naider F, Lennarz WJ (1983) Substrate recognition by oligosaccharyl transferase: inhibition of co-translational glycosylation by acceptor peptides. J Biol Chem 258:15255-15260.
Lingappa VR, Lingappa JR, Prasad R, Ebner KE, Blobel G (1978) Coupled cell-free synthesis, segregation, and core glycosylation of a secretory protein. Proc Natl Acad Sci USA 75:2338-2341
Marshall RD (1974) The nature and metabolism of the carbohydrate-peptide linkage of glycoproteins. Biochem Soc Symp 40:17-26
McCloskey MA, Troy FA (1980) Paramagnetic isoprenoid carrier lipids: Dispersion and dynamics in lipid membranes. Biochemistry 19:2061-2066
Milstein C, Brownlee G, Harrison T, Matthews MD (1972) A possible precursor of immunoglobulin light chains. Nature New Biol 239:117-120
Palmiter RD, Thibodeau SN, Rogers G, Boime I (1980) Cotranslational sequestration of egg white proteins and placental lactogen inside membrane vesicles. Anal NY Acad Sci 343:192-209
Parodi AJ, Leloir LF (1979) The role of lipid intermediates in the glycosylation of proteins in the eucaryotic cell. Biochim Biophys Acta 559:1-37
Perez M, Hirschberg CB (1985) Translocation of UDP-N-acetylglucosamine into vesicles derived from rat liver rough endoplasmic reticulum and Golgi apparatus. J Biol Chem 260:4671-4678
Perez M, Hirschberg CB (1986a) Transport of sugar nucleotides and adenosine 3'-phosphate 5'-phosphosulfate into vesicles derived from the Golgi apparatus. Biochim Biophys Acta 864:213-222
Perez M, Hirschberg CB (1986b) Topography of glycosylation reactions in the rough endoplasmic reticulum membrane. J Biol Chem 261:6822-6830
Pless DD, Lennarz WJ (1977) Enzymatic conversion of proteins to

glycoproteins. Proc Natl Acad Sci USA 74:134-138
Redman CM (1969) Biosynthesis of serum proteins and ferritin by free and attached ribosomes of rat liver. J Biol Chem 244:4308-4315
Ronnett GV, Lane MD (1981) Post-translational glycosylation-induced activation of aglycoinsulin receptor accumulated during tunicamycin treatment. J Biol Chem 256:4704-4707
Rothblatt JA, Meyer DI (1986) Secretion in yeast: translocation and glycosylation of prepro-α-factor in vitro can occur via an ATP-dependent post-translational mechanism. EMBO Journal 5:1031-1036
Rothman JE, Lodish HF (1975) Synchronised transmembrane insertion and glycosylation of a nascent membrane protein. Nature (London) 269:775-780
Sharma CB, Lehle L, Tanner W (1981) N-Glycosylation of yeast proteins. Eur J Biochem 116:101-108
Sheares BT, Robbins PW (1986) Glycosylation of ovalbumin in a heterologous cell: Analysis of oligosaccharide chains of the cloned glycoprotein in mouse L cells. Proc Natl Acad Sci USA 83:1993-1997
Snider MD, Robbins PW (1982) Transmembrane organization of protein glycosylation. Membrane oligosaccharide-lipid is coated on the luminal side of microsomes from Chinese hamster ovary cells. J Biol Chem 257:6796-6801
Snider MD, Rogers OC (1984) Transmembrane movement of oligosaccharide-lipids during glycoprotein synthesis. Cell 36:753-761
Snider MD, Sultzman LA, Robbins PW (1980) Transmembrane location of oligosaccharide-lipid synthesis in microsomal vesicles. Cell 21:385-392
Struck DK, Lennarz WJ (1980) The role of saccharide lipids in glycoprotein synthesis. In: Lennarz WJ (ed) The Biochemistry of Glycoproteins and Proteoglycans. Plenum Publishing Co., New York, p 35
Struck DK, Lennarz WJ, Brew K (1978) Primary structural requirements for the enzymatic formation of the N-glycosidic bond in glycoproteins. Studies with α-lactalbumin. J Biol Chem 253:5786-5794
Tucker P, Pestka S (1977) De novo synthesis and glycosylation of the MOPC-46B mouse immunoglobulin light chains in cell-free extracts. J Biol Chem 252:4474-4486
Valtersson C, van Duyn G, Verkleij AJ, Chojnacki T, deKruijff B, Dallner G (1985) The influence of dolichol, dolichol esters, and dolichyl phosphate on phospholipid polymorphism and fluidity in model membranes. J Biol Chem 260:2742-2751
Waechter CJ, Lennarz WJ (1976) The role of polyprenol-linked sugars in glycoprotein synthesis. Annu Rev Biochem 45: 95-112
Waters MG, Blobel G (1986) Secretory protein translocation in a yeast cell free system can occur posttranslationally and requires ATP hydrolysis. J Cell Biol 102: 1543-1550
Welply JK, Shenbagamurthi P, Lennarz WJ, Naider F (1983) Substrate recognition by oligosaccharyl transferase: studies on glycosylation of modified Asn-X-Thr/Ser tripeptides. J Biol Chem 258:11856-11863
Welply JK, Shenbagamurthi P, Naider F, Park HR, Lennarz WJ (1985) Active site-directed photoaffinity labeling and partial characterization of oligosaccharyltransferase. J Biol Chem 260:6459-6465
Welply JK, Kaplan HA, Shenbagamurthi P, Naider F, Lennarz WJ (1986) Studies on properties of membrane-associated oligosaccharyltransferase using an active site-directed photoaffinity probe. Arch Biochem Biophys 246: 808-819
Wieland F, Heitzer R, Schaefer W (1983) Asparaginylglucose: Novel type of carbohydrate linkage. Proc Natl Acad Sci USA 80:5470-5474

Sequence Determinants of Protein Sorting Into and Across Membranes

Gunnar von Heijne
Research Group for Theoretical Biophysics
Department of Theoretical Physics
Royal Institute of Technology
S-100 44 Stockholm
Sweden

NATO ASI Series, Vol. H 16
Membrane Biogenesis. Edited by J. A. F. Op den Kamp

Introduction

My intentions with this contribution are twofold: to describe briefly some methods of computer-aided sequence analysis in molecular biology, and to review the most important results obtained through such theoretical studies in the related fields of protein sorting and membrane protein biogenesis.

The methodological background is simple enough: since the arrival of rapid DNA sequencing techniques in the mid 1970s, an overwhelming wealth of sequence data is now available in computerized data banks. Our abilities to analyze and extract meaningful information from these banks have lagged far behind, however, and very little work in this area, beyond the development of homology-search and alignment algorithms, has met with general recognition in the basically lab-oriented world of molecular biology. For a general overview of the field of theoretical sequence analysis, see von Heijne (1987).

Proteins, being in general more structurally complex than polynucleotides, have proved particularly difficult to analyze theoretically. However, protein sorting has emerged as an area where the "molecular language" is particularly simple since the sorting "signals" can be recognized and studied already on the primary sequence level; probably, these signals interact weakly if at all with the "passenger" proteins, and fold to a large degree independently of these. Without the complication of non-local folding interactions, relatively unsophisticated methods of sequence analysis are sufficient for defining the main determinants or sequence patterns involved in many of the sorting processes.

From the biological point of view, correct and efficient protein sorting is one of the major demands put on all living cells, prokaryotic as well as eukaryotic. With the exception of a handful of mitochondrial and chloroplast proteins, Nature's solution to this problem has been to make all proteins in one subcellular compartment, the cytoplasm, and to provide those proteins destined for other compartments with specific sorting signals that can be recognized by "transporter" proteins in the membrane(s) surrounding the target organelle. Thus, transiently attached N–terminal "signal peptides" serve as the initial address labels for the secretory pathway; N–terminal "targeting sequences" have been found on proteins destined for mitochondria and chloroplasts; and internal "nuclear localization signals" apparently direct nuclear proteins through the nuclear membrane. The basic design principles of such signals, the "least common denominators", as they appear from computer analysis of large samples of known signals, will be discussed at length below.

Many integral membrane proteins are also provided with normal sorting signal such as secretory signal peptides, but are only partially translocated across the target membrane. "Stop transfer" or "anchor" sequences downstream of the sorting signal somehow block transmembrane transport, thus leaving the polypeptide chain spanning the membrane. In the case of "polytopic" integral membrane proteins, *i.e.* proteins that weave back and forth across the membrane many times, patterns detected through sequence analysis seem to suggest that the amino acid composition not only of the spanning segments proper, but also of the loops connecting the transmembrane segments, are important determinants of the final topology.

Some Methods of Theoretical Sequence Analysis

Data Banks and Downloading of Information

Obviously, easy access to sequence data is the first requirement of theoretical sequence analysis. As is well known, the EMBL and GenBank DNA sequence data banks are the main sources of DNA and RNA sequences. Being essentially identical in content, they now store more than 10^7 bp of sequence information, in addition to bibliographical data and notes on coding/noncoding parts of the sequences, known mutations, putative open reading frames, etc.

Protein sequences can be obtained from the EMBL and GenBank entries by automatic translation of the relevant region(s). This is a rather cumbersome process, and it is often easier to extract the sequences from a protein sequence data bank, such as the one maintained by the National Biomedical Research Foundation. This has the added advantage that any data obtained by protein rather than DNA sequencing is automatically included, which may be important when the precise N-terminal residue need to be known, or when questions of post-translational processing in general are studied. Many entries in the protein sequence data banks are only known from translation of the corresponding DNA sequence, however.

Most users of these data banks will not access them directly, but will only experience them "embedded" in a larger software package. This makes various kinds of searching easy, and also allows analyses such as hydrophobicity plots or secondary structure predictions to be made on-line.

Those of us who are interested in developing our own sequence analysis software often find it convenient to move, or "download", the sequences we want from the main data bank to our own computer, where we can work with them more easily. Microcomputers are in fact ideal for much of this work, as a lot of the programming done today does not require the power and speed of a main-frame computer.

Hydrophobicity Analysis

Hydrophobicity analysis is central to all theoretical work on membrane proteins and protein targeting. It is a commonplace observation in undergraduate chemistry that "like dissolves like", i.e. that molecules tend to prefer neighbors of their own kind to strangers with dissimilar chemical characteristics. In aqueous solution, apolar molecules or parts of molecules tend to aggregate such that the total apolar surface area exposed to water is minimized. Likewise, the structure of a globular protein is to a large measure dictated by the requirement that its polar and apolar residues should be respectively exposed on its surface and buried in its interior as far as possible.

The intuitive concept of a "hydrophobic bond" between apolar molecules, or a "hydrophobic effect" that accounts for such diverse chemical phenomena as the stability of lipid bilayers and the folding of large proteins, lies behind much of our present understanding of the physical chemistry of the living cell; yet, hydrophobicity has proven very difficult to quantify unambiguously. Clearly, it must in the final analysis be related to quantum mechanical properties of both solvent and solute (e.g. van der Waals interactions, hydrogen bonds, electrostatic energies) as well as to entropic effects stemming mainly from the relatively well-ordered water layer that forms around any apolar surface - all difficult problems in their own right that are far from completely understood.

Fortunately, chemists and, perhaps even more, molecular biologists are pragmatic people. There are scores of hydrophobicity scales in the literature, scales that have either been optimized with some given application in mind, or that have been derived from various experimental or theoretical approaches. A recent compilation lists no fewer than 46 scales (Cornette *et al.,* 1987). Basically, three different ways of constructing hydrophobicity scales have been tried: experimentally measured partition equilibria of amino acids between polar and non-polar solvents or between solvent and vapor; statistically derived tendencies for a given kind of residue to be found in buried or exposed locations in known protein structures; and theoretical optimizations to produce scales that maximize one or other discriminatory parameter (*e.g.* the hydrophobic moment of known amphiphilic helices). Given the rather unsophisticated methods available for hydrophobicity analysis, the choice of scale is of minor importance in most instances.

Hydrophobicity Plots & Hydrophobic Moment Calculations

A hydrophobicity plot is simply a profile of the local hydrophobicity *vs.* position in the protein sequence. To reduce noise, some kind of averaging is necessary. The size of the averaging window should be chosen according to the problem at hand; thus, if one is searching for potential membrane spanning segments in a membrane protein a window of around 20 residues (the typical length of a transmembrane α–helix) is appropriate, whereas a shorter window (5–7 residues) is used when looking for likely surface-exposed (often highly charged or polar) regions. A hydrophobicity plot of the MalF protein from the inner membrane of *E.coli* is shown in Fig.1.

In this example, all 8 membrane spanning segments known to be present in the protein are easily identified. This is not always the case, however, and a conservative rule is to make a definite prediction only when the mean hydrophobicity measured on the so–called Engelman-Steitz scale with a 19–residues window is either >1.4 (membrane spanning segment) or <0.9 (exposed segment) (von Heijne, 1986a).

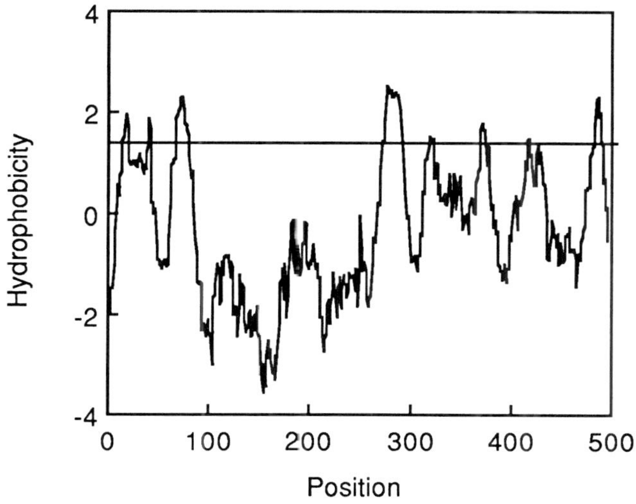

Fig.1. Hydrophobicity plot of MalF. Window size=19.

Hydrophobicity analysis can be extended to search not only for unbroken stretches of hydrophobic or hydrophilic residues but also to detect periodicities in the distribution of such residues. Any periodic secondary structure (i.e. an α-helix or a ß-sheet) can be characterized by the number of residues per turn m, or alternatively by the angle δ at which the side chains of successive residues project from the backbone: $\delta = 2\pi/m$. Thus, for an α-helix $\delta=100^o$, a 3_{10} helix has $\delta=120^o$, and a ß structure is characterized by δ=160-180

For any value of δ one can define a hydrophobic moment vector (Eisenberg et al., 1984):

$$\mathbf{m} = \Sigma h_i \mathbf{s_i}$$

where h_i is the hydrophobicity of residue i and $\mathbf{s_i}$ is a unit vector pointing from the helical axis towards the α-carbon of residue i. The sum is taken over a given window (typically 11 or 18 residues long) that is moved successively along the sequence.

In most applications only the absolute value of **m** is of interest; this is easily computed as:

$$m = \{[\Sigma h_i \sin(\delta i)]^2+[\Sigma h_i \cos(\delta i)]^2\}^{1/2}$$

A large value of m for a given δ indicates that polar and apolar residues tend to segregate to opposite sides of the periodic structure in the segment under consideration. Thus, plots of m versus the position of the window, or even contour-plots of m=f(δ) versus sequence position (Finer-Moore & Stroud, 1984), can be used to detect elements of "amphiphilic" secondary structure.

It is also possible to analyze each helical face separately. Vogel & Jähnig (1986) calculate the mean hydrophobicity of one side of a 20-residue α-helix with residue i in the middle from the formula:

$$Ha(i) = 1/10[h(i\pm8)+h(i\pm7)+h(i\pm5)+h(i\pm4)+3/4h(i\pm3)+1/2h(i\pm1)+h(i)]$$

where h(i) is the hydrophobicity of residue i. In an amphiphilic helix Ha will fluctuate wildly between neighboring positions, as one goes from the apolar to the polar and back to the apolar face. The advantage of this kind of analysis is that the hydrophobicity of the apolar face as well as the hydrophilicity of the polar face will be immediately apparent from the plot of Ha versus sequence position; in a plot of m, only the difference between the two sides is shown. On the other hand, the Vogel-Jähnig plot easily becomes rather 'cluttered', and is not amenable to smoothing procedures.

Weight-matrix predictions

The weight-matrix approach is often used for locating putative regulatory sites in DNA sequences. The idea is to use the incidence of the four bases in each position in a set of aligned sequences of known function (ribosome binding sites, say) as a template, which is then moved along the sequence to be tested. The degree of match between the template and the test sequence is calculated for each position, and displayed graphically. When the template is based on a relatively well–conserved consensus sequence, this method is quite reliable. Below, an application of this technique to the problem of detecting secretory signal peptides and predicting the site of cleavage between the signal peptide and the mature protein will be described.

Protein Sorting Signals

Secretory Signal Peptides

Proteins destined for secretion, in prokaryotes as well as in eukaryotes, are made initially with a short amino-terminal extension some 15–25 residues long. The presence of such a signal (or leader) peptide is necessary (but not always sufficient) for secretion to occur. In eukaryotic cells, the emergence of a nascent signal peptide from the ribosome somehow induces the binding of a ribonucleoprotein particle, the signal recognition particle or SRP, to the ribosome. Upon SRP binding, translation of the nascent chain slows down or stops. When the ribosome eventually finds an export site on the rough endoplasmic reticulum (ER) membrane, the membrane bound SRP receptor (or Docking Protein) catalyzes the liberation of SRP and lifts the translational block, whereupon secretion into the lumen of the ER takes place (Rapoport,1986). The biochemistry of the process is less well characterized in prokaryotes, but a number of genes have been shown to be required for efficient secretion through the cytoplasmic membrane.

As the number of known signal peptides has grown (my collection now counts some 500 eukaryotic and over 100 prokaryotic sequences), more and more refined statistical studies of their common design principles have become possible. Basically, three distinct regions can be defined: a positively charged amino terminal region (n-region), a central apolar region (h-region), and a more polar carboxy terminal region (c-region) leading up to the cleavage site, Fig.2.

```
 -   ++                             ↓
MDYYRKYAAIFLVTLSVFLHVLHSA...
<--  n  --><--    h     -><- c -><--  mature -->
```

Fig.2. The signal peptide from HCG, a human secretory protein

Although the basic design is the same from bacteria to man, some subtle differences between signal peptides from prokaryotes and eukaryotes have been found. The n-region has on average one more Arg or Lys residue and is on average one residue longer in the prokaryotic sequences (von Heijne, 1984). The fraction of all sequences with a given net n-region charge becomes very similar in the two samples if an additional positive charge is added to all eukaryotic sequences; it may be that the unformylated free amino group on the initiator Met in the eukaryotic sequences provides the "missing" charge, Fig.3.

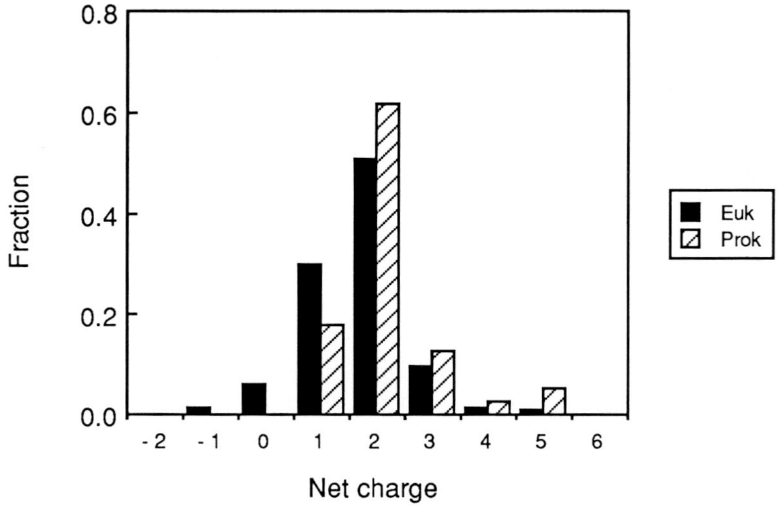

Fig.3. Distribution of the net charge in the n-region of samples of eukaryotic and prokaryotic signal peptides. One positive charge has been added to all eukaryotic sequences.

The h-regions in prokaryotic and eukaryotic signal peptides are also similar, except that the eukaryotic ones tend to have a slightly higher mean hydrophobicity per residue. No fine-structure inside the h-region has been found so far; rather, this region seems to be best described as a random assortment of apolar residues (von Heijne,1985). The boundary between the h- and c-regions, although not always immediately discernible, tends to fall between positions -7 and -6 (counting from the cleavage site) in prokaryotic sequences, and between -6 and -5 in eukaryotic sequences (von Heijne,1985).

The carboxy terminal c-region, finally, has some very strong amino acid preferences in the immediate neighborhood of the cleavage site. Indeed, all naturally occurring cleavage sites obey the "(-3,-1)-rule": the residue in position -1 must be small and uncharged (Ala, Ser, Gly, Cys, Thr, Gln, Pro), and the residue in position -3 must not be bulky or charged (Phe, His, Tyr, Trp, Asn, Gln, Asp, Glu, Arg, Lys) (von Heijne,1983; Perlman & Halvorson,1983), Fig.4.

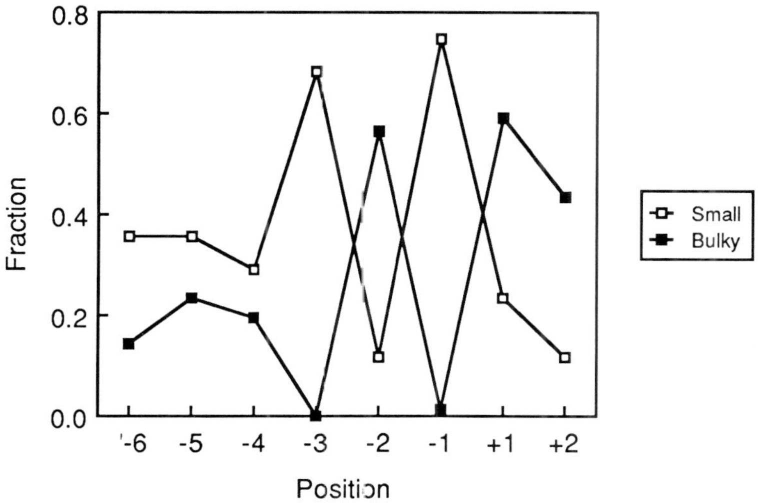

Fig.4. Fraction of small and bulky+charged residues in the c-region in a sample of eukaryotic signal peptides

Based on the statistical analysis of signal peptides, a weight–matrix method that can accurately discriminate between true signal peptides and amino terminal regions from cytoplasmic proteins, as well as predict the most likely site of cleavage between a signal peptide and the mature protein, has been developed (von Heijne,1986b). Basically, a large sample of signal peptides were aligned from their cleavage sites, and the number of residues of type a in position i, N(a,i), was counted for positions -13 to +2. A weight matrix W(a,i) was then constructed by first dividing all counts by the abundance of the particular residue in a control sample of soluble proteins, N(a), and then taking the natural logarithms of these quotients:

$$W(a,i) = \ln\left[\frac{N(a,i)}{N(a)}\right]$$

The most probable cleavage site in a new signal peptide can now be identified by scanning the sequence with the weight matrix and summing the weights for each position i of the weight matrix over the sequence, i.e. by calculating:

$$S(i) = W(a_{i-12}, i-12) + W(a_{i-11}, i-11) + \ldots + W(a_{i+2}, i+2)$$

where the summation window extends from position i-12 to position i+2. The site with the highest S(i)-value is predicted as the most likely cleavage site. In practice, >95% of all signal peptides can be told apart from >95% of amino terminal regions from cytoplasmic proteins on the basis of their maximal S-values, and the correct cleavage site is found in 75-80% of all cases.

Mitochondrial Targeting Peptides

Import of proteins into mitochondria is also dependent on a (often cleavable) amino terminal targeting (or transit) peptide. Comparative sequence analysis, as well as experimental data, indicate that a common theme in this case is a propensity to fold into an amphiphilic α–helix with one apolar and one polar, positively charged face. A plot of the mean hydrophobic moment m calculated as a function of δ, the angle between two successive residues, for a sample of mitochondrial targeting sequences, as well as for a sample of known surface-active amphiphilic helical peptides, is shown in Fig.5 (von Heijne, 1986c). Note that both curves peak around 90°-100°, i.e. close to the value expected for an ideal amphiphilic α-helix. Control samples of randomly scrambled targeting sequences and amino terminal sequences from cytoplasmic proteins have no similar peaks (not shown).

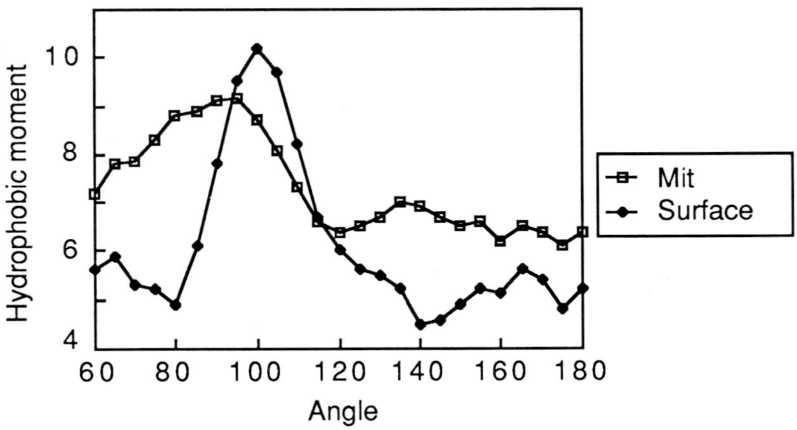

Fig.5. The mean hydrophobic moment for a sample of 23 mitochondrial targeting sequences and a sample of known amphiphilic peptides.

The mitochondrial targeting peptides are cleaved by a soluble protease located in the matrix of the mitochondrion, but the sequence patterns specifying the site of cleavage have not yet been identified.

Topogenic Signals in Membrane Proteins

Integral membrane proteins are bound to membranes by virtue of interactions between their hydrophobic transmembrane segments and the fatty acyl core of the bilayer. Each membrane protein has a unique orientation, yet different proteins span the membrane with a bewildering variety of topologies (Wickner & Lodish, 1985). Over the past few years the primary sequences of a large number of membrane proteins have been determined, but the sequence characteristics that determine the transmembrane topology of a protein still remain rather vaguely defined.

The ability to predict (i) the topology, and (ii) the fully folded structure from the primary sequence is mainly limited by our ignorance of the way membrane proteins are inserted into their target membrane. "Monotopic" membrane proteins with only one apolar membrane spanning segment generally are made with a cleavable amino terminal signal sequence that initiates translocation of the first part of the chain across the membrane. The internal apolar membrane spanning segment presumably halts the translocation process, and anchors the protein to the membrane in its final topology.

The biogenesis of "polytopic" membrane proteins, i.e. proteins that weave back and forth across the membrane one or more times, can be envisaged as proceeding by two mechanistically very different routes: either by an amino- to carboxy-terminal succession of "start" and "stop" events involving the normal secretory machinery (Blobel,1980), or by an insertion mechanism where neighboring apolar segments pair up and penetrate the membrane as "helical hairpins" (Engelman & Steitz,1981). In the first model, the topology is essentially determined by the apolar spanning segments alone (once the first translocation step is under way, the whole topology is determined), whereas the topology in the helical hairpin model is an outcome of kinetic competition between the formation and insertion of all possible helical hairpins in a process where the characteristics of the polar segments connecting the apolar stretches could be decisive.

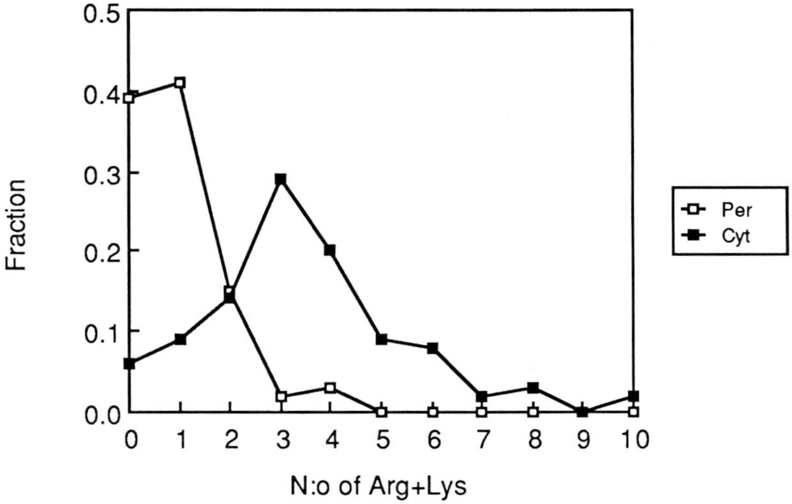

Fig.6. Fractions of all periplasmic and cytoplasmic loops with a given number of Arg+Lys residues in a sample of 20 bacterial inner membrane proteins.

To look for correlations between sequence patterns and topology, I collected data on the amino acid composition of known periplasmic and cytoplasmic connecting loops in inner membrane proteins from bacteria (von Heijne,1986a). A very strong effect soon became apparent: loops exposed on the periplasmic side contain very few positively charged residues ($f_{Arg+Lys}$=4%), whereas the cytoplasmic loops are enriched for these same residues ($f_{Arg+Lys}$=16%). As shown in Fig.6, very few periplasmic loops have more than 2 Arg or Lys residues. Indeed, all of the inner membrane proteins analyzed in this study, as well as integral membrane proteins from a number of membranes in eukaryotic cells (plasma membrane, mitochondrial inner membrane, chloroplast thylakoid membrane; von Heijne & Gavel, in preparation), show a nice pattern of alternating (Arg+Lys)–rich and (Arg+Lys)–poor domains, with the highly charged loops always following a "positive inside"-rule and remaining on the side of the membrane where the protein was synthesized, Fig.7.

Interestingly, in long periplasmic loop (>70 residues) no reduction in the frequency of Arg+Lys was seen, suggesting that short and long loop are translocated across the membrane by different routes. In particular, we hypothesize that long loops utilize the normal secretory machinery, whereas short loops for some reason cannot trigger this system but must insert "spontaneously".

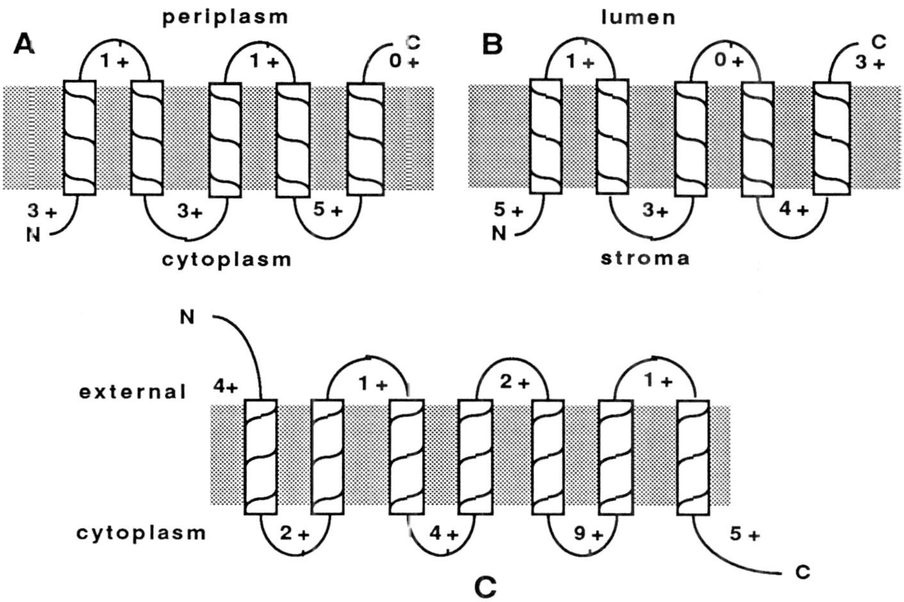

Fig.7. Distribution of Arg+Lys residues in (A) a bacterial inner membrane protein (R. viridis reaction center L-subunit), (B) a thylakoid membrane protein (the 32kd protein), and (C) a eukaryotic plasma membrane protein (Drosophila rhodopsin).

A Worked Example: Leader Peptidase from *Escherichia coli*

A good model system for studying sequence patterns important for the membrane assembly of polytopic membrane proteins is the inner membrane protein leader peptidase from *E. coli*. This protein has two apolar regions (Wolfe et al., 1983). The positions of the two hydrophobic (H1, H2) and two polar (P1, P2) regions of leader peptidase is shown in Fig.8. Segment H1 (residues 1-22) is entirely apolar and is probably oriented with the amino terminus facing the periplasm. The P1–domain (residues 23-61) is extremely polar with 10 basic and 5 acidic residues out of a total of 39, and protrudes into the cytoplasm. The second apolar domain H2 (residues 62-76) spans the membrane, leaving the large carboxy terminal domain P2 in the periplasm (Moore & Miura, 1987).

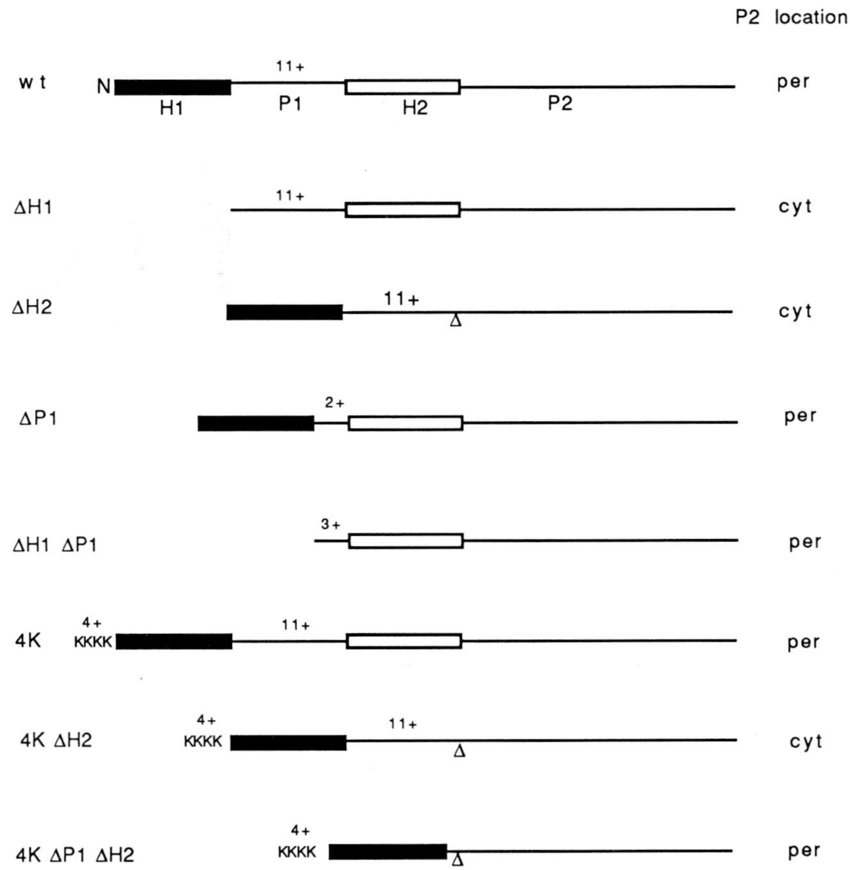

Fig.8. Leader peptidase mutants. Arg+Lys residues are indicated.

A number of insertions and deletions have been introduced into this protein, and their effects on its membrane assembly (i.e. the efficient translocation of P2 into the periplasm) have been studied. These constructions clearly show that the polar flanking sequences are as important as the apolar regions. Briefly, when either H1 or H2 is deleted (constructions ΔH1 and ΔH2, Fig.8) no assembly is seen (von Heijne et al.,1987; Dalbey & Wickner,1987). H2, when fused to the amino terminus of mature M13 coat protein or OmpA, supports their translocation across the inner membrane, albeit at a reduced rate (Dalbey et al., 1987). By this criterion, H2 is an internal, uncleaved signal peptide. By adding 4 lysines to the amino terminus of the protein, H1 can also be turned into an efficient signal peptide, but only if most of the downstream polar region P1 is absent (4KΔP1ΔH2 vs. 4KΔH2; von Heijne et al.,1987).

One possible interpretation of these results is that the topological properties of a given apolar region is determined by the incidence of positively charged residues in its flanking regions - the "positive inside"-rule. Thus, in the wild type protein, H1 inserts with its uncharged amino terminus facing the periplasm, leaving H2, which is preceded by a highly basic domain, to function as an internal, uncleaved signal peptide initiating the translocation of the P2 domain into the periplasm. The ΔH2 mutant lacks the internal signal peptide, and P2 is not translocated. In ΔH1, the polar domain P1 apparently blocks the signalling function of H2, perhaps by folding into a structure that disturbs the interaction between H2 and the secretory machinery; note that when P1 is also removed, H2 again becomes functional (ΔH1ΔP1). In the wild type, H1 seems to act as a "hydrophobic helper" for H2, needed to overcome the block imposed by P1.

H1, which normally does not act to initiate translocation of its immediate downstream sequence, can nevertheless be turned into an efficient uncleaved signal peptide (4KΔP1ΔH2). However, when the highly basic P1 region is present, no translocation is seen (4KΔH2). P1 can thus be regarded as a "translocation poison" sequence; a sequence that blocks the action of an otherwise functional upstream signal peptide. This interpretation is also consistent with the phenotype of the 4K-mutant, where H2 rather than H1 acts as the translocation signal, even though H1 has the N-terminal charge properties of a signal peptide and is closer to the amino terminus of the protein.

Conclusion

As I have tried to show in this paper, theoretical sequence analysis has been a surprisingly powerful tool for discovering sequence patterns involved in protein translocation into and across membranes. Clearly, statistical analyses of this kind cannot prove but only suggest hypotheses. Considering the enormous effort put into the experimental investigation of protein-membrane interactions, however, first trying to read the "message in the sequences" rather than relying on mere intuition when embarking on a new line of experimentation may be worthwhile, even if it means having to deal with computers.

References

Blobel,G. (1980) *Intracellular Protein Topogenesis*. Proc.Natl.Acad.Sci.USA 77, 1496-1500.

Cornette,J.L., Cease,K.B., Margalit,H., Spouge,J.L., Berzofsky,J.A., & DeLisi,C. (1987) *Hydrophobicity Scales and Computational Techniques for Detecting Amphipathic Structures in Proteins*. J.Mol.Biol. 195, 659-685.

Dalbey,R.E., Kuhn,A., & Wickner,W. (1987) J.Biol.Chem.in press.

Dalbey,R.E., & Wickner,W. (1987) *Leader Peptidase of Escherichia coli: Critical Role of a Small Domain in Membrane Assembly*. Science 235, 783-787.

Engelman,D.M., & Steitz,T.A. (1981) *The Spontaneous Insertion of Proteins Into and Across Membranes: The Helical Hairpin Hypothesis*. Cell 23, 411-422.

Eisenberg,D., Schwarz,E., Komaromy,M., & Wall,R. (1984) *Analysis of Membrane and Surface Protein Sequences with the Hydrophobic Moment Plot.* J.Mol.Biol. 179, 125-142.

Finer-Moore,J., & Stroud,R.M. (1984) *Amphipathic Analysis and Possible Formation of the Ion Channel in an Acetylcholine Receptor.* Proc.Natl.Acad.Sci.USA 81, 155-159.

Moore,K., & Miura,S. (1987) J.Biol.Chem. in press

Perlman,D., & Halvorson,H.O. (1983) *A Putative Signal Peptidase Recognition Site and Sequence in Eukaryotic and Prokaryotic Signal Peeptides.* J.Mol.Biol. 167, 391-409.

Rapoport,T.(1986) *Protein Translocation Across and Integration Into Membranes* CRC Crit.Rev.Biochem. 20, 73-137.

Vogel,H., & Jähnig,F. (1986) *Models for the Structure of Outer-Membrane Proteins of Escherichia coli Derived from Raman Spectroscopy and Prediction Methods.* J.Mol.Biol. 190, 191-199.

von Heijne,G. (1983) *Patterns of amino Acids near Signal-Sequence Cleavage Sites.* Eur.J.Biochem. 133, 17-21.

von Heijne,G. (1984) *Analysis of the Distribution of Charged Residues in the N-terminal Region of Signal Sequences: Implications for Protein Export in Prokaryotic and Eukaryotic Cells.* EMBO J. 3, 2315-2318.

von Heijne,G. (1985) *Signal Sequences: The Limits of Variation.* J.Mol.Biol. 184, 99-105.

von Heijne,G. (1986a) *The Distribution of Positively Charged Residues in Bacterial Inner Membrane Proteins Correlates with the Trans-Membrane Topology.* EMBO J. 5, 3021-3027.

von Heijne,G (1986b) *A New Method for Predicting Signal Sequence Cleavage Sites.* Nucl.Acids Res. 14, 4683-4690.

von Heijne,G. (1986c) *Mitochondrial Targeting Sequences May Form Amphiphilic Helices.* EMBO J. 5, 1335-1342.

von Heijne,G. (1987) *Sequence Analysis in Molecular Biology: Treasure Trove or Trivial Pursuit?* Academic Press, New York.

von Heijne,G., Wickner,W., & Dalbey,R. (1987) *The Cytoplasmic Domain of Escherichia coli Leader Peptidase is a Translocation Poison Sequence.* Submitted.

Wickner,W., & Lodish,H. (1985) *Multiple Mechanisms of Protein Insertion Into and Across Membranes.* Science 230, 400-407.

Wolfe,P.B., Wickner,W., & Goodman,J. (1983) *Sequence of the Leader Peptidase Gene of Escherichia coli and the Orientation of Leader Peptidase in the Bacterial Envelope.* J.Biol.Chem. 258, 12073-12080.

COMPONENTS INVOLVED IN PROTEIN TRANSLOCATION ACROSS THE MEMBRANE OF THE ENDOPLASMIC RETICULUM.

Bernhard Dobberstein
European Molecular Biology Laboratory
Meyerhofstr. 1
Postfach 102209
6900 Heidelberg

Proteins which are synthesized in the cytoplasm of eucaryotic cells can have different destinations. They can remain in the cytoplasm or be transported to a particular organelle such as the endoplasmic reticulum (ER), the mitochondria or the nucleus. The ER is the entry site to the secretory pathway for secretory, lysosomal and for those membrane proteins that become resident of the ER, the Golgi or the plasma membrane. Proteins which enter this pathway have signal sequences that guide them to the ER membrane. A typical signal sequence is about 20 amino acid residues long and its main distinctive feature is a core of hydrophobic amino acid residues. On the luminal side of the membrane the signal sequence of most presecretory proteins is cleaved by signal peptidase.

At least two cytoplasmic components are involved in the specific targeting of nascent secretory and membrane proteins to the ER, the signal recognition particle (SRP) and the docking protein (DP) or SRP receptor (Fig. 1). SRP binds to the signal sequence of nascent polypeptides and can arrest further synthesis at different sites (Walter and Blobel, 1981 a and b; Lipp et al. 1987). The arrest is relieved after interaction of SRP with the DP in the ER membrane (Meyer et al. 1982; Gilmore et al. 1982a and b). Elongation then proceedes and translocation of the nascent polypeptide across the membrane starts. This sequence of events has been established by using cell-free systems assembled from components derived from wheat germ and various mammalian cells or tissues (see Hortsch and Meyer, 1986; Walter and Lingappa, 1986).

Not all secretory proteins strictly depend on SRP and DP for their membrane insertion. Some small polypeptides, at least when tested in cell-free systems, can be inserted into ER membranes in the absence of SRP and DP (Schlenstedt and Zimmermann, 1987). Membrane translocation of these peptides requires ATP.

Although cotranslational translocation seems to be the rule for most eucaryotic secretory and membrane proteins, some exceptions exist. Alpha factor of the yeast strain Saccharomyces Cerevisiae can be translocated posttranslationally across yeast membranes (Rothblatt and Meyer, 1986; Hansen et al. 1986; Waters and Blobel, 1986). A cytoplasmic component, distinct from SRP, is required for translocation of alpha factor (Waters et al. 1986).

NATO ASI Series, Vol. H16
Membrane Biogenesis. Edited by J. A. F. Op den Kamp

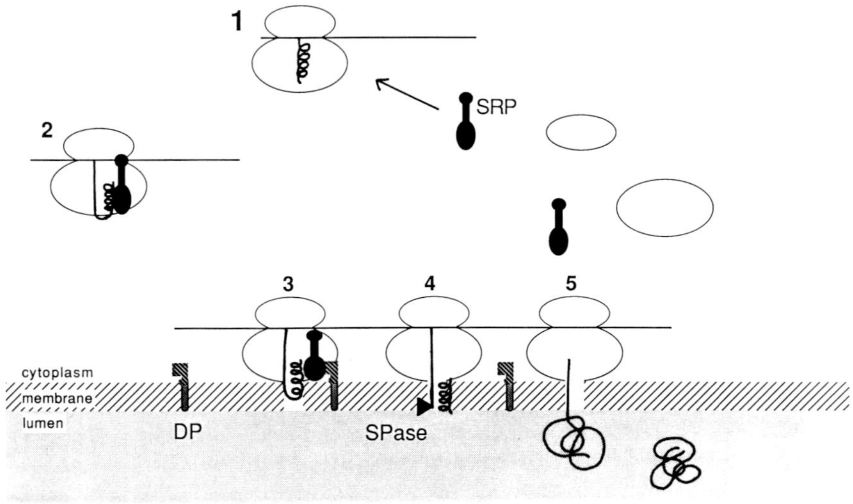

Figure 1
Steps in the translocation of a secretory protein across the membrane of the ER.
1. Start of translation of the m RNA coding for a secretory protein on a free cytoplasmic ribosome. 2. SRP interacts with the signal sequence in the nascent polypeptide chain. 3. Ribosomes which have bound SRP contact the ER membrane. 4. The nascent polypeptide chain is translocated across the membrane and the signal sequence is cleaved off. 5. When translation is completed, the secretory protein accumulates in the lumen of the ER. The ribosome dissociates from the membrane and can start a new round of translation.
SRP: signal recognition particle; DP: docking protein; SPase: signal peptidase.

In this review I will describe the components involved in the translocation of secretory and membrane proteins across mammalian ER membranes and discuss the functions of these components. For a description of other aspects of the secretory process see the following reviews: Hortsch and Meyer, 1986; Walter and Lingappa, 1986; Oliver,1985; Rapoport and Wiedmann, 1985; von Heijne, 1985; Wickner and Lodish,1985.

Isolation of components involved in protein targeting to the ER membranes

A prerequisite for the isolation of components involved in protein translocation across ER membranes was the development of an in vitro test system. Such a system consists of a protein synthesizing extract and purified microsomal membranes across which nascent polypeptides are translocated (Blobel and Dobberstein, 1975). Radioactively labelled secretory or membrane proteins are synthesized in the presence of microsomal vesicles. Protein translocation is determined by treatment of the vesicles with protease. Proteins that have not been transported into the vesicles are digested by the added protease while those which have been translocated are protected from the digestion.

The microsomal membranes can be depleted of components present on their surface and then be tested for their ability to translocate nascent or completed polypeptides. Such an approach was successfully applied in the detection and isolation of signal recognition particle (SRP) and docking protein (DP).

When microsomal membranes, usually from dog pancreas, are washed with 0.5M KCl they loose their capacity to translocate secretory proteins when tested in the wheat germ cell-free system. If however the salt extract is recombined with the membranes, the translocation activity is restored (Warren and Dobberstein, 1978). It was found that the active component in the salt extract is the SRP (Walter and Blobel, 1980).

Treatment of microsomal membranes with low amounts of elastase and high concentrations of salt led to the removal of the cytoplasmically exposed domain of another component involved in translocation, the DP (Walter et al 1979; Meyer and Dobberstein, 1980; Meyer et al. 1982; Gilmore et al. 1982 a and b).

Structure of SRP

From the salt extract of dog pancreas microsomal membranes SRP can be purified by chromatography on ω-aminopentylagarose and DEAE-Sepharose and centrifugation through a sucrose density gradient. SRP is an 11S cytoplasmic ribonucleoprotein particle. It consists of six nonidentical polypeptide chains of 9, 14, 19, 54, 68 and 72 kDa and one molecule of a 7S RNA (Fig. 2). Both, the RNA and the proteins are required for restoring translocation activity to salt-extracted microsomes in vitro. Electron microscopic analysis revealed, that SRP is a rod-shaped particle 5-6 nm wide and 23-24nm long (Andrews et al. 1985).

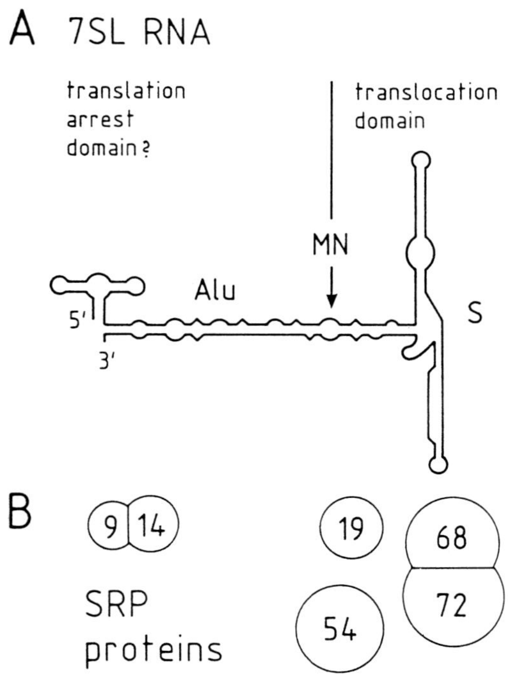

Figure 2
Structural components of the SRP
A: Proposed secondary structure of 7SL RNA of SRP. (redrawn from Zwieb, 1985)
Alu: Alu repetitive sequences at the 5´ and 3´end of 7SL RNA. S: the central 7SL RNA specific segment. MN: sites for cleavage by micrococcal nuclease.
B: The proteins of the SRP. Their estimated molecular weights are given in kilo dalton. The 9/14 kDa heterodimer binds to the Alu-like parts of 7SL RNA while the 19, 54 and 68/72 kDa proteins bind to the central S segment.

7SL RNA

The 7S RNA in SRP, also called 7SL, comprizes about 300 nucleotides. Alu-like sequences are found within the 100 nucleotides of the 5´ end and the 40 nucleotides of the 3´ end. The central 160 nucleotides, called S fragment, are unique for the 7SL RNA (Ullu et al. 1982). The secondary structure of the human 7SL RNA has been established by using specific

nucleases and the compensatory base change approach (Gundelfinger et al.1984, Zwieb, 1985). The basic structural features of 7SL RNA are a central rod formed by pairing of the bases between positions 48 and 118 with those between 233 and 299, flanked by two small stem-loop structures at one end and two larger loop structures at the other. The small stem-loops are formed by the 44 bases at the 5′ end and the two larger ones by bases 119 to 232. The small stem-loops and most of the rod are formed by the Alu segments, the larger loops by the central S segment. Linker insertion analysis has revealed, that 7S RNA can exist in alternative conformations. It has been proposed, that these alternative conformations might play a role in the function of SRP (Zwieb and Ullu, 1986).
Limited digestion of the 7S RNA in SRP with micrococcal nuclease leads to two subparticles of uneven size. The small one contains the paired Alu segments from the 5′ and 3′ end of the RNA and the larger one the central S fragment. The 9 and 14 kDa proteins are found bound to the Alu segments and the 19, 54, 68 and 72 kDa proteins to the central S segment (Gundelfinger et al. 1983). Siegel and Walter (1985) showed, that the large SRP subparticle containing the S segment and the four large proteins can still promote secretory protein translocation across microsomal membranes, but can no longer trigger an arrest of presecretory protein synthesis. From this it has been concluded, that it is the Alu-like RNA in SRP and the 9/14 kDa proteins that provide the particle with the elongation-arresting activity (Siegel and Walter, 1986). The results also imply, that the elongation arrest is not a prerequisite for protein translocation across ER membranes (Siegel and Walter, 1985).

SRP proteins

The proteins in SRP are required for the membrane translocation activity of the particle, as shown by several lines of evidence. Alkylation with N-ethylmaleimide inactivates the particle (Walter and Blobel, 1980). Similarly, antibodies against the 54, 68 and the 72 kDa proteins have been shown to neutralize SRP activity in vitro (Walter and Blobel, 1983).
SRP can be disassembled into its native RNA and protein components by unfolding the particle with EDTA and dissociating of the components on polycationic matrices (Walter and Blobel, 1983). Isolated SRP proteins are inactive in promoting translocation of secretory proteins across ER membranes. However, when recombined with 7SL RNA in the presence of magnesium, the proteins reassemble stoichiometrically with 7SL RNA and form fully active SRP (Walter and Blobel, 1983). A stepwise removal of the proteins from the 7SL RNA has been achieved by incubating SRP in 2 M KCl. After this treatment, heterodimers of the 9/14

kDa and of the 68/72 kDa proteins are released from the RNA. (Scoulica, et al. 1987). From reconstitution studies it has been shown, that the 9/14, the 19 and the 68/72 kDa proteins bind directly to the 7S RNA whereas the 54 kDa protein requires for its binding to the RNA the presence of the 19 kDa protein (Walter and Blobel, 1983).

The domain structure of the SRP proteins has been probed by using mild elastase treatment and protein-specific antibodies . A 55 kDa domain can be cleaved from the 72 kDa protein and a 35 kDa domain from the 54 kDa protein. Release of these domains leads to inactivation of the particle (Scoulica et al., 1987).

Using a photocrosslinking approach it has been shown, that the signal sequence of nascent preprolactin interacts with the 54 kDa SRP protein (Kurzchalia, et al. 1986, Krieg et al. 1986, Wiedman, et al. 1987).

Docking protein (SRP receptor)

The site where free and ribosome-bound SRP interacts with the ER membrane is the docking protein (DP) or SRP receptor. Cell fractionation and immunocytochemical analysis indicated that DP is essentially restricted to the rough ER (Hortsch et al., 1985; Hortsch and Meyer, 1985).

DP is an oligomeric protein complex that contains two distinct polypeptide chains, DPα and DPβ (Tajima et al. 1986). DPα was originally purified as the 52 kDa elastase fragment which reconstitutes the translocation activity of proteolyzed and high salt-treated rough microsomes (Meyer and Dobberstein, 1980). The 52 kDa fragment was purified and antibodies were raised against it. Using these antibodies it was shown, that the intact, functional molecule has an apparent molecular weight of 73 kDa (Meyer and Dobberstein, 1980). DPα was also characterized by its ability to bind to SRP and release the SRP-mediated translation arrest (Gilmore et al., 1982b).

The complete primary structure of dog DPα has been deduced from the complementary DNA sequence (Lauffer et al. 1985). DPα contains 638 amino acid residues and has a calculated molecular weight of about 70kDa. This is in good agreement with its apparent molecular weight as estimated by SDS-PAGE. The fragment which can be liberated from DPα by treatment with elastase comprizes 487 amino acid residues (52 kDa) and contains the C-terminus. The fragment which interacts with the membrane comprizes 151 amino acids and contains the N-terminus. No distinct hydrophobic region, as is typical for membrane-spanning proteins, can be detected in DPα. The molecule is basic in character (calculated pI=9.86) and charged amino acid residues appear largely clustered at the site where elastase cleaves the molecule. Mixed-charge amino acid clusters are also found in regions of ribosomal proteins and aminoacyl-tRNA synthetases that are thought to be involved in binding of these proteins to RNA molecules. Lauffer et al (1985) suggested, that the region, where elastase cleaves DPα, may function in nucleic acid binding. It is still unclear, whether the DPα itself spans the membrane or whether it is held in the membrane by interaction with a membrane-integrated protein, the DPβ.

The DPβ has long escaped detection as it comigrated with proteolytic fragments generated from the DPα subunit (Tajima et al. 1986). DPβ is a 30 kDa membrane protein which can bind directly to SRP. The DPα and DPβ subunit have not yet been separated under nondenaturing conditions and it therefore remains unclear whether one subunit alone can functionally interact with the SRP. Both subunits are present in the ER membrane in about equimolar amounts. Proteolysis releasing the 52 kDa fragment from the DPα subunit leaves the DPβ subunit intact and membrane associated. Therefore, it is likely that DPβ provides the binding site for the 18 kDa fragment of DPα that remains bound to the membrane after elastase treatment (Tajima et al. 1986).

Signal Sequence Receptor

The signal sequence is not only recognized by its cytoplasmic receptor, the SRP, but most likely also by protein(s) in the membrane of the ER (Prehn, et al. 1981; Austen and Ridd,1983). A good candidate for this receptor has recently been identified by using a very elegant cross-linking approach (Wiedmann et al. 1987). A photoactivatable cross-linking reagent was incorporated into the signal sequence of nascent preprolactin. After binding to the ER membrane the nascent chain was found to interact with a glycosylated 35 kDa membrane protein called signal sequence receptor (SSR) (Wiedmann et al 1987). Wiedmann et al. speculated that SSR could mediate the interaction of the signal sequence with the lipid bilayer or induce the formation of a channel through which the nascent polypeptide could traverse the membrane.

Postulated Componts Involved in Translocation across the Membrane

After the nascent secretory polypeptide is targeted to the endoplasmic reticulum its translocation through the membrane might be mediated by other protein components or the nascent chain might directly partition into the lipid bilayer. Direct partitioning into the lipid bilayer was proposed based largely on theoretical considerations (von Heijne and Blomberg, 1979; Engelman and Steitz, 1981). The involvement of protein components is suggested from experiments in which an amino-terminal segment of a secretory protein was reacted with microsomal membranes. It was found, that the puromycin-released peptide interacted with the membrane and could only be released after treatment with protein denaturants such as 4 M urea or alkaline pH (Gilmore and Blobel, 1985).

Using short polypetides the requirements for the binding of nascent polypeptides extruding from the ribosome to the ER membrane were investigated (Conolly and Gilmore, 1986). It was found, that the high salt resistent binding of the nascent polypeptide chains to the ER membrane requires GTP but not GTP hydrolysis. The actual transport across the membrane was neither dependent on ribonucleotide hydrolysis nor on continued polypeptide chain elongation.

Signal Peptidase

The signal sequence of most presecretory proteins is cleaved during translocation of the protein across the microsomal membrane (Blobel and Dobberstein, 1975). The enzyme responsible for this cleavage has been termed signal peptidase (Jackson and Blobel, 1977). Signal peptidase is an integral membrane protein (Lively and Walsh, 1983). Its activity can be monitored after detergent solubilization using a posttranslational cleavage assay (Jackson and Blobel, 1977). Since it cannot be inactivated by proteolysis of intact microsomal vesicles, it was concluded that the active center of signal peptidase is located on the luminal side of the ER membrane (Jackson and Blobel, 1977). Several procedures have been described by which signal peptidase can be enriched to various degrees. (Jackson and Blobel, 1980; Mollay et al. 1982; Hikawa et al. 1985; Fujimoto et al. 1985; Evans et al. 1986). Evans et al (1986) purified a complex composed of six polypeptides of 25, 23, 22, 18, and 12 kDa which showed signal peptidase activity. The 22 and 23 kDa subunits were shown to be glycoproteins based an their sensitivity to endoglycosidase H and their ability to bind concanavaline A. It was suggested, that only one subunit of the complex carries out signal peptide cleavage and that the others might be involved in nascent chain translocation across the microsomal membranes (Evans et al. 1986)

The Postulated Ribosome Receptor

Several lines of evidence demonstrate the existence of a specific ribosome receptor in the rough ER membrane. Ribosomes can be detached from the ER membrane by treatment with puromycin at a high salt concentration. They bind back to the membrane at low ionic strength (Kreibich, et al. 1983). This suggests, that eucaryotic ribosomes are bound to the ER membrane not only via the nascent polypeptide chain, but also by an electrostatic interaction. A specific interaction of ribosomes with the ER membrane is also suggested from experiments showing, that a secretory protein synthesized on Escherichia coli ribosomes was not translocated across microsomal membranes from dog pancreas. However, when the same protein is synthesized on mammalian ribosomes it is translocated across these membranes (Ibrahimi and Fuchs, 1987).

Based on cell fractionation and cross-linking studies, Kreibich and his collaborators proposed two integral rough ER-specific membrane proteins as the binding sites for ribosomes. They named these proteins ribophorin I and II (Kreibich, et al. 1978). Ribophorins are found nearly exclusively in the rough ER in a stoichiometric relationship with the number of membrane-bound ribosomes (Marcantonio, et al. 1982). They remain associated with the ribosomal fraction when microsomal membranes are solubilized with nonionic detergents (Kreibich et al. 1978). However, two recently obtained results make it unlikely, that ribophorins are involved in binding of ribosomes to the ER membrane. 1. Mild proteolysis of microsomes eliminated ribosome binding without affecting ribophorins (Hortsch et al., 1986). 2. Biochemical and sequence analysis of ribophorins revealed, that only rather small domains of these proteins are exposed on the cytoplasmic side of the ER membrane and that the major portion of these proteins is exposed on the luminal side of the membrane (Fig. 3). This would suggest, that the main function of ribophorins is probably performed on the luminal and not on the cytoplasmic side of the membrane (Crimaudo et al. 1987).

In conclusion, very little is known about the postulated ribosome receptor in the ER membrane.

Conclusions

Two distinct steps can be distinguished in the translocation of proteins across the membrane of the ER: i) targeting of the nascent polypeptide to the membrane and ii. the actual translocation across the membrane. SRP and DP, the two components so far isolated and characterized, are both involved in targeting. It is still unclear whether the actual translocation of the polypeptide across the membrane is also mediated by protein components or whether it occurs directly through the lipid bilayer. Several genetic and biochemical approaches are currentely used in several laboratories to identify further components involved in protein translocation across the ER membrane.

Literature

Andrews, D. W., Walter, P. and Ottensmeyer, F.P. (1985) Structure of the signal recognition particle by electron microscopy. Proc. Natl. Acad. Sci. USA, 82: 785-789

Blobel, G. and Dobberstein, B. (1975) Transfer of proteins across membranes. II. Reconstitution of functional rough microsomes from heterologous components. J. Cell Biol. 67: 852-862

Crimaudo, C., Hortsch, M., Gausepohl, H. and Meyer, D. I. (1987) Human ribophorins I and II: the primary structure and membrane topology of two highly conserved rough endoplasmic reticulum-specific glycoproteins. EMBO J. 6: 75-82

Engelman D. M. and Steitz, T. A. (1981) The spontaneous insertion of proteins into and across membranes: The helical hairpin hypothesis. Cell, 23: 411-422

Evans, E. A., Gilmore, R. and Blobel, G. (1986) Purification of microsomal signal peptidase as a complex. Proc. Natl. Acad. Sci. USA, 83: 581-585

Fujimoto, Y., Watanabe, Y., Uchida, M. and Ozaki, M. (1984) Mammalian signal peptidase: Partial purification and general characterization of the signal peptidase from microsomal membranes of porcine pancreas. J. Biochem. 96, 1125-1131

Gilmore, R., Blobel, G. and Walter, P. (1982a) Protein translocation across the endoplasmic reticulum. I. Detection in the microsomal membrane of a receptor for the signal recognition particle. J. Cell Biol. 95: 463-469

Gilmore, R., Walter, P. and Blobel, G. (1982b) Protein translocation across the endoplasmic reticulum. II. Isolation and characterization of the signal recognition particle receptor. J. Cell Biol. 95: 470-477

Gilmore, R. and Blobel, G. (1985) Translocation of secretory proteins across the microsomal membrane occurs through an environment accessible to aqueous perturbants. Cell, 42: 497-505

Gundelfinger, E.D., Di Carlo, M., Zopf, D. and Melli, M. (1984) Structure and evolution of the 7SL RNA component of the signal recognition particle. EMBO J. 3: 2325-2332

Gundelfinger, E.D., Krause, E., Melli, M. and Dobberstein, B. (1983) The organization of the 7SL RNA in the signal recognition particle. Nucleic Acids Res. 11: 7363-7374

Hansen, W., Garcia, P. D. and Walter, P. (1986) In vitro protein translocation across the yeast endoplasmis reticulum: ATP-dependent posttranslational translocation of the prepro-alpha-factor. Cell, 45: 397-406

Hikawa, A., Hashimoto, M., Horigome, T., Omata, S. and Sugano, H. (1985) Properties of rat liver signal peptidase reconstituted into liposomes. J.Biochem. 97:105-112

Hortsch, M. and Meyer, D. I. (1985) Immunochemical analysis of rough and smooth microsomes from rat liver. Segregation of docking protein in rough membranes. Eur. J. Biochem. 150: 559-564

Hortsch, M., Griffiths, G. and Meyer, D. I. (1985) Restriction of docking protein to the rough endoplasmic reticulum: immunocytochemical localization in rat liver. Eur. J. Cell Biol. 38: 271-279

Hortsch, M. and Meyer, D.I. (1986) Transfer of secretory proteins through the membrane of the endoplasmic reticulum. Intern. Rev. Cytol. 102: 215-242

Hortsch, M., Avossa, D. and Meyer, D. I. (1986) Characterization of secretory protein translocation: Ribosome - membrane interaction in endoplasmic reticulum. J. Cell Biol. 103: 241-253

Ibrahimi, I. and Fuchs, E. (1987) Nascent secretory polypeptides synthesized on Escherichia coli ribosomes are not translocated across mammalian endoplasmic reticulum. J. Bacteriol. 169: 1603-1610

Jackson, R. C. and Blobel, G.(1977) Post-translational cleavage of presecretory proteins with an extract of rough microsomes from dog pancreas containing signal peptidase activity. Proc. Natl. Acad. Sci. USA 74: 5598-5602

Jackson, R.C., Walter,P. and Blobel, G.(1980) Secretion requires a cytoplasmically disposed sulphydryl of the RER membrane. Nature, 286: 174-176.

Kreibich, G., Czako-Graham, M., Grebenau, R., Mok, W., Rodriguez-Boulan, E. and

Sabatini, D.D. (1978) Characterization of the ribosomal binding site in rat liver rough microsomes: Ribophorins I and II, two integral membrane proteins related to ribosome binding. J. Supramolec. Structure 8: 279-302

Kreibich,G., Marcantonio, E. E. and Sabatini, D. D. (1983) In: Methods Enzymology, Academic Press, New York (S. Fleischer and B. Fleischer, eds.), 96: 520-530

Krieg, U. C., Walter, P. and Johnson, A. E. (1986) Photocrosslinking of the signal sequence of nascent preprolactin to the 54-kilodalton polypeptide of the signal recognition particle. Proc. Natl. Acad. Sci. USA 83: 8604-8608

Kurzchalia, T. V., Wiedmann, M., Girshovich, A.S., Bochkareva, E. S., Bielka, H. and Rapoport, T.A. (1986) The signal sequence of nascent preprolactin interacts with the 54 K polypeptide of the signal recognition particle. Nature 320:634-636

Lauffer, L.,Garcia, P.D., Harkins, R.N., Coussens, L., Ullrich, A. and Walter, P. (1985) Topology of signal recognition particle receptor in endoplasmic reticulum membrane. Nature, 318: 334-338

Lipp, J. and Dobberstein, B. (1986) Signal recognition particle-dependent membrane insertion of mouse invariant chain: A membrane-spanning protein with a cytoplasmically exposed amino terminus. J. Cell Biol., 102: 2169-2175

Lipp, J., Dobberstein, B. and Haeuptle, M-T. (1987) Signal recognition particle arrests elongation of nascent secretory and membrane proteins at multiple sites in a transient manner. J. Biol. Chem. 262: 1680-1684

Lively, M. O. and Walsh, K.A. (1983) Hen oviduct signal peptidase is an integral membrane Protein. J. Biol. Chem. 258: 9488-9495

Marcantonio, E. E., Amar-Costesec, A. and Kreibich, G. (1982) Segregation of the polypeptide translocation apparatus to regions of the endoplasmic reticulum containing ribophorins and ribosomes. II. Rat liver microsomal subfractions contain equimolar amounts of ribophorins and ribosomes. J. Cell Biol. 99: 2254-2259

Meyer, D. I. and Dobberstein, B. (1980) Identification and characterization of a membrane component essential for the translocation of nascent proteins across the membrane of the endoplasmic reticulum. J. Cell Biol. 92: 503-508

Meyer, D. I., Krause, E. and Dobberstein, B. (1982) Secretory protein translocation across membranes - the role of the docking protein. Nature, 297: 647-650

Mollay, C., Vilas, U. and Kreil, G. (1982) Cleavage of honeybee prepromelittin by an endoprotease from rat liver microsomes: Identification of intact signal peptide. Proc. Natl. Acad. Sci. USA 79: 2260-2263

Oliver, D. (1985) Protein secretion in Escherichia coli. Ann. Rev. Microbiol. 39: 615-648

Prehn, S., Nürnberg, P. and Rapoport, T. A. (1981) A receptor for signal elements of secretory proteins in rough endoplasmic reticulum membranes. FEBS Letters, 123: 79-84.

Rapoport, T.A. and Wiedmann, M. (1985) Application of the signal hypothesis to the incorporation of integral membrane proteins. Current topics in membranes and transport. 24: 1-63

Rothblatt, J.A. and Meyer, D. I. (1986) Secretion in yeast: reconstitution of the translocation and glycosylation of alpha-factor and invertase in a homologous cell-free system. Cell, 44: 619-628

Schlenstedt, G. and Zimmermann, R. (1987) Import of frog prepropeptide Gla into microsomes requires ATP but does not involve docking protein or ribosomes. EMBO J. 6: 699-703

Scoulica, E., Krause, E., Meese, K. and Dobberstein, B. (1987) Disassembly and domain structure of the proteins in the signal-recognition particle. Eur. J. Biochem. 163: 519-528

Siegel, V. and Walter, P. (1985) Elongation arrest is not a prerequisite for secretory protein translocation across the microsomal membrane. J. Cell Biol. 100: 1913-1921

Siegel. V. and Walter, P. (1986) Removal of the Alu structural domain from signal recognition particle leaves its protein translocation activity intact. Nature 320: 81-84

Tajima, S., Lauffer, L., Rath, V. L. and Walter, P. (1986) The signal recognition particle receptor is a complex that contains two distinct polypeptide chains. J. Cell Biol. 103: 1167-1178

Ullu, E., Murphy, S. and Melli, M.(1982) Human 7SL RNA consists of a 140 nucleotide middle-repetitive sequence inserted in an ALU sequence. Cell, 29: 195-202

von Heijne, G. and Blomberg, C. (1979) Trans-membrane translocation of proteins: The direct transfer model. Eur. J. Biochem. 97: 175-181

von Heijne, G. (1985) Structural and thermodynamic aspects of the transfer of proteins into and across membranes. Current topics in membranes and transport, 24: 151-179

Walter, P., Jackson, R. C., Marcus, M. M., Lingappa, V. R. and Blobel, G. (1979) Tryptic dissection and reconstitution of translocation activity for nascent presecretory proteins across microsomal membranes. Proc. Natl. Acad. Sci. USA, 76: 1795-1799

Walter, P. and Blobel, G. (1980) Purification of a membrane-associated protein complex required for protein translocation asross the endoplasmic reticulum. Proc. Natl. Acad. Sci. USA 77: 7112-7116

Walter, P. and Blobel, G. (1981a) Translocation of proteins across the endoplasmic reticulum II. Signal recognition protein (SRP) mediates the selective binding to microsomal membranes of in-vitro-assembled polysomes synthesizing secretory protein. J. Cell Biol. 91: 551-556

Walter, P. and Blobel, G. (1981b) Translocation of proteins across the endoplasmic reticulum. III. Signal recognition protein (SRP) causes signal sequence-dependent and site-specific arrest of chain elongation that is released by microsomal membranes. J. Cell Biol. 91: 557-561

Walter, P. and Blobel, G. (1983) Disassembly and reconstitution of signal recogniton particle. Cell, 34: 525-533

Walter, P. and Lingappa, V. R. (1986) Mechanism of protein translocation across the endoplasmic reticulum membrane. Ann. Rev. Cell Biol. 2: 499-516

Warren, G. and Dobberstein, B. (1978) Protein transfer across microsomal membranes reassembled from separated membrane components. Nature, 273: 569-571

Waters, M. G. and Blobel, G. (1986) Secretory protein translocation in a yeast cell free system can occur posttranslationally and requires ATP hydrolysis. J. Cell Biol.102: 1543-1550

Waters, M. G., Chirico, W. J. and Blobel, G. (1986) Protein translocation across the yeast microsomal membrane is stimulated by a soluble factor. J. Cell Biol. 103: 2629-2636

Wickner, W.T. and Lodish, H.F. (1985) Multiple mechanisms of protein insertion into and across membranes. Science 230: 400-407

Wiedmann, M., Kurzchalia, T. V., Bielka, H. and Rapoport, T. A. (1987) Direct probing of the interaction between the signal sequence of nascent preprolactin and the signal recognition particle by specific cross-linking. J. Cell Biol. 104: 201-208

Wiedmann, M., Kurzchalia, T. V., Hartmann, E. and Rapoport, T.A. (1987) A signal sequence receptor in the endoplasmic reticulum membrane Nature, 328: 830-833

Zwieb, C. (1985) The secondary structure of the 7SL RNA in the signal recognition particle: functional implications. Nucleic acids Res. 13: 6105-6124

Zwieb, C. and Ullu, E. (1986) Identification of dynamic sequences in the central domain of 7SL RNA. Nucleic Acids Res. 14: 4639-4657

IMPORT OF SMALL SECRETORY AND PLASMA MEMBRANE PROTEINS INTO THE ENDOPLASMIC RETICULUM

Richard Zimmermann, Maria Sagstetter, Gabriel Schlenstedt, Hans Wiech, Birgitta Kaßeckert and Günter Müller
Institut für Physiologische Chemie, Physikalische Biochemie und Zellbiologie
Universität München
Goethestr. 33
D-8000 München 2
Federal Republic of Germany

SUMMARY

We are employing precursors of small secretory and plasma membrane proteins as tools for defining the different stages in the import of proteins into the endoplasmic reticulum (ER). The precursor proteins that we selected are M13 procoat protein, the precursor of a bacterial plasma membrane protein, and the precursors of two eucaryotic secretory proteins, honeybee prepromelittin and frog prepropeptide GLa. Our experimental systems involves in vitro systems, suitable for translation as well as import of the translation products into microsomes. The following stages in import can be resolved: i) Specific association of a precursor protein with the ER-membrane; ii) Insertion of a precursor protein into the ER-membrane; and iii) Assembly of a protein into the ER-membrane and transport of a protein across the ER-membrane, respectively. We present a working model for the import of small precursor proteins into ER.

NATO ASI Series, Vol. H16
Membrane Biogenesis. Edited by J. A. F. Op den Kamp

INTRODUCTION

Every polypeptide has a specific function as well as a unique functional location, i.e. an intra- or extracellular location where it fulfils its function (Zimmermann, 1986; Zimmermann and Meyer, 1986). There are two facts which turned the latter notion into a central problem in cell biology: i) There is essentially only one site of protein synthesis, the cytoplasm, but there are many different potential functional locations, the cell organelles and the extracellular space; ii) The site of synthesis is separated from these locations by phospholipid bilayers. Therefore, there must exist mechanisms which guarantee the specific transport of proteins across membranes. Membrane assembly of proteins can be visualized as a similar problem.

About fifteen years ago a first hypothesis was put forward which tried to explain specific import of proteins into the endoplasmic reticulum (ER), the "signal hypothesis" (for review see Zimmermann and Meyer, 1986). According to this hypothesis there are specific signals present in the proteins to be imported and the signals, typically aminoterminal and transient, guide the nascent polypeptides and the translating ribosomes to pores in the ER membrane. The tight interaction between ribosomes and pores and the resulting coupling of translation and import were suggested to prevent folding of a particular protein before it had reached its functional location. Therefore, the mechanism was termed cotranslational. On first sight the identification of a system which apparently decodes the signals, the signal recognition particle (SRP)/docking protein-system, seemed to provide support for this hypothesis.

However, during the last year the whole concept has been seriously questioned by data published by several laboratories. The novel themes seem to be that there does not have to be a coupling between translation and import and that folding of precursor proteins in the cytoplasm is prevented by the action of different cytoplasmic systems, at least one of which depends on ATP.

We are employing precursors of small secretory and membrane proteins as tools for defining the different stages in the import of proteins into the endoplasmic reticulum. The logic behind this approach is: i) that these precursor proteins have typical signal sequences, are processed by signal peptidase and thus presumably share some steps of the import pathway with larger precursor proteins, and ii) that these proteins may be able to bypass certain steps, obligatory for larger precursor proteins, possibly because of a lower tendency for folding into structures, incompatible with membrane insertion or transport. The precursor proteins, we selected, are M13 procoat protein (73 amino acid residues), the precursor of a bacterial plasma membrane protein, and the precursors of two eucaryotic secretory proteins, honeybee prepromelittin (70 amino acid residues) and frog prepropeptide GLa (64 amino acid residues) (Figure 1). An additional reason for choosing these precursor proteins is that they could be expected to be imported posttranslationally just because of their size and, therefore, to give a technical advantage: Approximately 40 amino acid residues of a nascent polypeptide are shielded within a protease-inaccessible domain of the ribosome. This does not leave much time between the point where a 20 to 23 residues long signal peptide emerges from the ribosome and becomes free to interact with a receptor or a membrane and the point of termination of protein synthesis.

```
                       ↓
 +-    +                - --  +           -                   +  ++   +
MKKSLVLKASVAVATLVPMLSFAAEGDDPAKAAFNSLQASATEYIGYAWAMVVVIVGATIGIKLFKKFTSKAS
    *** * **** **  * ***     * ***  * * *   *  *** ***** * * * **  *   *

 +                     - -   - - - - - - -       +             ++++
MKFLVNVALVFMVVYISYIYAAPEPEPAPEPEAEADAEADPEAGIGAVLKVLTTGLPALISWIKRKRQQG
   ** ***** ** *  * **     *    * * * *   * * *** **   * *** **

  +                 -     - -----++  +    +      +  +   +   ++-
MYKQIFLCLIIAALCATIMAEASAFADADEDDDKRYVRGMASKAGAIAGKIAKVALKALGRRDS
    *** ****** * * * * *** *        *   *  * ***  ** *** **
```

Figure 1. Primary Structure of M13 Procoat Protein, Honeybee Prepromelittin and Frog Prepropeptide GLa.
The cleavage site for signal peptidase is indicated by an arrow. Positively (+) and negatively (-) charged amino acid residues are indicated as well as hydrophobic ones (*).

RESULTS

Our experimental system involves in vitro systems, isolated from E.coli, wheat germs or rabbit reticulocytes, for translation of native mRNAs or in vitro-transcripts and ER-derived vesicles, isolated from dog pancreas. We assay association of small precursor proteins with microsomes in binding experiments where precursors are incubated with microsomes at low temperatures and microsomes are reisolated by gradient centrifugation or gel filtration. Membrane insertion is typically assayed as removal of the signal peptide by signal peptidase on the luminal side of the membrane. The standard assays for assembly or transport are testing sequestration, i.e. protection against externally added protease in the absence of detergent but sensitivity in the presence of detergent, and subfractionation of microsomes at neutral and alkaline pH-values.

Specific Association of a Precursor Protein with the ER-Membrane

The prediction for a specific association of a precursor with its target membrane is that the precusor is soluble and is able to expose its signal peptide. The classic model for specific association of precursors with microsomes involves SRP and docking protein. The small precursors, however, do not use the SRP/docking protein-system (Watts et al, 1983; Zimmermann and Mollay, 1986; Schlenstedt and Zimmermann, 1987). Therefore, we concluded that the SRP/docking protein-system has a role in keeping the signal peptide exposed while the small precursors can do so because of intrinsic properties (Müller and Zimmermann, 1987a).

A set of typical experiments on the docking protein-independence of prepromelittin processing is shown in Figures 2 to 4: While even the lowest concentration of trypsin used for pretreatment of microsomes is sufficient to completely destroy docking protein and, accordingly, to inactivate the microsomes with respect to their ability to process pre-kappa light chain, there is no effect on the ability of such pretreated microsomes to process prepromelittin (for further details refer to Zimmermann and Mollay, 1986 and Schlenstedt and Zimmermann, 1987).

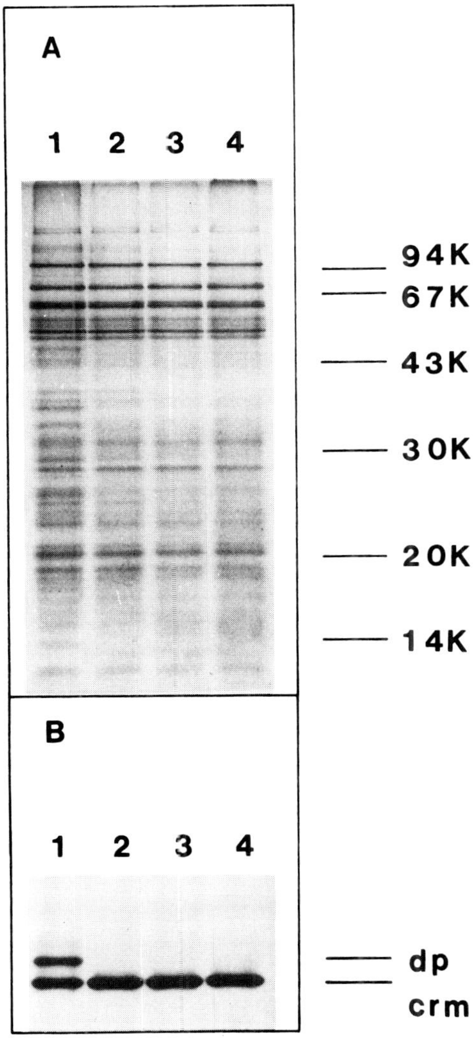

Figure 2. Effect of Trypsin Treatment of Microsomes on the Total Protein Pattern and on Docking Protein.
Four aliquots of dog pancres microsomes were incubated without trypsin (mock-treatment) (lane 1) or in the presence of trypsin at final concentrations of 10 (lane 2), 20 (lane 3) or 50 µg/ml (lane 4). After incubation for 60 min at 0 °C soybean trypsin inhibitor (twofold molar excess) and PMSF (1 mM) were added and the incubation was continued for 5 min at 0 °C. For analysis of the total protein pattern (A) and the docking protein content (B) two aliquots of each sample were run on SDS polyacrylamide gels (17.5 %). One set of samples was analysed by staining with Coomassie Brilliant Blue (A). The other set was electrophoretically transferred onto nitrocellulose and then probed with a rabbit antiserum directed against docking protein (dp) and with 14C-protein A (B). The relevant portion of an autoradiograph of this blot is shown. The antiserum crossreacts with a luminal protein (crm).

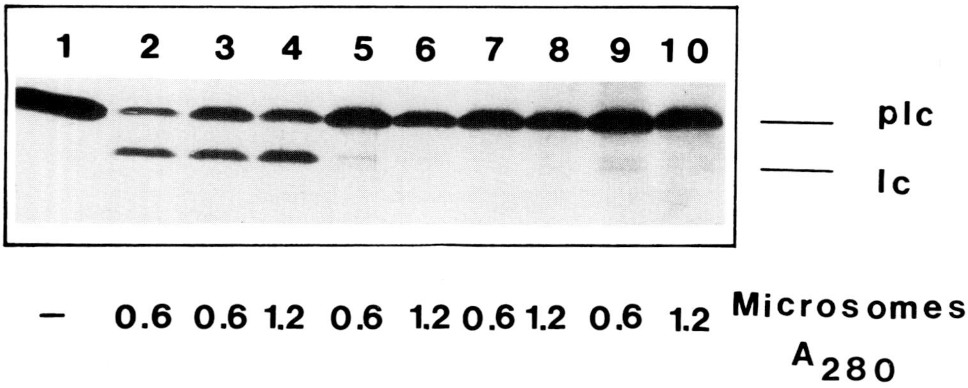

Figure 3. Effect of Trypsin Treatment of Microsomes on Processing of Pre-kappa-Light Chain.
Pre- kappa-light chain (plc) was synthesized in rabbit reticulocyte lysates for 60 min at 37 °C in the absence of membranes (lane 1) or in the presence of dog pancreas microsomes, which were either not treated prior to their inclusion in the translation reactions (lane 2), mock-treated (lanes 3 and 4) or treated with trypsin and trypsin inhibitors as described in the legend to Figure 2. The trypsin concentrations during the treatment were 10 µg/ml (lanes 5 and 6), 20 µg/ml (lanes 7 and 8) and 50 µg/ml (lanes 9 and 10). The relevant portion of a fluorograph after electrophoretic analysis is shown.

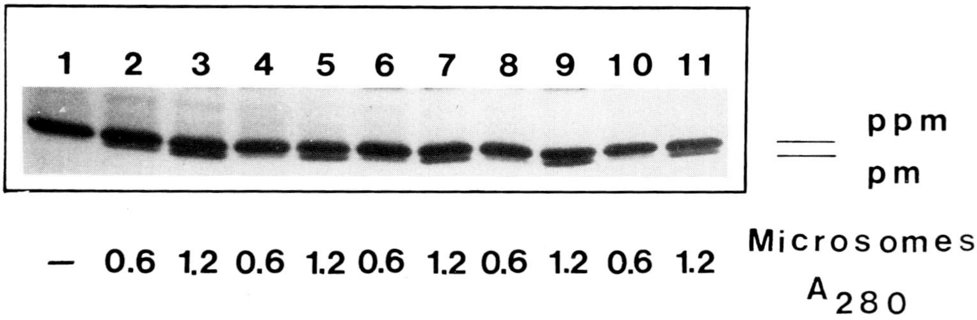

Figure 4. Effect of Trypsin Treatment of Microsomes on Processing of Prepromelittin.
Prepromelittin (ppm) was synthesized in rabbit reticulocyte lysates for 60 min at 37 °C in the absence of membranes (lane 1) or in the presence of dog pancreas microsomes, which were either not treated prior to their inclusion in the translation reactions (lanes 2 and 3), mock-treated (lanes 4 and 5) or treated with trypsin and trypsin inhibitors as described in the legend to Figure 2. The trypsin concentrations during the treatment were 10 µg/ml (lanes 6 and 7), 20 µg/ml (lanes 8 and 9) and 50 µg/ml (lane 10 and 11). The relevant portion of a fluorograph after butanol extraction and electrophoretic analysis is shown.

Insertion of a Precursor Protein into the ER-Membrane

The insertion of a precursor into the hydrophobic core of its target bilayer can be expected to be the second step in the import pathway and to precede membrane assembly as well as completion of translocation, i.e. transport. For the classic examples of proteins to be imported into mirosomes the ribosome in collaboration with a ribosome receptor at the surface of the microsomal membrane seem to be involved at this step. Again, however, the small precursors do not show such a requirement (Schlenstedt and Zimmermann, 1987; Wiech et al., 1987; Müller and Zimmermann, 1987b). On the other hand, we observed a requirement for the hydrolysis of ATP and a so far unknown cytosolic component in these cases. We concluded from these data that the ribosome in the case of large precursor proteins and the ATP-dependent component in general have a role in keeping precursor proteins competent for membrane insertion, i.e. preventing an unfavourable conformation. This notion is supported by our observation that the processing of small precursor proteins are inhibited at this step by formation of intramolecular disulfide bridges (Müller and Zimmermann, 1987b).

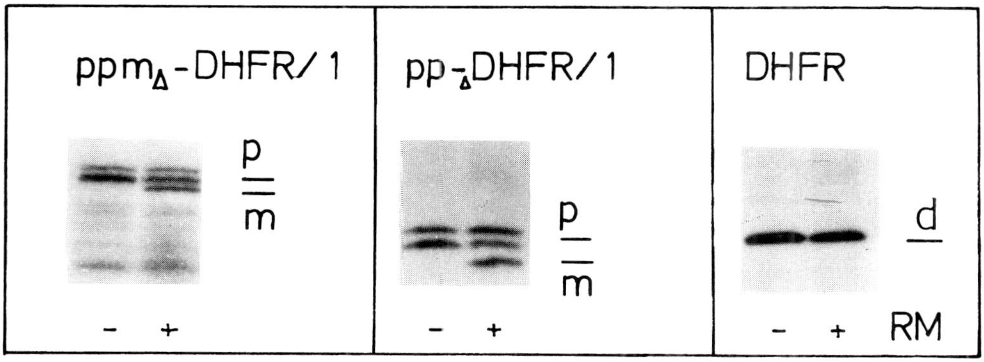

Figure 5. Processing of Hybrid Proteins between Prepromelittin and Dihydrofolate Reductase by Microsomes.
Hybrid Proteins between prepromelittin and dihydrofolate reductase, one containing prepromelittin with a minor deletion at the carboxy terminus and full length dihydrofolate reductase (ppm$_\Delta$-DHFR/1) and the other one containing prepromelittin without the melittin domain and dihydrofolate reductase with a large amino terminal deletion (pp-$_\Delta$DHFR/1), and dihydrofolate reductase (DHFR, d), respectively, were synthesized in rabbit reticulocyte lysates for 30 min at 37 °C in the absence or presence of microsomes (RM). The relevant portion of a fluorograph after electrophoretic analysis is shown.

The following set of experiments points to a role of the ribosome in the case of large, SRP and docking protein-dependent, precursor proteins (Figures 5 to 8): While methotrexate is able to bind to a hybrid protein consisting of prepromelittin and dihydrofolate reductase in the absence of microsomes and to induce a native conformation in the dihydrofolate reductase domain (dd), it cannot do so in the presence of microsomes and, accordingly, does not interfere with import (for further details refer to Müller and Zimmermann, 1987a and b).

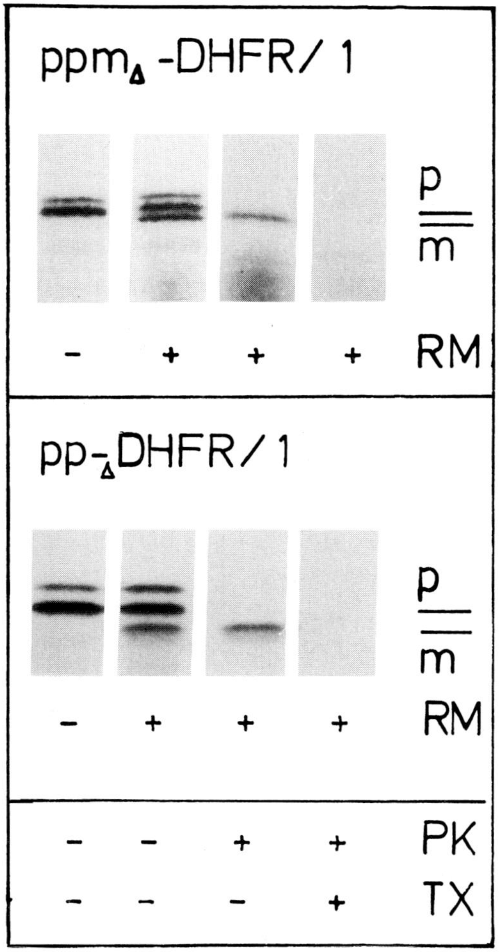

Figure 6. Import of Hybrid Proteins between Prepromelittin and Dihydrofolate Reductase into Microsomes. Hybrid proteins between prepromelittin and dihydrofolate reductase were synthesized in rabbit reticulocyte lysates for 30 min at 37 °C in the absence or presence of microsomes (RM). Afterwards each sample was divided into three aliquots. One aliquot was incubated in the absence of proteinase K (PK), the second one in the presence of proteinase K (50 µg/ml) and the third one in the presence of both, Triton X-100 (TX) and proteinase K, for 60 min at 0 °C. The relevant portion of a fluorograph after electrophoretic analysis is shown.

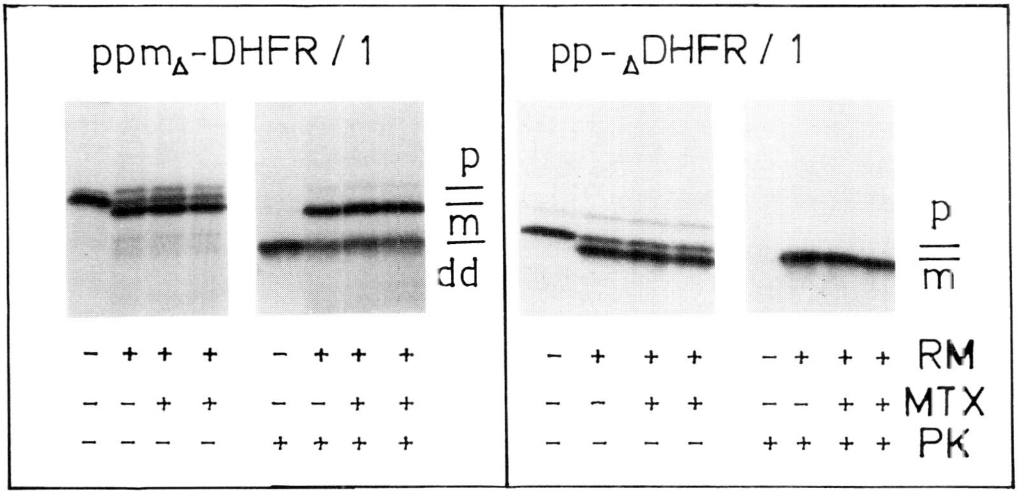

Figure 7. Effect of Methotrexate on the Import of Hybrid Proteins between Prepromelittin and Dihydrofolate Reductase into Microsomes.
Hybrid proteins between prepromelittin and dihydrofolate reductase were synthesized in rabbit reticulocyte lysates for 30 min at 37 °C in the absence of methotrexate (MTX) or in the presence of methotrexate (10 µM or 100µM) in the absence or presence of microsomes (RM). Afterwards each sample was divided into two aliquots. One aliquot was incubated in the absence of proteinase K (PK), the other one in the presence of proteinase K (50 µg/ml) for 60 min at 0 °C. The relevant portion of a fluorograph after electrophoretic analysis is shown.

Assembly of a Protein into the ER-Membrane and Transport of a Protein across the ER-Membrane

Membrane assembly of a membrane protein and completion of translocation of a soluble protein, respectively, can be seen as the final step of an import pathway. In the case of our small precursor proteins we were able to suggest a component of the microsomal membrane to be involved in transport of promelittin (Mollay and Zimmermann, 1986). However, not all small precursor proteins depend on this component (Schlenstedt and Zimmermann, 1987; Müller and Zimmermann, 1987a).

A typical experiment on the involvement of a protease-sensitive component of the microsomal membrane on the sequestration of promelittin is shown in

Figure 9: While trypsin pretreatment of microsomes does not inactivate the microsomes with respect to prepromelittin processing (see above), it completely inhibits such pretreated microsomes with respect to promelittin sequestration (for further details refer to Zimmermann and Mollay, 1986 and Müller and Zimmermann, 1987a).

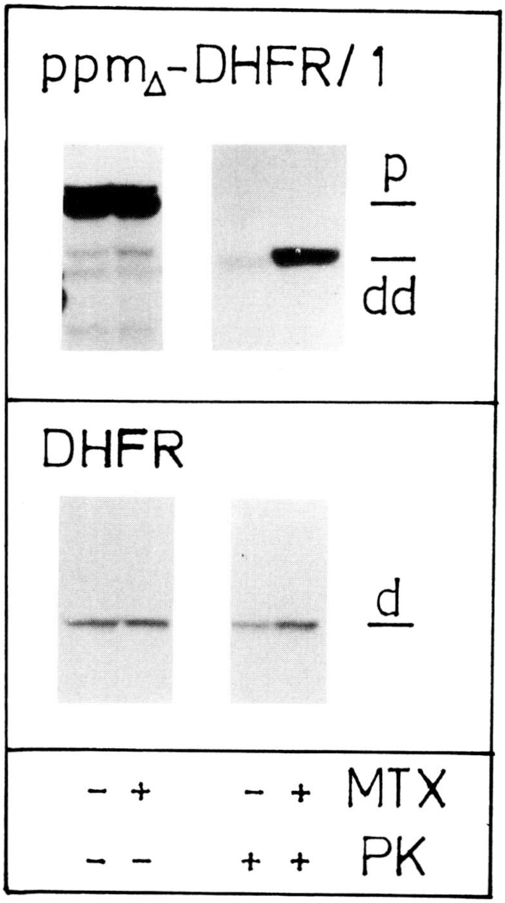

Figure 8. Effect of Methotrexate on the Folding of Hybrid Proteins between Prepromelittin and Dihydrofolate Reductase.
A hybrid protein between prepromelittin and dihydrofolate reductase (ppm$_\Delta$-DHFR/1) and dihydrofolate reductase (DHFR, d), respectively, were synthesized in rabbit reticulocyte lysates for 30 min at 37 °C in the absence of methotrexate (MTX) or in the presence of methotrexate (100 µM). Afterwards each sample was divided into two aliquots. One aliquot was incubated in the absence of proteinase K (PK), the other one in the presence of proteinase K (50 µg/ml) for 60 min at 0 °C. The portion of a fluorograph is shown.

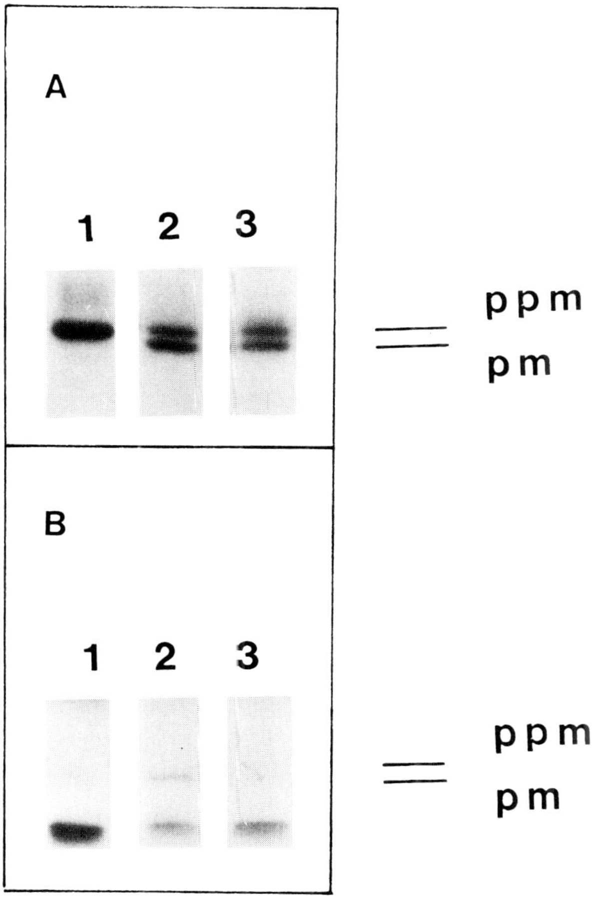

Figure 9. Effect of Trypsin Treatment of Microsomes on Processing of Prepromelittin and Sequestration of Promelittin.
Prepromelittin (ppm) was synthesized in rabbit reticulocyte lysates for 60 min at 37 °C in the absence of membranes (lanes 1) or in the presence of dog pancreas microsomes, which were either not treated prior to their inclusion in the translation reactions (lanes 2), or treated with trypsin and trypsin inhibitors as described in the legend to Figure 2 at a concentration of 10 µg/ml (lanes 3). Afterwards each sample was divided into two aliquots. One aliquot was incubated in the absence of proteinase K (A), the other one in the presence of proteinase K (B) for 60 min at 0 °C. The relevant portion of a fluorograph after electrophoretic analysis and butanol extraction is shown.

DISCUSSION

We propose the following mechanism for the import of small secretory and membrane proteins into the endoplasmic reticulum (Figure 10): Precursor proteins with a content of less than 80 amino acids are released from the ribosome before an interaction between either the signal peptide and signal recognition particle or the ribosome and the ribosome receptor can occur (Schlenstedt and Zimmermann, 1987; Müller and Zimmermann, 1987a; Wiech et al., 1987). Their primary structures, however, have evolved in a way which allows them to stay competent for membrane association and insertion without the aid of these components (Müller and Zimmermann, 1987a). Possibly, a loop is formed by the mature part, thereby bringing the amino- and carboxy-terminal parts in close proximity of each other. Ionic interactions may stabilize this structure, the presence of unbalanced charges in these positions may act destabilizing (Müller and Zimmermann, 1987a and b). Even disulfide bridges are allowed at this point, although, they probably never occur in vivo. This structure can associate with the microsomal membrane. It is presently unclear, whether a signal receptor is involved at this step or whether it is a pure protein-lipid interaction. We assume that an unfolding, i.e. destabilization of this structure, has to take place before membrane insertion can occur (Müller and Zimmermann, 1987b). Typically this is catalysed by a cytoplasmic protein which depends on the hydrolysis of ATP for its action (Wiech et al., 1987). Strikingly, the protein without any charged amino acid residues at the mature termini does not depend on this activity and disulfide bridges cannot be dissociated (Müller and Zimmermann, 1987b). Insertion may occur in form of a hairpin structure between the signal peptide and the mature amino terminus. This event leads to exposure of the cleavage site to the luminal side of the membrane and typically results in processing by leader peptidase. We assume that this takes place in a lipid environment because we observed the same characteristics for insertion into detergent micelles (Müller and Zimmermann, 1987b; Wiech et al, 1987). The following membrane assembly or transport of the mature proteins may occur in one or more steps and may or may not involve membrane components. We have defined one protease sensitive component which seems to be necessary in order to expell the hydrophobic melittin domain within promelittin out of the membrane (Zimmermann and Mollay, 1986; Müller and Zimmermann, 1987a). Accordingly, proteins which do not contair a similar domain do not depend on this component. The alternatives fcr the action of such a component are still the ones proposed by either the signal hypothesis or the membrane triggered folcing hypothesis or something in between. There may be transient proteinaceous pores involved or just the phospholipids, possibly in form of a transient bilayer distortion. Or there may be membrane proteins which do not form a pore but trigger or / and control the bilayer distortion.

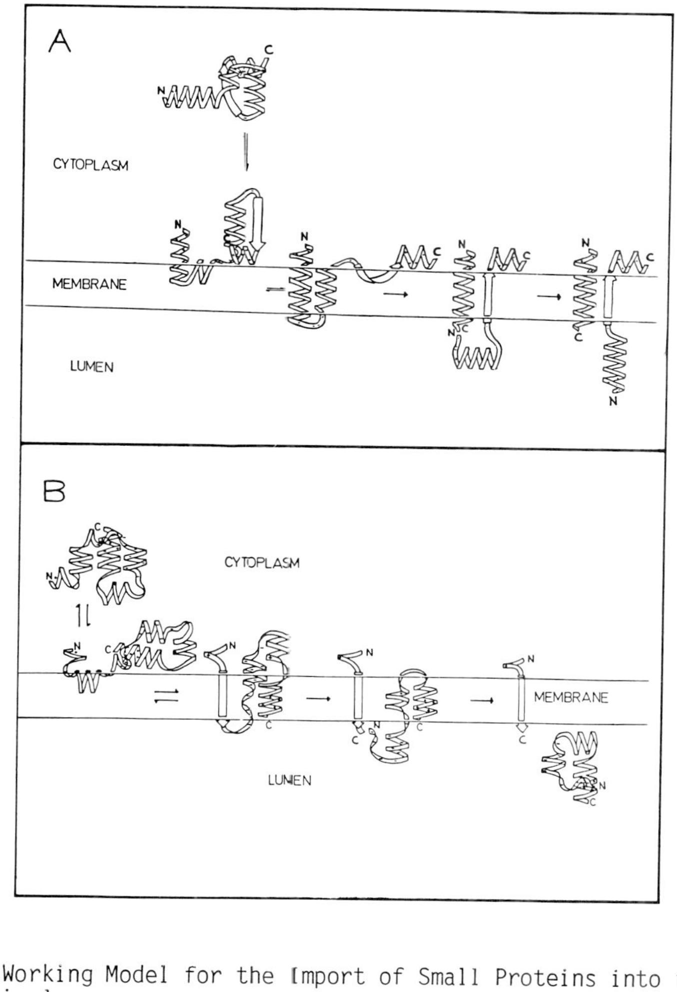

Figure 10. Working Model for the Import of Small Proteins into the Endoplasmic Reticulum.
Models for the assembly of M13 coat protein into microsomal membranes (A) and the transport of small secretory proteins, in this case promelittin into the ER-lumen (B), respectively, are shown. Charged amino acids are indicated.

ACKNOWLEDGEMENTS

We are thankful to Dr. Günther Kreil for mRNA and cDNA coding for prepromelittin and prepropeptide GLa, to Dr. William Wickner for DNA coding for procoat protein, and to Drs. Hermann Bujard and Dietrich Stueber for cDNA coding for dihydrofolate reductase. This work was supported by grants Zi234/2-1 and Zi234/2-2 from the 'Deutsche Forschungsgemeinschaft' and by grant B10 from the 'Sonderforschungsbereich 184: Molekulare Grundlagen der Biogenese von Zellorganellen'.

REFERENCES

Müller G, Zimmermann R (1987) Import of honeybee prepromelittin into the endoplasmic reticulum: structural basis for independence of SRP and docking protein. EMBO J 6: 2099-2107.

Müller G, Zimmermann R (1987) Import of SRP and docking protein independent precursor proteins into the endoplasmic reticulum: ATP requirement and effect of disulfide bond formation. Submitted for publication.

Schlenstedt G, Zimmermann R (1987) Import of frog prepropeptide GLa into microsomes requires ATP but does not involve docking protein or ribosomes. EMBO J 6: 699-703.

Watts C, Wickner W, Zimmermann R (1983) M13 procoat and pre-immunoglobulin share processing specificity but use different membrane receptor mechanisms. Proc Natl Acad Sci USA 80: 2809-2813.

Wiech H, Sagstetter M, Müller G, Zimmermann R (1987) The ATP requiring step in assembly of M13 procoat protein into microsomes is related to preservation of transport competence of the precursor protein. EMBO J 6: 1011-1016.

Zimmermann R (1986) Import of proteins into mitochondria. In: Strauss A W, Boime I, Kreil G (eds) Protein compartmentalization. Springer, Berlin Heidelberg New York, pp 119-136.

Zimmermann R, Meyer D I (1986) 1986: a year of new insights into how proteins cross membranes. Trends Biochem Sci 11: 512-515.

Zimmermann R, Mollay C (1986) Import of honeybee prepromelittin into the endoplasmic reticulum: requirements for membrane insertion, processing and sequestration. J Biol Chem 261: 12889-12895.

BIOGENESIS AND MEMBRANE TOPOLOGY OF OUTER MEMBRANE PROTEINS IN *ESCHERICHIA COLI*

Jan Tommassen
Department of Molecular Cell Biology
State University of Utrecht
Transitorium 3
Padualaan 8
3584 CH UTRECHT
The Netherlands

NATO ASI Series, Vol. H16
Membrane Biogenesis. Edited by J. A. F. Op den Kamp

INTRODUCTION

The outer membrane of *E. coli* and its proteins

The cell envelope of *E. coli* consists of two membranes, the cytoplasmic membrane and the outer membrane, which are separated by the peptidoglycan-containing periplasm (Fig. 1). The outer membrane is an asymetrical membrane, consisting of lipopolysaccharide (LPS) and phospholipids in the outer and inner monolayer, respectively. It is covalently attached to the peptidoglycan via a lipoprotein. For a detailed description of the molecular architecture of the outer membrane, the reader is refered to a review by Lugtenberg and Van Alphen (1983).

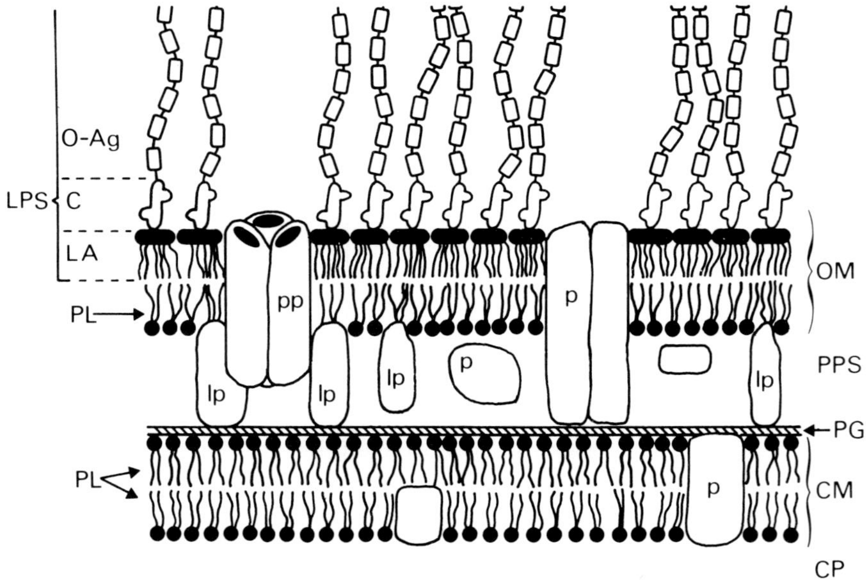

Fig. 1: Model of the cell envelope of Enterobacteriaceae. The cytoplasm (CP) is surrounded by the cytoplasmic membrane (CM), the peptidoglycan layer (PG), the periplasm (PPS) and the outer membrane (OM). The cytoplasmic membrane consists of phospholipids (PL) and proteins (p). The outer membrane consists of phospholipids (PL), lipopolysaccharide (LPS) and proteins. The most abundant outer membrane proteins are lipoprotein (lp), pore proteins (pp) and OmpA protein (p). The LPS consists of three moieties, namely lipid A (LA), the core (C) and the O-antigen (O-Ag). The O-antigen is lacking in *E. coliK-12*.

The outer membrane enables the bacteria to live in the colon by creating a barrier for harmful compounds like bile salts and enzymes. A number of abundantly expressed proteins makes the membrane permeable to nutrients. These proteins, designated as porins, form channels or pores in the outer membrane through which small hydrophilic solutes with molecular weights up to about 600 can pass (for reviews, see Nikaido and Vaara 1985; Benz 1985; Nakae 1986). When grown under standard laboratory conditions, *E. coli* K12 produces two distinct porins, namely OmpF and OmpC. The synthesis of another pore protein, PhoE, is induced when cells are grown under P_i-limitation (Overbeeke and Lugtenberg 1980; Tommassen and Lugtenberg 1980). Although the pores formed by OmpC, OmpF and PhoE are general pores which facilitate the diffusion of small hydrophilic solutes, the PhoE pores are particularly efficient for phosphate-containing and other negatively charged solutes (Korteland et al. 1982; Benz et al. 1984), whereas the OmpF and OmpC pores are more efficient for cationic solutes (Benz et al. 1985).

In addition to the porins, the outer membrane contains proteins which facilitate the uptake of specific solutes. One of these proteins is LamB protein. By the presence of a specific binding site (Ferenci et al. 1980; Benz et al. 1986) this protein allows the uptake of maltodextrins consisting of up to seven glucose residues (Wandersman et al. 1979). Such molecules are too large to pass through the porin pores. The synthesis of LamB protein is induced by the presence of maltose or maltodextrins in the growth medium (Schwartz 1967).

Another abundant outer membrane protein is OmpA protein. No pore function has been assigned to this protein. Together with the previously mentioned lipoprotein, OmpA protein is somehow involved in maintaining the structural integrity of the outer membrane and the rod shape of the cell (Sonntag et al. 1978).

The structural genes encoding the porins (Inokuchi et al. 1982; Overbeeke et al. 1983; Mizuno et al. 1983), OmpA protein (Beck and Bremer 1980), LamB protein (Clément and Hofnung 1981) and several other outer membrane proteins have been sequenced. The porins are mutually very homologous (Fig. 2), but they share hardly any homology with LamB or OmpA. Only a few regions with weak homology between these different proteins were discerned (Nikaido and Wu 1984) and a role of these regions in protein targeting has been implicated. In the following section, I will describe experiments which were designed to identify intragenic export information in outer membrane proteins. In one peculiar aspect the mentioned outer membrane

```
                                         a
OmpF            V L   AVGL   F KGNGENSYGGN  MT A L        45
PhoE  AEIYNKDGNKLDVYGKVKAMHYMSDNASKD*****GDQSYIRFGF        40
OmpC   V         L    DGL   F   KDV            T M L        40

                                  b
           SD      Q  YN Q  NSEG DAQ GN*                     87
      KGETQINDQLTGYGRWEAEFAGN***KAESDTAQQKTRLAFAG            80
          VT       Q  YQIQ     S  NENNS*W  V                 79

                                              c
        A V        Y VV  ALGY    L      *T YS                126
      LKYKDLGSFDYGRNLGALYDVEAWTDMFPEFGGDSSAQTD               120
       FQ V        Y VV   TS    VL      *TYGS                118

                                                   d
      D FVG VG V      SN    LV     FAV  L    *           TA  166
      NFMTKRASGLATYRNTDFFGVIDGLNLTLQYQGKNEN**************[R]D*V  160
         QQ GN F            LV     FAV       G PSGEGFTSGVTNNG   A  172

                                              e
      RRS    V G IS EYE ** G V    GAA    L E A P             204
      KKQNGDGFGTSLTYDFGGSDFAISGAYTNSDRTNEQN*LQS               200
      LR     V G I   YE ** G G  ISS K  DA  TAAY               211

                                                   f
      L N  K  Q                 AN G  NA    NKFTNTS           250
      [R]GTGKRAEAWATGLKYDANNIYLATFYSETRKMTPIT******[G]GF        240
      I N D   TYTG              AQ TQ YNA RVG     SLGW        252

                                              g
          DVLL                 IA TK  A  V       V           290
      ANKTQNFEAVAQYQFDFGLRPSLGYVLSKGKDIE*[G]IGDED              280
         A                    A LQ     NLGR YD               293

                                                 h
          FE             TY    I   I       GVGS ***          330
      LVNYIDVGATYYFNKNMSAFVDYKINQLDSDNKLN****INNDD            320
      ILK V              TY        L  *   QFTRDAG   T N       336

      T    IV                                                 340
      IVAVGMTYQF                                              330
         L LV                                                 346
```

Fig. 2: Comparison of the amino acid sequences of the porins PhoE (middle line), OmpF (upper line) and OmpC (bottom line). Amino acids in OmpF and OmpC are only indicated when they differ from PhoE. The proteins are aligned to give maximal homology. A * indicates a deletion of one amino acid residue. The regions with the most pronounced differences between the proteins are indicated and numbered a-h. The residues Arg-158, Arg-201, Gly-238 and Gly-275 of PhoE which were shown to be cell surface-exposed, are boxed.

proteins are similar: in spite of the fact that they are integral membrane proteins, they lack segments of hydrophobic amino acids, long enough to span the membrane. In the last section of this chapter, I will describe how such apparently hydrophilic proteins can exist in the hydrophobic environment of the membrane. Emphasis will be layed on PhoE protein, which is the model system in our laboratory.

INTRAGENIC EXPORT INFORMATION IN OUTER MEMBRANE PROTEINS

Role of the signal sequence

Since protein synthesis takes place in the cytoplasm, outer membrane proteins have to pass the cytoplasmic membrane to reach their destination. Like other exported proteins, outer membrane proteins are synthesized in a precursor form with an N-terminal signal sequence. Since such a signal sequence is present in all exported proteins, it is likely to play an important role in the export process. Indeed, mutations in the signal sequence, e.g. of LamB protein (Emr et al. 1980), have been described which prevent export.

Is a signal sequence sufficient to mediate translocation? In eukaryotes, most proteins which have to be transported from the cytoplasm into cellular organelles are synthesized as precursors. There N-terminal peptide extensions appear to contain all the information which is required for transport. Thus a normally cytosolic protein can be transported into mitochondria when hooked to the presequence of a mitochondrial protein (see e.g. Hurt et al. 1985; Horwich et al. 1985). The situation appears to be different in case of bacterial signal sequences. The cytoplasmic protein β-galactosidase was not exported from the cytoplasm when fused to the signal sequence of LamB protein (Moreno et al. 1980). Thus, either the signal sequence does not contain all the information which is required for protein export, or β-galactosidase contains sequences or conformations which prevent export. This question will be addressed later on.

Does the signal sequence determine the ultimate localization of an exported protein? In eukaryotes, the N-terminal peptide extensions of organellar proteins not only contain the information required for import into these organelles, but also for sorting to different compartments within the organelles (Hurt and Van Loon, 1986). For instance, replacement of the transit-peptide of plastocyanin, which is normally transported to the thylakoid lumen of chloroplasts, by the transit-peptide of ferredoxin, which is located in the stroma of the chloroplasts, results in accumulation of the protein in the stroma (Smeekens et al. 1986). In *Escherichia coli*, a multivariate data analysis revealed that signal peptide amino acid sequences contain information related to the final protein localization (Sjöström et al. 1987). However, experimental evidence argues against a role of the signal sequence in determining the ultimate localization of an exported protein. Replacement of the signal sequence of PhoE protein by the one of the periplasmic enzyme β-lactamase did not affect the localization of the protein (Tommassen et al. 1983). Similarly, OmpF protein was normally translocated to the outer membrane with the signal sequence of an inner membrane protein (Jackson et al. 1985).

In summary, the signal sequence is required for protein export but may not be sufficient for the transport process. In addition, the rest of the protein determines the ultimate localization of the exported protein.

Identification of additional export and sorting signals by use of gene fusions

Moreno et al. (1980) showed that β-galactosidase is not transported out of the cytoplasm by the signal sequence of LamB protein. An interpretation of this result is that additional information which is essential for protein export, is present in the mature portion of LamB. If this interpretation is correct, an obvious way to identify this additional export information is the construction of hybrid genes in which *lacZ*, the structural gene for β-galactosidase is fused to larger portions of the *lamB* gene. Indeed, it was found that such hybrid proteins co-fractionated with the outer membrane, provided that they contained a sufficiently large portion of LamB protein (Silhavy et al. 1977; Hall et al. 1982). Analysis of the hybrid genes suggested that an essential export signal is located between residues 27 and 39 of mature LamB protein and that a sorting signal that specifies outer membrane localization is present between residues 39 and 49 (Benson et al. 1984). Interestingly, the latter region was identified by Nikaido and Wu (1984) as a region with homology to OmpA and the porins, suggesting the existence of a common sorting signal in outer membrane proteins. However, the results obtained with this approach, depended upon the passenger protein used in the fusions. When the signal sequence of the periplasmic enzyme alkaline phosphatase was replaced by the signal sequence and the first 60 amino acids of LamB protein, the hybrid protein ended up in the periplasm (Hoffman and Wright 1985). This argues against the presence of a sorting signal for outer membrane localization between residues 39 and 49 of LamB.

The cause of the discrepancies in the results obtained with different gene fusions probably lies in the techniques used to localize the hybrid proteins at the subcellular level. It appears that standard cell fractionation techniques are not reliable when applied to localize genetically engineered proteins. This was first observed in case of a PhoE-LacZ fusion protein (Tommassen et al. 1985). The PhoE moiety of this hybrid consists of 300 out of the 351 amino acids of the PhoE precursor. As was found for the LamB-LacZ fusions, induction of the synthesis of the PhoE-LacZ fusion was lethal to the cells and led to the accumulation of precursors of other exported proteins. Cell fractionation experiments suggested that the protein was efficiently translocated to the outer membrane. However, when ultrathin cryosections of the cells were labelled with antibodies directed against β-galactosidase or PhoE protein and with protein A-gold complexes, the gold-label was almost exclusively found in the cytoplasm (Tommassen et al. 1985). The latter result was confirmed by protease-accessibility experiments (Tommassen and De Kroon 1987). The vast majority of hyrid protein turned out to be inaccessible to trypsin when whole cells or spheroplasts were treated with trypsin. Thus, the hybrid protein is present in the cytoplasm, presumably as aggregates which co-fractionate with the outer membrane. However, also apparently inner-membrane-spanning forms of the hybrid protein could be detected. A minor quantity of hybrid protein was accessible to trypsin from the periplasmic side of the membrane, leaving degradation products with apparent molecular weights of 116000 (which is identical to the molecular weight of β-galactosidase) and 79000. These degradation products reacted with anti-β-galactosidase serum, but not with anti-PhoE protein serum. Our interpretation of these results is that export, initiated at the PhoE moiety of the hybrid protein, hampers at or in the β-galactosidase moiety. Thus, the export sites in the membrane become blocked. This leads to the accumulation of precursors of other exported proteins and is lethal to the cells. Also hybrid protein which is synthesized after the export-sites have been blocked, accumulates in the cytoplasm in the form of aggregates.

Since cell fractionation experiments appeared to be unreliable for the subcellular localization of the PhoE-LacZ hybrid protein, we also re-investigated the localization of LamB-LacZ fusion proteins which were reported to co-fractionate with the outer membrane (Voorhout, De Kroon, Leunissen-Bijvelt, Verkleij and Tommassen, manuscript submitted for publication). We observed in ultrathin cryosections that induction of the synthesis of the hybrid proteins leads to the appearance of membrane-like structures within the cytoplasm in over 50% of the cells (Fig. 3). Immuno-gold labelling experiments showed that the hybrid proteins were associated to these membrane-like structures.

In summary, when β-galactosidase is fused to N-terminal portions of

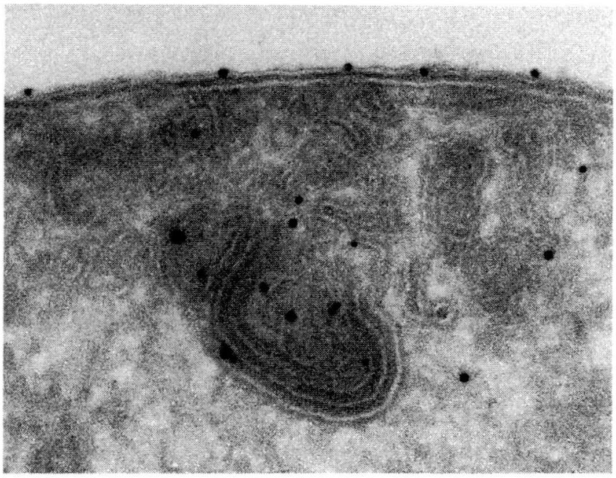

Fig. 3: Appearance of membrane-like structures in the cytoplasm of cells, induced for the synthesis of a LamB-LacZ hybrid protein. The synthesis of the hybrid protein in strain pop3299 was induced by adding maltose to the growth medium. Ultrathin cryosections of the cells were incubated with anti-LamB protein serum and protein A-gold complexes.

outer membrane proteins PhoE or LamB, the hybrid proteins are not exported. They accumulate as aggregates in the cytoplasm or in intra-cytoplasmic membranes which co-fractionate with the outer membrane. Although this could mean that essential export information is located close to the C-termini of these outer membrane proteins, the interpretation that β-galactosidase contains sequences or conformations which are incompatible with export seems to be more likely. If this is indeed the case, the use of these gene fusions will not lead to the identification of intragenic export information in outer membrane proteins.

Internal deletions in outer membrane protein genes

As described in the previous section, the disadvantage of using gene fusions for the identification of export and sorting signals in a protein is, that the passenger protein may influence the localization of the hybrid protein. This problem is circumvented by making deletions internal to the protein under study, followed by the subcellular localization of the truncated proteins. This approach was followed for PhoE protein (Bosch et al. 1986). A series of deletions in the *phoE* gene was created with the aid of exonuclease *Bal*31. Together, the overlapping deletions covered the complete gene, except for the DNA encoding the signal sequence and the first eleven animo acids. Pulse-label and pulse-chase experiments showed that all mutant proteins were processed. Since processing takes place at the outside surface of the inner membrane all the mutant proteins are therefore most likely exported from the cytoplasm. Two mutant proteins could not be localized in the cell because they were rapidly degraded by cellular proteases. Immunogold labelling on ultrathin cryosections demonstrated that the other mutant proteins accumulated in the periplasm. Therefore, no parts of PhoE protein seem to be essential for export through the cytoplasmic membrane except for the signal sequence (a possible role for the first eleven amino acids of the mature protein could be ruled out in other experiments described below). Furthermore, it appears that protein conformation, rather than a specific sorting signal is required for outer membrane localization.

Similar conclusions were reached when a series of overlapping *ompA* mutations was analysed (Freudl et al. 1985). Interestingly, some of the OmpA mutant proteins were detected in the outer membrane in immunocytochemical experiments. In one of these, the region with homology to LamB which was proposed to function as a common sorting signal, was removed. Rasmussen and Silhavy (1987) described small internal deletions in the corresponding region of LamB. The region between amino acids 27-49, which was presumed to contain export and sorting signals, could be removed without any effect on the rate of processing of the mutant proteins. However, the rate of processing was affected by deletions which extend amino terminally from amino acid 28 of mature LamB. This suggested that the first 28 amino acids of mature LamB are required for rapid and efficient export from the cytoplasm. However, an alternative interpretation is that only the conformation of the processing-site is affected. Also for PhoE protein, a series of deletions close to the N-terminus was created (Bosch et al. manuscript in preparation). In this series of mutations, the signal sequence and the first amino acid of mature PhoE were not affected, whereas the second residue, Glu, was changed in Asp. A deletion which removed amino acids 3-11 did not affect the rate of processing of the mutant protein, but processing was severely disturbed in case amino acids 3-18 were removed. This suggests the presence of an export signal between residues 12 and 18. However, a mutant protein in which amino acids 3-24 were deleted was normally processed and even incorporated into the outer membrane. This result suggests that the deletion Δ3-18 does not remove an essential export signal, but rather affects the conformation of the processing site.

To conclude, it appears that there are no essential export signals present in outer membrane proteins, except for the signal sequence. The mature parts of the proteins only play a passive role: they must be compatible with export. This conclusion is consistent with the very recent observation the the signal sequence of OmpA protein is sufficient to mediate the translocation of large fragments of protein 37 of phage T4 (MacIntyre et al. 1987). Probably, protein folding makes polypeptides incompatible with transport. It has been shown that the precursor of the periplasmic maltose-binding protein is competent for export only before it folds into its final stable tertiary conformation (Randall and Hardy 1986). Thus, the observation that β-galactosidase can not be exported can be explained by assuming that the growing chains of the protein rapidly fold in the cytoplasm to form export-incompatible domains.

Model for outer membrane protein translocation

The observation that many mutant PhoE and OmpA proteins with internal deletions accumulate in the periplasmic space suggests that the wild-type proteins are also transiently present in this compartment during translocation. This suggestion is underscored by the observation that overproduction of these proteins results in the accumulation of the proteins in the periplasm (Leunissen et al. manuscript in preparation; Freudl et al. 1986). In both cases, the periplasmic form is distinguishable from the correctly assembled protein. PhoE protein is present in the outer membrane in a trimeric form and antibodies directed against this trimeric form do not recognize the denatured form of the protein. The PhoE protein which accumulated in the periplasm, was only recognized by antibodies raised against the denatured form of the protein (Leunissen et al. manuscript in preparation). OmpA is a "heat modifiable" protein, i.e. its electrophoretic mobility in SDS-polyacrylamide gels is dependant on the solubilization temperature. The heat modifiability is caused by a conformational change. The overproduced OmpA protein which accumulated in the periplasm did not display this heat modifiable character (Freudl et al. 1986). However, after purification heat modifiability was observed upon addition of LPS.

An outer membrane protein translocation model that accounts for these results is the following: the presence of a signal sequence is sufficient to mediate export through the inner membrane as long as the remainder of the protein is not folded. At the periplasmic side of the membrane, the precursor is processed. The processed polypeptide is then recognized by LPS which causes a conformational change (trimerization; heat modifiability) and forces the protein into the outer membrane.

TOPOLOGY OF OUTER MEMBRANE PROTEINS

Structure of outer membrane proteins

The usual approach to predict the structure of a membrane protein is to search for stretches of approximately 20 hydrophobic amino acids that would form a membrane-spanning α-helix (Kyte and Doolittle 1982). However, inspection of the sequences of the porins (Fig. 2) reveals the presence of a large percentage of hydrophilic amino acids and the absence of long segments of hydrophobic residues. This may be related to the fact that transport to the outer membrane involves translocation across the inner membrane. The presence of long hydrophobic segments would probably block this process (Davis and Model 1985). Nevertheless, the porins span the outer membrane in a compact form with very small domains extending into the aqueous phase and in their native trimeric form they have a surface that is very hydrophobic (Garavito et al. 1983).

Using infrared absorption and high-angle X-ray diffraction, Kleffel et al. (1985) showed that the majority of OmpF is arranged in anti-parallel β-pleated sheet structure, with the β-strands oriented nearly normal to the membrane plane. Vogel and Jähnig (1986) confirmed this arrangement by using Raman spectroscopy. They showed that also LamB and the N-terminal 177 amino-acids of OmpA are similarly arranged (The remainder of the 325 residues long OmpA protein is known to extend in the periplasm; Schweizer et al. 1978). Thus, it appears that there is a common folding pattern for outer membrane proteins, which is substantially different from most other integral membrane proteins.

Identification of cell-surface exposed amino acids in PhoE protein

One approach to study the path of PhoE protein in the membrane is to pinpoint amino acid residues which are cell-surface exposed. This was achieved by selecting for mutants which produce a PhoE protein with altered cell-surface related properties. Bacteriophage TC45, which uses PhoE protein as its receptor, was used for this purpose (Korteland et al. 1985). TC45 resistant mutants were selected which still produced PhoE protein. Nucleotide sequence analysis of the mutations showed that the Arginine in position 158 of mature PhoE was changed into Histidine. Since the mutant proteins were functional as pores, they were normally incorporated into the membrane. Thus, it appears that Arg^{158} is cell-surface exposed and involved in TC45 receptor function.

In an alternative method, we employed monoclonal antibodies directed against the cell-surface exposed part of PhoE (Van der Ley et al. 1985). These monoclonal antibodies recognize the native trimeric structure of the protein, but not the denatured protein. Mutants could be selected which fail to bind the antibodies by using the antibody-dependant bactericidal action of the complement system (Van der Ley et al. 1986). Among the mutants obtained which still produce PhoE, the following amino acid substitutions were observed: $Arg^{201} \rightarrow$ His, $Arg^{201} \rightarrow$ Cys, $Gly^{238} \rightarrow$ Ser, $Gly^{275} \rightarrow$ Ser and $Gly^{275} \rightarrow$ Asp. Since the mutant proteins formed functional pores and conferred TC45 sensitivity, no gross conformational alterations had occurred. Therefore, the residues Arg^{201}, Gly^{238} and Gly^{275} are most likely cell-surface exposed and directly involved in the binding of the antibodies.

A model for the folding of PhoE protein

The amino acids Arg^{158}, Arg^{201}, Gly^{238} and Gly^{275} which have been demonstrated to be cell surface-exposed, show a remarkably regular spacing, being approximately 40 amino acids apart in the primary sequence. This distance is sufficient to span the membrane twice in β-sheet conformation. Furthermore, these residues are located in regions where PhoE notably differs from OmpF and OmpC (Fig. 2) and in hydrophilic maxima of the protein (Fig. 4).

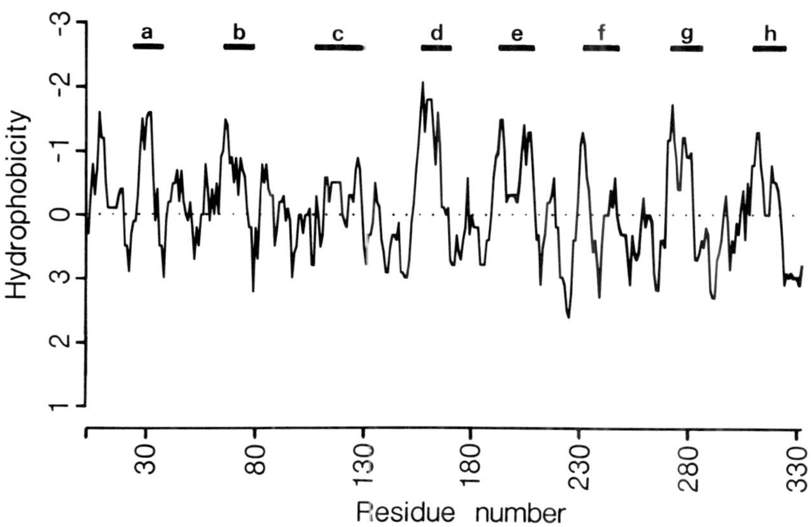

Fig. 4: Smoothed hydrophilicity profile of PhoE protein. The hydrophilic maxima, indicated as a-h, correspond to the hypervariable regions indicated in Fig. 2.

These properties are to be expected for cell surface-exposed regions. In total, eight such variable, hydrophilic regions can be discerned (numbered a-h in Fig. 2 and 4) which are all separated by approximately 40 amino acids. We suppose that these eight regions are all surface-exposed and that the membrane is crossed twice between each exposed region in β-sheet conformation (Fig. 5) (Van der Ley et al. 1986). Inspection of the sequences of the 16 predicted membrane-spanning segments revealed that most of them display an amphipathic character when present in β-sheets of 9-10 amino acids long (Table 1) (Van der Ley and Tommassen 1987). The high hydrophobicity of one side of the presumed β-strands is also observed in the corresponding segments of OmpC and OmpF (Table 1). Remarkable is the high frequency of occurrence of a tyrosine residue as the fourth hydrophobic residue in the β-strands. This is highly conserved in the other porins. It seems likely that the hydrophobic sides of the amphipathic β-sheets are facing outward to the lipid bilayer and that the other side is facing inward to the inside of the pore or to other parts of the protein. Finally, it should be noted that secondary structure predictions show that regions, likely to form reverse turns, are present in parts of PhoE, OmpF and OmpC which, according to the model, are located at the periplasmic side or the cell surface, and hardly in the membrane-spanning fragments.

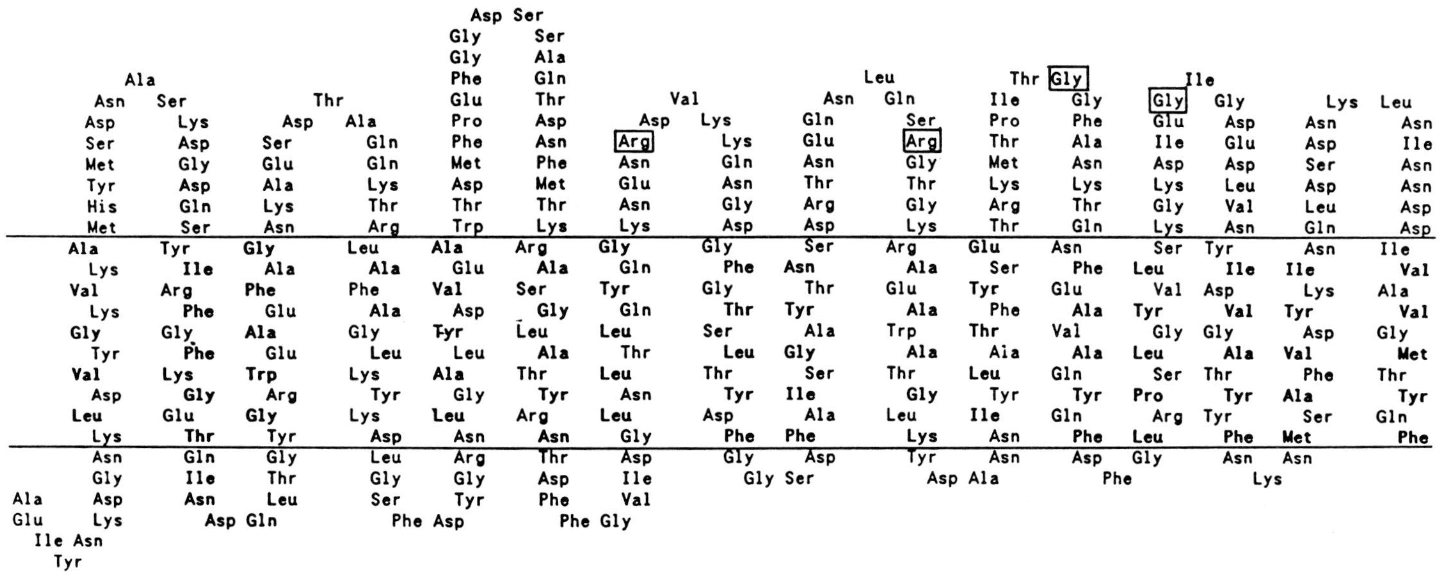

Fig. 5: Model for the topology of PhoE protein in the outer membrane. The amino acids Arg-158, Arg-201, Gly-238 and Gly-275 which were shown to be cell surface-exposed, are boxed. The membrane-spanning segments are as indicated in Table 1. Amino acids at the more hydrophobic side of a potential β-sheet are depicted in bold face.

Tabel 1: Potentially membrane-spanning segments of PhoE protein

Segment	Sequence[a]	Average hydrophobicity[b] a	b	C	F
10 - 19	lys *leu* asp *val* tyr *gly* lys *val* lys *ala*	-1.03	*+0.86*	*+0.83*	*+0.74*
35 - 44	tyr *ile* arg *phe* gly *phe* lys *gly* glu *thr*	-0.81	*+0.84*	*+0.66*	*+0.66*
53 - 62	tyr *gly* arg *trp* glu *ala* glu *phe* ala *gly*	-0.63	*+0.72*	*+0.68*	*+0.64*
76 - 84	leu *ala* phe *ala* gly *leu* lys *tyr* lys	-0.05	*+0.64*	*+0.87*	*+0.64*
94 - 103	asn *leu* gly *ala* leu *tyr* asp *val* glu *ala*	-0.18	*+0.73*	*+0.50*	*+0.54*
126 - 135	arg *ala* ser *gly* leu *ala* thr *tyr* arg	-0.85	*+0.50*	*+0.46*	*+0.61*
144 - 153	gly *leu* asn *leu* thr *leu* gln *tyr* gln *gly*	-0.41	*+0.78*	*+0.81*	*+0.81*
167 - 176	gly *phe* gly *thr* ser *leu* thr *tyr* asp *phe*	-0.03	*+0.73*	*+0.69*	*+0.69*
181 - 190	*phe* ala *ile* ser *gly* ala *tyr* thr *asn* ser	+0.17	*+0.51*	*+0.85*	*+0.79*
206 - 213	arg *ala* glu *ala* trp *ala* thr *gly* leu	-0.29	*+0.58*	*+0.25*	*+0.22*
220 - 228	asn *ile* tyr *leu* ala *thr* phe *tyr* ser	+0.37	*+0.66*	*+0.83*	*+0.83*
246 - 255	asn *phe* glu *ala* val *ala* gln *tyr* gln *phe*	-0.43	*+0.78*	*+0.78*	*+0.84*
259 - 268	*leu* arg *pro* ser *leu* gly *tyr* val *leu* ser	-0.27	*+0.71*	*+0.33*	*+0.26*
284 - 293	tyr *ile* asp *val* gly *ala* thr *tyr* tyr *phe*	+0.01	*+0.91*	*+0.85*	*+0.87*
297 - 306	*met* ser *ala* phe *val* asp *tyr* lys *ile* asn	-0.43	*+0.80*	*+0.66*	*+0.66*
321 - 330	ile *val* ala *val* gly *met* thr *tyr* gln *phe*	+0.32	*+0.85*	*+0.93*	*+1.00*

a) Amino acid residues on the most hydrophobic side of a potential β-sheet are shown in italics.

b) Average hydrophobicity of the amino acids on each side of the potential β-sheets; (a) hydrophilic side; (b) hydrophobic side; (C) and (F) hydrophobic sides of the corresponding fragments in OmpC and OmpF, respectively. The hydrophobicity values are those of the normalized consensus scale of Eisenberg et al. (1984).

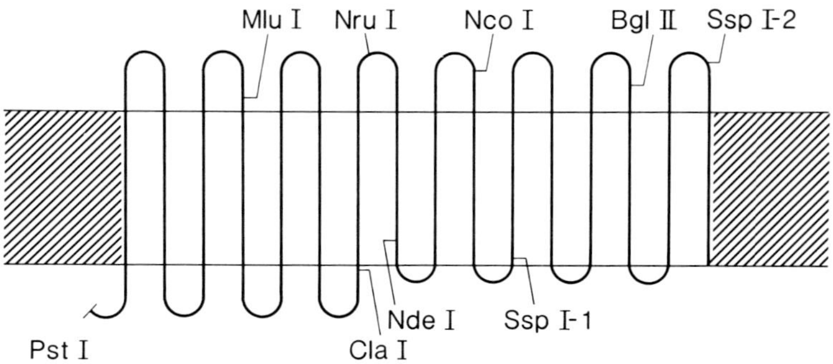

Fig. 6: Positions of the restriction enzyme recognition sites in the *phoE* gene, relative to the folding model of PhoE protein shown in Fig. 5. All the recognition sites are present in the wild-type *phoE* gene, except for the *Nru*I site which was created by oligonucleotide-directed site-specific mutagenesis (Agterberg et al. 1987b) and the *Nco*I site which was found to be present in one of the mutants with altered antibody binding properties.

Evidence for the PhoE model

Two lines of evidence provided support for the PhoE model presented in Fig. 5. First, small oligonucleotide insertions were created in the restriction sites indicated in Fig. 6 (Bosch and Tommassen 1987; Agterberg et al. 1987a, 1987b; Bosch et al. unpublished observations). Except for one out of two different insertions in the *Cla*I-site which will not be discussed further, the mutations did not interfere with the assembly of the proteins in the membrane, since functional pores were formed. Insertions of 3-4 amino acids in regions located at or close to the periplasmic side of the membrane (i.e., insertions in the restriction sites *Pst*I, *Cla*I, *Nde*I and *Ssp*I-1) did not disturb any cell surface related function. Similar insertions in the *Mlu*I and *Nru*I sites disturbed the binding of TC45, showing that the corresponding regions of the protein are indeed surface-exposed and involved in phage binding. Similarly, insertions in the *Nru*I, *Bgl*II and *Ssp*I-2 sites interfered with the binding of one or several monoclonal antibodies. Oligonucleotides encoding a viral antigenic determinant were inserted in the *Nru*I and *Nco*I sites. Intact *E. coli* cells expressing the mutant proteins were able to bind a monoclonal antibody directed against the virus, providing direct evidence for cell surface-exposure of the corresponding regions. Finally, the porins when imbedded in the membrane, are not attainable to most proteases. OmpC protein contains a "natural insertion" of 14 amino acids in the cell surface-exposed region d (Fig. 2). Morona et al. (1985b)

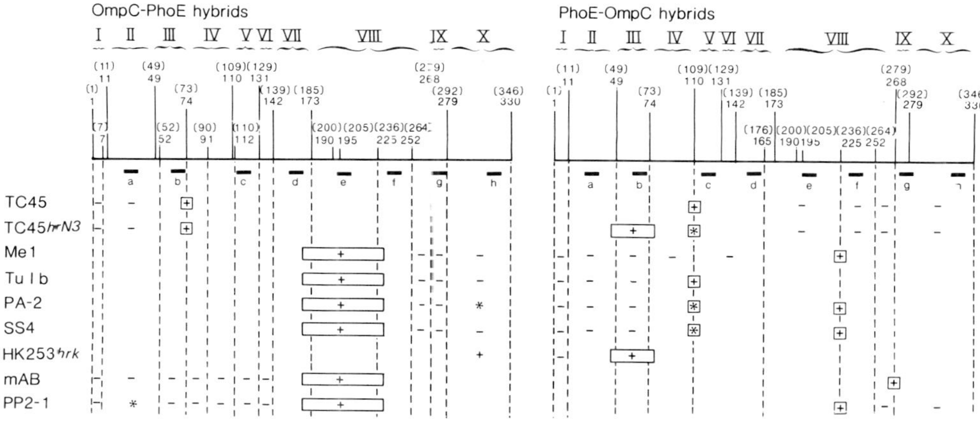

Fig. 7: Functional analysis of OmpC-PhoE and PhoE-OmpC hybrids. Hybrid genes in which N-terminal (OmpC-PhoE hybrids) or C-terminal (PhoE-OmpC hybrids) portions of *phoE* are replaced by the corresponding regions of *ompC* were obtained by making use of *in vivo* recombination between these homologous genes. The locations of the fusion sites were determined by restriction enzyme mapping and, in some cases, by sequence analysis. The numbers correspond to the relative positions of the restriction sites or the sequenced fusion joints in the animo acid sequence of PhoE or, between brackets, of OmpC. The bars, numbered a-h, correspond to regions of PhoE that are proposed to be cell surface-exposed. mAB stands for nine different monoclonal antibodies which recognize cell surface-exposed regions of PhoE. PP2-1 is another monoclonal antibody with different binding properties. Analysis of the OmpC-PhoE hybrids showed that the N-terminal 52 residues of PhoE protein can be replaced by the corresponding part of OmpC without effect on the binding of PhoE-specific phages TC45 and TC45*hr*N3. Thus, these 52 residues are not required for the binding of these phages (denoted as -). A hybrid in which 91 amino acids of PhoE were replaced did not function as phage receptor anymore. Thus, residues between positions 52 and 91 and thus most likely exposed region b, are involved in the phage receptor function (denoted as +). Analysis of the PhoE-OmpC hybrids showed that amino acids C-terminal to position 165 of PhoE could be replaced by the corresponding region of OmpC without effect on the phage receptor function. Amino acids between residues 74 and 165 appear to be involved in the binding of TC45 (most likely exposed regions c and/or d). This region is also involved in the binding of phage TC45*hr*N3, since replacement by the corresponding region of OmpC affects, but not completely destroys, the phage receptor function (denoted as *). The results shown for the binding of OmpC-specific phages and PhoE specific monoclonal antibodies have to be interpreted in a similar way.

showed that OmpC could be cleaved by externally added proteinase K and that cleavage occurred in this region.

A second line of evidence was provided by the construction and analysis of a series of hybrid genes in which parts of *phoE* are replaced by the corresponding parts of *ompC*. These hybrid genes were created by employing *in vivo* recombination between these homologous genes (Tommassen et al.

1985; Van der Ley et al. 1987). The proteins encoded by the hybrid genes were normally assembled into the outer membrane since they formed functional pores. They were characterized for functions and properties in which the native PhoE and OmpC proteins differ, such as the binding of PhoE and OmpC specific phages and PhoE specific monoclonal antibodies. For the assignment of roles in TC45 and monoclonal antibody binding to specific exposed regions, the results of these studies (Fig. 7) are in complete agreement with the results described above. Furthermore, they allow the assignment of the binding of OmpC specific phages to different exposed regions in OmpC. Finally, they provide evidence for the cell surface-exposure of region a of PhoE protein, since amino acids between residues 7 and 53 are involved in the binding of monoclonal antibody PP2-1.

From the combined results of all the different approaches, evidence has been obtained for the exposure at the cell surface of seven out of eight regions which were predicted to be exposed. Only evidence for the exposure of region c is lacking up to now. It should be noted that of the regions a-h, this is the less variable region when compared to OmpC and OmpF (Fig. 2) and its hydrophilicity is less pronounced (Fig. 4).

Topology of other outer membrane proteins

OmpA-specific bacteriophages were employed to select for mutants which produce an OmpA protein with altered cell surface-related properties (Morona et al. 1984; 1985a). Sequence analysis of the mutant alleles showed that the mutations were located in four regions, centered around amino acid residues 25, 70, 110 and 154. Interestingly, these regions are separated by approximately 40 residues in the primary sequence. Comparison of the sequences of the *ompA* genes of several enterobacterial species showed that the regions around residues 25, 70 and 110 are hypervariable (Braun and Cole 1984). A model for the folding of OmpA protein was proposed according to which the four regions around residues 25, 75, 110 and 154 are surface-exposed, with eight membrane-spanning fragments (Morona et al. 1984). Vogel and Jähnig (1986) showed that these predicted membrane-spanning β-sheets have an amphipathic character. Thus, the proposed model for OmpA is very similar to the model proposed for the porins.

The analysis of mutants, resistant to LamB-specific phages (Clément et al. 1983; Charbit et al. 1984; Gehring et al. 1987) or resistant to LamB-specific monoclonal antibodies and complement (Desaymard et al. 1986) suggested cell-surface exposure of the following five amino acids or amino acid regions of the 421 residue long LamB protein: 18, 148-164, 245-259, 333 and 382-401. Mapping of protease-sensitive sites confirmed the cell surface-exposure of the second and the third of these regions, suggested that the fifth region is larger and extends to residue 370 and suggested that another region around residue 203 is also surface-exposed (Schenkman et al. 1984). Inspection of a hydrophobicity profile shows that these regions are located in hydrophilic maxima. Additional hydrophilic maxima are observed around residues 53, 300 and a less pronounced one around residue 105. If also these hydrophilic maxima are exposed, LamB protein contains a total of 9 exposed regions, containing residues 18,53,105,151-164,203,245-259,300,333 and 370-401. Again, these regions are regularly spaced, being approximately 40 amino acids apart in the primary sequence. This suggests a folding model with 18 membrane-spanning segments. Inspection of the regions supposed to contain membrane spanning regions, reveals the presence of at least 13 segments that would have a pronounced hydrophobic side when arranged in a β-sheet conformation (not shown).

Conclusions

Outer membrane proteins with a totally different primary structure appear to be arranged in the membrane in a rather similar way. The cell surface-exposed regions are the hydrophilic maxima of the protein, and these regions are separated in the primary structure by approximately 40 amino acids. The membrane-spanning regions form β-sheets of 9-10 residues long. Many of these β-sheets have a pronounced hydrophobic side. According to secondary structure predictions β-turns are generally present in the exposed regions (at the cell surface and at the periplasmic side) but hardly or not in the membrane-spanning fragments. By applying these criteria, it should be possible to predict the topology of any outer membrane protein of which the sequence is available. When the sequences of two related proteins are available, an additional criterium can be added, since the surface-exposed regions of the outer membrane proteins are hypervariable.

REFERENCES

Agterberg M, Adriaanse H, Tommassen J (1987a) Use of outer membrane protein PhoE as a carrier for the transport of a foreign antigenic determinant to the cell surface of *Escherichia coli*K12. Gene in press

Agterberg M, Benz R, Tommassen J (1987b) Insertion mutagenesis on a cell surface exposed region of outer membrane protein PhoE of *Escherichia coli*K-12. Eur J Biochem in press

Beck E, Bremer E (1980) Nucleotide sequence of the gene *ompA* coding the outer membrane protein II* of *E. coli*K-12. Nucl Acids Res 8:3011-3024

Benson SA, Bremer E, Silhavy TJ (1984) Intragenic regions required for LamB export. Proc Natl Acad Sci USA 81:3830-3834

Benz R (1985) Porin from bacteria and mitochondrial outer membranes. Crit Rev Biochem 19:145-190

Benz R, Darveau RP, Hancock REW (1984) Outer membrane protein PhoE from Escherichia coli forms anion selective pores in lipid bilayer membranes. Eur J Biochem 150:319-324

Benz R, Schmid A, Hancock REW (1985) Ion selectivity of gram-negative bacterial porin. J Bacteriol 162:722-727

Benz R, Schmid A, Nakae T, Vos-Scheperkeuter GH (1986) Pore formation by LamB of *Escherichia coli* in lipid bilayer membranes. J Bacteriol 165:978-986

Bosch D, Leunissen J, Verbakel J, De Jong M, Van Erp H, Tommassen J (1986) Periplasmic accumulation of truncated forms of outer-membrane PhoE protein of *Escherichia coli*K-12. J Mol Biol 189:449-455

Bosch D, Tommassen J (1987) Effects of linker insertions on the biogenesis and functioning of the *Escherichia coli* outer membrane pore protein PhoE. Mol Gen Genet 208:485-489

Braun G, Cole ST (1984) DNA sequence analysis of the *Serratia marcescens ompA* gene: Implication for the organisation of an enterobacterial outer membrane protein. Mol Gen Genet 195:321-328

Charbit A, Clément J-M, Hofnung M (1984) Further sequence analysis of the phage lambda receptor site. Possible implications for the organization of the LamB protein in *E. coli*K12. J Mol Biol 175:395-401

Clément J-M, Hofnung M (1981) Gene sequence of the λ receptor, an outer membrane protein of E. coli K12. Cell 27:507-514

Clément J-M, Lepouce E, Marchal C, Hofnung M (1983) Genetic study of a membrane protein: DNA sequence alterations due to 17 *lamB* point mutations affecting adsorption of phage lambda. EMBO J 2:77-80

Davis NG, Model P (1985) An artificial anchor domain: hydrophobicity suffices to stop transfer. Cell 41:607-614

Desaymard C, Débarbouillé M, Jolit M, Schwartz M (1986) Mutations affecting antigenic determinants of an outer membrane protein of *Escherichia coli*. EMBO J 5:1383-1388

Eisenberg D, Schwarz E, Komaromy M, Wall R (1984) Analysis of membrane and surface protein sequences with the hydrophobic moment plot. J Mol Biol 179:125-142

Emr SD, Hedgpeth J, Clément J-M, Silhavy TJ, Hofnung M (1980) Sequence analysis of mutations that prevent export of λ receptor, an *Escherichia coli* outer membrane protein. Nature 285:82-85

Ferenci T, Schwentorat M, Ullrich S, Vilmart J (1980) Lambda receptor in the outer membrane of *Escherichia coli* as a binding protein for maltodextrins and starch polysaccharides. J Bacteriol 142:521-526

Freudl R, Schwarz H, Klose M, Movva NR, Henning U (1985) The nature of information, required for export and sorting, present within the outer membrane protein OmpA of *Escherichia coli*K-12. EMBO J 4:3593-3598

Freudl R, Schwarz H, Stierhof Y-D, Gamon K, Hindennach I, Henning U (1986) An outer membrane protein (OmpA) of *Escherichia coli*K-12 undergoes a conformational change during export. J Biol Chem 261:11355-11361
Garavito RM, Jenkins J, Jansonius JN, Karlsson R, Rosenbusch JP (1983) X-ray diffraction analysis of matrix porin, an integral membrane protein from *Escherichia coli* outer membranes. J Mol Biol 164:313-327
Gehring K, Charbit A, Brissaud E, Hofnung M (1987) Bacteriophage λ receptor site on the *Escherichia coli*K-12 LamB protein. J Bacteriol 169: 2103-2106
Hall MN, Schwartz M, Silhavy TJ (1982) Sequence information within the *lamB* gene is required for proper routing of the bacteriophage λ receptor protein to the outer membrane of *Escherichia coli*K-12. J Mol Biol 156:93-112
Hoffman CS, Wright A (1985) Fusions of secreted proteins to alkaline phosphatase: an approach for studying protein secretion. Proc Natl Acad Sci USA 82:5107-5111
Horwich AL, Kalousek F, Mellman I, Rosenberg LE (1985) A leader peptide is sufficient to direct mitochondrial import of a chimeric protein. EMBO J 4:1129-1135
Hurt EC, Pesold-Hurt B, Suda K, Oppliger W and Schatz G (1985) The first twelve amino acids (less than half of the presequence) of an imported mitochondrial protein can direct mouse cytosolic dihydrofolate reductase into the yeast mitochondrial matrix. EMBO J 4:2061-2068
Hurt EC, Van Loon APGM (1986) How proteins find mitochondria and intramitochondrial compartments. Trends Biochem Sci 6: 262-266
Inokuchi K, Mutoh N, Matsuyama S, Mizushima S (1982) Primary structure of the *ompF* gene that codes for a major outer membrane protein of *Escherichia coli*K-12. Nucl Acids Res 10:6957-6968
Jackson ME, Pratt JM, Stoker NG, Holland IB (1985) An inner membrane protein N-terminal signal sequence is able to promote efficient localisation of an outer membrane protein in *Escherichia coli*. EMBO J 4:2377-2383
Kleffel B, Garavito RM, Baumeister W, Rosenbusch JP (1985) Secondary structure of a channel-forming protein: porin from *E. coli* outer membranes. EMBO J 4:1589-1592
Korteland J, Overbeeke N, De Graaff P, Overduin P, Lugtenberg B (1985) Role of the Arg-158 residue of the outer membrane PhoE pore protein of *Escherichia coli*K12 in bacteriophage TC45 recognition and in channel characteristics. Eur J Biochem 152:691-697
Korteland J, Tommassen J, Lugtenberg B (1982) PhoE protein pore of the outer membrane of *Escherichia coli*K12 is a particularly efficient channel for organic and inorganic phosphate. Biochim Biophys Acta 690:282-289
Kyte J, Doolittle RF (1982) A simple method for displaying the hydropathic character of a protein. J Mol Biol 157:105-132
Lugtenberg B, Van Alphen L (1983) Molecular architecture and functioning of the outer membrane of *Escherichia coli* and other gram-negative bacteria. Biochim Biophys Acta 737:51-115
MacIntyre S, Freudl R, Degen M, Hindennach I, Henning U (1987) The signal sequence of an *Escherichia coli* outer membrane protein can mediate translocation of a not normally secreted protein across the plasma membrane. J Biol Chem 262:8416-8422
Mizuno T, Chou M-Y, Inouye M (1983) A comparative study on the genes for three porins of the *Escherichia coli* outer membrane. DNA sequence of the osmoregulated *ompC* gene. J Biol Chem 258:6932-6940
Moreno F, Fowler AV, Hall M, Silhavy TJ, Zabin I, Schwartz M (1980) A signal sequence is not sufficient to lead β-galactosidase out of the cytoplasm. Nature 286:356-359

Morona R, Klose M, Henning U (1984) *Escherichia coli* K-12 outer membrane protein (OmpA) as a bacteriophage receptor: analysis of mutant genes expressing altered proteins. J Bacteriol 159:570-578

Morona R, Krämer C, Henning U (1985a) Bacteriophage receptor area of outer membrane protein OmpA of *Escherichia coli* K-12. J Bacteriol 164:539-543

Morona R, Tommassen J, Henning U (1985b) Demonstration of a bacteriophage receptor site on the *Escherichia coli* K12 outer-membrane protein OmpC by use of a protease. Eur J Biochem 150:161-169

Nakae T (1986) Outer-membrane permeability of bacteria. Crit Rev Microbiol 13:1-62

Nikaido H, Vaara M (1985) Molecular basis of bacterial outer membrane permeability. Microbiol Rev 49:1-32

Nikaido H, Wu HCP (1984) Amino acid sequence homology among the major outer membrane proteins of *Escherichia coli*. Proc Natl Acad Sci USA 81:1048-1052

Overbeeke N, Bergmans H, Van Mansfeld F, Lugtenberg B (1983) Complete nucleotide sequence of *phoE*, the structural gene for the phosphate limitation inducible outer membrane pore protein of *Escherichia coli*K12. J Mol Biol 162:513-532

Randall LL, Hardy SJS (1986) Correlation of competence for export with lack of tertiary structure of the mature species: a study in vivo of maltose-binding protein in E. coli. Cell 46:921-928

Rasmussen BA, Silhavy TJ (1987) The first 28 amino acids of mature LamB are required for rapid and efficient export from the cytoplasm. Genes Developm. 1:185-196

Schenkman S, Tsugita A, Schwartz M, Rosenbusch JP (1984) Topology of phage λ receptor protein. Mapping targets of proteolytic cleavage in relation to binding sites for phage or monoclonal antibodies. J Biol Chem 259:7570-7576

Schwartz M (1967) Sur l'existence chez *Escherichia coli* K12 d'une régulation commune à la biosynthèse de récepteurs du bacteriophage λ et au métabolisme du maltose. Ann Inst Pasteur Paris 113:685-704

Schweizer M, Hindennach I, Garten W, Henning U (1978) Major proteins of the *Escherichia coli* outer cell envelope membrane. Interaction of protein II* with lipopolysaccharide. Eur J Biochem 82:211-217

Silhavy TJ, Shuman HA, Beckwith J, Schwartz M (1977) Use of gene fusions to study outer membrane protein localization in *Escherichia coli*. Proc Natl Acad Sci USA 74:5411-5415

Sjöström M, Wold S, Wieslander Å, Rilfors L (1987) Signal peptide amino acid sequences in *Escherichia coli* contain information related to final protein localization. A multivariate data analysis. EMBO J 6:823-831

Smeekens S, Bauerle C, Hageman J, Keegstra K, Weisbeek P (1986) The role of the transit peptide in the routing of precursors toward different chloroplast compartments. Cell 46:365-375

Sonntag I, Schwarz H, Hirota Y, Henning U (1978) Cell envelope and shape of *Escherichia coli*: multiple mutants missing the outer membrane lipoprotein and other major outer membrane proteins. J Bacteriol 136:280-285

Tommassen J, De Kroon T (1987) Subcellular localization of a PhoE-LacZ fusion protein in *E. coli* by protease accessibility experiments reveals an inner membrane spanning form of the protein. FEBS Lett, in press

Tommassen J, Leunissen J, Van Damme-Jongsten M, Overduin P (1985a) Failure of *E. coli* K-12 to transport PhoE-LacZ hybrid proteins out of the cytoplasm. EMBO J 4: 1041-1047

Tommassen J, Lugtenberg B (1980) Outer membrane protein e of *Escherichia coli* K-12 is co-regulated with alkaline phosphatase. J Bacteriol 143:151-157

Tommassen J, Van der Ley P, Van Zeijl M, Agterberg M (1985b) Localization of functional domains in *E. coli* K-12 outer membrane porins. EMBO J 4:1583-1587

Tommassen J, Van Tol H, Lugtenberg B (1983) The ultimate localization of an outer membrane protein of *Escherichia coli* K-12 is not determined by the signal sequence. EMBO J 2:1275-1279

Van der Ley P, Amesz H, Tommassen J, Lugtenberg B (1985) Monoclonal antibodies directed against the cell-surface-exposed part of PhoE pore protein of the *Escherichia coli* K-12 outer membrane. Eur J Biochem 147:401-407

Van der Ley P, Burm P, Agterberg M, Van Meersbergen J, Tommassen J (1987) Analysis of structure-function relationships in *Escherichia coli* K12 outer membrane porins with the aid of *ompC-phoE* and *phoE-ompC* hybrid genes. Mol Gen Genet, in press

Van der Ley P, Struyvé M, Tommassen J (1986) Topology of outer membrane pore protein PhoE of *Escherichia coli*. Identification of cell surface-exposed amino acids with the aid of monoclonal antibodies. J Biol Chem 261:12222-12225

Van der Ley P, Tommassen J (1987) PhoE protein structure and function. In: Torriani-Gorini A, Rothman FG, Silver S, Wright A, Yagil E (eds) Phosphate metabolism and cellular regulation in microorganisms. American Society for Microbiology, Washington, DC, in press

Vogel H, Jähnig F (1986) Models for the structure of outer membrane proteins of *Escherichia coli* derived from Raman spectroscopy and prediction methods. J Mol Biol 190:191-199

Wandersman C, Schwartz M, Ferenci T (1979) *Escherichia coli* mutants impaired in maltodextrin transport. J Bacteriol 140:1-13

WHAT CAN WE LEARN FROM COLICINS ABOUT THE DYNAMICS OF INSERTION AND TRANSFER OF PROTEINS INTO AND ACROSS MEMBRANES.

Claude J. LAZDUNSKI
Centre de Biochimie et de Biologie Moléculaire du C.N.R.S.
31 Chemin Joseph Aiguier, B.P. 71
13402 Marseille Cedex 9
France

Introduction

New insights into how proteins cross membranes are beginning to emerge. Although there is no universal mechanism, recent results suggest that there are common features of protein translocation. Three steps usually are involved : i) the association of the protein with receptors on the correct membrane, ii) the translocation through the membrane and iii) covalent modification and folding on the opposite membrane surface. A requirement which has now been well established is that the unfolded state of the polypeptide chain to be transferred has to be maintained. By unfolded we mean that this polypeptide has not attained its final most stable conformation although it has obviously acquired some secondary structure. Therefore, an essential feature of translocation may have to do with preventing the folding of the protein into a tertiary structure approximating that of the mature protein. This feature has been demonstrated in mitochondria with dihydrofolate reductase fused with the cytochrome oxydase subunit IV (Eilers and Schatz, 1986), in *E. coli* with the phosphate binding protein (Pagès et al., 1984) and the maltose binding protein (Randall and Hardy, 1986) and in the rough endoplasmic reticulum with preprolactin by Maher and Singer (1986).
What prevents folding ? Although there is as of yet no clear answer it seems that the components of the export machinery such as the ribosomes and the Signal Recognition Particle (Walter and Blobel, 1982) and some unknown protein in mitochondria (Zimmerman, 1986) may be involved. Post-translational transfer across the endoplasmic reticulum membrane, in contrast to the situation in bacteria or mitochondria, does not require a membrane potential but does require ATP hydrolysis. At present the energy of translocation is the "black box" of the process and there is an urgent need to understand this.

Why colicins ?

Colicins are toxic proteins produced by and active against *E. coli* and closely related bacteria.

NATO ASI Series, Vol. H16
Membrane Biogenesis. Edited by J. A. F. Op den Kamp

They are produced in large amounts and, in general, are secreted across the cell envelope to the extracellular medium. They can adsorb to specific receptors located at the external surface of the outer membrane of sensitive cells, and are then translocated to their specific targets within these cells. Colicins were discovered a long time ago and were studied intensively until the sixties (Frederick, 1957). After a period of relative neglect, they have now emerged as powerful model systems to investigate the mechanism and the energetics of protein insertion into and across membranes.

In general, colicins are large molecules of molecular weights between 40,000 and 80,000. Although they are secreted to the extracellular medium their polypeptide chains, in contrast to those of periplasmic proteins, like alkaline phosphatase, do not contain disulfide bridges. This bias suggests that the latter might not be compatible with their functions, which are to get out of the colicinogenic cells and to penetrate into sensitive cells.

Colicins are encoded by plasmids that confer upon their hosts the property of being insensitive (one says immune) to the colicin produced. Immunity proteins are generally of low molecular weight (less than 20,000). The largest group of colicins comprises those which can form voltage-dependent channels in membranes, thereby destroying the cell's energy potential. This group includes colicins A, B, E1, Ia, Ib, K and N. Immunity to pore-forming colicins may either prevent the colicin from spanning the cytoplasmic membrane, or prevent the pore from opening (Konisky, 1982).

Most of the plasmids which encode colicins, also encode a small protein called "lysis protein" or "bacterial release protein", whose expression is required for colicin release (Pugsley, 1984; Cavard et al., 1985; de Graaf and Oudega, 1986). The term lysis protein must be qualified since it has been shown that the release of colicin A is unaffected in *E. coli* mutants which display decreased levels of autolytic enzyme activities (Howard et al., 1987). Furthermore the ultrastructure of colicinogenic cells is not grossly altered following full induction and colicin release (Cavard et al., 1984).

Here I will review some of the recent work devoted to pore-forming colicins and their immunity proteins and will elaborate some insights of general interest for membrane biology that have resulted from that work. The reader is referred to earlier reviews for wider aspects of colicinogeny (Reeves, 1972; Holland, 1975; Konisky, 1982).

Functional Organization of Colicin Operons

Lysis protein genes form with colicin genes an operon regulated by the SOS response and therefore repressed by the LexA protein (Little and Mount, 1982). The organization of the 3 genes *caa, cai* and *cal* encoding respectively colicin A (Caa), the immunity protein (Cai) and the pColA-lysis protein (Cal) has been analysed in detail (Lloubès et al., 1986). The genes *caa* and *cal* form an operon, whereas *cai* is located between these two genes and transcribed in the opposite direction from its own promoter (Fig. 1). Downstream of *caa*, there is a *rho*-independent terminator which arrests most of the transcripts initiated at the promoter of the *caa-cal* operon (Lloubès et al., 1986). This explains the coordinated expression of Caa and Cal and the fact, previously reported (Cavard et al., 1985), that much more colicin A than Cal protein is produced after induction. This feature may be significant since it is essential that a cell, once committed to the normally suicidal process of colicin production and release, should maximise the amount of colicin it produces.

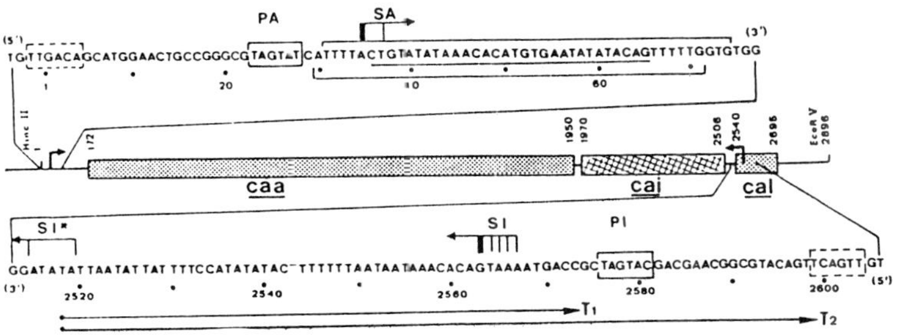

Fig. 1 : Organization of the *caa-cal* operon and the *cai* gene.
The nucleotide sequences for the promoters are presented : the recognition and binding sequences for RNA polymerase are boxed with dotted lines (-35 sequence) and solid lines (Pribnow box). PA and SA indicate the Pribnow box and transcriptional start site for the *caa-cal* operon while PI and SI indicate those for the *cai* gene. SI* indicates the start site for *cai in vitro*. Arrows above the start sites indicate the direction of transcription, the bold line indicates the major band detected by Sl mapping. In the case of SI*, one of the nucleotides within the bracket corresponds to the start site. The binding site for LexA in the operator region for *caa* is indicated by brackets overlining the protected region for the sense-strand and brackets underlining the sequence (not represented) for the anti-sense strand. The two overlapped consensus "SOS sequences" are underlined. The sequence is numbered taking as the +1 nucleotide the cleavage site for *Hinc*II. Coding sequences for *caa*, *cai* and *cal* are indicated by stippled and hatched boxes. Numbers above the boxes indicate the first and last nucleotides of the coding sequence, and arrows indicate the direction of transcription. Transcripts of *caa* and *caa-cal* are indicated as well as the terminators T1 and T2.

Colicin synthesis

Colicins are produced in free polysomes in the cytoplasm (Varenne et al., 1981) and not in a precursor form (Yamada et al., 1982; Morlon et al., 1983; Varley and Boulnois, 1984).
A useful feature of this system is provided by the fact that their polypeptide chain elongation is

quite discontinuous and discrete intermediates can be easily demonstrated (Varenne et al., 1982). This feature is not particular to colicins but is general to all proteins. It is due to ribosomes pausing in mRNA regions containing codons recognized by rare iso-tRNAs (Varenne et al., 1984). It so happens that colicin genes contain a high proportion of such codons (Varenne et al., 1984). By using pulse and chase labelling, we have shown that as long as less than 450 amino acids of the 592 residue of colicin A polypeptide chain have been assembled, the nascent chains are soluble and have no affinity for membranes. Beyond 450 residues the nascent chain acquires a COOH-terminal region, comprising a 48 amino acid hydrophobic stretch, which is involved in pore-formation (Pattus et al., 1985). At this point, the nascent chains acquire two new properties. Firstly, they become competent for membrane insertion (Varenne et al., 1984), and secondly they can form stable dimers of colicin A (Knibiehler and Lazdunski, 1987). The nascent chain intermediates resulting from pause sites and the dimer form are shown in figure 2. As soon as they are secreted to the extracellular medium, the high dilution from 0.7μm^3 (intracellular volume) to the "pacific ocean" (the extracellular medium) appears to promote a slow conformational change by which the polypeptide loses its competence for membrane insertion and the ability to form dimers. In this process, the COOH-terminal hydrophobic stretch most probably becomes masked in the interior of the protein.

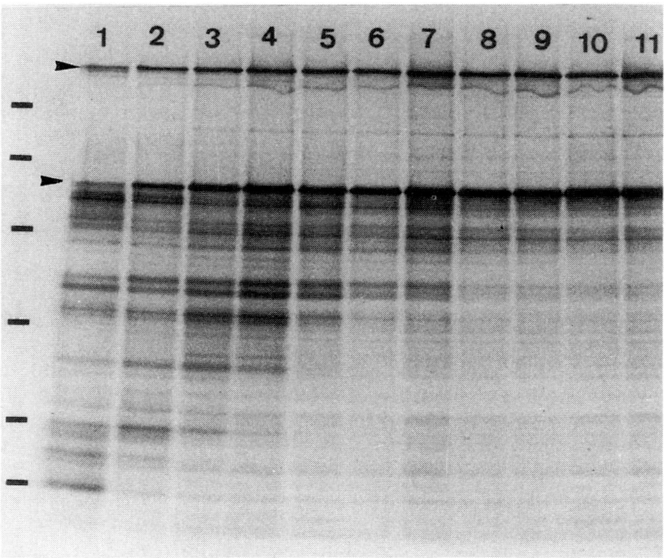

Fig. 2 : Elongation intermediates of colicin A polypeptide.
Samples from 10 sec pulse-labelled cells (lane 1) were removed at different times of chase (lanes 2 to 11). Samples were solubilized, immunoprecipitated and analyzed on a SDS/polyacrylamide gel : lane 2, 10 sec; lane 3, 20 s; lane 4, 30 s; lane 5, 40 s; lane 6, 50 s; lane 7, 60 s; lane 8, 70 s; lane 9, 80 s; lane 11, 100 s. The migrations of colicin A monomer and dimer are indicated by arrowheads.

Colicin secretion

Using transposon insertion mutagenesis or via deletion of the lysis gene, strains deficient in the production of cal protein can be obtained and these cells do not release colicin A, thus demonstrating the role of this protein in colicin release. We have observed that a critical concentration of Cal is required to trigger colicin release and, deletion of the T1 terminator causes an early decrease in the turbidity of the culture with concomitant loss of viability (data not shown).

Although cells which are excreting colicin do not show signs of extensive desintegration (Cavard et al., 1984), they do exhibit a number of features which indicate that envelope functions are affected. They are no longer able to accumulate labelled substrates, they release ions from the cytoplasm, and are sensitive to sucrose, which suggest that the inner membrane has become permeable to small molecules (Pugsley and Schwartz, 1984). These effects can be almost completely overcome by adding 20 mM Mg^{2+} to the medium. This treatment also prevents the decline in culture turbidity but was reported to have no effect on colicin export (Pugsley and Schwartz, 1984; de Graaf and Oudega, 1986). Some colicins are released as a complex with their immunity protein. However, it has been demonstrated that the cloacin DF13 immunity protein was also produced and excreted in the absence of cloacin molecules (de Graaf and Oudega, 1986).

```
-1  ↓ +1
AlaCysGlnValAsnAsnValArgAspThrGlyGlySerValSerProSer
SerIleValThrGlyValSerMetGlySerAspGlyValGlyAsnPro (33)
```

Fig. 3 : Amino acid sequence of wild-type Cal. In the AK31 mutant Cal, the underlined alanine (-1) has been substituted by a proline and in the AL16 mutant, the underlined cysteine (+1) has been substituted by a threonine. The arrow indicates the site of cleavage. The mature Cal polypeptide chain is numbered 1 to 33.

The structures of the known bacterial release proteins feature a high degree of homology (Cavard et al., 1985; de Graaf and Oudega, 1986). They all contain a signal peptide with a sequence, at the cleavage site, resembling the consensus lipoprotein modification sequence described by Wu and et al. (1983). The polypeptide chain of the mature form is small, comprising 33 amino acids for the colicin A lysis protein (Cal) and 28 amino acids for the colicin E1, E2, E3 and cloacin DF13 lysis proteins.

For Cal, we have demonstrated that the precursor must be modified by a lipid before it can be processed and that the maturation is prevented by globomycin, an inhibitor of signal peptidase II (Cavard et al., 1987). Using oligonucleotide-directed mutagenesis, the alanine (mutant

AK31) and cysteine (mutant AL16) residues in the -1 and +1 positions of the cleavage site were replaced by proline and threonine residues, respectively, in two different constructs (Fig. 3). Both substitutions prevented the normal modification and cleavage of the parent protein (Fig. 4). It has also been suggested that the C-terminal nine amino acids are not essential for the quasi-lysis phenotype caused by the ColE2 lysis protein (Toba et al., 1986).

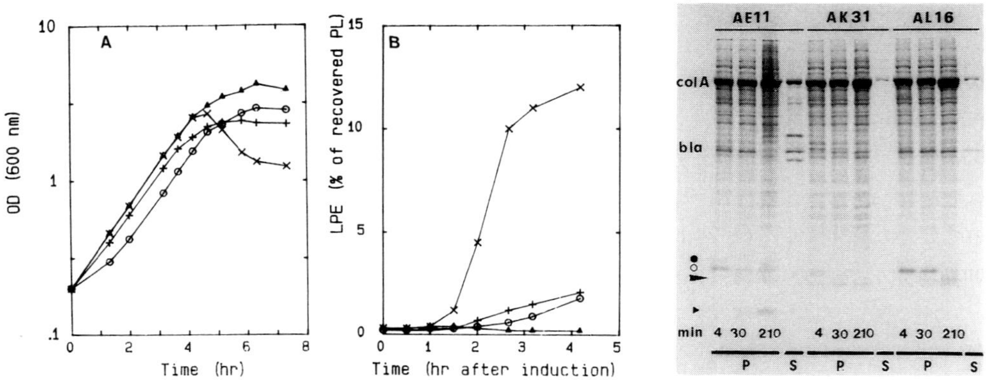

Fig. 4 : Effect of mitomycin C (MTC) induction on growth and LPE content of cells containing wild-type and mutant Cal proteins. The growth curve (A) and the amount of LPE as a proportion of total phospholipids (B) are presented. The cells were grown to an optical density (OD) at 600 nm of approximately 1 and were either not induced or induced with MTC. The phospholipids were extracted and analyzed after further incubation. The plasmids and the induction conditions of the cells are as follows : X, W3110 (pAE11) wild-type induced; , W3110 (pAE11) non induced; + W3110 (pAK31) induced; o W3110 (pAL16) induced. (C) The modification and processing of wild type and mutant Cal proteins, as well as the release of colicin A were analyzed. The cells were induced with MTC for 90 min before labelling with S35-methionine. Then, at various times (as indicated), culture samples were centrifuged, the pellets (P) and supernatants (S) were analyzed by SDS/PAGE. A fluorogram is presented. The various Cal protein derivatives are indicated : •, Cal modified precursor form; o, Cal unmodified precursor form; ➤Cal mature form; ▸ Cal signal peptide. The position of colicin A is indicated.

Although the detailed mechanism of release of colicins is not yet known, the increase in envelope permeability observed late after MTC induction may be attributed, in part or totally, to the lysis protein-dependent activation of the detergent-resistant phospholipase A present in the outer membrane (Pugsley and Schwartz, 1984; Luirink et al., 1985).

The marked activation of the detergent resistant phospholipase A observed with wild-type Cal was not observed with Cal mutants AK31 and AL16. Both Cal mutants were also defective for the secretion of colicin A (Fig. 4) (Cavard et al., 1987). The increase in lysophosphatidyl ethanolamine (LPE), the product of the action of phospholipase A on phosphatidylethanolamine, the major membrane phospholipid, was not observed (Fig. 4). LPE may thus act as a membrane perturbant which could alter the permeability properties of the envelope and allow colicin to leak out of the cell. In addition, a subset of at least 20 proteins is also released to the extracellular medium (Fig. 4). Electron microscope studies with immunogold labelling of colicin A on cryosections of *pldA* and *cal* mutants indicated that the colicin remains in the cytoplasm and is not transferred to the periplasmic space (Cavard et al., 1987). These results demonstrated that Cal must be modified and processed to activate the detergent-resistant phospholipase A and to promote release of colicin A. However, so far the mechanism of transfer of colicins across the inner membrane remains poorly understood. In particular, it is not known if any interaction between a given colicin and its lysis protein is involved, or if any specific regions or specific interactions between regions of the colicin polypeptide chain are required for this transfer. We have now addressed this question by constructing deletion mutants of colicin A and by a new approach using a cassette which contains a stop codon, a Shine-Dalgarno sequence and an initiation codon to separate the NH_2- and COOH-terminal regions of colicin A. Together, these deletions span the region from amino acid 15 to the end of the protein (Fig. 5).

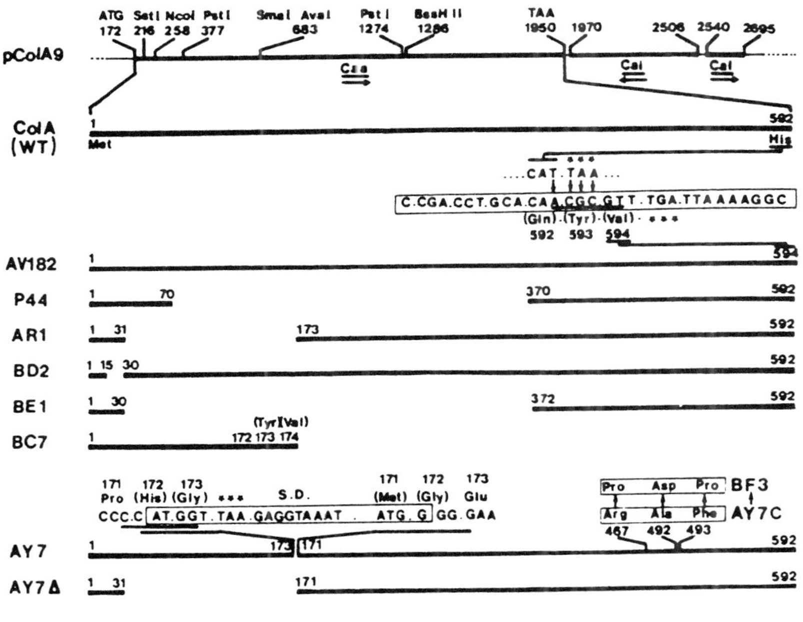

Fig.5 : Construction of recombinant plasmids. A part of the plasmid pColA9 is shown, as well as the constructions derived from it. The restriction sites are indicated above the number of the cleavage site. The oligonucleotides used for the constructions are boxed and the restriction sites created are underlined. Stars represent stop codons, S.D. the Shine-Dalgarno sequence. The amino acids in brackets were created during the construction. AY7 represents two proteins AY7N and AY7C. BF3 corresponds to the three mutations introduced into AY7C (Baty et al., 1987a). AY7Δ contains the first 31 amino acids of AY7N (called AY7ΔN) and the AY7C protein.

None of these regions was found to be required for extracellular release or had any effect on the efficiency of this process. As well, the NH_2-terminal and central plus COOH-terminal domains could be demonstrated to be released to the same extent when produced as separate polypeptides as when produced as linked ones. The introduction into the COOH-terminal domain of mutations promoting cytoplasmic aggregation (Baty et al., 1987a) prevented its release but had no effect on the secretion of the NH_2-terminal polypeptide (Baty et al., 1987b). These results demonstrated that no specific interaction between the NH_2- and the COOH-terminal regions of the colicin A polypeptide chain is involved in the release of colicin A. We are led to conclude that there is no topogenic export signal in the polypeptide chain of colicin A involved in the release mechanism. Thus the process is non-specific with respect to the colicin itself and depends solely on the expression of the colicin A lysis protein.

Colicin receptors and entry into sensitive cells

Once they are released from producing cells, colicins can bind to specific receptors on the surface of susceptible cells (Kadner et al., 1979). These receptors are comprised of outer membrane proteins, called porins, that allow the diffusion of low molecular weight nutrients across the outer membrane. These porins have been parasitized by bacteriophages and colicins which use them as receptors. The colicin-receptor interaction is energy-independent and can occur below the transition temperature of the membrane. The receptors for pore-forming colicins are shown in Table 1.

Table 1 : Receptors for pore-forming colicins.

Colicin	Genetic locus	Molecular weight	Other ligands or Nutrients	Reference
A	*btuB OmpF*	60.000 37.000	Vit. B12 various <700	 Cavard et al.
B	*fepA (cbr)*	81.000	Enterochelin	Pugsley and Reeves
E1	*btuB*	60.000	vit. B12	Sabet & Schnaitman
Ia Ib	*cit*	74.000	iron accumulation	Konisky and Cowell
K	*tsx*	27.000	Nucleosides	Hantke
N	OmpF	37.000	Various < 700	Pugsley

The expression of most of these receptor proteins is regulated by growth conditions. Many of the colicin receptors have been shown to be involved in outer membrane-mediated nutrient uptake. Thus, the polypeptide that serves as the receptor for colicins E1, and A (BtuB) functions in the uptake of vitamins B12, whereas the colicin K receptor serves as a specific diffusion pathway for nucleosides. Several colicin receptors are involved in iron uptake, serving as siderophore-binding proteins. These important physiological functions exert selective pressure for the maintenance of active receptors on the surface of sensitive organisms.
The molecular events between the initial adsorption to receptors and final interaction with a particular cellular target is not known for any colicin. However, there are indications that such translocation may be energy dependent.
According to its activity spectrum against a variety of mutants, a particular colicin can be unambiguously assigned to one of two groups. The type "B" colicins (B, Ia, Ib, V, D and M) are inactive against strains that have a lesion in the *tonB* gene, but active against strains mutant in tolA or tolB genes whereas the type "A" colicins (A, E1, K) show the opposite specificity. This distinction is independent of mode of action, but is thought to reflect two different modes of colicin uptake. Although the exact function of the *tonB* gene product has not been determined, available information has led to the consensus that it plays a role in mediating transfer of outer membrane-bound nutrients (such as vitamin B12 and siderophores) or type "B" colicins from their respective surface receptors to the cytoplasmic membrane (for a review, see Konisky, 1982). The receptors can be by-passed since osmotic shock alleviates the need for receptors in cells challenged with colicins E3, M or E1 (Braun et al., 1980; Uratani et al., 1981). Under these conditions, the TonB function is no longer needed suggesting its involvement in the translocation across the outer membrane. It is thought that the TonB protein is somehow involved in the energy-coupling process between inner and outer membranes. The product of the gene *exbB* is required for the vitamin B12 and iron transport and also for the uptake of various colicins (Pugsley and Reeves, 1975).
A homology region [(Glu or Asp)-Thr-Val-Ile-Val] at the amino-terminal end of the mature form common to all TonB-dependent receptor proteins has been demonstrated (Sauer et al., 1987). The TonB protein was found in both the cytoplasmic and outer membrane and it has been proposed that this protein might form intermembrane bridges thus regulating the activity of outer membrane proteins according to the energy state of the cytoplasmic membrane (Sauer et al., 1987).
For group "A" colicins, the possible involvement of new gene products for uptake in addition to TolA and TolB, has been suggested. Two open reading frames ,tolQ and tolR proximal to the TolAB region have been described. The proteins coded by this region comprise a system which allows transport of large molecules such as colicins or filamentous phage DNA, into or across the bacterial membrane (Sun and Webster, 1986, 1987). It has been proposed that

colicins E1, A, K and N as well as the gene III protein of f1 phage, all of which feature a glycine-rich region at the amino terminus, may be interacting with the protein products of *tolQRA* gene cluster. Ohno-Iwashita and Imahori (1982) showed that the carboxy two-thirds of colicin E1 can kill osmotically shocked *btuB* cells but not intact *btuB* cells. They suggested that the amino-terminal portion of colicin E1 is required for its transport through the outer membrane after the receptor binding step. It is also of interest to point out that only sensitive cells with an energized membrane can be killed. This requirement applies not only to the formation and opening of channels for pore-forming colicins (see further) but also to colicin E3, which has its target in the cytoplasm (Jetten and Jetten, 1975). Therefore, a potential across the inner membrane is probably also required for the transport of *TonB* independent colicins across the outer membrane.

Although the comparison might be fallacious it is tempting to suggest that colicins conform to the rules that apply to protein transfer across membranes in mitochondria, the rough endoplasmic reticulum membrane and bacteria in general. The three steps, association with a receptor on the correct membrane, translocation, and folding on the opposite membrane surface also apply to colicins. The energy requirement for colicin uptake may be related to unfolding which could explain the lack of disulfide bridges in their polypeptide chains. This might hold true for both *TonB* dependent and *TonB* independent colicins.

Binding of colicins to lipid in model membrane systems

To each of the three stages, receptor-binding, translocation and insertion, a specific domain of the colicin polypeptide chain is associated. The amino-terminal region appears to be required for translocation, the central region is required for receptor binding, and the COOH-terminal region is involved in channel formation (Dankert et al., 1982; Martinez et al., 1983; Suit et al., 1983; Cavard et al., 1986). Pore-forming colicins are highly elongated molecules with axial ratios between 10 and 20 (Konisky, 1982).

Despite the results concerning the *tol* genes and the *ton* B gene discussed above, the mechanics of colicin uptake are poorly understood. This is a difficult problem since the number of colicin molecules entering a cell is below the level of detection and a single colicin molecule is theoretically sufficient to kill a sensitive cell. To circumvent this problem, we have studied the interactions of colicin A with lipids in model membrane systems. Colicin A can insert into phospholipid monolayers, although at neutral pH the affinity is not high. In contrast, below pH 5.5, this affinity is considerably increased and colicin A becomes a highly penetrating protein, as strong as the lytic peptides mellitin and cardiotoxin (Pattus et al., 1983). These results were fully confirmed using unilamellar vesicles. At acid pH, not only could colicin A bind efficiently to liposomes but it could also promote their aggregation and fusion (Pattus et al., 1985). In order to be inserted into the membrane below pH 5.5, the colicin A polypeptide chain must switch from a water-soluble conformation to one that is more stable in the membrane. The polypeptide chain of pore-forming colicins contains a large hydrophobic stretch of 35 to 50 residues in the COOH-terminal region (Ebina et al., 1981; Morlon et al., 1983; Mankovich et al., 1986; A. Pugsley, personal communication). Putative membrane spanning hydrophobic helices of about 20 residues have been predicted in this region for colicin E1 (Davidson et al., 1984) and colicin A (Pattus et al., 1985). It is thus tempting to hypothesize that either the formation of these helical hairpins or their exposure, is favored at acid pH (below pH 5). The spontaneous insertion of this exposed hydrophobic helical hairpin into the membrane would be energetically favored as pointed out by Engelman and Steitz (1981). With regard to this point, it is of interest to mention the existence of a high affinity lipid binding site in colicin A. Below pH 3, a colicin A tetramer is spontaneously formed (P. Sauve, personal communication). This micelle-like protein aggregate probably arises after unmasking of putative hydrophobic and amphipatic α-helices (Baty et al., 1987). The tetramer can be dissociated (into monomers) in the presence of an excess of lipid. These monomers were shown to be associated to 22 lipid monomer molecules, suggesting that the high affinity lipid binding site is involved in tetramer assembly. Colicin A mutants altered in the COOH-terminal region have been found to spontaneously aggregate in the cytoplasm of producing cells (Baty et al., 1987). This result also suggests that a rather discrete conformational change might be needed to expose the high affinity lipid binding site of colicin A.

A bank of monoclonal antibodies directed against epitopes distributed all along the polypeptide chain of colicin A has been obtained (Cavard et al., 1985). The role of the COOH-terminal region for membrane binding has been directly demonstrated using these antibodies (M. Frenette, unpublished result).

In view of all these results, it is tempting to suggest that the interaction with the receptor and then eventually with accessory chromosomally-encoded gene products, TolA, TolB, TolQR (Kadner et al., 1979; Sun and Webster, 1987) assumed to be required for the translocation process might favor the conformational change exposing the putative helical hairpin. Translocation also requires cell respiration and proton extrusion through the membrane (Kadner et al., 1979). This is known to induce a steep pH gradient from a more acidic surface towards the bulk in energy transducing membranes (Prats et al., 1986) and this acidification of the surface of the membrane would be expected to favor colicin insertion.

Mechanism of action of channel-forming colicins

What is the structure and action of colicin A and other pore-forming colicins in the cytoplasmic membrane of sensitive cells ? To investigate this point, planar lipid bilayers have been used (Schein et al., 1978; Weaver et al., 1981; Martinez et al., 1983; Cleveland et al., 1983; Davidson et al., 1984). With this system, known voltages can be applied across the membrane and the polypeptide insertion can be probed through ion transfer (conductance). Colicins A, E1, K, Ia and Ib can form voltage-dependent single channels as demonstrated by the stepwise increase in conductance observed after injection of very low concentrations of colicin (Fig. 6).

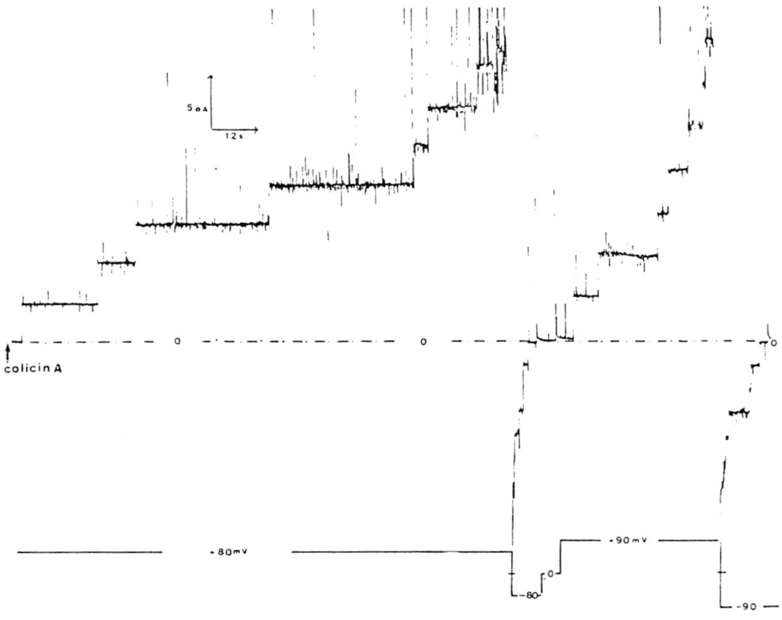

Fig. 6 : Stepwise current increase after injection of colicin A (0.1 ng/ml). Soybean phospholipid planar bilayers separating two aqueous compartments (10 mM Tris-acetate buffer pH 7.0, containing 1 M KCl, 5 mM CaCl2) were formed by transformation of liposomes to planar bilayers as described (Martinez et al., 1983). A voltage of + 90 mV was applied before the addition of colicin A, after the insertion of 8 channels, the applied voltage was reversed. After a few seconds, the current reached the bare membrane current value. Reversing the voltage to + 90 mV induced the opening of the pores, the applied voltage was then again reversed. (This figure has been kindly provided by F. Pattus).

These channels are functionally similar to the sodium and potassium channels of nerve membranes, although their ion selectivity is not marked (Raymond et al., 1985). The COOH-terminal domain of pore-forming colicins is responsible for the channel formation (Martinez et al., 1983; Cleveland et al., 1983; Davidson et al., 1984). This domain of about 20

kd can in general be easily isolated by limited proteolytic digestion and by itself can form ion channels in planar lipid bilayers. For colicin A, this protein domain has been crystallised and its three-dimensional structure at 2.7 Å resolution should be available in the near future (Tucker *et al.*, unpublished result). However this will be the water-soluble structure since the protein has been crystallized from 2.3 M-amonium sulphate aqueous buffer and we also would like to know the structure after insertion into the membrane. Two possible structures have been proposed for the channel. In one, the 20 kd polypeptide becomes stably inserted, spanning the membrane several times (Davidson et al., 1984; Pattus et al., 1985). In the other, only part of the polypeptide is stably inserted, the rest of the polypeptide chain flipping in and out of the membrane, thereby providing voltage-dependent gating for the channel (Raymond et al., 1986). At present it is not possible to rule out any model although recent results support the latter. Circular dichroism studies indicated that the COOH-terminal domain contains a high proportion of α-helical structure (Davidson et al., 1984; Pattus et al., 1985). By analogy with the structures proposed for other ion channels (Stroud and Finner Moore, 1985; Henderson and Unwin, 1975), this suggests that the pore might be formed as a bundle of α-helices crossing the bilayer. Such helices should be either hydrophobic or amphipatic, having their non-polar faces oriented toward the lipid and their polar faces oriented toward the aqueous lumen of the pore. By using secondary structure prediction methods, 6 α-helices have been predicted in the COOH-terminal domain of colicins A, E1 and Ib and, among these putative helices, only the 3 COOH-terminal ones appear to be able to span the membrane according to the hydrophobic moment method of Eisenberg (Pattus et al., 1985).
However, based on conductance measurements, the diameter of the pores should be at least 8 Å (Raymond et al., 1985) and one can calculate that at least 6 α-helices are required to form such a pore. It has thus been proposed that the pore may be formed by a dimer or a trimer of colicins A, E1 or Ib (Pattus et al., 1985). Studies using site-directed mutagenesis, which allowed the isolation of 45 different mutants in the COOH-terminal domain of colicin A, supported the dimer model (Baty et al., 1987). These studies also suggested that subtle conformational changes caused by some of the mutations in the putative α-helices lead to the aggregation of colicin within the cytoplasm of producing cells. Other mutations led to alterations of the voltage-dependent channel properties in terms of conductance, stability, voltage switching, etc... (Baty et al., 1987). These mutants will prove very useful since we can now begin to relate defined amino acid modifications with defined pore properties.

Colicin immunity.

Immunity to channel-forming colicins is provided by polypeptides of 11 kd to 18 kd that span the cytoplasmic membrane 3 to 4 times (Fig. 7) (Goldman et al., 1985; Geli et al., 1986; Mankovich et al., 1986).

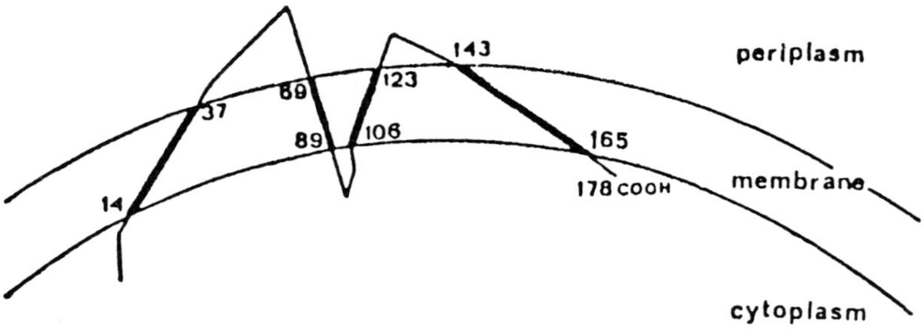

Fig. 7 : Hypothetical topology of the colicin A immunity polypeptide chain in the cytoplasmic membrane. This topology has been probed using various techniques (Geli et al., 1987).

They specifically interact with the COOH-terminal domains of colicins and they confer upon cells protection against the colicin they produce, but not against heterologous colicins with identical modes of action (Konisky, 1982). It is important to point out that they are required to protect the cells against the effects of external colicin, presumably because colicin in the cytoplasm of the colicinogenic cell cannot act correctly since the polarity of the transmembrane energy potential is opposite to that required (Geli et al., 1986). Studies using site-directed mutagenesis have suggested that interactions of putative periplasmic loops of the colicin A immunity polypeptide chain may be involved in the colicin-immunity protein interaction (Geli et al., 1986).

The amino terminal region of the immunity protein, which is in the cytoplasm, is neither required for insertion nor for function. This has been demonstrated by substituting the first 13 amino acid residues of the immunity protein by the 170 NH_2-terminal residues of colicin A. The expression of the hybrid protein obtained is inducible and confers protection to otherwise sensitive cells (Geli et al., 1987). Since this protein has a defined orientation, is easily assayable, and since we have now constructed recombinant plasmids encoding the immunity protein under the control of inducible promoters (Geli et al., 1987), this system is suitable for studying the mechanism and the energetics of polypeptide insertion into the cytoplasmic membrane, the role of accessory proteins such as secA and prlA (Michaelis and Beckwith, 1982), the energetic requirements, etc... Such studies are in progress.

A summing up.

The study of colicin release from producing cells has revealed a novel mechanism of secretion. Instead of a built-in "tag", such as a signal peptide containing information for secretion, the mechanism employs coordinate expression of a small protein which causes an increase in the envelope permeability, resulting in the release of the colicin as well as other proteins.
On the other hand, the mechanism of entry of colicins into sensitive cells involves the same three stages of protein translocation that have been demonstrated for various cellular organelles. They first interact with receptors located at the surface of the outer membrane and are then transferred across the cell envelope in a process that requires energy and depends upon accessory proteins (TolA, TolB, TolC, Fii) which might play a role similar to that of the secretory apparatus of eukaryotic and prokaryotic cells. At this point, the type of colicin described in the review interacts specifically with the inner membrane to form an ion channel. The pore-forming colicins are isolated as soluble proteins and yet insert spontaneously into lipid bilayers. The three-dimensional structures of some of these colicins should soon become available and site-directed mutagenesis studies have now provided a large number of modified polypeptides. Their use in model systems, particularly those in which the role of transmembrane potential can be tested for polypeptide insertion, constitutes a powerful handle with which to improve our understanding of the dynamics of protein insertion into and across membranes. In addition, their immunity proteins, which only exist in one state (membrane inserted) will also contribute to such an understanding.

Acknowledgements

The co-workers associated with the work described here which has been carried out in my laboratory are : Drs D. Baty, D. Cavard, V. Geli, S.P. Howard, M. Knibiehler, R. Lloubès, J. Morlon and S. Varenne. This work was supported by the CNRS, the INSERM (CRE n° 861020), the direction des Recherches Etudes et Techniques and the Fondation pour la Recherche Médicale. I am grateful to S.P. Howard for careful reading and to M. Payan for preparing the manuscript.

REFERENCES

Baty D, Knibiehler M, Verheij H, Pattus F, Shire D, Bernadac A, Lazdunski C (1987a) Site directed mutagenesis of the COOH-terminal region of colicin A : effect on secretion and voltage-dependent channel activity. Proc Natl Acad Sci USA 84:1152-1156

Baty D, Lloubès R, Geli V, Lazdunski C, Howard SP (1987b) (to be published) Extracellular release of colicin A is non-specific. EMBO J

Blobel G (1980) Intracellular protein topogenesis. Proc Natl Acad Sci USA 77:1496-1500

Braun V, Frana J, Hantke K, Schaller K (1980) Penetration of colicin M into cells of *E. coli.* J Bacteriol 142:162-168

Cavard D, Bernadac A, Pages J-M, Lazdunski C (1984) Colicins are not transiently accumulated in the periplasmic space before release from colicinogenic cells. Biol Cell 51:79-86

Cavard D, Lloubès R, Morlon J, Chartier M, Lazdunski C (1985) Lysis protein encoded by plasmid ColA-CA31. Gene sequence and export. Mol Gen Genet 199:95-100

Cavard D, Crozel V, Gorvel J-P, Pattus F, Baty D, Lazdunski C (1986) A molecular, genetic and immunological approach to the functioning of colicin A, a pore-forming protein. J Mol Biol 187:449-459

Cavard D, Baty D, Howard SP, Verheij HM, Lazdunski C (1987) Lipoprotein nature of the colicin A lysis protein. Effect of amino acid substitutions at the modification and processing site. J Bacteriol 169:2187-2194

Cleveland M, Slatin J, Finkelstein A, Levinthal C (1983) Structure-function relationships for a voltage-dependent ion channel : Properties of COOH-terminal fragments of colicin E1. Proc Natl Acad Sci USA 80:3706-3710

Dankert JR, Uratani Y, Grabau C, Cramer WA, Hermondson M (1982) On a domain structure of colicine E1. J Biol Chem 257:3857-3863

Davidson VL, Brunden KR, Cramer WA, Cohen FS (1984) Studies on the mechanism of action of channel-forming colicins using artificial membranes. J Membrane Biol 79:105-118

De Graaf FK, Oudega B (1986) Production and release of cloacin DF13 and related colicins. Curr Top in Microbiol and Immun 125:183-205

Ebina Y, Nakazawa (1983) Cyclic AMP-dependent and rho-dependent termination of colicin E1 gene transcription. J Biol Chem 258:7072-7078

Eilers M, Schatz G (1986) Binding of a specific ligand inhibits inport of a purified precursor protein into mitochondria. Nature 322:228-232

Engelman DM, Steitz TA (1981) The spontaneous insertion of proteins into and across membranes. Cell 23:411-418

Fredericq P (1957) Colicins Ann Rev Microbiol 11:9-22

Geli V, Baty D, Crozel V, Morlon J, Lloubès R, Pattus F, Lazdunski C (1986) A molecular genetic approach to the functioning of the immunity protein to colicin A. Mol Gen Genet 202:455-460

Geli V, Baty D, Lazdunski C (1987) (to be published) Use of a foreign epitope as a "tag" for the localization of minor proteins within a cell : the case of the immunity protein to colicin A. Proc Natl Acad Sci USA

Goldman K, Suit JL, Kayalar C (1985) Identification of the plasmid-encoded immunity protein for colicin E1 in the inner membrane of *Escherichia coli*. FEBS Lett 190:319-323

Hantke K (1976) Phage T6-colicin K receptor and nucleoside transport in *E. coli*. FEBS Lett 70:109-112

Henderson R, Unwin PN (1975) Three dimensional model of purple membrane obtained by electron microscopy. Nature 257:28-32

Holland IB (1975) Physiology of colicin action. Adv Microb Physiol 12:429-438

Howard SP, Leduc M, Van Heijenoort J, Lazdunski C (1987) (to be published) Lysis and release of colicin A in colicinogenic antolytic deficient *Escherichai coli*. FEMS Lett

Jetten AM, Jetten MER (1975) Energy requirement for the initiation of colicin action in *E. coli*. Biochim Biophys Acta 387:12-22

Kadner RJ, Jr Bassford PJ, Pugsley AP (1979) Colicin receptors and mechanisms of colicin uptake. Zentrablatt für Bakteriologie, Parasitenkunde, Infektionskrankheiten und Hygiene 244:90-104

Knibiehler M, Lazdunski C (1987) (to be published) Conformation of colicin A : possible difference between cytoplasmic and extracellular polypeptide chain. FEBS Lett 216:183-189.

Konisky J (1982) Colicins and other bacteriocins with established modes of action. Ann Rev Microbiol 36:125-144

Little JW, Mount DW (1982) The SOS regulatory system of *Escherichia coli*. Cell 29:11-22

Lloubès R, Baty D, Lazdunski C (1986) The promoters of the genes for colicin production, release and immunity in the ColA plasmid : effects of convergent transcription and LexA protein. Nucl Acids Res 14:2621-2636

Luirink J, Van der Sande C, Tommassen J, Veltkamp E, de Graaf FK, Oudega B (1985) Mode of action of protein H encoded by plasmid CloDF13 : effects of culture conditions and of mutations affecting phospholipase A activity on excretion of cloacin DF13 and on growth and lysis of host cells. J Gen Microbiol 132:825-834

Maker PA, Singer SJ (1983) Disulfide bonds and the translocation of proteins across membranes. Proc Natl Acad Sci USA 83:9001-9005

Mankovich JA, Hsu CH, Konisky J (1986) DNA and amino acid sequence analysis of structural and immunity genes of colicins Ia and Ib. J Bacteriol 168:228-236

Martinez MC, Lazdunski C, Pattus F (1983) Isolation, molecular and functional properties of the C-terminal domain of colicin A. EMBO J 2:1501-1507

Michaelis S, Beckwith J (1982) Mechanism of incorporation of cell-envelope protein in *E. coli*. Ann Rev Microbiol 36:435-465

Morlon J, Lloubès R, Chartier M, Bonicel J, Lazdunski C (1983) Complete nucleotide sequence of the structural gene for colicin A, a gene translated at non-uniform rate. J Mol Biol 170:271-285

Ohki M, Doi O, Nojima S (1972) Mutant of *Escherichia coli* K12 deficient for detergent-resistant phospholipase A. J Bacteriol 110:864-869

Ohno-Iwashita Y, Imajori K (1982) Assignment of the functional loci in the colicin E1 molecule by characterization of its proteolytic fragments. J Biol Chem 257:6446-6451

Pagès J-M, Anba J, Bernadac A, Shinagawa H, Nakata A, Lazdunski C (1984) Normal precursors of periplasmic proteins accumulated in the cytoplasm are not exported post-translationally in *E. coli*. Eur J Biochem 143:499-505

Pattus F, Martinez MC, Dargent B, Cavard D, Verger R, Lazdunski C (1983) Interaction of colicin A with phospholipid monolayers and liposomes. Biochemistry 22:5698-5703

Pattus F, Cavard D, Crozel V, Baty D, Adrian M, Lazdunski C (1985) pH-dependent membrane fusion is promoted by various colicins. The EMBO J 4:2469-2474

Pattus F, Heitz F, Martinez C, Provencher SW, Lazdunski C (1985) Secondary structure of the pore-forming colicin A and its C-terminal fragment : experimental facts and structure prediction. Eur J Biochem 152:681-689

Prats M, Tessié J, Toccane JF (1986) Lateral proton conduction at lipid-water interfaces and its implications for the chemiosmotic-coupling hypothesis. Nature 322:756-758

Pugsley AP, Reeves P (1975) Iron uptake in colicin B resistant mutants of *E. coli* K12. J Bacteriol 127:218-228

Pugsley AP, Schwartz M (1983) Expression of a gene in a 400 bp fragment of colicin plasmid ColE2-P9 is sufficient to cause host cell lysis. J Bacteriol 156:109-114

Pugsley AP (1984) Genetic analysis of VolN plasmid determinants for colicin production release and immunity. J Bacteriol 158:523-529

Pugsley AP, Schwartz M (1984) Colicin E2 release : lysis, leakage or secretion ? Possible role of a phospholipase. EMBO J 3 : 2392-2397

Randall LL, Hardy SJ (1986) Correlation of competence for export with lack of tertiary structure of the mature species : a study *in vivo* of maltose-binding protein in *E. coli*. Cell 46:921-928

Raymond L, Slatin SL, Finkelstein A (1985) Channels formed by colicin E1 in planar lipid bilayers are large and exhibit pH-dependent ion selectivity. J Membrane Biol 84:173-181

Raymond L, Slatin SL, Finkelstein A, Lin QR, Levinthal C (1986) Gating of a voltage-dependent channel (colicin E1) in planar lipid bilayer : translocation of regions outside the channel-forming domain. J Membrane Biol 92:255-268

Reeves P (1972) The bacteriocins, London : Chapman and Hall

Sabet SF, Schnaitman CA (1971) Localization and solubilization of colicin receptors. J Bacteriol 108:422-430

Sauer M, Hantke K, Braun V (1987) Ferric coprogen receptor *Fhu*E of *E. coli* : processing and sequence common to all TonB-dependent outer membrane receptor proteins. J Bacteriol 169:2044-2049

Schein SJ, Kagan BL, Finkelstein A (1978) Colicin K acts by forming voltage-dependent channels in phospholipid bilayer membranes. Nature 276:159-163

Stroud RM, Finer-Moore R (1985) Acetylcholine receptor, structure function and evolution. Ann Rev Cell Biol 1:317-351

Suit JL, Fan ML, Sabik JS, Lebarre R, Luria SE (1983) Alternative forms of lethality in mitomycin C-induced bacteria carrying ColE1 plasmids. Proc Natl Acad Sci USA 80:579-583

Sun TP, Webster RA (1986) *Fii*, a bacterial locus required for filamentous phage infection and its relation to colicin-tolerant *tol*A and *tol*B. J Bacteriol 165:107-115

Sun TP, Webster RE (1987) (to be published) Nucleotide sequence of a gene cluster involved in the entry of "E" colicins and the single stranded DNA of infecting filamentous phage into *E. coli*. J Bacteriol

Toba MH, Masaki H, Takahisa O (1986) Primary structures of the ColE2-P9 and ColE3-CA38 lysis genes. J Biochem 99:591-596

Tucker A, Pattus F, Tsernoglou D (1986) Crystallization of the C-terminal domain of colicin A carrying the voltage-dependent pore activity of the protein. J Mol Biol 190:133-134

Uratani Y, Kageyama M (1977) A fluorescent probe response to the interaction of Pyocin R1 with sensitive cells. J Biochem 81:333-341

Varenne S, Cavard D, Lazdunski C (1981) Biosynthesis and export of colicin A in *Citrobacter freundii CA31*. Eur J Biochem 116:615-620

Varenne S, Knibiehler M, Cavard D, Morlon J, Lazdunski C (1982) Variable rate of polypeptide chain elongation for colicins A, E2 and E3. J Mol Biol 159:57-70

Varenne S, Buc J, Lloubès R, Lazdunski C (1984) Translation is a non uniform process. Effect of tRNA availability on the rate of elongation of nascent polypeptide chains. J Mol Biol 180:549-576

Varley JM, Boulnois GJ (1984) Analysis of a cloned colicin Ib gene : complete nucleotide sequence and implication for regulation of expression. Nucl Acids Res 12:6727-6739

Walter P, Blobel G (1982) Signal recognition particle contains a 7 s RNA essential for protein translocation across the endoplasmic reticulum. Nature 299:691-698

Watson RJ, Lau PC, Vernet T, Visentin LP (1984) Characterization and nucleotide sequence of a colicin-release gene in the *hic* region of plasmid ColE3-CA38. Gene 29:175-184

Weaver C, Kagan B, Finkelstein A, Konisky J (1981) Mode of action of colicin Ib : formation of ion-permeable membrane channels. Biochim Biophys Acta 645:137-142

Wu HC, Tokunaga M, Tokunaga H, Hayashi S, Giam CZ (1983) Post-translational modification and processing of membrane lipoproteins in bacteria. J Cell Biochem 22:161-171

Yamada M, Ebina Y, Miyata T, Nakazawa T, Nakazawa A (1982) Nucleotide sequence of the structural gene for colicin E1 and predicted structure of the protein. Proc Natl Acad Sci USA 79:2827-2831

Zimmermann R (1986) in Protein Compartmentalization. Strauss A, Bovine I, Kreil G (ed). Springer-Verlag Berlin

BACTERIAL PROTEIN TRANSLOCATION

Elliott Crooke and William Wickner
Molecular Biology Institute and
Department of Biological Chemistry
School of Medicine
University of California
Los Angeles, CA 90024
USA

Introduction

Protein translocation is a process common to all cells. Both prokaryotes and eukaryotes are divided into distinct compartments, each containing a unique set of proteins important to specific cellular functions. Except for a few mitochondrial and chloroplast proteins, almost all proteins are synthesized in the cytoplasm; therefore many proteins must cross at least one membrane to reach their final destination within the cell. The eukaryotic endoplasmic reticulum and mitochondria, the bacterial plasma membrane and various secreted proteins have been used as model systems for extensive research on protein translocation. From these studies certain common themes for membrane assembly have emerged (Wickner and Lodish, 1985). In summary (i) translocation can occur independently of protein synthesis, (ii) the information for translocation is contained throughout the protein, not just in the leader (signal) sequences, (iii) translocation requires energy, either as a high energy phosphate, a membrane electrochemical potential, or both. Still, the molecular mechanisms for translocation remain largely unknown. To gain further understanding of these problems, we have focused our efforts on studying membrane assembly and protein export in *E. coli*, exploiting the sophistication of bacterial genetic engineering and the relative ease of cellular fractionation and biochemistry of bacteria.

NATO ASI Series, Vol. H16
Membrane Biogenesis. Edited by J. A. F. Op den Kamp

Genetic Analysis of Membrane Assembly Elements in Procoat and Leader Peptidase

As models for protein membrane assembly we have utilized two secreted proteins; procoat, the precursor to the M13 coat protein, and leader peptidase, an integral plasma membrane protein of E. coli. Procoat is comprised of two hydrophobic segments (one in the leader sequence and one in the mature region) separated by an acidic sequence, with basic regions at the amino and carboxy termini (Wickner, 1983). Leader peptidase contains three amino terminal hydrophobic domains, a polar cytoplasmic sequence between the first and second hydrophobic domains, and a large, polar, perplasmic region (Wolfe, et al. 1983; Moore and Miura, 1987). Our research is aimed at identifying and characterizing the required assembly domains within these proteins. We have invesitgated the relevant features of both apolar and polar sequences. Using molecular genetics we have discovered: (i) a novel secretion element, termed the "hydrophobic helper," in both leader peptidase (Dalbey, et al. 1987) and procoat (Kuhn, et al. 1986a), (ii) the presence of an uncleaved internal signal sequence in leader peptidase (Dalbey and Wickner, 1987), (iii) the importance of polar segments as topogenic elements in membrane assembly. The charged carboxy terminus of procoat, which never leaves the cytoplasm, is vital for membrane insertion (Kuhn, et al, 1986b). Also, the polar cytoplasmic domain of leader peptidase appears to act as a "translocation poison" sequence (von Heijne, et al, unpublished data). The context-dependent roles of these various secretion elements in membrane assembly, specifically for leader peptidase, are described elsewhere in this publication (G. von Heijne).

Trigger Factor and the Reconstitution of In Vitro ProOmpA Translocation

Recently, an *in vitro* bacterial protein translocation system was developed (Rhoads et al, 1984; Muller and Blobel, 1984; Chen, et al, 1985); precursor proteins synthesized *in vitro* can assemble posttranslationally into inverted bacterial plasma membrane vesicles in an ATP and membrane electrochemical potential dependent fashion (Chen and Tai, 1985; Geller et al, 1986). We have started to fractionate and reconstitute this reaction to define which components are necessary for translocation.

Specifically, our strategy has been to purify the precursor protein proOmpA from the in vitro synthesis reaction to determine what, if any, other soluble components are required for subsequent translocation (Crooke and Wickner, 1987). One step in the purification involves eluting the proOmpA from an anti-OmpA antibody column with 8 M urea. Further purification of the proOmpA by anion exchange chromatography in the presence of 8 M urea results in radiochemically pure proOmpA that is close to homogeneous. Following the removal of the urea by dialysis, the renatured purified proOmpA is incompetent for translocation into inverted vesicles. Inclusion of a cytoplasmic extract in the translocation reaction does not restore translocation competence to the proOmpA. However, if the cytoplasmic extract is mixed with the proOmpA prior to the removal of the urea, the resulting renatured proOmpA is competent, with an efficiency for translocation equal to the proOmpA in the initial synthesis reaction. We believe that a soluble factor, termed "trigger factor," is involved in triggering the folding of proOmpA into a form which is competent for membrane assembly as the urea is removed.

Evidence exists that suggests that trigger factor and proOmpA form a complex: 1) Prior to purification, proOmpA sediments as a high molecular species, 2) Chemical crosslinking experiments indicate that proOmpA is involved in a heterologous complex (Crooke and Wickner, unpublished data), and 3) treatment of competent proOmpA with chaotropic agents renders it incompetent for translocation unless it is subsequently incubated with cytoplasmic extract. A folding function for trigger factor is supported by studies showing that assembly-competent proOmpA (renatured in the presence of trigger factor) is more sensitive to limited proteolytic digestion than assembly-incompetent proOmpA (Brundage, Crooke, and Wickner, unpublished data).

We are currently attempting to purify trigger factor and to isolate its gene so that we can test the molecular mechanisms of trigger factor and its in vivo role in protein folding and translocation.

References

Chen, L, Rhoads D, Tai PC (1985) Alkaline phosphatase and OmpA protein can be translocated posttranslationally into membrane vesicles of Escherichia coli. J Bact 161:973-980

Chen L, Tai PC (1985) ATP is essential for protein translocation into Escherichia coli membrane vesicles. Proc Natl Acad Sci USA 82:4384-4388

Crooke E, Wickner W (1987) Trigger Factor: A soluble protein which folds pro-OmpA into a membrane assembly competent form. Proc Natl Acad Sci USA 84:5216-5220

Dalbey R, Kuhn A, Wickner W (1987) The internal signal sequence of Escherichia coli leader peptidase is necessary, but not sufficient, for its rapid membrane assembly. J Biol Chem (in press)

Dalbey R, Wickner W (1987) Characterization of the internal signal-anchor domain of Escherichia coli leader peptidase. J Biol Chem (in press)

Geller BL, Movva R, Wickner W (1986) Both ATP and the electrochemical potential are required for optimal assembly of pro-OmpA into Escherichia coli inner membrane vesicles. Proc Natl Acad Sci USA 83:4219-4222

Kuhn A, Kreil G, Wickner W (1986a) Both hydrophobic domains of M13 procoat are required to initiate membrane insertion. EMBO J 5: 3681-3685

Kuhn A, Wickner W, Kreil G (1986b) The cytoplasmic carboxy terminus of M13 procoat is required for the membrane insertion of its central domain. Nature 322:335-339

Moore K, Muira S (1987) A small hydrophobic domain anchors leader peptidase to the cytoplasmic membrane of Escherichia coli. J Biol Chem 262:8806-8813

Muller M, Blobel G (1984) In vitro translocation of bacterial proteins across the plasma membrane of Escherichia coli. Proc Natl Acad Sci USA 81:7421-7425

Rhoads DB, Tai PC, David BD (1984) Energy-requiring translocation of the OmpA protein and Alkaline phosphatase of Escherichia coli into inner membrane vesicles. J Bact 159:63-70

Wickner W (1983) M13 coat protein as a model of membrane assembly. TIBS 8:90-92

Wickner W, Lodish HF (1985 Multiple mechanisms of protein insertion into and across membranes. Science 230:400-407

Wolfe PB, Wickner W, Goodman JM (1983) Sequence of the leader peptidase gene of Escherichia coli and the orientation of leader peptidase in the bacterial envelope. J Biol Chem 258:12073-12080

THE USE OF HYBRID PROTEINS IN THE STUDY OF PROTEIN TARGETING SIGNALS

Anthony P.Pugsley
Institut Pasteur
25 rue du Dr.Roux
75724 Paris Cedex 15
France

Introduction. A hybrid protein is a complex of two or more polypeptide sequences or fragments thereof which would not normally be associated, but are coupled together either by fusing the genes which encode them (gene fusions) or by chemically cross-linking the purified component parts. One of the major applications of hybrid proteins has been in the study of protein targeting. This article reviews basic concepts behind the use of hybrid proteins, the ways in which they can be constructed, specific details of their applications, and their short-comings.

Non-cytoplasmic proteins have specific targeting signals. A basic assumption common to all studies on protein targeting is that the sequences of proteins other than those which remain at their site of synthesis (usually the cytoplasm) contain one or more signals which target them to their final destination. There are four potential types of targeting signals:

1. **Routing signals,** which direct polypeptides into a targeting pathway without necessarily influencing their final destination along the pathway. Typical examples of routing signals are signal peptides of precursors to proteins targeted into the secretory pathway in eukaryotes or prokaryotes, and transit peptides and presequences which target nuclear-encoded protein precursors to chloroplasts and mitochondria respectively.
2. **Retention or salvage signals,** which prevent a polypeptide from travelling along the full length of a targeting pathway eg, the C-terminal peptide K-D-E-L which enables certain secretory proteins to be retained in or recycled back to the lumen of the RER, rather than continuing through the Golgi apparatus and on to the cell surface (see below).
3. **Sorting signals** are important in branched targeting pathways. Examples are the mannose-6-phosphate residues found on soluble lysosomal proteins of higher eukaryotes, and the signal close to the N-terminus which targets proteins to the yeast vacuole. Another example may be the amino-terminal sorting signals which redirect some proteins to the mitochondrial intramembranous space after they have been imported into the mitochrondrial matrix.

NATO ASI Series, Vol. H16
Membrane Biogenesis. Edited by J.A.F. Op den Kamp

4. Transmembrane topology signals, which determine whether and how a polypeptide will adopt a transmembrane configuration.

Targeted proteins will often contain several types of targeting signals, not all of which will be easily-identifiable, linear sequences of amino acids such as signal peptides or RER retention signals.

Passenger or tracer proteins. Hybrid proteins have been extensively employed in the search for targeting signals. The basic principle is to couple parts of the sequence of a targeted polypeptide to a second sequence from a polypeptide devoid of such signals, and then to determine whether the passenger protein follows the same targeting pathway and reaches the same final destination as the polypeptide from which the putative signal was taken. The choice of passenger protein is obviously critical for the success of these experiments. Passenger proteins should be easy to identify; indeed many studies on protein targeting have been greatly facilitated simply by coupling targeting signals from a polypeptide which, for reasons of poor antigenicity or the absence of enzymatic activity, is difficult to detect, to a more readily-identifiable polypeptide. Ideally, the passenger protein should have an enzymatic activity which can be detected by histological methods, facilitating ready detection of the hybrid polypeptide during its construction and, later, _in situ_ or during cell fractionation. In some cases, the fusion of an epitope to the targeted protein may suffice to enable the protein to be detected more easily, especially if monoclonal antibodies or other epitope-specific antibodies are available (Munro and Pelham, 1984). Most hybrid proteins are the products of genetic engineering. The gene for the passenger protein should, therefore, be easy to manipulate, either by _in vitro_ recombinant genetic techniques or by random, transposon-mediated _in vivo_ insertion mutagenesis (see below). The hybrid gene usually consists of the sequence coding for (part of) the targeted polypeptide at the 5' end, and for the passenger protein at the 3' end. Consequently, replacement of the natural N-terminus of the passenger protein by sequences from the targeted protein should not affect its enzymatic or other properties. In most cases, the passenger protein should not have its own endogenous targeting signals, although it may be desirable to study the combined effects of different endogeneous and exogenous signals. The passenger should also not form strong secondary or tertiary structures which might impede its movement through membranes. Although normally cytoplasmic proteins would seem ideal for use as passengers, experience shows that many of them contain sequences such as normally cryptic transmembrane segments of 12 or more hydrophobic or neutral residues which prevent their translocation through certain lipid bilayer membranes. It is often advantageous to screen or select for targeting of the passenger protein. This may be done by choosing passenger proteins which are only active once they reach their target, or are toxic unless correctly targeted. Finally, as recently demonstrated in elegant studies by Eilers and Schatz (1986), targeting signal-passenger protein hybrids are often relatively easy to purify using affinity chromatography, which could be very useful for *in vitro* experiments with purified components. In practice, the choice of passenger protein, the most commonly-used of which are listed in Table 1, is often a compromise between these various features.

Table 1. Enzymes and other proteins with potential applications as tracers in protein targeting studies.

a,+, targeted correctly in at least some cases; -, not targeted/incorrectly targeted; NT, not tested.
b, toxic to producing cells.
c, in-frame gene fusions in bacteria or other organisms may confer antibiotic resistance.
d, B.Doberstein and I.Palva personal communications.
e, *in vitro* using dog pancreas microsomes.
f, selection for in-frame fusions may be possible in bacteria devoid of dihydrofolate reductase activity.
g, lacking endogenous signal peptide.

References: A,Benson *et al*, 1985; B,Emr *et al*., 1984; C,Douglas *et al*., 1984; D,Hall *et al*., 1984; E,Manoil and Beckwith, 1986; F,van den Broeck *et al*., 1985; G, Lipp and Dobberstein, 1986. H,Hurt *et al*., 1985; I,Kalderon *et al*., 1984; J,Grayhed and Inouye, 1984; K,Simon *et al*., 1987; L,Davey *et al*., 1985; M,Emr *et al*., 1983; N,Emr *et al*., 1986; O,Johnson *et al*., 1987.
(This reference list is meant as a guide and is not exhaustive; further details are given in the text).

Table 1.

Protein	Normal location	Size	Active form	Simple screen select	Simple assay	Examples of use in potentially targeted protein[a]/reference					
						Bacterial export	RER sec. route	Mitochondria	Chloroplast	Nucleus	Vacuole/ lysosome
E.coli β-galactosidase	cytoplasm	117kD	tetramer	+	+	-(A)[b]	-(B)	+(C)	NT	+(D)	NT
E.C.alkaline phosphatase[g]	periplasm	50kD	dimer with intramol S-S and Zn^{++}	+	+	+(E)	NT	NT	NT	NT	NT
Neomycin phosphotransferase (bacteria)	cytoplasm	28kD		+[c]	-	NT	NT	NT	+(F)	NT	NT
Cloramphenicol transacetylase (bacterial)	cytoplasm	24kD	tetramer	+[c]	-	-[b,d]	+(G)[a]	NT	NT	NT	NT
Dihydrofolate reductase (mouse)	cytoplasm	18kD	monomer	-[f]	-	-[b,d]	NT	+(H)	NT	NT	NT
Pyrivate kinase (chicken)	cytoplasm	57kD		-	-	NT	NT	NT	NT	+(I)	NT
β-lactamase (bacterial)[g]	periplasm	32kD	monomer	+[c]	+	+(J)	+(K)	NT	NT	NT	NT
α-globin (monkey red blood cells)	cytoplasm	15kD	monomer	-	-	NT	+(K)	NT	NT	+(L)	NT
Invertase (yeast)[g]	secret/ cytoplasm	60 kD	dimer/ octamer	+	+	NT	+(M)	+(N)	NT	NT	+(O)

Construction of gene fusions. As already mentioned, most hybrid proteins used in targeting studies are the products of gene fusions constructed *in vitro*. A simple method for creating in-frame fusions at different sites stretching into the gene for a targeted protein from the 3' end is illustrated in Fig.1.

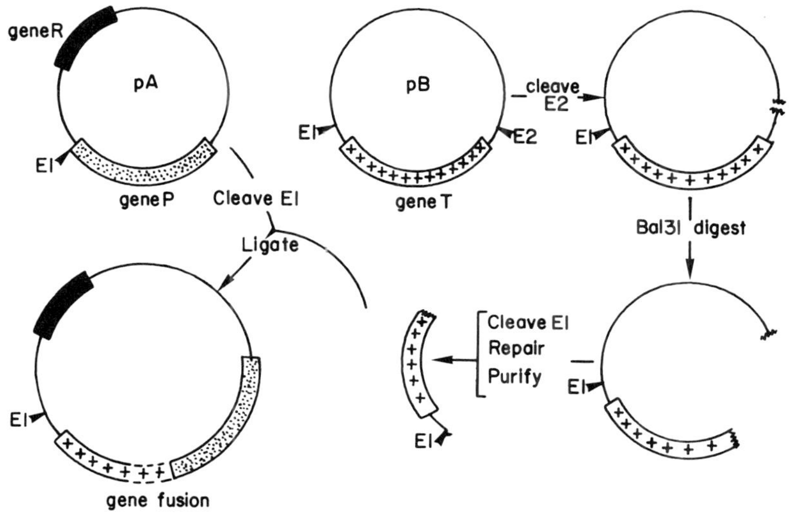

Fig.1. Use of cloned genes for a targeted protein (gene T), and for a passenger protein (gene P) to create gene fusions coding for hybrid polypeptides containing various lengths of the targeted protein. Gene P is carried by a multiple copy number plasmid (A) with a selectable antibiotic resistance marker (R). Gene P is devoid of translation initiation signals (ribosome binding site and ATG) and has a unique blunt-end restriction endonuclease site (E1) at its extreme 5' end. Gene T is also carried by a multiple copy number plasmid (B). It is bordered by unique restriction endonuclease sites E1 (5' of its promoter) and E2 (3' of gene T). Plasmid B is linearised with enzyme E2 and deleted with nuclease Bal31 which chews into the 3' end of gene T, and then digested with restriction enzyme E1. The resulting DNA is treated with Klenow fragment of DNA polymerase I to fill in the ragged ends generated by Bal31, and the deleted gene fragments are purified and ligated into plasmid A linearised at the E1 site. 50% of the inserted fragments will be in the same orientation as gene P, and of these 33% should be in-frame and encode a hybrid polypeptide. Several variations on this basic scheme are possible.

Alternatively, fragments of the gene for the targeted protein can be fused into the gene for the passenger protein, or DNA for the passenger protein may be inserted into that for the targeted protein such they are in-frame at both extremities (sandwich fusion). An alternative method for creating gene fusions is to use a transposon carrying a 5' truncated gene for the passenger protein at one of its extremities. Natural transposons such as Tn5 or the DNA of bacteriophage Mu insert more-or-less at random into any cloned DNA in plasmids in *Escherichia coli*. It is therefore relatively easy to obtain a bank of insertions of Tn5 or Mu derivatives carrying, for example, truncated genes for alkaline phosphatase (AP) or β-galactosidase (βGAL), in a gene for a targeted protein (Manoil and Beckwith, 1986 Castihlo *et al*, 1984,).

Hybrid proteins containing routing signals. Routing signals of most bacterial and eukaryotic secretory proteins and of all nuclear-encoded chloroplast and mitochondrial proteins are located at the extreme N-terminus, probably protruding from the surface of the folded polypeptide (see below). Simple gene fusions constructed as outlined in Fig.1 are, therefore, particularly well-suited to the study of this type of targeting signal.

1. Bacterial proteins

The identification of the first routing signal, the signal peptide of eukaryotic secretory proteins, pre-dates the use of gene fusions . According to the signal hypothesis, the signal peptide (or the related signal and insertion sequences) is the only routing signal directing proteins across the RER membrane. Signal peptides, therefore, might be able to facilitate the translocation of **any** polypeptide sequence into the lumen of the RER. The proposal was first tested in the bacterium *E.coli* at a time when the extent to which the signal peptide hypothesis could be applied to bacterial protein export had not been evaluated. These first experiments tested the possibility that the normally cytoplasmic enzyme βGAL could be converted into an exported protein by fusing a bacterial signal peptide to its N-terminus. The observation by Moreno *et al* (1980) that a hybrid βGAL derivative carrying the 24-residue signal peptide plus 15 N-terminal residues from the outer membrane LamB protein remained almost entirely in the cytoplasm was originally treated with disbelief by the hard-line protagonists of the signal peptide hypothesis. The explanation favoured by Moreno *et al* was that successful entry of at least the LamB protein into the bacterial secretory pathway required more than 15 residues of the mature LamB protein **in addition** to the signal peptide.

Indeed, it is now known that E.coli will initiate the export of βGAL hybrids carrying slightly more LamB sequence information (Benson et al, 1984) or signal peptides and the amino termini of a variety of other exported proteins. However, **none** of these hybrids are completely translocated across the cytoplasmic membrane, although some of them may be processed (Rasmussen et al, 1984). Instead, the hybrid polypeptides seem to remain anchored in the cytoplasmic membrane, preventing processing of other exported proteins (Fig.2), and causing the cells to die and usually to lyse (eg, Ito et al, 1981; Michaelis et al, 1986; Benson et al, 1984; d'Enfert and Pugsley, 1987). The reasons for these effects have not been thorougly investigated, but several groups have suggested that βGAL cannot be translocated through the cytoplasmic membrane, and therefore remains jammed in the estimated 20,000 specific sites involved in protein export (Ito et al, 1981). This may be because the βGAL part of the hybrid oligomerises before it is translocated, or because βGAL has one or more sequences which are intrinsically unable to cross the cytoplasmic membrane. Interestingly, the βGAL sequence includes a potential membrane anchor or "stop transfer" signal around residue 100 which might prevent translocation. The membrane-perturbing effects of accumulating βGAL sequences in the cytoplasmic membrane, and failure to export enzymes essential for cell wall synthesis would both be expected to result in lysis.

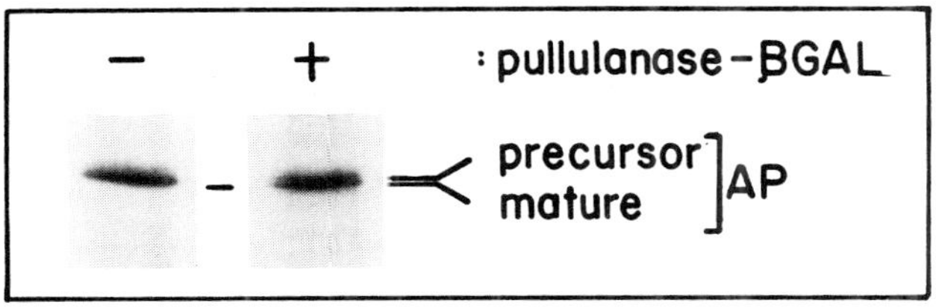

Fig.2. Accumulation of proalkaline phosphatase (AP) in cells producing a pullulanase-βGAL hybrid which eventually causes cell death and lysis (d'Enfert and Pugsley, 1987)

Although the results of studies with βGAL hybrids proved disappointing in some respects, their toxicity and their extremely low of activity when anchored in the membrane (presumably caused by incorrect oligomerisation) have been extremely useful in selecting mutations affecting protein export. Most notably, intragenic suppressors of low enzymatic activity or toxicity almost invariably affected the DNA encoding the signal peptide part of the hybrid. When the mutated DNA was recombined into the gene coding for the authentic exported protein, the resulting signal peptides exhibited low export proficiency. The mutations were sequenced and were found to affect the hydrophobic core of the signal peptide, usually causing substantial deletions or the insertion of polar or charged residues (Fig.3).

```
LamB wild type       MMITLRKLPLAVAVAAGVMSAQAMA↓V
point mutations                  REE  R
deletions                   ←——→
                            ←—·······—→

MBP  wild type        MKIKTGARILALSALTTMMFSASALA↓K
point mutations               P    E K RR
deletion                        ←—····—→

AP wild type              MKQSTIALALLPLLFTPVTKA↓R
point mutations                  QEH    R
                                        Q
deletion                         ←→
```

Fig.3. Sequence changes resulting from mutations in signal peptides for three exported proteins of E.coli selected for their ability to prevent the lethality of signal peptide-βGAL hybrids. LamB is the outer membrane maltoporin and MBP and AP are periplasmic maltose binding protein and alkaline phosphatase respectively. When recombined into the wild type gene, these mutations cause a 70-99% drop in protein export efficiency; polar residue insertions generally having a lower effect than deletions or charged residue insertions. The hydrophobic core is underlined and the cleavage site is indicated by ↓. Data compiled from Briggs and Gierasch (1986), Bankaitis and Bassford (1984), Benson *et al* (1985) and Michaelis *et al* (1986), and references therein.

Extragenic suppressors were also obtained, and a complex selection and counter-selection procedure was developed to identify genes whose products might interact with signal peptides during protein export (Benson *et al*, 1985).

The question of whether or not cytoplasmic proteins can cross the bacterial cytoplasmic membrane could be addressed using other passenger proteins. We are currently attacking this question through the use of another *E.coli* cytoplasmic enzyme, amylomaltase (AM). This enzyme has the advantage of being smaller than β-galactosidase and of being monomeric.

Furthermore, it should be possible to design a procedure for screening or selecting for export of AM, which is essential for growth on maltose. Meanwhile, Plückthun and Knowles (1987) have reported that triosephosphate isomerase, a cytoplasmic protein from chicken, can be exported into the E.coli periplasm when fused to a fragment of bacterial pro-β-lactamase (βLAC) including the signal peptide and a few residues of the mature polypeptide, but not when only the signal peptide is used. Other attempts to export cytoplasmic proteins such as mouse dihydrofolate reductase (DHFR) or bacterial chloramphenicol transacetylase (CAT) by fusing them to a bacterial signal peptide have not been successful; indeed these hybrids were also toxic to the producing cells, possibly because they too blocked the export pathway (Table 1). The toxicity of proamylase-CAT hybrids in B.pumilis (Palva et al, 1987) might offer a means of selecting export-defective mutants in this bacterium.

2. Eukaryotic secretory proteins

Given the weight of evidence that βGAL cannot cross the E.coli cytoplasmic membrane when fused to a bacterial signal peptide, it was not too surprising that hybrids obtained by fusing βGAL to varying lengths of the precursor of the yeast secreted protein invertase were not be secreted by yeast cells. Instead, all of the hybrids, including one containing only the signal peptide and 41 residues from the mature invertase amino terminus, remained trapped in the ER, where they were core glycosylated (Emr et al, 1984). Even the longest hybrid, which contained 90% of the invertase polypeptide, did not progress beyond the ER to the Golgi. These hybrids may have been translocated into the lumen of the ER, rather than remaining jammed in the membrane, because they were glycosylated and did not block protein secretion. If this is correct, it may point to a fundamental difference between protein translocation across the RER and bacterial cytoplasmic membranes. Signal peptides and signal sequences have been also shown to catalyse the in vitro transfer of other cytoplasmic proteins across the ER membrane of dog pancreas microsomes (Table 1 and Eskridge and Shields, 1986). It therefore seems likely that translocation of these hybrids could occur in vivo, but it remains unlikely that they would progress beyond the ER to the Golgi and on to the cell surface.

Are signal peptides necessarily N-terminal? Several workers have proposed that the signal peptide must protrude from the surface of the polypeptide in order to interact with the membrane or components of the secretory apparatus. However, three recent gene fusion studies indicate that the signal peptide can be preceeded by relatively long foreign sequences without greatly affecting protein translocation across the bacterial cytoplasmic or RER membranes (Hayashi et al, 1985; Pohler et al, 1986; Simon et al, 1987), although other studies produced the contrary result (eg, Finidori et al, 1987).

3. Mitochondrial and chloroplast proteins.

Other examples of βGAL being translocated across membranes have been reported with yeast mitochondria. Douglas et al (1984) found that βGAL coupled to the N-terminal 350 amino acids of the 509-residue F1 ATPase β subunit was not only translocatead into the mitochondrion, as judged by protease protection tests, but also interfered with mitochondrial respiratory functions, indicating that it probably interacted with other ATPase components in the inner membrane. Emr et al (1986) subsequently showed that a maximum of 169 amino acids of the ATPase subunit sufficed for mitochondial targeting of βGAL. Shorter hybrids were not delivered to the mitochondrion, possibly because they were unstable, but as few as 29 amino-terminal residues of the ATPase subunit targeted the normally secreted invertase (lacking its own signal peptide) into mitochondria. Similarly, Hase et al (1986) found that the N-terminal 21 amino acids of the yeast mitochondrial 70 KDa outer membrane protein could direct βGAL into the mitochondrial matrix.

These experiments demonstrate that even bulky, potentially oligomeric polypeptides can be translocated across the two mitochondrial membranes. Among the other examples of organelle-specific targeting of normaly cytoplasmic proteins listed in Table 1, the work on the uptake of DHFR by yeast and by Neurospora mitochondria deserves special mention. Studies by Schatz and his colleagues showed that as few as 12 amino acids from the N-terminus of a mitochondrial protein presequence could direct DHFR into the mitochondrion. However, presequence-mediated DHFR import was blocked in the presence of methotrexate, which binds to DHFR and induces an apparently irreversible conformational change. Eilers and Schatz (1986) interpreted these results as showing that proteins with a certain degree of secondary structure cannot be imported into mitochondria, implying that the import of βGAL described above probably occurred before the enzyme oligomerised. However, Pfanner et al (1987) have recently proposed that NTP hydrolysis may be important for an enzyme which unfolds proteins, including hybrids containing normally excluded sequences such as DHFR, before or as they are translocated across the mitochondrial envelope.

Another important result obtained through studies with DHFR hybrids was that DHFR itself contains a cryptic mitochondrial routing sequence which is normally buried within the folded molecule. However, when this sequence was fused to the N-terminus of a complete DHFR molecule, the hybrid, which contained no mitochondrial protein sequences what-so-ever, was routed to the mitochrondrion (Hurt and Schatz, 1987). These studies have now been extended to show that the structural and sequence constraints on mitochondrial routing signal are so low that stretches of DNA coding for such sequences are extremely common in a normal bacterial genome. Gene fusions constructed by joining these sequences to a sequence for a protein devoid of a presequence encode mitochondrially-targeted hybrids (Baker and Schatz, 1987).

4. Nuclear proteins.

Other examples of a routing signals which can translocate βGAL across a membrane are the nuclear targeting signals of yeast nuclear proteins (Hall et al, 1984). Similary, nuclear targeting of βGAL and of cytoplasmic pyruvate kinase was used to define a short sequence, P-K-K-K-R-K-V, in SV40 virus large T antigen as a nuclear routing signal (Kalderon et al, 1984). Replacement of the second lysine residue by a threonine completely prevent nuclear targeting of the hybrids. One of the routing signals identified by Hall et al in the yeast MAT a2 protein is N-terminal, whereas that which Kalderon et al identified in the large T antigen, and a second T antigen nuclear targeting signal subsequently identified using pyruvate kinase gene fusions (Richardson et al, 1986) are close to the centre of the polypeptide. Furthermore, micro-injection studies using E.coli cytoplasmic galactokinase fusions indicated that the nuclear targeting signal of Adenovirus E1a protein consists primarily or exclusively of the last five amino acids (K-R-P-R-P, Lyons et al, 1987). Nuclear targeting is clearly very different from the targeting of secretory or mitochondrial proteins, since the polypeptides cross the double nuclear membrane via large, apparently non-specific pores. These pores seem to exclude only relatively large polypeptides, but these can be translocated into the nucleus if coupled to nuclear routing signals. However, the relative importance of nuclear routing signals and nuclear retention signals (such as DNA binding sites) is still unclear. The first might lead the polypeptide into the nucleus more efficiently than simple diffusion through the pores, possibly by specific recognition by pore components, but the latter might be important in preventing the protein from leaking back into the cytoplasm.

Although genetically-engineered hybrid proteins have been useful in the study of nuclear protein targeting, chemically cross-linked polypeptides (Goldfarb et al, 1986) and nuclear proteins coupled to inert particles have also proven particularly useful. Until recently, this approach was almost entirely restricted to the study of nuclear protein targeting because it demands that polypeptides be injected directy into the cell, a technique which is more easily applied to amphibian oocytes, which are large, easy to obtain and have a well defined nucleus, than to other cell types. Perhaps most impressive among these studies is that Feldherr et al (1984) who showed that gold particles with a diameter larger than that of proteins normally excluded by nuclear pores, could be transported into the Xenopus oocytes when coated with the nuclear protein nucleoplasmin.

Another interesting feature of nuclear protein targeting is that some proteins can "ride piggy-back" into the nucleus. For example, Gritz et al (1985) showed that yeast ribosomal protein RP15A-βGAL hybrids containing the first 64 residues of RP15A were not nuclear-targeted unless wild type RP15A protein was also made by the cells, and piggy-back uptake of micro-injected monoclonal antibodies to a nuclear protein was reported by Madsen et al (1986). Interestingly, some proteins may enter the nucleus from a pre-existing cytoplasmic pool at different stages in cell development (De Robertis, 1983), a phenomenon which might be explained by the synthesis of specific piggy-back carriers only when nuclear targeting of particular proteins or groups of proteins is required.

Hybrid proteins with retention signals. Retention or salvage signal, which prevent proteins from being transported to the end of a particular pathway, are clearly important in targeting pathways in which individual steps are reversible. Proteins targeted to organelles such as mitochondria are likely to remain there because vesicle movement is not involved in the targeting pathway. However, proteins in the ER, Golgi cisternae, lysosomes, vacuoles etc. could escape via vesicles which are continually shuttling back and forth along the pathway, and nuclear proteins could leak out through the pores. The only retention signal which has been studied in detail in K-D-E-L found in some luminal proteins of the endoplasmic reticulum (Munro and Pelham, 1987). This sequence was genetically fused to the end of the sequence for secreted chicken lysozyme, which consequently remained in the lumen of the ER.

Hybrid proteins with sorting signals. Sorting signals are important in branched protein targeting pathways. It was recently reported that as few as 50 N-terminal residues from the precursor of the yeast vacuolar protein carboxypeptidase Y (CPY) can direct the normally secreted invertase into the vacuole. This region of CPY includes a 20-residue signal peptide and the vacuolar targeting signal (Johnson et al, 1987). Gene fusion technology would seem particularly well suited to further studies of lysosomal and other vacuolar sorting signals. Furthermore, it cannot necessarily be assumed that secreted or plasma membrane proteins complete their passage through the secretory pathway, rather than being shunted to lysosomes/vacuoles, by default; some of them may also have secretory sorting signals which operate during or before they leave the Golgi complex. One example of the sorting of secretory proteins is in the distinction between the constitutive and inducible branches at a late stage of the secretory pathway. Moore and Kelly (1986) used a gene fusion approach to determine which of the two groups of secretory proteins required a specific sorting signal to be transported to the cell surface by their appropriate pathway. They found that hybrids containing a C-terminally-truncated version of the constitutively-secreted G protein of vesicular stomatitis virus (lacking the transmembrane and cytoplasmic domains) fused to the C terminus of human growth hormone, which is secreted by the inducible pathway, were packaged into dense secretory granules typical of the inducible pathway. This indicates that sorting into this pathway probably involves specific sorting signals, or at least that constitutive pathways sorting signals in G protein can be overridden by those in human growth hormone.

Stretches of protein sequence which anchor polypeptides in membranes, which may be regarded as sorting signals or as retention signals, can also be studied by gene fusion technology. Related to these are the sorting signals which are proposed to distinguish between periplasmic and outer membrane proteins in Gram-negative bacteria. Early studies seemed to indicate that outer membrane LamB protein of E.coli contained sorting signals which could direct βGAL into the outer membrane (Benson et al 1984). In fact, it now seems clear that at most only a small proportion of the hybrid molecules actually reached the cell surface, and that Benson et al erroneously interpreted results from cell fractionation experiments (see article by Tommassen in this edition), which highlights one of the potential difficulties of using hybrids in the study of protein targeting.

Hybrid proteins with membrane topology signals. One of the most fascinating recent applications of gene fusion technology has been its use to study membrane protein topology. Three passenger proteins, βGAL, alkaline phosphatase (AP) and β-lactamase (βLAC) are particularly useful for the study of bacterial cytoplasmic membrane proteins. As already mentioned, βGAL apparently cannot cross the cytoplasmic membrane, and has very low specific activity when it is forced into the membrane by fusion to sequences which are normally exported beyond the cytoplasmic membrane. In the case of type III integral inner membrane proteins, which cross the membrane several times, βGAL hybrids with high specific activity can only be obtained with gene fusions in which the fusion point is within a sequence coding for a cytoplasmic loop of the polypeptide. Conversly, alkaline phosphatase is only active when it is translocated through the cytoplasmic membrane. Thus, hybrids between type III membrane proteins and AP which have high specific activity must have the AP sequence fused to periplasmic loop of the polypeptide (Manoil and Beckwith, 1986). Finally, these features of βGAL and AP hybrids are combined in βLAC hybrids. Cells carrying hybrid polypeptides with βLAC fused to either cytoplasmic or periplasmic loops of a type III membrane protein will be penicillin resistant, but the cytoplasmic loop fusions exhibit only low-level resistance because penicillin only comes into contact with the βLAC from a small number of cells in the colony which lyse, whereas periplasmic loop fusions will confer high level penicillin resistance (Broome-Smith and Spratt, 1986 and A.Edelman, personal communication). It is easy to see how these characteristics could be used to build up a picture of the topology of a type III membrane protein (Fig.4).

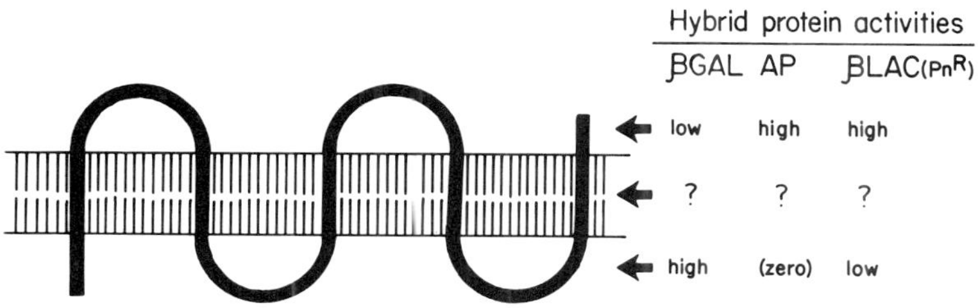

Fig. 4. Use of βGAL, AP and βLAC fusions to determine the topology of a type III cytoplasmic membrane protein of E.coli. Pn^R, penicillin resistance. The upper side of the membrane is the periplasmic face.

Hybrid proteins with protein modification signals. Hitherto in this discussion, I have tended to imply that targeting signals are usually specific, more-or-less well-defined stretches of amino acids within the protein sequence. This is clearly not always the case, however, since there are examples of covalent modification (eg, glycosylation, acylation) affecting protein targeting, and of transient modifications (eg, phosphorylation of secretory, mitochondrial and chloroplast proteins) thought to be important in transmembrane movement. Post-translational modification must itself depend on signals within the polypeptide, however, and these can also be located and studied through the use of gene fusions. One example of this is in the study of fatty acylation of bacterial lipoproteins. Early protein chemistry studies indicated that the N-terminal cysteine residue was the site at which the major E.coli lipoprotein (LPP) was fatty acylated. This cysteine residue is only exposed at the N terminus when the signal peptide is cleaved off, and it is now known that esterification of the cysteine is a prerequisite for signal peptide cleavage. To see whether a signal peptide and the relatively well-conserved sequence around the cysteine residue was all that was needed for fatty acylation, Grayheb et al (1984) fused DNA coding for the LPP signal peptide plus 9 N-terminal residues of LPP to βLAC (devoid of its own signal peptide). The resulting hybrid was fully fatty acylated and was processed by lipoprotein signal peptidase. We have obtained similar results through the study of another lipoprotein, pullulanase, coupled to βGAL or AP (d'Enfert and Pugsley, 1987). Thus, fatty acylation of bacterial lipoproteins is probably completely independent of the sequences which follow the modification/processing site.

Gene fusions in biotechnology. An expanding area of interest in the use of gene fusions is in promoting the secretion or export (periplasmic localization) of heterologous proteins, chiefly in bacteria and yeasts but also in mammalian cell culture lines. In its simplest form, this technology consists of fusing DNA coding for a bacterial or yeast signal peptide to DNA coding for another proteins (eg, a growth hormone or other commercially-important polypeptide) and expressing the fused gene such that the polypeptide is processed by signal peptidase and exported to the periplasm (E.coli) or into the medium (B.subtilis or yeasts). Although this system has been shown to work on numerous occasions, it is often necessary to include parts of the mature sequence in the hybrid to ensure efficient export/secretion, and signal peptide cleavage may not occur at the required site. In these cases, more sophisticated gene fusions including DNA coding for a "bridge" between the bacterial/yeast sequences and the hormone sequences, are required. These bridges contain sites for specific cleavage by reagents such as those listed in Table 2, allowing the two halves of the molecule to be separated and the important part to be purified (the sequence should obviously not occur elsewhere in the sequence of the hybrid polypeptide). Purification of the hybrid polypeptide may also be facilitated by the presence of sequences which will eventually be removed from the final product. For example, hybrid proteins containing the IgG-binding domain of the secreted Staphylococcus aureus protein A can be easily purified on immunoglobulin affinity adsorption columns. The protein A sequence is then cleaved off by specific reagents and separated from the remaining parts of the hybrid by a second chromatography step on the same affinity matrix (Nilsson et al, 1985).

Table 2. Agents which can be used to induce specific cleavage between carrier and passenger sequences in secreted or exported hybrid proteins.

Site	Agent	Site	Agent
Met↓X	cyanogen bromide	Gly-Asp-Asp-Asp↓X	enterokinase
Asp↓Pro	formic acid	Ile-Glu-Gly-Arg↓X	factor X
Asn↓Gly	hydroxylamine	Pro-Val↓Gly-Pro	collagenase
Trp↓X	bromosuccinimide	Lys↓X	Pseudomonas protease Ps-1
Cys↓X	nitro-thiocyano-benzoate	Pro-Phe-Arg↓X	plasma kallekrein
Ala Gly Ala Ser-X-Gly↓X Leu Ser Val Ile	signal (leader) peptidase	Thr-Pro-Pro↓Thr- Pro-Ser-Pro↓Ser- Thr-Pro-Pro↓Thr- Pro-Ser-Pro-Ser	IgA proteases
Gly-Pro-Arg↓X	thrombin	Arg-Lys↓X	Yeast KEX2 protease

X represents almost any amino acid, ↓ indicates cleavage site.

Another interesting development in the application of gene fusion technology is the localization of foreign epitopes to the surface of bacterial cells by fusing them into cell surface loops of type III outer membrane proteins (Charbit et al, 1986). This development could be particularly useful in designing live vaccine strains to raise antibodies against such epitopes as key sequences from viral surface proteins. However, there appear to be limitations on the length of foreign sequence which can be inserted into these loops. We are now developing an alternative procedure which may allow longer sequences to be exposed on the surface of E.coli cells by fusing them to the C-terminus of pullulanase, an enzyme which is itself apparently fully exposed on the cell surface where it is anchored by covalently-linked fatty acyl chains. We have already shown that pullulanase can act as a carrier to target AP to the E.coli cell surface (d'Enfert and Pugsley, 1987; see article by d'Enfert et al, this volume).

Concluding remarks. In the past ten years we have witnessed increasing use of gene fusion technology in studies on protein targeting. The results can often provide useful indications as the nature and location of different targeting signals within a polypeptie sequence, but, strictly-speaking, such signals can only be fully characterized in the context of the polypeptide of which they are part. Nevertheless, we expect to see hybrid proteins being put to further use, particularly with the development of novel passenger proteins such as firefly luciniferase (Keller et al, 1987), and increasing use of gene fusions in biotechnology. Further effort must now be devoted to identifying cell components interacting with targeting signals in order to learn more about protein routing, sorting and membrane topogenesis.

References

Baker A, Schatz G (1987) Sequences from a prokaryotic genome or the mouse dihydrofolate reductase gene can restore the import of a truncated precursor protein into yeast mitochondria. Proc Natl Acad Sci USA 84: 3117-3121.

Bankaitis VA, Bassford PJ Jr (1984) The synthesis of export-defective proteins can interfere with normal protein export in Escherichia coli. J Biol Chem 259:12193-12200.

Benson SA, Bremer E, Silhavy TJ (1984) Intragenic regions required for LamB export. Proc Natl Acad Sci USA 81:3830-3834.

Benson SA, Hall MN, Silhavy TJ (1985) Genetic analysis of protein export in Escherichia coli. Ann Rev Biochem 54:101-134.

Briggs MS, Gierasch LM (1986) Molecular mechanisms of protein secretion: the role of the signal sequence. Advances in Protein Research 32: 109-180.

Van den Broeck G, Timko MP, Kausch AP, Cashmore AR, van Montagu M, Herrera-Estrella L (1985) Targeting of a foreign protein to chloroplasts by fusion to the transit peptide from the small subunit of ribulose 1,5-biphosphate carboxylase. Nature 313:358-365.

Broome-Smith JK Spratt BG (1986) A vector for the construction of translational fusions to TEM β-lactamase and the analysis of protein export signals and membrane protein topology. Gene 49:341-349.

Castihlo BA, Olfsen P, Casadaban MJ (1984) Plasmid insertion mutagenesis and lac gene fusions with mini-Mu bacteriophage transposons. J.Bacteriol. 158:488-495.

Charbit A, Boulain JC, Ryter A, Hofnung M (1986) Probing the topology of a bacterial membrane protein by genetic insertion of a foreign epitope: expression at the cell surface. EMBO J 5:3029-3037.

Davey J, Dimmock NJ, Colman A (1985) Identification of the sequence responsible for nuclear accumulation of the influenza virus nucleoprotein in Xenopus oocytes. Cell 40:667-675.

De Robertis EM (1983) Nucleocytoplasmic segregation of proteins and RNAs. Cell 32:1021-1025.

Douglas MG, Geller BL, Emr,SD (1984) Intracellular targeting and import of an F1-ATPase β-subunit-β-galactosidase hybrid protein into yeast mitochondria. Proc Natl Acad Sci USA 81:3983-3387.

Eilers M, Schatz G (1986) Binding of a specific ligand inhibits import of a purified precursor protein into mitochondria. Nature 322:228-232.

Emr SD, Schekman R, Flessel MC, Thorner J (1983) An MFα1-SUC2 (α-factor-invertase) gene fusion for study of protein localization and gene expression in yeast. Proc Natl Acad Sci USA 80:7080-7084.

Emr SD, Shauer I, Hansen W, Esmon P, Schekman R. (1984) Invertase-β-galactosidase hybrid proteins fail to be transported from the endoplasmic reticulum in Saccharomyces cerevisiae. Mol Cell Biol 4: 2347-2355.

Emr SD, Vassarotti A, Garrett J, Geller BL, Takeda M, Douglas MG (1986) The amino terminus of the yeast F1 ATPase β-subunit precursor functions as a mitochondrial import signal. J Cell Biol 102: 523-533.

d'Enfert C, Pugsley AP (to be published). A gene fusion approach to the study of pullulanase export and secretion in Escherichia coli Mol Microbiol 1.

Eskridge EM, Shields D (1987) The NH_2 terminus of preproinsulin directs the translocation and glycosylation of a bacterial cytoplasmic protein by mammalian microsomal membranes. J Cell Biol 103:2263-2272.

Feldherr CM, Kallenbach E, Schultz N (1984) Movement of a karyophilic protein through the nuclear pores of oocytes J Biol Chem 99:2116-2222.

Finidori J, Rizzolo L, Gonzalez A, Kreibach G, Adesnik m, Sabatini DD (1987) The influenza haemagglutinin insertion signal is not cleaved and does not halt translocation when presented to the endoplasmic reticulum as part of a translocating polypeptide. J Cell Biol 104: 1075-1714.

Goldfarb DS, Gariepy J, Schoolnik G, Kornberg RD (1986) Synthetic peptides as nuclear localization signals. Nature 322:641-644.

Grayheb J, Inouye M (1984) Nine amino acids at the NH2-terminal of lipoprotein are sufficient for its modification, processing and localization in the outer membrane of Escherichia coli. J Biol Chem 259:463-467.

Gritz L, Abovich N, Teem JL, Rosbach M (1985) Posttranscriptional regulation and assembly into ribosomes of a Saccharomyces cerevisiae ribosomal protein-β-galactosidase fusion. Mol Cell Biol 5:3436-3442.

Hall MN, Hereford L, Herskowitz I (1984) Targeting of E.coli β-galactosidase to the nucleus in yeast. Cell 36:1057-1065.

Hase T, Nakai M, Matsubara H (1986) The N-terminal 21 amino acids of a 70 kDA protein of the yeast mitochondrial outer membrane direct E.coli β-galactosidase into the mitochondrial matrix in yeast cells. FEBS Letts 197:199-203.

Hayashi S, Chang S-Y, Chang S, Giam C-Z, Wu HC (1985) Modification and processing of internalized signal sequences of prolipoprotein in Escherichia coli and Bacillus subtilis. J Biol Chem 260:5753-5759.

Hurt EC, Müller U, Schatz G (1985) The first twelve amino acids of a yeast mitochondrial outer membrane protein can direct a nuclear-encoded cytochrome oxidase into the mitochondrial inner membrane. EMBO J 4:3509-3518.

Hurt EC, Schatz G (1987) A cytosolic protein contains a cryptic mitochondrial targeting signal. Nature 325:499-503.

Ito K, Bassford PJ Jr, Beckwith J (1981) Protein localization in E.coli: is there a common step in the secretion of periplasmic and outer membrane proteins? Cell 24:707-717.

Johnson LM, Bankaitis VA, Emr SD (1987) Distinct sequence determinants direct intracellular sorting and modification of a yeast vacuolar protease. Cell 48:875-885.

Kalderon D, Robert BL, Richardson WD and Smith AE (1984) A short amino acid sequence able to specify nuclear location. Cell 39:499-509.

Keller GA, Gould S, Deluca M, Subramani S (1987) Firefly luciniferase is targeted to peroxisomes in mammalian cells. Proc Natl Acad Sci USA 84: 3264-3268.

Lipp J, Dobberstein B (1986) Signal recognition particle-dependent membrane insertion of a mouse invariant chain: a membrane-spanning protein with a cytoplasmically exposed amino terminus. J.Cell Biol 102: 2169-2175.

Lyons RH, Ferguson BQ and Rosenberg M (1987) Pentapaptide nuclear localization signal in Adenovirus Ela. Mol Cell Biol 7:2451-2456.

Manoil C, Beckwith J (1986) A genetic approach to analyzing membrane protein topology. Science 233:1403-1408.

Masden P, Nielsen S, Celis JE (1986) Monoclonal antibody specific for human nuclear proteins IEF 8Z30 and 8Z31 accumulates in the nucleus a few hours after cytoplasmic microinjection of cells expressing these proteins. J Cell Biol 103: 2083-2089.

Michaelis S, Hunt JF, Beckwith J (1986) Effects of signal sequence mutations on the kinetics of alkaline phosphatase export to the periplasm in Escherichia coli. J Bacteriol 167:160-167.

Moore HP, Kelly RB (1986) Re-routing of a secretory protein by fusion with human growth hormone sequences. Nature 321:443-446.

Moreno F, Fowler AV, Hall M, Silhavy TJ, Zabin I, Schwartz M (1980) A signal sequence is not sufficient to lead β-galactosidase out of the cytoplasm. Nature 286: 356-359.

Munro S and Pelham HRB (1984) Use of peptide tagging to detect proteins expressed from cloned genes: deletion mapping functional domains of Drosophila hsp70. EMBO J 3:3087-3093.

Munro S and Pelham HRB (1987) A C-terminal signal prevents secretion of luminal ER proteins. Cell 48:899-907.

Nilsson B, Holmgreen E, Josephson S, Gatenbeck S, Philipson L and Uhlen M (1985) Efficient secretion and purification of human insulin-line growth factor I with a gene fusion in Staphylococci. Nucl Acids Res 13: 1151-1162.

Palva I, Sibakov M, Kallio P, Nyberg K, Simoren M (1987) Secretion of proteins in Bacilli. Proc Fifth International Symposium on the Genetics of Industrial Microorganisms. eds Alacevis M, Hranufi D, Toman Z. Pliva, Zagreb, Yugoslavia. p 421.

Pfanner N, Tropschug M, Neupert W (1987) Mitochondrial protein import: nucleoside triphosphates are involved in conferring import competence to precursors. Cell 49:815-823.

Plüchthun A, Knowles JR (1987) The consequence of stepwise deletions from the signal-processing site of β-lactamase. J Biol Chem 262:3951-3957.

Pohler J, Meyer TF, Manning PM (1986) Serological properties and processing in Escherichia coli K12 of OmpV fusion proteins of Vibrio cholera. Mol Gen Genet 205:501-506.

Rasmussen BA, Bankaitis VA, Bassford PJ Jr (1984) Export and processing of MalE-LacZ hybrid proteins in Escherichia coli. J Bacteriol 160:612-617.

Richardson WD, Robert BL, Smith AE (1986) Nuclear location signals in polyoma virus large-T. Cell 44:77-85.

Simon K, Perera E, Lingappa VR (1987) Translocation of globin fusion proteins across the endoplasmic reticulum membrane in Xenopus laevis oocytes. J.Cell Biol 104: 1165-1172.

PULLULANASE: A NEW SPECIFIC SECRETION PATHWAY IN *ESCHERICHIA COLI*

Christophe d'Enfert, Isabelle Reyss, Antoinette Ryter and Anthony P. Pugsley

Unités de Génétique Moléculaire et de Microscopie Electronique
Institut Pasteur
25, rue du Docteur Roux
75724 Paris Cedex 15
France

Bacterial proteins which are exported to the periplasm or outer membrane (Gram-negative bacteria) or into the medium (Gram -positive bacteria) are invariably synthesized as precursors from which the signal peptide is removed during translocation of the polypeptide through the cytoplasmic membrane (Oliver, 1985). Translocation across the cytoplasmic membrane occurs by a process similar to that by which proteins are localized to the lumen of the endoplasmic reticulum in the first stage of the secretory pathway in eukaryotic cells (Wickner and Lodish, 1985). Outer membrane proteins of Gram-negative bacteria seem to adopt a conformation which allows them to insert into the outer membrane, rather than having specific outer membrane targeting signals (Freudl *et al*, 1985; Bosch *et al*, 1986).

The translocation of proteins across the outer membrane of Gram-negative bacteria and their release into the medium (**secretion**) appears to occur by unique pathways which are specific for a particular secreted protein or group of proteins (Pugsley and Schwartz, 1985; Pugsley, 1988). This idea relies essentialy on the fact that when structural genes for normally secreted proteins from different Gram-negative species are expressed in *Escherichia coli*, their products are not released into the culture medium, except when autocatalytic cleavage is sufficient to ensure their release from the outer membrane (Pohlner *et al*, 1987; Yanigida *et al*, 1986). Moreover, the cloning of additional genetic information into *E.coli* may allow the specific release of a particular protein, as shown for *E.coli* hemolysin (Mackman *et al*, 1986) and *Klebsiella pneumoniae* pullulanase (d'Enfert

NATO ASI Series, Vol. H16
Membrane Biogenesis. Edited by J. A. F. Op den Kamp

et al, 1987b).

Pullulanase is an enzyme that hydrolyzes α(1-6) bonds in glucose polymers such as amylopectin, glycogen or pullulan, a linear polysaccharide composed of maltotriose units (Bender and Wallenfels, 1961). The enzyme is produced by *K.pneumoniae* growing in the presence of maltose (Hope and Dean, 1974). During exponential growth, pullulanase is localized to the outer membrane, with its active site exposed to the culture medium (Wallenfels *et al*, 1966; Wöhner and Wöber, 1978; Michaelis *et al*, 1985; d'Enfert *et al*, 1987a). Pullulanase is progressively released after the end of the exponential growth phase and is found in an aggregated state in the culture medium (Michaelis *et al*, 1985; Pugsley *et al*, 1986). This short review describes our studies on pullulanase secretion as a model system for the mechanisms by which a protein can cross the outer membrane and be released into the culture medium.

The pullulanase structural gene is part of the maltose regulon.

The structural gene for pullulanase (*pulA*) has been cloned from different *K.pneumoniae* strains (W70, Takizawa and Murooka, 1984; ATCC15050, Michaelis *et al*, 1985; UNF5023, d'Enfert *et al*, 1987b) and expressed in *E.coli*. In all three cases, pullulanase production by *E.coli* was shown to be induced by growth in the presence of maltose, as is also the case in *K.pneumoniae.* Moreover, pullulanase synthesis depends of the *malT* product, the activator protein of the *E.coli* maltose regulon (Michaelis *et al*, 1985; Schwartz, 1987), indicating that *pulA* is part of the maltose regulon. By sequencing the 5' region of the *pulA* gene of *K.pneumoniae* ATCC15050, Chapon and Raibaud (1985) detected two divergent promoters, both of which included a putative MalT binding site [the conserved hexanucleotide 5'-GGA(T/G)GA-3' present at least once in front of all MalT-regulated promoters]. One MalT box was located in front of *pulA* and the other in front of *malX*, a gene whose product and function are unknown. Putative MalT boxes were also detected in the two other sequenced *pulA-malX* intergenic regions; those from strains W70 (Katsugari *et al*, 1987) and UNF5023 (d'Enfert *et al*, 1987b), the latter of which contained four rather than two copies of the hexanucleotide.

Pullulanase is a lipoprotein synthesized with a signal peptide.

The sequence of the 5' region of *pulA* indicated the presence of a signal peptide (Fig.1). This signal peptide has a cleavage site which is typical of bacterial lipoproteins, whose N-terminal cysteine residue is modified by three fatty acids (Wu and Tokunaga, 1986).

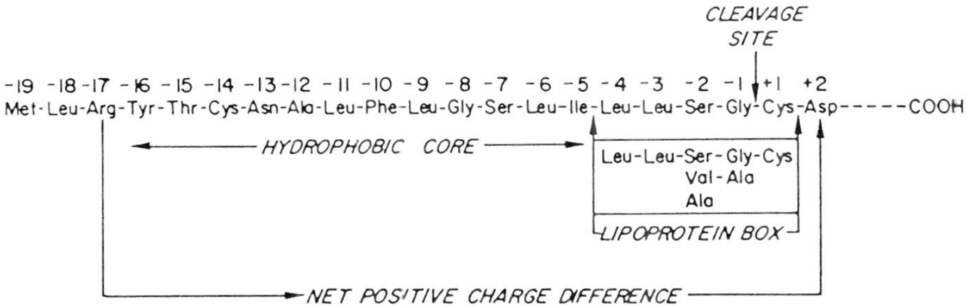

Figure 1: Amino-acid sequence of the N-terminal region of the pullulanase signal peptide from *K.pneumoniae* ATCC15050 and localization of its functional domains. The cleavage site is indicated by an arrow and is compared with the consensus lipoprotein cleavage site ("lipoprotein box").

Pugsley *et al* (1986) confirmed that pullulanase is synthesized as a precursor by expressing the *pulA* gene in an *E.coli* strain carrying a *malE-lacZ* gene fusion whose maltose-induced expression causes the accumulation of the precursors of exported proteins (see article by Pugsley, this volume). Moreover, they showed that expression of *pulA* in the presence of globomycin, a specific inhibitor of the lipoprotein signal peptidase, resulted in the accumulation of a pullulanase precursor. Finally, pullulanase could be labelled with ^{3}H-palmitate, confirming the lipoprotein nature of the mature protein produced in *K.pneumoniae* and *E.coli*. We have recently shown that a hybrid pullulanase-alkaline phosphatase containing only the first 11 residues of pullulanase can also be fatty-acylated (d'Enfert and Pugsley, 1987).

Cell-surface localization of pullulanase in *K.pneumoniae*

Post-translational modification of pullulanase by fatty-acids is followed in *K.pneumoniae* by its translocation to the cell-surface. Membrane associated pullulanase sediments with outer membrane vesicles in isopycnic sucrose density gradients (Wöhner and Wöber, 1978; Michaelis *et al*, 1985; d'Enfert *et al*, 1987a), but these results must be interpreted with caution in view of the lipoprotein nature of pullulanase. The following additional data are, however, consistent with cell surface localization of pullulanase :

1) Pullulan, a large polysaccharide which is unable to cross the outer membrane, has free access to the active site of pullulanase in whole cells (Wallenfels *et al*, 1966; Michaelis *et al*, 1985; d'Enfert *et al*, 1987a).

2) Pullulanase is degraded by proteinase K without the need to permeabilize the outer membrane (d'Enfert *et al*, 1987a).

3) Pullulanase can be detected on the cell surface by immunofluorescence (d'Enfert *et al*, 1987a) or by immuno-electron-microscopy (d'Enfert *et al*, 1987b) using antiPulA serum.

Pullulanase is not automatically exposed on the cell surface.

Expression of the *pulA* gene in *E.coli* did not allow pullulanase exposition or secretion (Michaelis *et al*, 1985; Takizawa and Murooka, 1984, 1985). These results suggested either that the *K.pneumoniae* outer membrane was significantly different to that of *E.coli*, which is unlikely considering the close relationship of the two species, or that *E.coli* does not produce specific pullulanase secretion functions. This second hypothesis was confirmed by expressing the *pulA* gene in *K.pneumoniae* under *lac* promoter control (d'Enfert *et al*, 1987a). In this case, pullulanase was synthesized in the absence of maltose, but was not exposed or secreted. This indicated that maltose-regulated genes were required for cell-surface localization of pullulanase, and that these genes were not present in *E.coli*.

Cloning of a region of the *K.pneumoniae* chromosome coding for production, exposition and secretion of pullulanase.

We speculated that a gene in the MalT-regulated transcription unit located near *pulA* could be involved in pullulanase exposition and secretion. Therefore, we cloned a large chromosomal DNA fragment from *K.pneumoniae* encompassing *pulA* and the region flanking it. A plasmid carrying a 23 kb DNA fragment from *K.pneumoniae* UNF5023 allowed *E.coli* to produce, expose and secrete pullulanase.

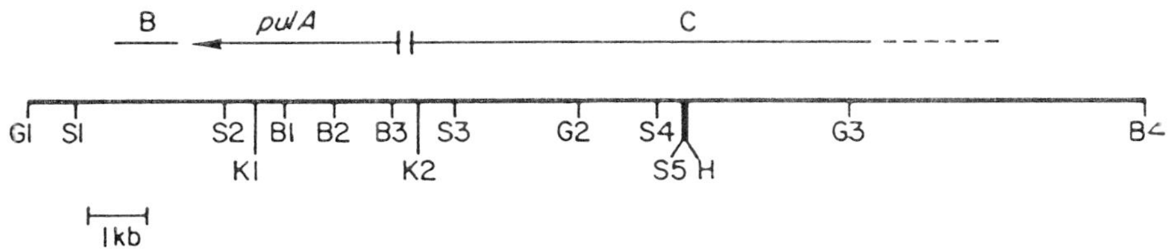

Figure 2: Restriction map of the 18.8 kb fragment of the *K.pneumoniae* UNF5023 chromosome coding for production, exposition and secretion of pullulanase. Cleavage sites are indicated for the following restriction enzymes and are numbered from left to right : B, *Bam*HI; G, *Bgl*II; H, *Hin*dIII; K, *Kpn*I; S, *Sma*I.

By subcloning, we showed that an18.8 kb region of this insert carried *pulA* and the genes required for exposition and secretion (Figs 2 and 3). Secretion of pullulanase by *E.coli* was obtained when this region was carried either on a multiple copy number plasmid or on the *E.coli* chromosome, without overt lysis, suggesting that secretion by *E.coli* was specific (Fig.4)

Tn*5* and miniMudI mutagenesis of this 18.8 kb fragment showed that two regions are required for exposition and secretion. Region B is located 3' of *pulA* (Fig.2). Three lines of evidence indicate that it forms a *pulA*-independant transcription unit :

1) a Tn*5* insertion which did not affect pullulanase production or exposition/secretion was found between *pulA* and region B as defined by other Tn*5* inserts (d'Enfert *et al*, 1987b).

2) pullulanase exposition and secretion was abolished when this region was deleted (d'Enfert and Pugsley, manuscript in preparation).

3) the effect of Tn*5* insertions in region B was abolished by introducing a plasmid carrying the 2.5 kb *Sma*I fragment (S1-S2; Fig.2) encompassing region B (d'Enfert and Pugsley, manuscript in preparation).

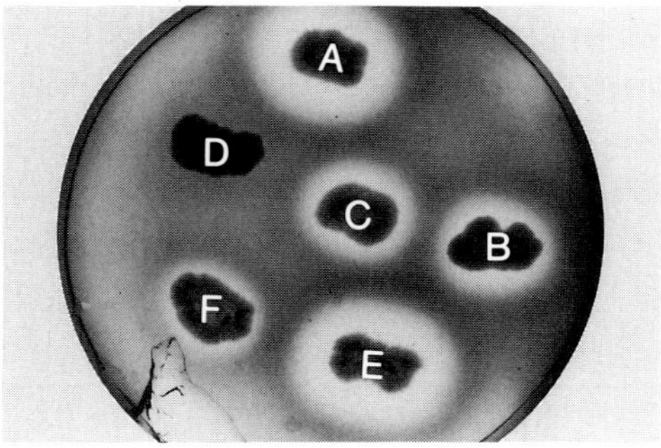

Figure 3: Plate-test for pullulanase production and exposition/ secretion. Plates containing L agar with 0.01% red pullulan (reactive red 120 coupled to pullulan) were inoculated with *E.coli* carrying cloned fragments of *K.pneumoniae* DNA and incubated for 2 days at 30°C. Clear zones around the bacterial growth are regions of pullulan hydrolysis. A, pBR322::*pulABC*; B, *pulABC* in chromosome; C, *pulAB* in chromosome plus pBR322::*pulC*; D, as C but region C is absent; E, pACYC184::*pulAB* plus pBR322::*pulC*; F, as E but region C is absent.

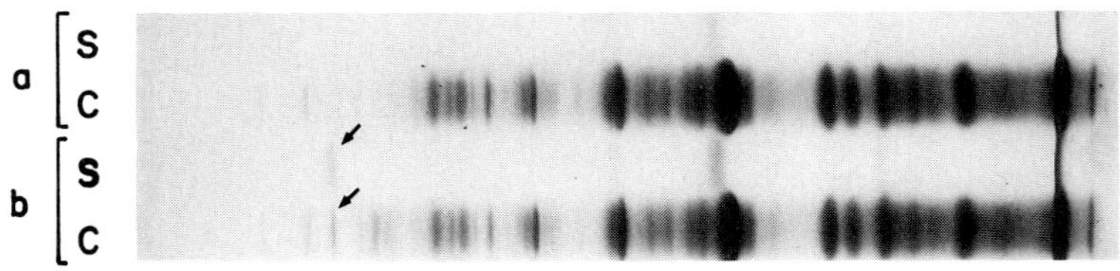

Figure 4: SDS-PAGE of cell-associated (C) and secreted (S) proteins of *E.coli* K-12 devoid of *pul* genes (a) and the same strain carrying *pulABC* on a pBR322 derivative (b). Cells were grown to the end of exponential growth phase in minimal medium containing 0.1% maltose before sampling. Pullulanase is indicated by the arrow, and the direction of electrophoresis is from left to right.

Region C is located 5 to *pulA* and is entirely contained within the 12.7 kb *Bam*HI fragment (B3-B4; Fig.2). A large set of Tn*5* insertions was obtained in this region, all of which blocked pullulanase exposition **and** secretion. By constructing a series of operon fusions in which various lengths of the C region extending 3' of the *Bam*HI site B3 (Fig.2) were fused to the *malPQ* operon, we showed that region C constitutes an unique operon of at least 7.5 kb whose transcription is under the control of the MalT-regulated promoter *malXp*. This confirmed that at least one gene involved in pullulanase exposition/secretion is part of the maltose regulon. The absence of transposon insertions specifically affecting pullulanase secretion suggests that this phenomenon could be a passive event, resulting from the exposition of the pullulanase on the cell surface. However, this could also be due to a polar effect of transposon insertions in secretion genes on downstream exposition genes in the *malX* operon.

Pullulanase secretion.

Pullulanase is slowly released into the medium by *K.pneumoniae* after the end of exponential growth (Michaelis *et al*, 1985). This pattern of release was also manifested by *E.coli* cells carrying the 18.8 kb fragment on their chromosome. However, pullulanase seemed to be secreted during exponential growth when *pulA* and the exposition/secretion genes were present in *E.coli* on a multiple copy number plasmid (d'Enfert *et al*, 1987b). Pugsley *et al* (1986) showed that extracellular pullulanase was also fatty-acylated. This result was confirmed by studying pullulanase released by *E.coli* (d'Enfert *et al*, 1987b). Therefore, pullulanase release does not result from proteolytic cleavage of the N-terminal region which is probably anchored in the membrane by the fatty-acyl chains.

Approximately 50% of pullulanase in the medium can be pelleted by ultracentrifugation (Pugsley *et al*, 1986). When the sedimented material was examined by immunoelectron-microscopy, some of the vesicle-like structures were seen to have reacted with antiPulA serum and protein A-gold, whereas other had not (Fig.5). This preliminary result suggests that the cells were releasing two kinds of vesicles: outer membrane vesicles which cells released irrespective of pullulanase production and secretion, and pullulanase-enriched vesicles or micelles (Pugsley *et al*, 1986, 1987). We have previously suggested that vesiculation or micellation of pullulanase-enriched regions of the outer membrane may occur spontaneously when the local pullulanase concentration reaches a critical level (Pugsley *et al*, 1987), but further work is required to confirm this hypothesis.

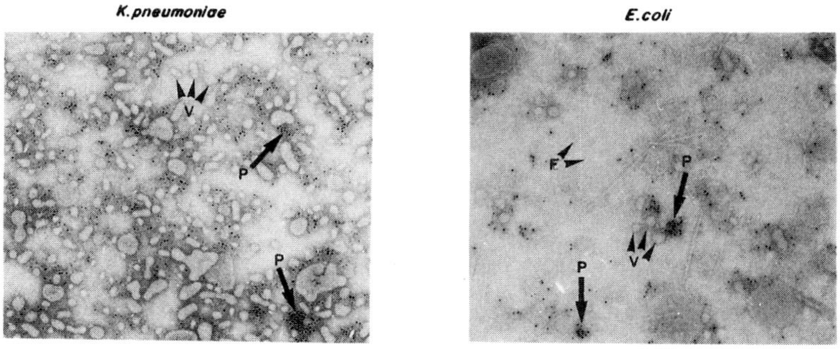

Figure 5: Analysis by immunoelectron-microscopy of sedimented material from the medium of maltose-induced cultures of *K.pneumoniae* and of *E.coli* carrying the *pulABC* region in a pBR322 derivative. Pullulanase was immunodetected using antiPulA serum and protein A-gold labeling. Arrows indicate outer membrane vesicles (V) and pullulanase-enriched vesicles or aggregates (P). Fimbriae are labelled F.

Perspectives

The next stage in our studies will be to determine the structure and role of the proteins encoded by the B and C regions of the pullulanase gene complex. Several different approaches are being used. The entire 18.8 kb DNA fragment is being sequenced. Several of the gene products have already been identified by using *in vitro* and *in vivo* expression systems, and there are already strong indications that some of them are synthesized with signal peptides, leading to the prediction that they are in the periplasm or outer membrane. We have also constructed a series of *phoA* (alkaline phosphatase) gene fusions in the C region. Some of these gene fusions encode hybrid proteins with alkaline phosphatase activity, indicating that they are translocated across the cytoplasmic membrane (see article by Pugsley, this volume). However, we do not yet know whether all of the genes in the B and C regions are actually required for pullulanase exposition/ secretion. Random linker mutagenesis is therefore being used to obtain a collection of non-polar mutations in the C-region, and deletion mutagenesis is being applied to the B region. Once the specific genes/proteins involved in exposition/secretion are identified, we intend to select mutations affecting these functions in the hope of identifying both the specific sorting (secretion) signals in the pullulanase polypeptide, and the components of the secretion pathway with which they interact.

References

Bender H, Wallenfels K (1961) Untersuchungen an pullulan. Biochem Z 334:79-95

Bosch D, Leunissen J, Verbakel J, de Jong M, van Erp H, Tomassen J (1986) Periplasmic accumulation of outer membrane PhoE protein of *Escherichia coli* K-12. J Mol Biol 189:449-455

Chapon C, Raibaud O (1985) Structure of two divergent promoters located in front of the gene encoding pullulanase in *Klebsiella pneumoniae* and positively regulated by the *malT* product. J Bacteriol 164:639-645

d'Enfert C, Chapon C, Pugsley AP (1987a) Export and secretion of the lipoprotein pullulanase in *Klebsiella pneumoniae*. Mol Microbiol 1:107-116

d'Enfert C, Ryter A, Pugsley AP (to be published) Cloning and expression in *Escherichia coli* of the *Klebsiella pneumoniae* genes for production, suface localization and secretion of the lipoprotein pullulanase. EMBO J

d'Enfert C, Pugsley AP (to be published) A gene fusion approach to the study of pullulanase export and secretion in *Escherichia coli*. Mol Microbiol

Freudl R, Schwartz H, Klose M, Movva NR, Henning U (1985) The nature of information, required for export and sorting, present within the outer membrane protein OmpA of *Escherichia coli* K-12. EMBO J 4:3595-3598

Hope GC, Dean ACR (1974) Pullulanase synthesis in *Klebsiella (Aerobacter) aerogenes* strains growing in continuous culture.

Biochem J 144:403-411
Katsugari N, Takizawa N, Murooka Y (1987) Entire nucleotide sequence of the pullulanase gene of *Klebsiella aerogenes* W70. J Bacteriol 169:2301-2306
Mackman N, Nicaud JM, Gray L, Holland IB (1986) Secretion of haemolysin by *E.coli*. Curr Topics Microbiol Immunol 125:159-181
Michaelis S, Chapon C, d'Enfert C, Pugsley AP, Schwartz M (1985) Characterization and expression of the structural gene for pullulanase, a maltose-inducible secreted protein of *Klebsiella pneumoniae*. J Bacteriol 164:633-638
Oliver D (1985) Protein secretion in *Escherichia coli*. Ann Rev Microbiol 39:615-648
Pohlner J, Halter R, Beyreuther K, Meyer TF (1987) Gene structure and extracellular secretion of *Neisseria gonorrhoeae* IgA protease. Nature 325:458-462
Pugsley AP (to be published) Getting out of Coli: protein secretion across the Gram-negative outer membrane. In: Das RC, Robbins PW (eds) Protein transfer and organelle biogenesis. Academic Press, Orlando
Pugsley AP, Schwartz M (1985) Export and secretion of proteins by bacteria. FEMS Microbiol Rev 32:3-38
Pugsley AP, Chapon C, Schwartz M (1986) Extracellular pullulanase of *Klebsiella pneumoniae* is a lipoprotein. J Bacteriol 166:1083-1088
Pugsley AP, d'Enfert C, Cole ST (1987) Protein secretion by *Escherichia coli*: colicins and pullulanase. In: Proceedings of the fifth International Symposium on the Genetics of Industrial Microorganisms, vol B. p 431
Schwartz M (1987) The maltose regulon. In: Neidhardt FC, Ingraham JL, Low KB, Schaechter M, Umbarger HE (eds) *Escherichia coli* and *Salmonella typhimurium*. Cellular and molecular biology. American Society for Microbiology, Washington DC, p 1482
Takizawa N, Murooka Y (1984) Intergenic transfer of the pullulanase gene between *Klebsiella aerogenes* and *Escherichia coli* by *in vivo* genetic manipulation. Agric Biol Chem 48:1451-1458
Takizawa N, Murooka Y (1985) Cloning of the pullulanase gene and overproduction of pullulanase in *Escherichia coli* and *Klebsiella pneumoniae*. Appl Environ Microbiol 49:294-298
Wallenfels K, Bender H, Rasched JR (1966) Pullulanase from *Aerobacter aerogenes*: production in a cell-bound state. Purification and properties of the enzyme. Biochem Biophys Res Commun 22:254-261
Wickner WT, Lodish HF (1985) Multiple mechanisms of protein insertion into and across membranes. Science 230:400-407
Wöhner G, Wöber G (1978) Pullulanase, an enzyme of starch catabolism, is associated with the outer membrane of *Klebsiella*. Arch Microbiol 116:303-310
Wu HC, Tokunaga M (1986) Biogenesis of lipoproteins in bacteria. Curr Topics Microbiol Immunol 125:127-157
Yanigida N, Ouozumi T, Beppu T (1986) Specific excretion of *Serratia marcescens* protease through the outer membrane of *Escherichia coli*. J Bacteriol 166:937-944

A GENETIC ANALYSIS OF PULLULANASE EXPORT FROM *KLEBSIELLA AEROGENES*

M. G. Kornacker[1], A. Boyd[2] and G. S. Plastow[1]

[1]The Leicester Biocentre
University of Leicester
University Road
Leicester, LE1 7RH
Great Britain

INTRODUCTION

The enzyme pullulanase (EC 3.2.1.41) is produced by various genera of Gram-positive and Gram-negative bacteria. This enzyme is defined by its ability to hydrolyze $\alpha(1-6)$ bonds of $\alpha(1-4)$, $\alpha(1-6)$ glucans. Pullulanases of approximately 140 kd (Eisele et al. 1972) are produced and secreted by strains of the Gram-negative bacterium *Klebsiella aerogenes*. The amounts synthesized and proportions secreted are strain-dependent. In this report we summarize data obtained from the analysis of *Klebsiella aerogenes* strain K21. This strain secretes 50% of pullulanase made into the culture medium. The remaining 50% is located at the cell surface.

We have confirmed that pullulanase is a lipoprotein (see also Pugsley et al. 1986) and show that the site of lipid modification is the extreme N-terminus of pullulanase. We present genetic evidence that pullulanase export requires extra gene functions and that most, if not all, of these gene functions are linked to the structural gene for pullulanase (*pul*A).

[2]Dept. of Biochemistry, University of Edinburgh, Medical School, George Square, Edinburgh, United Kingdom.

NATO ASI Series, Vol. H16
Membrane Biogenesis. Edited by J. A. F. Op den Kamp

R E S U L T S   A N D   D I S C U S S I O N

PULLULANASE IS A LIPOPROTEIN

The structural gene for pullulanase (*pulA*) has been cloned from strain K21 as a 6kb *EcoRI* fragment and partially sequenced (GSP, unpubl. results). The nucleotide sequence predicts the presence of a 19 amino acid signal sequence with the C-terminal residues Leu-Ser-Gly. This sequence is followed by a Cys residue suggesting that pullulanase may be a lipoprotein. Lipoproteins are a small subset of bacterial proteins that are covalently modified by the addition of a diacylglyceride to the side-chain and (after removal of the signal sequence) an acyl chain to the N-terminus of the Cys residue (Braun 1975). They are found among Gram-negative and Gram-positive bacteria and have in common a consensus signal sequence cleavage site of (Leu-Ser/Val/Ala-Ala/Gly-Cys) (Pugsley and Schwartz 1985).

H_2N-TERMINUS OF PULLULANASE: MLRYTCNALFLGSLILLSG CDNS---CO_2H

H_2N-TERMINUS OF B-LACTAMASE: MSIQHFRVALIPFFAAFCLPVFA HPET---CO_2H

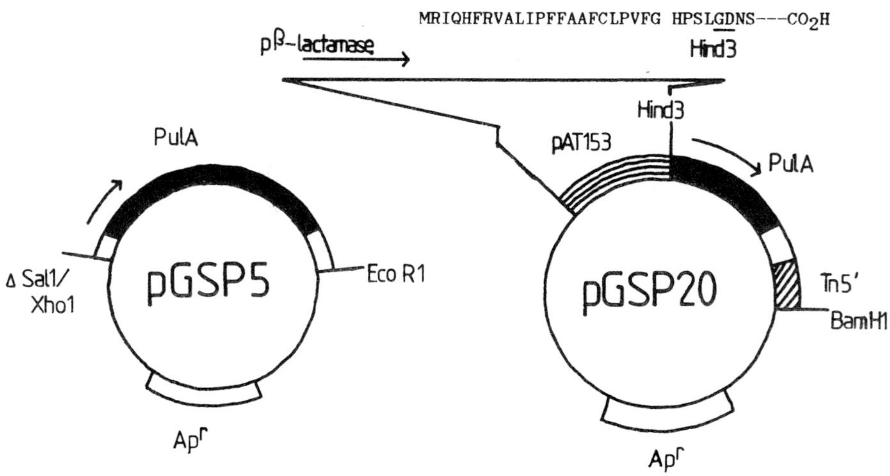

Fig. 1: Structures of pGSP5 and pGSP20. pGSP5 encodes the *pulA* gene and pullulanase from this strain would be expected to incorporate label. pGSP20 specifies pullulanase except that DNA encoding its signal sequence and the adjacent Cys residue has been replaced largely by DNA encoding a modified version of the signal sequence of the βlactamase of pAT153 (Kadonaga et al. 1984). Since the missing Cys residue is the site for attachement of the lipid, incorporation of the lipid is not expected to occur.

As a first step in the verification of the suspected lipoprotein nature of pullulanase from strain K21, cultures of *E. coli* strain NM522 containing either plasmid pGSP5 or plasmid pGSP20 were labelled by the addition of [9,10(n)-^{3}H]-palmitic acid. The signal sequence and Cys residue of *pulA* on pGSP5 have been replaced in the hybrid pullulanase gene carried by pGSP20 (fig. 1).

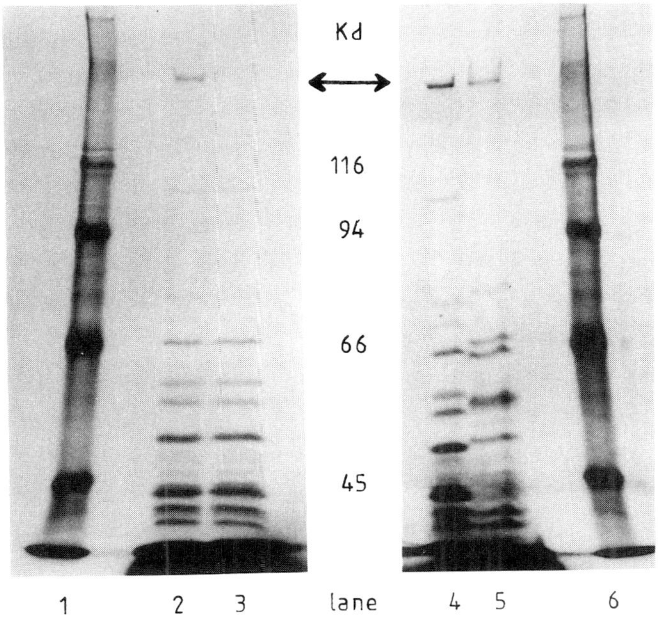

Fig. 2: Labelling of pullulanase with [9,10(n)-^{3}H]-palmitic acid. For details see text. Lanes: 1,6: molecular weight markers; 2,4: cell lysate from NM522(pGSP5); 3: cell lysate from NM522(pGSP20); 5: cell lysate from K21. Lane 3 carries five times as much pullulanase as lane 2. Note other lipoproteins (Ichihara et al. 1981). Arrows indicate the position of pullulanase.

As expected, label is only incorporated into pullulanase expressed from pGSP5 (fig. 2, lane 2). Thus, the site of lipid modification is the extreme N-terminal region of pullulanase, most likely the Cys residue at the N-terminus of mature pullulanase. The lipoprotein nature of pullulanase has also been confirmed for strain K21 itself (fig. 2, lane 5).

Evidence that pullulanase from *Klebsiella pneumoniae* strain ATTC 15050 is a lipoprotein has been published recently (Pugsley et al. 1986). The signal sequence of pullulanase from strain K21 is identical to pullulanase of strain ATTC 15050 (Chapon et al. 1985) and the restriction maps of the two genes are nearly identical. A further pullulanase gene, from *Klebsiella aerogenes* strain W70, has recently been cloned and sequenced (Katsuragi et al. 1987). This gene has a different restriction map; however, comparison of the predicted N-terminal amino acid sequences of the pullulanases from strains W70 and K21 shows very strong homology. (Katsuragi et al and GSP, unpubl. results). Their signal sequences differ at two amino acid positions but their signal peptidase cleavage sites are identical. Katsuragi et al have demonstrated that the pullulanase from strain W70 is also a lipoprotein.

SECRETED PULLULANASE FROM STRAIN K21 IS NOT ALWAYS MODIFIED BY LIPID

The only other specifically secreted bacterial lipoproteins characterised so far, are three penicillinases from Gram-positive bacteria (Nielsen et al. 1982). These are released from the cell surface into the culture medium by proteolytic removal of an N-terminal segment whose attached lipid presumably anchors the protein to the cell surface. The released form of the protein therefore lacks the lipid. Release of pullulanase from the cell surface of strain K21 may occur by an identical mechanism. To establish if this was the case a culture of strain K21 was labelled with [9,10(n)-^{3}H]-palmitic acid and pullulanase from cell lysates and culture supernatants of both log phase and overnight stationary phase cultures was analyzed for incorporation of label. As shown in fig. 3, cell-associated pullulanase is lipid modified (lanes 1 and 4). However, in contrast to pullulanase in overnight culture supernatants (lane 3), there is no apparent labelling of secreted pullulanase in late log culture supernatants (lane 2). This suggests that there may be growth-phase dependent release of two forms of pullulanase, one of which lacks the lipid.

There is no evidence for lysis in overnight cultures which might account for the presence of the lipoprotein form of pullulanase in the culture medium. In the case of strain ATTC15050 Pugsley et al (1986) have reported that only the lipoprotein-form of the enzyme is secreted, and have suggested that this material enters the culture medium by a mechanism involving the release of pullulanase-enriched membrane vesicles from the cell surface at the end of log phase. As there is no evidence for the existence of additional pullulanase structural genes in strain K21 (MGK, GSP, results not shown) which could direct the synthesis of secreted pullulanase not entering the lipoprotein pathway, we suggest that an additional secretion mechanism exists in this strain. By analogy with *Bacillus* penicillinases (Nielsen et al. 1982)) this would involve proteolytic cleavage to release pullulanase from the cell surface. The involvement of protease(s) in pullulanase secretion has also been proposed by other workers (Ohba et al. 1982). The absence of any such protease(s) in some strains of *Klebsiella* (Pugsley AP. pers. comm.) might explain the difference between our results and those of Pugsley et al. Further experiments are now in progress to confirm the predicted existence of a mixed population of pullulanase molecules in overnight culture supernatants.

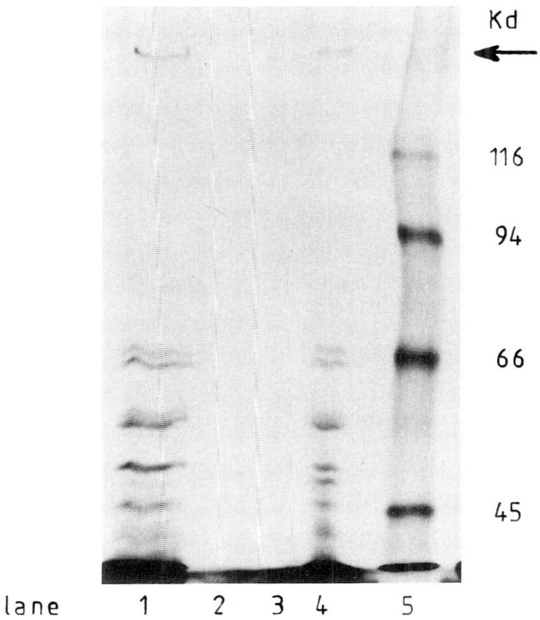

Fig 3: $[^3H]$-palmitate labelling of late log ond overnight culture of *Klebsiella aerogenes* strain K21. Lanes: 1: late log cell lysate; 2: late log culture supernatant; 3: overnight culture supernatant; 4; overnight cell lysate; 5: molecular weight markers. Each lane carries equal amounts of pullulanase. The arrow indicates the position of pullulanase.

ANALYSIS OF PULLULANASE EXPORT MUTANTS OF KLEBSIELLA AEROGENES

Isolation of secretion-defective Tn10-insertion mutants of *Klebsiella aerogenes* strain K21.

When the *pul*A gene is expressed in *E. coli* none of the pullulanase synthesized is localized at the cell surface. Instead, pullulanase is distributed between the cytoplasmic membrane and the inner face of the outer membrane, suggesting that *E. coli* lacks function(s) required to export pullulanase across the outer membrane to the cell surface and probably also into the culture medium (GSP, unpubl. results; Takizawa et al. 1985). In order to identify and characterize any such functions strain K21 was subjected to transposon mutagenesis using λ_{840}::Tn10 (Way et al. 1984). Since strain K21 is λ-insensitive, a plasmid carrying the *E. coli lamB* gene (which encodes a λ-receptor) was initially conjugated into K21 essentially as described by Salmond et al (1986). Strains harbouring this plasmid could now be infected with λ_{840}. Since phage λ cannot be stably maintained in *Klebsiella*, mutants resulting from transposition of Tn10 into the chromosome of K21 could therefore be isolated by selecting for the Tn10-encoded tetracycline resistance determinant. Approximately 5000 such tetracycline-resistant colonies were screened for the failure to secrete pullulanase into the medium by using a haloassay (Morgan et al. 1979). A total of 22 clones, representing at least 17 independent Tn10-insertions as judged by Southern blot analysis (see below), failed to secrete pullulanase. All of these mutants still synthesized pullulanase.

Properties of Tn10-insertion mutants of strain K21.

All of the Tn10-insertion mutants that were isolated fail to release pullulanase into the culture medium. However, they can be divided into two classes with respect to the amount of pullulanase synthesized and the proportion of pullulanase found on the cell surface (table 1). Whereas class 1 mutants appear to be defective in their ability to localize pullulanase to the cell surface and to secrete it into the culture medium, class 2 mutants only appear to be defective in the latter step. The results suggest that separate functions are required for translocation of pullulanase to the cell surface and its subsequent release into the culture medium. Class 2 mutants are further distinguished by the lack or reduced expression of a major 48 kd protein as judged by SDS-PAGE analysis, by their failure to utilize maltose to any significant extent and by their slow growth on minimal medium containing pullulan or maltose as the sole carbon source. The 48kd protein is exclusively located in the outer membrane of *Klebsiella* (data not shown) and may be a release function for pullulanase or, alternatively, it may be involved in maltose utilisation. In the latter case it may well correspond to the equivalent in *Klebsiella* of the maltose porin of *E. coli* (*lam*B). The phenotype of the class 2 mutants with respect to the defects in pullulanase release and maltose utilisation may be a consequence of polar effects on gene expression of the Tn10-insertions in these mutants. This would imply that gene(s) required for maltose utilisation are in the same operon as the putative pullulanase release function(s). However, there are alternative explanations such as inactivation of common regulatory genes by the Tn10-insertions in the class 2 mutants.

Table 1:

pullulanase	strain K21	class 1 mutants	class 2 mutants
% in culture medium	50-60	0	0
on cell surface	40-50	0-30	60-100
within cell	0	70-100	0-40
total activity relative to strain K21	100	30-60	10-20
growth on maltose or pullulan	+++	+++	+
Other			-lack or reduced expression of 48kd protein. -defect in maltose-utilisation

Mapping of Tn10-insertion mutants of strain K21

In order to determine if any pullulanase export functions are linked to the *pul*A gene, the 6kb *Eco*RI fragment encoding pullulanase (see above) was used as a probe in Southern blot analysis of genomic DNA isolated from Tn10-insertion mutants (fig. 4). This allowed the Tn10-insertions to be mapped with respect to known restriction sites inside and outside the 6kb *Eco*RI fragment. There were 13 insertions which mapped within the 6kb *Eco*RI fragment, with 5 insertions on either side of a unique *Bam*HI site (fig. 5). The remaining 3 insertions have not been mapped with respect to the *Bam*HI site. Calculation of the distances of the 10 Tn10-insertions from the *Eco*RI sites showed that none were in *pul*A itself. Since we have no evidence for the presence of other *pul*A genes in strain K21 which would explain the ability of the Tn10-insertion mutants to synthesize pullulanase, it seems that insertion into *pul*A, if it does occur, is detrimental to the cell. In order to map the remaining Tn10-insertions to regions adjacent to the 6kb *Eco*RI fragment, chromosomal DNA from strain K21 extending downstream and upstream of the *pul*A *gene*, was cloned into the positive-selection vector pKGW (Kuhn et al. 1986) as two *Bgl*II fragments of 6.5kb and 10kb, respectively (fig. 5). These were used as probes in mapping experiments. Five additional Tn10-insertions were found to map within the 10kb *Bgl*II fragment which extends upstream of *pul*A and these have been further mapped with respect to a *Bam*HI site within this fragment. Three insertions map outside both *Bgl*II fragments. Insertions have not been found in a 5kb region extending downstream of *pul*A from the *Eco*RI site 3' to *pul*A (fig. 5). The results suggest that DNA of not more than 1kb downstream of *pul*A is required for pullulanase export and agree with our finding that a Tn10-insertion downstream of *pul*A can be complemented by the 6kb *Eco*Rl fragment (preliminary date, results not shown). Assuming that the class 2 mutants (see above) represent adjacent insertions and since one class 2 mutant maps outside the 10kb *Bgl*II fragment, it appears that DNA of at least 10kb upstream of *pul*A is required for pullulanase export.

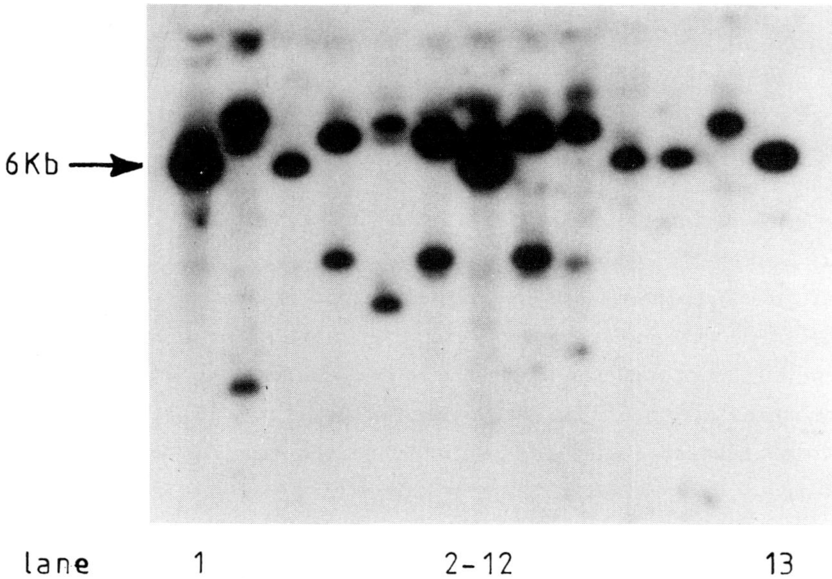

Fig 4: Determination of the positions of Tn10-insertions relative to *pul*A. *Eco*RI digested chromosomal DNA from Tn10-insertion mutants was hybridized in a Southern blot to a radiolabelled 6kb *Eco*RI fragment which carries the centrally-located *pul*A gene. Tn10 has a unique *Eco*RI site. The probe is therefore expected to hybridize to one or two bands depending on whether the Tn10-insertion is located outside or within the 6kb *Eco*RI fragment, respectively. Lanes: 1: strain K21; 2-12: Tn10-mutants; 13: probe.

The conclusion that some if not all the genes for pullulanase export are clustered around *pulA* is supported by the isolation (using the positive-selection vector pUN121, Nilsson et al. 1983) and expression in *E. coli* of the *Hind*III fragment which encompasses both *Bgl*II fragments mentioned above. In contrast to *E. coli* carrying the 6kb *Eco*RI fragment, in strains of *E. coli* expressing the *Hind*III fragment all of the cell-associated pullulanase synthesized is now located at the cell surface (preliminary results, data not shown). Up to 50% of pullulanase is released into the medium by what appears to be perturbation of the outer membrane rather than whole cell lysis as judged by the proportions of cytoplasmic and periplasmic marker enzymes which are released in parallel. It is not known whether this non-specific release is due to the *Hind*III fragment lacking further function(s) required for pullulanase release and/or whether it is a consequence of overproduction of some or all of these functions from the multicopy vector pUN121. However, it is consistent with the idea of pullulanase export functions being located in the outer membrane.

The possession of the *Hind*III fragment will allow us to further dissect the mechanism of pullulanase secretion at the DNA-level by complementation studies and at the protein level by identifying the number, sizes and cellular locations of pullulanase export functions.

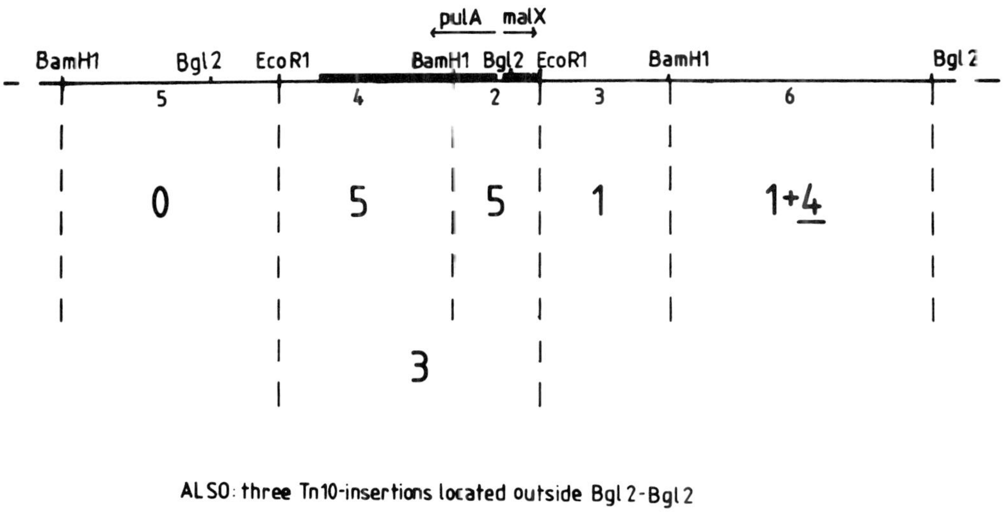

Fig 5: Diagram showing positions of Tn10-insertions relative to *pulA*. Numbers below arrows indicate number of Tn10-insertions mapped to within a region as indicated. Class 2 mutants are underlined.

REFERENCES

Braun V (1975) Covalent lipoprotein from the outer membrane of *E. coli*. Biochim Biophys Acta 415:335-377

Chapon C, Raibaud O (1985) Structure of two divergent promoters located in front of the gene encoding pullulanase in *Klebsiella pneumoniae* and positively regulated by the *malT* product. J Bacteriol 164:639-645

Eisele B, Rasched IR, Wallenfels K (1972) Molecular characterisation of pullulanase from *Aerobacter aerogenes*. Eur J Biochem 26:62-67

Ichihara S, Hussain M, Mizushima S (1981) Characterisation of new membrane lipoproteins and their precursors in *E. coli*. J Biol Chem 256:3125-3129

Kadonaga JT, Gautier AE, Straus DR, Charles AD, Edge MD, Knowles JR (1984) The role of the βlactamase signal sequence in the secretion of proteins by *E. coli*. J Biol Chem 259:2149-2154

Katsuragi N, Takizawa N, Murooka Y (1987) Entire nucleotide sequence of the pullulanase gene of *Klebsiella aerogenes* W70. J Bacteriol 169:2301-2306

Kuhn I, Stephenson F, Boyer HW, Greene PJ (1986) Positive-selection vectors utilizing lethality of *EcoR*1 endonuclease. Gene 44:253-263

Michaelis S, Chapon C, D'Enfert C, Pugsley AP, Schwartz M (1985) Characterisation and expression of the structural gene for pullulanase, a maltose-inducible secreted protein of *Klebsiella pneumoniae*. J Bacteriol 164:633-638

Morgan FJ, Adams KR, Priest FG (1979) A cultural method for the detection of pullulan-degrading enzymes in bacteria and its application to the genus *Bacillus*. J Appl Bacteriol 46:291-294

Nielsen JBK, Lampen JO (1982) Glyceride-cysteine lipoproteins and secretion by gram-positive bacteria. J Bacteriol 152:315-322

Nilsson B, Uhlen M, Josephson S, Gatenbeck S, Philipson L (1983) An improved positive-selection plasmid vector constructed by oligonucleotide mediated mutagenesis. Nucleic Acids Res 11:8019-8030

Ohba R, Ueda S (1982) An inductive effector in the production of extracellular pullulanase by *Aerobacter aerogenes*. Agric Biol Chem 46:2425-2431

Pugsley AP, Chapon C, Schwartz M (1986) Extracellular pullulanase of *Klebsiella pneumoniae* is a lipoprotein. J Bacteriol 166:1083-1088.

Salmond GPC, Hinton JCD, Gill DR, Perombelon MCM (1986) Transposon mutagenesis of *Erwinia* using phage λ vectors. Mol Gen Genet 203:524-528

Takizawa N, Murooka Y (1985) Cloning of the pullulanase gene and overproduction of pullulanase in *E. coli* and *Klebsiella aerogenes*. Appl Environ Microbiol 49:294-298

Way JC, Davis MA, Morisato D, Roberts DE, Kleckner N (1984) New Tn10 derivatives for transposon mutagenesis and for construction of *lacZ* operon fusions by transposition. Gene 32:369-379

CLONING OF *xcp* GENES POSSIBLY INVOLVED IN PROTEIN SECRETION IN *PSEUDOMONAS AERUGINOSA*.

M. BALLY, A. FILLOUX, M. MURGIER, B. WRETLIND*, A. LAZDUNSKI
Laboratoire de Chimie Bactérienne, C.N.R.S.
B.P. 71, 13277 Marseille Cedex 9
France

Introduction

New insights into how proteins cross membranes are beginning to emerge. Although there is no universal mechanism, recent results suggest that there are common features of protein translocation. Three steps usually are involved : i) the association of the protein with receptors on the correct membrane, ii) the translocation through the membrane and iii) covalent modification and folding on the opposite membrane surface. Secretion can be considered as a special case of protein export for which the final destination is extracellular. It has been mainly studied in Gram-positive organisms (i.e., *Bacillus* spp.). In Gram-negative bacteria, the problem is different, since secreted enzymes must be transferred across both the inner and outer membrane barriers. Among Gram-negative bacteria, *Escherichia coli* is not a very good secretory organism, whereas *Pseudomonas aeruginosa* secretes many proteins. Most strains of *Pseudomonas aeruginosa* produce at least two extracellular proteases, elastase and alkaline protease, an exotoxin A, a phospholipase C, an alkaline phosphatase and a lipase (Liu, 1974 ; Pavlovskis and Wretlind, 1982). Thus, *Pseudomonas aeruginosa* possesses machinery for the efficient release of exoproteins and one should keep in mind that different mechanisms for secretion of exoproteins may exist within the same organism.
In order to get a better understanding of secretion mechanisms, pleitropic mutations affecting protein release have been characterized. Then, plasmids complementing the different mutations have been isolated from a DNA genomic library of *Ps. aeruginosa*. The different gene products are under characterization.

*Department of Clinical Bacteriology, Danderyd Hospital, S-18288, Danderyd, Sweden

NATO ASI Series, Vol. H16
Membrane Biogenesis. Edited by J.A.F. Op den Kamp

Characterization of different types of mutants of Ps. aeruginosa defective in extracellular protein secretion

Pleiotropic mutants (xcp mutants) defective in extracellular protein secretion were isolated after treatment of Ps. aeruginosa PAO with a mutagen (EMS) (Wretlind and Pavlovskis, 1984 ; Filloux et al., 1987). The different mutants were unable to release at least four exoproteins (lipase, elastase, alkaline phosphatase, and phospholipase C) while alkaline protease was still secreted. The exoproteins appeared to be blocked in the periplasmic space. No difference in molecular weight was detected between cell-bound forms of elastase and alkaline phosphatase in the mutants and the corresponding extracellular forms from the wild type strain.

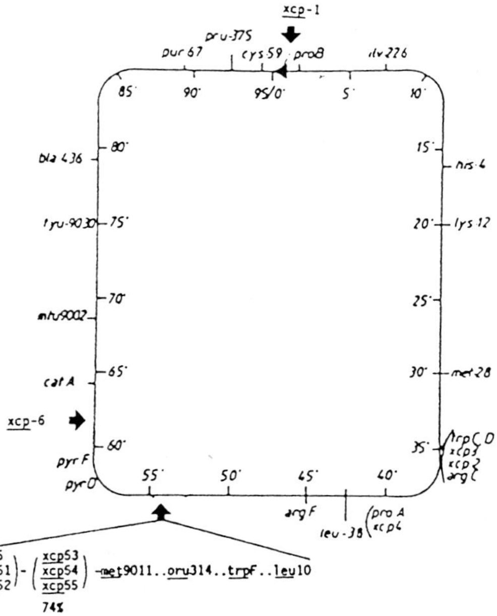

Fig. 1 : Map of P. aeruginosa PAO chromosome (from Holloway et al., 1982) and location of the different xcp mutations.
Genetic mapping was done to compare mutants xcp-54 and xcp-55 to those previously described (Wretlind and Pavlovskis, 1984). Transduction experiments with phage F116Lc4 showed that both mutations were located in the 55' region of the chromosome. They were both 78 % cotransducible with met-9011, the gene order being leu-10--met-9011--xcp-54/xcp-55. Those two mutations thus appear to be similar to xcp-53 which can be cotransduced at 74 % with met-9011. In contrast, other mutations (xcp-5, xcp-51, and xcp-52) are only cotransduced at 42-47 % with this marker.

Genetic mapping showed that the different mutations were located at different loci on the chromosome. So far, at least 6 secretion mutants were mapped at 55' on the chromosome (xcp-5-xcp-51... xcp-55).
Results based on transductional analysis suggest that two different loci are probably involved in protein secretion in this 55' region of the chromosome (Fig. 1).
Two other secretion mutants were mapped on the chromosome at 0' (xcp-1) and one mutant at 65' (xcp-6) (Wretlind and Pavlovskis, 1984).
Thus, although all the various secretion mutants feature similar phenotypes, different mutations are involved and are located at least at four different loci.

Cloning of *xcp* genes possibly involved in protein secretion.

In order to identify components of the export machinery of the cell, a genomic library containing *Sau*3A partial digests of *Pseudomonas aeruginosa* PAO1 DNA was constructed in cosmid vector pLAFR3 and propagated in *E. coli*. Recombinant plasmids were then transferred from *E. coli* to the different PAO secretion mutants by triparental mating, using the conjugative properties of pRK2013.
Approximately 2000 PAO-*xcp* exconjugants containing the pLAFR-PAO1 recombinant plasmids were examined for a secretory phenotype (zones of clearing on elastin-agar plates).
Using this approach, no convincing results have been obtained ; this was probably due to the instability of the DNA inserts on recombinant plasmids and to easier lysis of the cells containing those plasmids.
We thus had to use different stategies in order to clone successfully *xcp* genes.

a. Cloning of the *xcp*-1 gene.

A *Pseudomonas aeruginosa* transposon insertion mutant with defective release of several exoenzymes had been characterized (Lindgren and Wretlind, 1987). The Tn*5-751* insertion mutation was located in the previously described *xcp*-1 locus at 0 min on the chromosomal map.

The presence of an easily selectable antibiotic resistance marker in the xcp-1 locus offers the possibility of cloning the gene(s)involved in exoprotein secretion. Since the SalI sites in Tn5-751 lie between the Km^R and Tp^R determinants, it is possible to clone DNA adjacent to the transposon insertion by selecting for Km^R or Tp^R determinants (Rella et al., 1985). Those DNA fragments are then used as probes to screen the genomic bank. Eight cosmids have been selected. Two of them complemented the xcp-1 mutation and thus were believed to carry the xcp-1 gene.

To locate the xcp-1 gene on the cloned DNA fragment (30 kb) a number of deletion plasmids were constructed in E. coli. Each time, their ability to complement the xcp-1 mutation in PAO1 was determined, after a triparental mating. In this way, the xcp-1 gene has been located on a 2.9 kb PstI fragment which was then subcloned on a low copy number plasmid pRK310. The orientation of the xcp-1 gene transcription has been determined by insertion of a promoterless chloramphenicol acetyltransferase (CAT) cartridge. Moreover, this construction will allow us to study the xcp-1 promoter regulation by measuring CAT activity. Using a minicell system and the transposon TnphoA (Manoil and Beckwith, 1985), we are currently trying to characterize more precisely the xcp-1 gene product, particularly its cellular location.

b. Cloning of the xcp-54 and xcp-5 genes.

To isolate those genes, and more generally, all the chromosomal region located around those genes, we used a different strategy from that described above. From a different DNA genomic bank, we had previously cloned, on a 10 kb DNA insert, the neighbouring marker met-9011. By using this plasmid as a probe to screen the PAO cosmid bank, we intended to "walk" along the Ps. aeruginosa chromosome in the vicinity of the met-9011 gene to isolate overlapping DNA subclones which could harbor all the xcp genes located in this 55' region.

From genetic data, we knew that the oru-314 locus was mapped near met-9011, in the following order : pyrD... (xcp)...oru-314/met-9011/ trpF/leu-10. We first isolated a cosmid complementing oru-314. To our surprise, this cosmid did not complement any xcp mutants ; instead it complemented trpF and leu10. Thus, the true gene order in this area is

pyrD ..... (*xcp*) ..... *met*-9C11/*oru*314/*trp*F/*leu*10.
Finally, two other cosmids complementing *xcp*-54 have been isolated. Restriction mapping, subcloning and complementation analysis of a 30 kb DNA region is currently being done, with the aim to determine the organisation of the different loci involved in protein secretion in this region of the chromosome.

Perspectives

Very few secretion systems have been completly elucidated so far in Gram-negative bacteria (for a review see Pugsley and Schwartz, 1985). One of the best characterized is that of £-haemolysin in *E. coli* which comprises at least two genes involved in haemolysin secretion (Mackman et al., 1986). The extracellular secretion of *Neisseria gonorrhoeae* IgA protease appears to involve a carboxyl-terminal "helper" domain apparently required for secretion. This "helper" domain could serve as a pore for excretion of the protease domain through the outer membrane. Pullulanase is an extracellular catabolic enzyme produced by some strains of *Klebsiella pneumoniae*. This is a lipoprotein, found mainly in the outer membrane and subsequently released into the growth medium as a lipoprotein (Pugsley et al., 1986).
These results suggest that Gram-negative cells have evolved a wide variety of secretion systems which can be more or less specific.
In *Ps. aeruginosa*, different mechanisms for secretion of exoproteins appear to exist. Toxin A is made as a precursor with a classical signal peptide and is apparently co-translationally secreted directly into the medium (Lory et al., 1983). The specific secretion pathway of alkaline protease has not been elucidated at present (Wretlind and Pavlovskis, 1984 ; Filloux et al., 1987). Besides these two cases, there is also a general pathway for secretion of at least four proteins (alkaline phosphatase, phospholipase C, lipase and elastase). *xcp* mutations affect this pathway. We have now cloned different *xcp* genes. By characterizing the structure of the different *xcp* gene products and by determining their cellular location we hope to improve our understanding of this general secretion pathway in *Pseudomonas aeruginosa*.

REFERENCES

Filloux A, Murgier M, Wretlind B, Lazdunski A (1987) Characterization of two Pseudomonas aeruginosa mutants with defective secretion of extracellular proteins and comparison with other mutants. FEMS Microbiol Rev 40 : 159-163

Früh R, Haas D, Leisinger T (1985) Altered control of glutamate-dehydrogenases in ornithine utilization mutants of Ps. aeruginosa Arch Microbiol 141 : 170-176

Holloway BW, Crockett RJ (1982) Pseudomonas aeruginosa. Genet Maps. 2: 165-167

Lindgren V, Wretlind B (1987) Characterization of a Pseudomonas aeruginosa transposon insertion mutant with defective release of exoenzymes. J Gen Microbiol 133 : 675-681

Liu PV (1974) Extracellular toxins of Pseudomonas aeruginosa. J Infect Dis 130 : 94-99

Lory S, Tai PC, Davis BD (1983) Mechanism of protein excretion by Gram-negative bacteria : Pseudomonas aeruginosa exotoxin A. J Bacteriol 156 : 695-702

Mackman N, Nicaud JM, Gray L, Holland IB (1986) Secretion of haemolysin by E. coli. Curr Topics Microbiol Immunol Protein secretion and export in bacteria. Ed PC Tai and Wu H

Manoil C, Beckwith J (1985) ThphoA : A transposon probe for protein export signals. Proc Natl Acad Sci USA 82 : 8129-8133

Pavlovskis OR, Wretlind B (1982) Pseudomonas aeruginosa toxins in C.S.F. Easmon and J. Jeljaszewicz (ed.), Medical microbiology, vol.1. Academic Press, Inc., London

Pohlner J, Halter R, Beyreuther K, Meyer TF (1987) Gene structure and extracellular secretion of Neisseria gonorrhoeae IgA protease. Nature 325 : 458-462

Pugsley AP, Chapon C, Schwartz M (1986) Extracellular pullulanase of Klebsiella pneumoniae is a lipoprotein. J Bacteriol 166 : 1083-1088

Rella M, Mercenier A, Haas D (1985) Transposon insertion mutagenesis of Ps. aeruginosa with a Tn5 derivative : application to physical mapping of the arc gene cluster. Gene 33 : 293-303

Wretlind B, Pavlovskis OR (1984) Genetic mapping and characterization of Pseudomonas aeruginosa mutants defective in the formation of extracellular proteins. J Bacteriol 158 : 801-808

VACUOLE DIVISION AND INHERITANCE IN *SACCHAROMYCES CEREVISIAE*

Brenda Guthrie and William Wickner
Department of Biological Chemistry
and the Molecular Biology Institute
University of California
Los Angeles, California 90024 USA

Cell division requires the partitioning of cytoplasmic components to ensure daughter cell viability. Some organelles, such as ribosomes, may partition at cytokinesis simply by their abundant distribution throughout the cytoplasm (Birky,1983). Other organelles that are present singly or in low copy number must divide and segregate in a more regulated fashion (Warren,1985). In higher eukaryotic cells, the Golgi apparatus, endoplasmic reticulum, and nuclear membrane fragment at the start of mitosis (Burke et al.,1982, Lipski and Pagano,1985, Zeligs and Wollman, 1979, Gerace and Blobel,1980). This fragmentation event may enable the partitioning of organelle components into the daughter cell.

We have chosen to study the division and inheritance of the yeast vacuole. In G1 phase cells, the vacuole is present as a single large cytoplasmic organelle (Weimken et al.,1970, Weisman et al., 1987). The vacuole is readily visualized using a number of microscopic techniques (Weisman et al.,1987) and the availability of purified and cloned marker enzymes renders it amenable to biochemical study. The

NATO ASI Series, Vol. H16
Membrane Biogenesis. Edited by J. A. F. Op den Kamp

asymmetry of cell division in *Saccharomyces cerevisiae* provides an easy index of cell cycle position.

How does a daughter cell obtain its vacuole? Two possibilities seemed plausible. Perhaps the daughter cell vacuole is derived from parent vacuole components. Alternatively, since vacuolar components are derived from the Golgi (Stevens et al.,1982), the bud vacuole may be formed essentially *de novo* from Golgi derived vesicles. Both of these models suggested experiments to test these hypotheses.

Previous studies of vacuole morphology using synchronized cells suggest that the vacuole fragments into many small vesicles during bud emergence (Weimken et al., 1970). Some of these components are thought to migrate into the bud, where they fuse to form a single vacuole. Since synchronization may induce artificial physiological effects, we decided to study the vacuole morphology of asynchronous, exponentially growing cultures. Vacuoles were labeled with either of two fluorescent markers: quinacrine, a weak base which diffuses across membranes and concentrates in the acidic vacuole (Weisman et al.,1987), or FITC-dextran, a fluor which is endocytosed (Makarow,1986). Fluorescent and Nomarski optic photomicrographs were taken of fields of labeled cells. The cell cycle position was scored by comparing the diameter of the bud to that of the parent cell. Parental vacuole morphologies fell into two groups- a single major vacuole, or multiple vacuoles. The bud almost always contained a single vacuole. When parental vacuole morphology was compared to cell cycle position, a small increase in the number of multiple vacuoles was observed at the beginning of bud emergence (Weisman et al.,1987). This small increase did not seem to be an appreciable mechanism for vacuole partitioning.

To explore the inheritance in a slightly different way, we took advantage of the fluorescence found in *ade-2* cell vacuoles. This mutation in adenine biosynthesis causes an accumulation of a fluorescent substance in the vacuole when exogenous adenine becomes limiting. When cells with accumulated fluor are transferred to fresh media, the cells resume exponential growth. The inheritance of fluor by successive generations was measured using a fluorescence activated cell sorter. One-thirtieth of the fluorescence from the parent cells was found in the progeny cells. Vacuolar volume differences between parental and daughter cells contribute to this difference, and when this is taken into account, approximately 50% of the fluor is inherited by the daughter cells (Weisman et al.1987).

In order to further quantitate vacuolar inheritance, a biochemical method was developed. Yeast were covalently bound to polystyrene beads coated with Sanger's reagent. Bound yeast grow normally and release daughter cells into ythe surrounding media. Release of cells is growth dependent: no release is seen when bound yeast are treated with azide. Tethered yeast were pulse labaled with 35-S methionine, and chased with unlabeled methionine. Carboxypeptidase Y, a vacuolar marker enzyme, was immunoprecipitated from successive generations arising from the bound labeled yeast. Fifty percent of the labeled carboxypeptidase y was inherited by each generation (Bacallao and Wickner, submitted) in agreement with the fluorescence studies.

Although these experiments yielded insight into the amount of inheritance of vacuole components per generation, they did not suggest a mechanism by which this inheritance occurred. Since we saw no cell cycle dependent change in vacuole morphology, we searched for a more subtle inheritance process.

When yeast mate, the resultant zygote contains organelles from each parent in a common cytoplasm. Yeast with an unlabeled vacuole were mated with *ade-2* bearing a fluorescent vacuole. Early into the first zygotic budding cycle, the vacuoles remain intact and topologically distinct as visualized by Nomarski and fluorescence microscopy. Later into the budding cycle, the previously non-fluorescent vacuole acquires label and is visible using fluorescence microscopy, while still remaining separate from the labeled vacuole. Sometimes small fluorescent vesicles can be seen streaming throughout the cytoplasm into unlabeled areas (Weisman and Wickner, in prep). This vesicular communication defines a new pathway in organelle mediated traffic.

To determine whether cytoskeletal elements are involved in the morphology or inheritance of the vacuole, cells were treated with either of two drugs: nocodazole or MBC (methyl benzimidazole 2-yl carbamate). These drugs disrupt microtubules and cause growth arrest in the G2 portion of the cell cycle (Quinlan et al, 1980; Hasek et al., 1986). Vacuoles were visualized using quinacrine, FITC-dextran, and Nomarski optics. A surprising change in vacuole morphology was observed. Cells in the G2 portion of the cell cycle usually have a single or multiple vacuole morphology. Drug treated cell's vacuoles have an extremely vesiculated, or fragmented structure. (Guthrie and Wickner, submitted). A mutant in beta tubulin also exhibits this fragmented vacuole morphology, with a large percentage of cells bearing fragmented vacuoles exclusively in the bud (Guthrie and Wickner, submitted). These data provide evidence that microtubules may be involved in the maintenence of vacuolar morphology.

We have developed a model of vacuole inheritance based on these experiments. We propose that the vacuole is in a constant state of flux, with vesicular fission and fusion occurring throughout the cell cycle. Segregation of vacuole components could involve transport of these vesicles into the developing bud. These vesicles may also mediate communication between vacuoles in a common cytoplasm The fragmentation seen upon microtubule disruption implies that intact microtubules, or an intact microtubular network, may be needed for the fusion of vacuole derived vesicles.

References

Birky CW (1983) The partitioning of cytoplasmic organelles at cell division. Int Rev Cytol 15 49-89

Burke B, Griffiths G, Reggio H, Louvard D, Warren G (1982) A monoclonal antibody against a 135-kilodalton Golgi membrane protein. EMBO J 1:1621-1628

Gerace L, Blobel G (1980) The nuclear envelope is reversibly depolymerized during mitosis. Cell 19:277-287

Hasek J, Svoboda J, Streiblova E (1986) Immunofluorescence of the microtubular skeleton in growing and drug treated yeast protoplasts. Eur J Cell Biol 41:150-156

Lipski NG, Pagano RE (1985) A vital stain for the Golgi apparatus. Science 228:745-747

Makarow M (1985) Endocytosis in *Saccharomyces cerevisiae*: internalization of alpha amylase and fluorescent dextran into cells. EMBO J 4:1855-1860

Quinlan RA, Pogson CI, Gull K (1980) The influence of the microtubule inhibitor methyl benzimidazol 2-yl carbamate (MBC) on nuclear division and the cell cycle in *Saccharomyces cerevisiae.* J Cell Sci 46:341-352

Stevens T, Esmon B, Schekman R (1982) Early stages in the yeast secretory pathway are required for transport of carboxypeptidase y to the vacuole. Cell 30:439-448

Warren G (1985) Membrane traffic and organelle division. TIBS 439-443

Wiemken A, Matile P, Moor H (1970) Vacuolar dynamics in synchronously budding yeast. Arch Microbiol 70:89-103

Weisman LS, Bacallao R, Wickner W (1987) Multiple methods of visualizing the yeast vacuole permit evaluation of its morphology and inheritance during the cell cycle. J Cell Biol in press

Zeligs JD, Wollman SH (1979) Mitosis in rat thyroid cells in vivo I. Ultrastructural changes in cytoplasmic organelles during the mitotic cycle. J Ultrastruc Res 66:53-57

THE LACTOSE CARRIER OF E. COLI: PROTEASE DEPENDENT IN VIVO INACTIVATION.

Emanuel SHECHTER, Christine HOUSSIN, Alexandre GHAZI and Hélène THERISOD

Laboratoire des Biomembranes (UA 1116), Université de Paris-Sud,
91405 Orsay, France.

INTRODUCTION

Since its discovery in 1956 (Rickenberg et al., 1956), the β-galactoside permease (lactose carrier) of E. coli has been studied extensively in whole cells, as well as in membrane vesicles. Since the verification of the Mitchell hypothesis (Mitchell, 1963) that the mechanism of transport is a proton symport, this system has served as a model for the study of secondary active transport (Kaback, 1983; Rosen and Kaschket, 1978). More recently, the amino-acid sequence of the lactose carrier was deduced from the DNA sequence (Büchel et al., 1980) and the purification and functional reconstitution of the lactose carrier into proteoliposomes have been reported (Newman et al., 1981; Wright et al., 1982). Lately, extensive efforts have gone into the elucidation of the molecular mechanism of this lactose/proton symport using site-directed mutagenesis (see Kaback, 1987 for a review). In particular it was shown that one of the 4 histidines of the carrier (histidine 322) plays an important role in the coupling of lactose and proton transport (Padan et al., 1985; Pütner et al., 1986); it was also shown that glutamic acid 325, together with histidine 322, may participate in a charge relay system for proton transport (Carrasco et al., 1986). However, so far, and despite all efforts, the molecular mechanism remains to be elucidated.

A few years ago we have observed that lactose transport by whole cells may, under certain conditions, lead to an irreversible inactivation of the lactose carrier (Ghazi et al., 1983). We have pursued this study with two different objectives: 1) could the analysis of the structural modification responsible for the inactivation of the lactose carrier tell us something about the mechanism of lactose/proton symport; 2) could this inactivation be, in some way, part of the general process of membrane protein degradation. In particular, we have shown that the inactivation is strongly dependent upon the respiratory activity of the cell (Thérisod et al., 1985a), and that it is inhibited by serine protease inhibitors (Thérisod et al., 1985b), suggesting that the inactivation may be one way of controling sugar influx into the cell. More recently, we have purified and reconstituted into proteoliposomes the inactivated carrier and tried to identify the structural changes responsible for the inactivation. It is the whole of these data which we present here.

NATO ASI Series, Vol. H16
Membrane Biogenesis. Edited by J. A. F. Op den Kamp

MATERIAL AND METHODS

Growth conditions and cell treatment. Cells of E. coli ML308225 ($i^-z^-y^+a^+$) were grown in minimum medium A containing 0.4 % glycerol as the sole carbon source and harvested at an absorbance of 0.4 at 650 nm (1 cm path length). EDTA treatment of the cells was performed according to Booth et al., (1979).The cells were resuspendend in 10 mM Tris-HCl, 150 mM choline chloride, 1 mM KCl (pH 7.6) (tris-choline buffer) at 100 mg dry weight/mL and kept on ice.

Active transport of lactose in E. coli cells. For active transport of lactose, the cells were diluted in tris-choline buffer at 1 mg dry weight/mL and incubated in the presence of chloramphenicol with agitation for 2 min at 25 °C in the presence or absence of an exogenous energy source. ^{14}C lactose was then added at the desired final concentration. At given intervals, samples (100 µL) were removed, immediately diluted with 4 mL of tris-choline buffer, and filtered on Whatman glass microfiber filters (GF/F). The filters were washed twice with 4 mL of buffer and counted for radioactivity in a liquid scintillation counter. Uptakes were corrected by substracting blanks obtained under identical conditions, except that the cells were pretreated with 20 µM CCCP and 0.2 mM pCMBS. The internal lactose concentration was calculated assuming an internal cytoplasmic volume of 1 µL/mg dry weight (Booth et al., 1979).

Influx rate of lactose in E. coli cells. Cells were prepared and suspended in tris-choline buffer at a concentration of 1 mg dry weight/mL as described above. At time 0, non-radioactive lactose was added at the desired concentration. Samples (100 µL) were immediately withdrawn. To these, trace amounts of ^{14}C lactose were added at given times. The uptake of radioactive lacose was allowed to proceed for 15 s before dilution and filtration as described above (transport of lactose). The 15 s uptake of radioactive lactose reflects the rate of influx of non-radioactive lactose at the time of the addition of the radioactive solute.

Facilitated diffusion, efflux and exchange in E. coli cells. A concentrated cell suspension (100 mg dry weight/mL) in tris-choline buffer was incubated for 5 min with 30 µM CCCP. ^{14}C lactose was added at the desired concentration. For facilitated diffusion, aliquots (5 µL) were withdrawn as a function of time, diluted in 1 mL and filtered immediately as described for active lactose transport, except for the presence of pCMBS and CCCP in the dilution and washing buffers. For efflux, lactose was first allowed to equilibrate across the membrane for 30 min. Then, 5 µL aliquots were withdrawn, diluted in 1 mL tris-choline buffer and efflux was allowed to proceed for the desired amount of time. Filtration was performed as for facilitated diffusion. Exchange was performed as efflux, except for the presence of non-radioactive lactose in the dilution buffer.

Determination of $\Delta\Psi$ in E. coli cells. The membrane potential ($\Delta\Psi$) was determined by the accumulation of ^{14}C Ph_4P^+ (Ghazi et al., 1981). Cells (1 mg dry weight/mL) were incubated in tris-choline buffer. ^{14}C Ph_4P^+ was then added (final concentration: 10 µM). Samples (100 µL) were withdrawn as a function of time, diluted with 4 mL buffer and filtered on 0.45 µm EH filters (Millipore). The filters were washed once with 4 mL of buffer and counted for radioactivity. The uptake of Ph_4P^+ was corrected by substracting blanks obtained under identical conditions, except that the cells were pretreated with 50 µM CCCP, and the samples were filtered immediately after the addition of the radioactive probe. The accumulation of Ph_4P^+ reached a plateau within 3 to 5 min. $\Delta\Psi$ was calculated from this plateau value by the Nernst equation.

When $\Delta\Psi$ was determined as a function of time during lactose accumulation, a similar procedure was used. Ph_4P^+ was allowed to accumulate until the plateau was reached. Then, at time zero, non-radioactive lactose was added and samples (100 µL) were removed at given intervals, diluted, filtered and counted for radioactivity, as described above. Although the accumulation of Ph_4P^+ to its steady state level in response to $\Delta\Psi$ requires a few minutes (see above), we verified by the addition of CCCP, that the efflux of Ph_4P^+ to a new, lower, steady-state level of accumulation in response to a decrease in $\Delta\Psi$ is much more rapid and is accomplished within a few seconds.

Purification and reconstitution of the lactose carrier. The solubilization and purification were performed according to Newman et al., (1981). The purified lactose carrier solubilized in octyl-glucoside was reconstituted into proteoliposomes by detergent dilution according to Foster et al., (1982). The proteoliposomes were collected by high speed centrifugation and resuspended in a minimum volume of potassium phosphate buffer pH 7.5. The ratio of protein to lipid was around 1/200 (w/w). The internal volume of the proteoliposomes was calculated from the amount of ^{14}C lactose present inside the proteoliposomes at equilibrium under conditions where active transport was absent. It amounts to 1.25 µL per mg phospholipid and is in agreement with values reported by Garcia et al., (1983).

For some experiments, the lactose carrier was solubilized in C12E8. The exchange of octyl glucoside for C12E8 was performed according to Houssin et al., (1985).

Formation and determination of a potassium diffusion potential in proteoliposomes. A transmembrane potassium diffusion potential can be generated across proteoliposomes. Typically, a concentrated suspension of proteoliposomes prepared as described above, was sonicated in the presence of 20 µM valinomycin. The proteoliposomes were then diluted 1000 fold in 50 mM sodium phosphate buffer pH 7.5, or in control experiments, in 50 mM potassium phosphate buffer pH 7.5. The diffusion potential was monitored by following the quenching of a cyanine dye (diIC1(5)), a fluorescent membrane potential probe (Letellier and Shechter, 1979). Experimentally, the dye was solubilized in the dilution buffer (5 µM/mL). 1 mL of the solution was then transfered into a fluorescence cuvette and fluorescence of the dye was monitored at 640 nm (excitation: 600 nm). 1 µL of the concentrated proteoliposome suspension was added to the cuvette and the change in fluorescence was followed as a function of time.

Transport of lactose in proteoliposomes. For active transport, the concentrated proteoliposome suspension, prepared as described above in potassium phosphate buffer pH 7.5 was diluted 1000 fold into sodium phosphate buffer pH 7.5. ^{14}C lactose was added (0.5 mM final concentration). 100 µL aliquots were withdrawn at given times, diluted into 4 mL of the same buffer and filtered on 0.25 µm GSTF millipore filters. The filters were washed with 4 mL of the same buffer and counted for radioactivity.

For exchange, the concentrated proteoliposome suspension prepared as described above was first allowed to equilibrate for 1 hour with ^{14}C lactose (final concentration: 3 mM). Then, 5 µL aliquots were withdrawn, diluted in 1 mL potassium phosphate buffer containing 3 mM non radioactive lactose and exchange was allowed to proceed for the desired amount of time. Filtration and washing were performed as for active transport, except for the presence of pCMBS in the washing buffer.

Miscellaneous methods. High performance liquid chromatography and quantitative amino-acid analysis of the lactose carrier solubilized in C12E8 were performed as described in Houssin et al., (1985). SDS-PAGE was performed according to the method of Laemmli (1970).

RESULTS

Under certain experimental conditions, transport of lactose by E. coli cells results in its own impairment. As will become apparent later, this impairment is the result of an irreversible inactivation of the lactose carrier. The optimal conditions for this inactivation to take place are: aerated cell suspension, relatively large external lactose concentration (at or above the K(T) of transport, i.e., 0.5 mM (Ghazi and Shechter, 1981)), sligtly alkaline pH, low external K^+ concentration, pretreatment of the cells with EDTA. In addition and importantly, the inactivation is strongly enhanced by the presence of an exogenous energy source (glycerol, D-lactate, succinate, ascorbate-phenazine methosulfate) (Ghazi et al., 1983).

Effect of an exogenous energy source on lactose transport.

Figure 1 displays the time course of lactose transport under the optimal conditions mentioned above, in the presence and in the absence of glycerol.

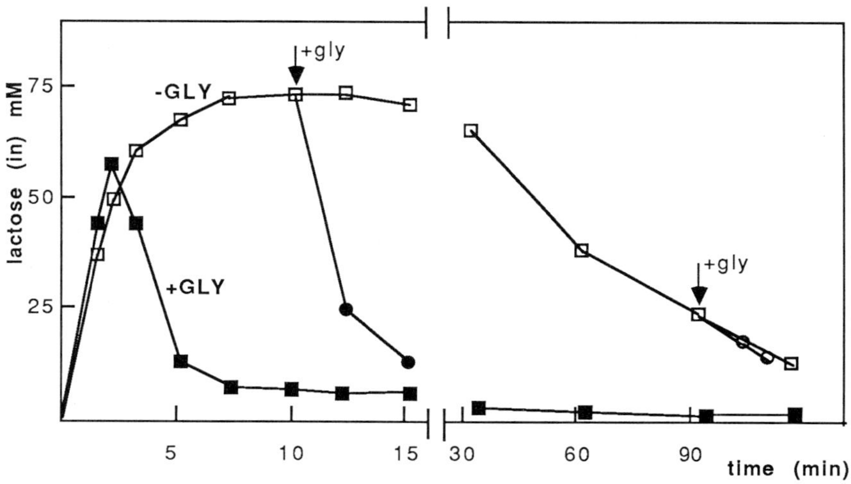

Figure 1. Time course of lactose uptake (0.5 mM external lactose concentration) as a function of time by E. coli cells (1 mg dry weight/mL) in the presence (■) and in the absence (□) of glycerol (0.4 %). Where indicated (arrows), glycerol (0.4 %) was added to aliquots of the cells which were accumulating lactose in the absence of glycerol (●,○).

In the presence of glycerol, the internal lactose concentration increased during the first 2 minutes of transport to a maximum of 60 mM (accumulation ratio: 120); thereafter it decreased and reached an apparent final steady state level of 5 mM after 10 minutes. In contrast, in the absence of an exogenous energy source, the impairment of lactose transport did not take place during the first 15 minutes of transport: within a few minutes, the internal lactose concentration reached its steady-state level (70 mM) and remained stable thereafter; addition of glycerol at this stage resulted in an immediate decrease of the internal lactose concentration. If transport was followed over a longer period of time, it was apparent that the absence of glycerol did not prevent the impairment of lactose transport but simply delayed it. After 2 hours of transport, the internal lactose concentration of cells which had been incubated in the absence of glycerol had decreased from its maximal value of 70 mM to a value of some 10 mM. This decrease was not the trivial result of a depletion of the endogenous energy sources as a consequence of the long incubation time; indeed, addition of external glycerol to the cells after 2 hour incubation did not restore an accumulation of lactose. The internal lactose concentration of cells which had been incubated with glycerol decreased further as a function of time and after 2 hours it was practically equal to the external lactose concentration (0.5 mM) (Thérisod et al., 1985a).

The enhanced impairment of lactose transport in the presence of glycerol was not the result of its metabolism. Identical results were obtained using other exogenous energy sources, i.e., succinate, D-lactate, and ascorbate-phenazine methosulfate. The data are shown in Figure 2. Addition of any one of these energy sources resulted in a rapid impairment of lactose transport. In particular, ascorbate-phenazine methosulfate is an artificial electron donnor which injects its electrons directly into the respiratory chain without prior transformation.

Increased respiratory activity is responsible for the enhanced impairment of lactose transport in the presence of an exogenous energy source.

The enhanced impairment of lactose transport in the presence of an external energy source as represented in Figures 1 and 2 was due to an increase in the respiratory activity of the cells (Thérisod et al., 1985a).

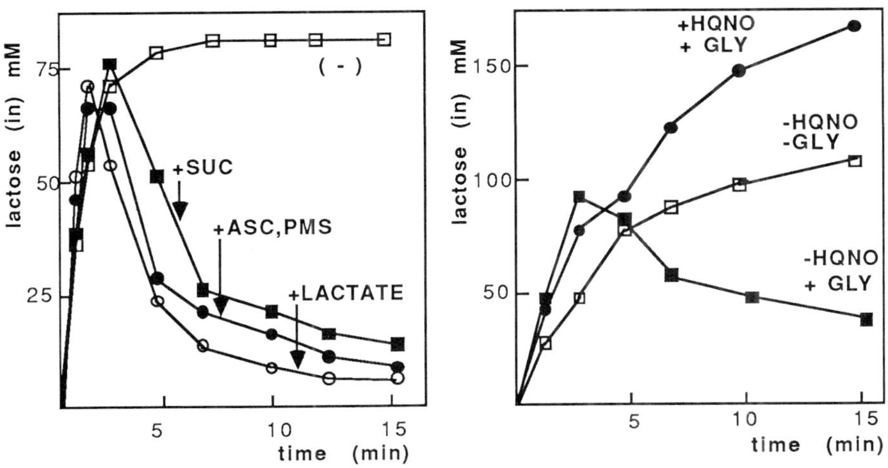

Figure 2 (left). Effect of various exogenous energy sources on the time course of lactose uptake (0.5 mM external lactose concentration) by E. coli cells (1 mg dry weight/mL). (□): no added energy source; (○): 20 mM D-lactate; (■): 10 mM sodium succinate; (●): 10 mM ascorbate-0.1 mM phenazine methosulfate.

Figure 3 (right). Effect of HQNO on the time course of lactose uptake (0.5 mM external lactose concentration) by E. coli cells (1 mg dry weight/mL). (■): uptake in the presence of 0.4 % glycerol; (●): uptake in the presence of glycerol and HQNO (0.25 mM); (□): uptake in the absence of glycerol and HQNO.

Indeed, addition of glycerol increased the rate of respiration of the cells by a factor of 3 (data not shown). Subsequent addition of HQNO (0.25 mM) or cyanide (0.2 mM), inhibitors of the respiratory chain, reduced the rate of respiration to the initial level (data not shown). The data in Figure 3 display the time course of lactose uptake in the absence of glycerol, in the presence of glycerol, in the presence of glycerol and HQNO (0.25 mM). Clearly in the presence of HQNO and glycerol the accumulation of lactose as a function of time is more pronounced than in the presence of glycerol alone or even than in the absence of glycerol. Similar results were obtained in the presence of cyanide (data not shown).

The impairment of lactose transport is the result of a decrease of the influx rate of transport.

The decrease of the internal lactose concentration reported in Figures 1 and 2 was the result of a decrease of the influx rate of lactose as it was accumulated. The data are shown in Figure 4. In the presence of glycerol the influx rate decreased steadily as lactose was accumulated during the 15 first minutes. In contrast, in the absence of glycerol, the influx rate remained constant during that period of time.

The impairment of lactose transport is accompanied by a collapse of $\Delta\Psi$.

The impairment of lactose transport (Figures 1 to 3) was accompanied by

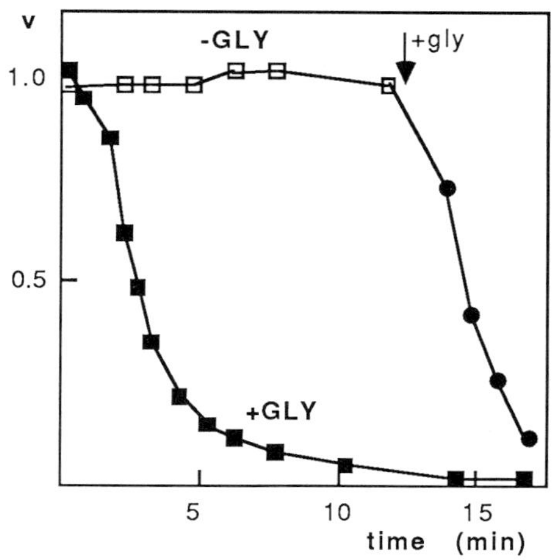

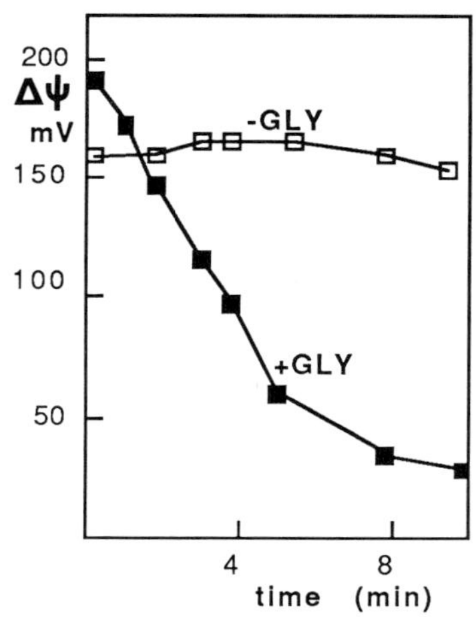

Figure 4 (left). Influx rate (v) of lactose (0.5 mM external lactose concentration) as a function of time during lactose accumulation in E. coli cells (1 mg dry weight/mL). (■): influx rate in the presence of glycerol (0.4 %); (□): influx rate in the absence of glycerol. Where indicated (arrow) glycerol was added to cells which had accumulated lactose in the absence of glycerol (●). The influx rates are given in arbitrary units and a value of 1 corresponds to 80 nmole/mg dry weight/min.

Figure 5 (right). Time course of membrane potential change ($\Delta\Psi$) during lactose accumulation (0.5 mM external lactose concentration) by E. coli cells (1 mg dry weight/mL) in the presence (■) and in the absence (□) of glycerol (0.4 %).

a decrease of the membrane potential ($\Delta\Psi$). At pH 7.6, $\Delta\Psi$ is the sole component of the proton motive force. In the absence of glycerol, $\Delta\Psi$ was 150 mV; addition of glycerol increased $\Delta\Psi$ to 180 mV. The data in Figure 5 display the time course of the change of $\Delta\Psi$ as lactose is being accumulated in the presence and in the absence of glycerol. In the absence of glycerol, $\Delta\Psi$ remained constant at 150 mV, at least during the 15 min of the experiment. In contrast, in the presence of glycerol and upon addition of lactose, $\Delta\Psi$ decreased steadily as lactose accumulated. The decrease was more or less pronounced depending on the cell preparations; the final value ranged from 140 mV to 40 mV.

We have performed similar experiments on cells devoid of the lactose permease (non-induced ML 30 strains). The presence of lactose and glycerol had no effect on $\Delta\Psi$ (data not shown). This demonstrates that the presence of the lactose carrier is required in order to observe a collapse of $\Delta\Psi$.

<u>Effect of other experimentals conditions on the impairment of lactose transport.</u>

We will not present the data related to the effect of external lactose concentration, external pH, external K^+ concentration, EDTA treatment, on the impairment of lactose transport and on the decrease of $\Delta\Psi$. As compared to the data presented in Figures 1 to 5, the impairment of lactose transport and the decrease of $\Delta\Psi$ were delayed (but not abolished) if the external lactose concentration was decreased, and/or if the external pH was decreased from pH 7.6 to pH 6, and/or if the external K^+ concentration was increased from 1 mM to 10 mM, and/or if the cells were not pretreated with EDTA.

<u>The impairment of lactose transport and the collapse of $\Delta\Psi$ are irreversible phenomena.</u>

The impairment of lactose transport and the concomitant decrease of $\Delta\Psi$ are irreversible phenomena. The evidences are presented in Figure 6.

Cells impaired for lactose transport were prepared in the following way. Cells (1 mg dry weight/mL) were incubated at 25 °C in tris-choline buffer in the presence of lactose (0.5 mM) and glycerol (0.4 %). This resulted in the impairment of lactose transport and in the collapse of $\Delta\Psi$. After 15 minutes, CCCP (30 μM final concentration) was added in order to collapse any remaining $\Delta\Psi$ and to allow for efflux of residual internal lactose. The suspension was incubated for 1 more min and centrifuged. The pellet was resuspended at 4 mg dry weight/mL in tris-choline buffer containing in addition 15 μM CCCP. After 1 minute, the suspension was centrifuged again and the pellet was finally suspended at 1 mg dry weight/mL in tris-choline buffer. Control cells, not impaired for lactose transport, were obtained in the same way except that the first incubation was performed for only 1 min and in the absence of glycerol. Transport and $\Delta\Psi$ were determined on both types of cells in the absence of an external energy source. The data are presented in Figure 6. Cells which had not been impaired for lactose transport were able to accumulate lactose and developed a $\Delta\Psi$ of 150 mV. In contrast, cells which had been impaired for lactose transport were unable to accumulate again lactose, even if replaced under conditions were lactose transport should take place. In addition, the cells were unable to generate a significant $\Delta\Psi$.

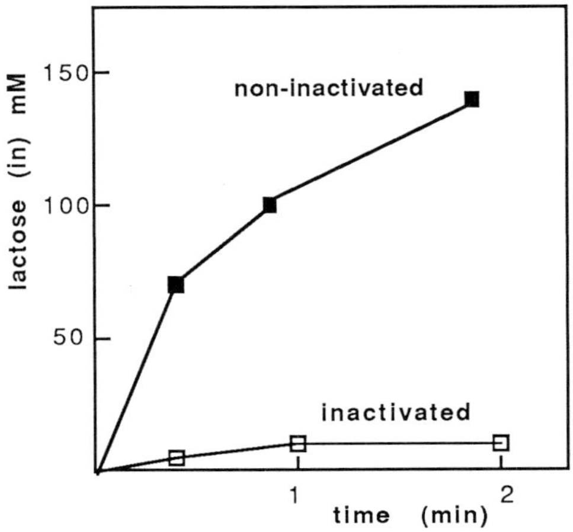

Figure 6. Time course of lactose uptake (0.5 mM external lactose concentration) in the absence of glycerol by E. coli cells (1 mg dry weight/mL) which had been previously inactivated. (■): control (non inactivated cells); (□): inactivated cells. (see text for details)

Impairment of lactose transport and collapse of $\Delta\Psi$ may take place even in the absence of lactose; effect of various galactosides.

The inability of the cells to accumulate lactose and to maintain the original high $\Delta\Psi$ may take place even in the absence of lactose uptake, although at a reduced extent and under a prolonged incubation. First, $\Delta\Psi$ and the ability of the cells to accumulate lactose (15 minutes uptake in the absence of glycerol) were determined on control aliquots before incubation. Then, cells (1 mg dry weight/mL) were incubated in tris-choline buffer under aerated conditions for 2 hours in the presence of lactose (0.5 mM) and in the absence of lactose. After the 2 hour incubation, the cells were washed twice to eliminate any accumulated lactose and resuspended in fresh aerated tris-choline buffer at 1 mg dry weight/mL. $\Delta\Psi$ and transport of lactose (15 min uptake) were determined on both types of cells in the absence of an external energy source. The data are presented in Table I. As expected from the irreversible nature of the inactivation, when the 2 hour incubation was performed in the presence of lactose, $\Delta\Psi$ decreased from 143 mV (before incubation) to a value below 40 mV. Also, before incubation the cells were able to accumulate lactose at a steady-state level of 75 mM (control); after the 2 hour incubation a maximal internal lactose concentration of 6 mM could be reached. When the 2 hour incubation was performed in the absence of lactose, the decrease of $\Delta\Psi$ and of the ability of the cells to accumulate lactose were less pronounced but still significant. $\Delta\Psi$ decreased from 143 mV to 110 mV; a maximal internal lactose concentration of only 30 mM could be reached. Interestingly, when the 2 hour incubation was performed in the presence of TDG , a competitive analog of lactose, rather than lactose itself, the cells lost, neither their ability to accumate lactose (72 mM), nor their capacity to maintain a high $\Delta\Psi$ (147 mV).

Table I. Effect of prolonged incubation (2 h) on the steady state level of internal lactose concentration(15 min uptake) and on the transmembrane electrical potential (see text for details).

	incubation time (h)	control 0	2	2	2	2	2	2
	addition (mM)	none	none	TDG (1)	α-NPG (1)	ONPG (1)	TMG (8)	lactose (8)
accumulation of lactose(mM)		75	30	72	72	60	11	6
$\Delta\Psi$(mV)		143	110	147	n.d.	n.d.	n.d.	40

The different behaviour of lactose and TDG led us to investigate the effects of various galactosides on the ability of the cells to accumulate lactose and to maintain a high $\Delta\Psi$ after a long term incubation. The galactosides fall into two classes: those which enhance the impairment of lactose transport and the decrease of $\Delta\Psi$ (TMG, lactose), those which prevent these phenomena (TDG, α-NPG, ONPG).

The different effects of the two classes of galactosides on lactose transport and on $\Delta\Psi$ strongly suggest that the impairment of lactose transport and the collapse of $\Delta\Psi$ which we report here are the result of a modification of the lactose permease itself. This modification results in an irreversible inactivation of the lactose permease. This inactivation is enhanced by one class of substrate; it is inhibited by a second class of substrate.

<u>The inactivated lactose carrier is still able to catalyze facilitated diffusion, efflux and exchange.</u>

It was of interest to determine wether or not the irreversibly inactivated lactose carrier was still able to catalyze facilitated diffusion, efflux and exchange. The lactose carrier was irreversibly inactivated as described above (see paragraph on "irreversiblility of the impairment of lactose transport"). Facilitated diffusion, efflux and exchange were determined on the cells having the inactivated carrier (inactivated cells) and on contro cells (non-inactivated cells) as described under "Material and methods". The data are shown in Figure 7.

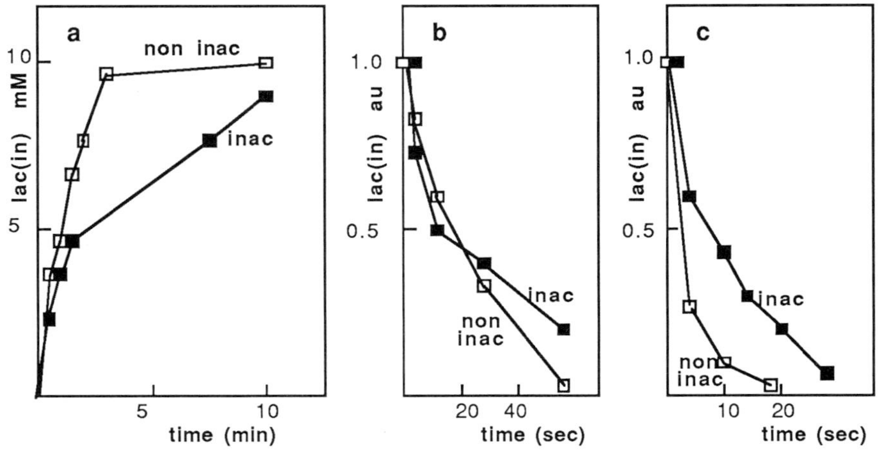

Figure 7. Facilitated diffusion (7a), efflux (7b) and exchange (7c) of lactose by inactivated (■) and non-inactivated (□) cells. For facilitated diffusion, the external lactose concentration was 10 mM and accumulation was followed as a function of time in concentrated cells (100 mg dry weight/mL). For efflux and exchange, the concentrated cells were first equilibrated with 10 mM radioactive lactose and then diluted 200 fold in buffer free of lactose and containing 10 mM non radioactive lactose, respectively. Efflux and exchange were followed by measuring the amount of radioactive lactose remaining in the cells as a function of time. For efflux and exchange, the data are presented as fractions of the internal lactose concentration relative to the initial internal lactose concentration.

Figure 7a displays the facilitated diffusion of lactose (external lactose concentration: 10 mM). Figures 7b and 7c display respectively the efflux and the exchange. Inactivated cells catalyzed facilitated diffusion, efflux and exchange. The time courses were slightly slower for the inactivated than for the non-inactivated cells.

A kinetic analysis was performed in the case of efflux. The corresponding Eadie-Hofstee plot is represented in Figure 8. The inactivation results in a significant increase of the K(T) of efflux from some 4 mM (non-inactivated cells) to 20 mM (inactivated cells); the V(max) are little changed. The K(T) of efflux of 4 mM for non-inactivated cells agrees reasonably well with values previously reported (2 mM) (Kaczorowski et al., 1979). A kinetic analysis could not be performed for facilitated diffusion and exchange. The time courses are too fast, the initial rates are linear for only a few seconds, and the precision in the measurements precludes a meaningful comparison between the kinetic parameters of exchange or facilitated diffusion of the inactivated and non-inactivated cells.

For inactivated cells, efflux occurs via a single component whose K(T) is very different from that of non-inactivated cells. It excludes the possibility that the inactivation affects only a minor fraction of the lac carrier.

Inhibitors of serine proteases protect against the inactivation of the lactose carrier.

The inactivation of the lactose carrier was protected by the presence of inhibitors of serine proteases (Thérisod et al., 1985b). Figure 9 displays the time course of lactose uptake in the absence of glycerol, in the presence of glycerol, in the presence of glycerol and either PMSF (5 mM) or TLCK (2.5 mM). Clearly, in the presence of these 2 inhibitors of serine proteases, the uptake in the presence of glycerol ressembled that in the absence of glycerol. We have checked that these inhibitors are without effect on the rate of respiration (data not shown). These inhibitors of serine proteases also suppressed the decrease of $\Delta\Psi$. In the presence of glycerol $\Delta\Psi$ decreased from 200 mV to 40 mV during the first 15 minutes of lactose uptake. In the presence of PMSF or TLCK the decrease was significantly smaller (from 160 mV to 145 mV and from 185 to 155 mV, respectively).

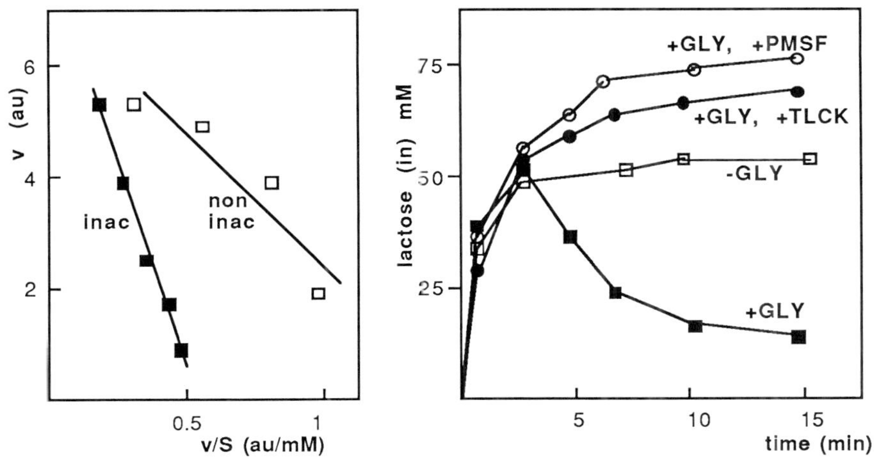

Figure 8 (left). Kinetic analysis of lactose efflux by inactivated (■) and non-inactivated (□) E. coli cells as a function of the initial internal lactose concentration. Cells (100 mg dry weight/mL) were equilibrated with radioactive lactose at a given well defined concentration (S ranging from 1 mM to 20 mM). Then, the cells were diluted 200 fold and initial rates of efflux (v) were determined over 5 sec intervals. The data are presented in the Eaddie Hofstee representation(v as a function of v/S). The slope is equal to -K(T) of efflux; the intercept with the ordinate axis is equal to the maximum rate of efflux at saturating internal lactose concentration.
Figure 9 (right). Time course of lactose uptake (0.5 mM external lactose concentration) by E. coli cells (1 mg dry weight /mL). (□): in the absence of glycerol and protease inhibitors; (■) in the presence of glycerol (0.4 %) and in the absence of protease inhibitors; (○): in the presence of glycerol and PMSF (8 mM); (●): in the presence of glycerol and TLCK (2.5 mM).

Purification and reconstitution of the inactivated carrier

The lactose carrier can be specically labeled with the photoaffinity reagent α-NPG (Kaczorowski et al., 1980). French pressed vesicles obtained from inactivated and non-inactivated cells were assayed for photoaffinity labeling with tritiated α-NPG. In both cases the same photoaffinity labeling was obtained indicating that the inactivation of the lactose carrier does not prevent binding of the substrate (data not shown). This was expected from the fact that inactivated cells were able to catalyze facilitated diffusion, efflux and exchange.

When vesicles photolabeled under those conditions were subjected to SDS-PAGE followed by fluorography, radioactivity migrated in both cases (inactivated and non inactivated vesicles) at 33 kD, which is the apparent molecular weight reported for the lactose carrier by SDS-PAGE. Thus, if the inactivation is the consequence of a proteolytic attack of the carrier, this proteolysis is necessarily limited to a few amino-acids at one end of the protein.

The lactose carrier can be purified to homogeneity, after solubilization in octylglucoside, and functionnaly reconstituted in liposomes according to the procedure of Newman et al., (1981). We have followed this procedure to purify and reconstitute in parallel the lactose carrier obtained from inactivated and non-inactivated cells. In order to follow the purification, in each case, labeled and unlabeled vesicles were mixed in a 3:97 ratio.

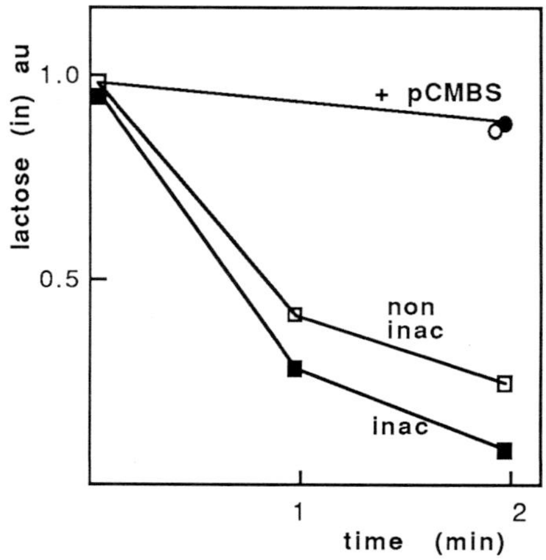

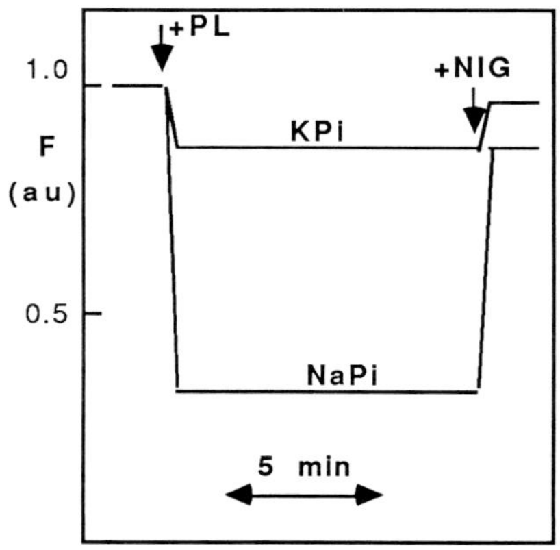

Figure 10 (left). Time course of lactose exchange by proteoliposomes prepared with inactivated(■) and non-inactivated (□) lactose carrier. The proteoliposomes were prepared as described in Material and Methods. They were equilibrated with 3 mM ^{14}C lactose and diluted 200 fold in a medium containing 3 mM non radioactive lactose. Exchange was followed by determining the internal concentration of radioactive lactose as a function of time. The data are presented as fractions of the internal lactose concentration relative to the initial internal lactose concentration. In control experiments, pCMBS (300 µM) (a thiol reagent which blocks the lactose carrier) was added 2 min before the dilution (●,○).

Figure 11(right). Determination by the fluorescence quenching of diIC1(5) of a potassium diffusion potential generated across the proteoliposomes (see Material and methods for experimental details). (PL): addition of 1 µL concentrated proteoliposomes in potassium phosphate to 1 mL of the dye solution in sodium phosphate (NaPi) or in potassium phosphate (KPi); (NIG): addition 10 min later of 10 µM nigericin.

In both cases, and at each step, the purification factor and the yield, calculated from the recovery of the photolabel were similar. The purified inactivated and non-inactivated lactose carrier were subsequently reconstituted into proteoliposomes. The inactivated carrier incorporated into the proteoliposomes to the same extent as the non-inactivated carrier.

We have assayed the two types of proteoliposomes for exchange of lactose. The proteoliposomes were loaded with 3 mM ^{14}C lactose and diluted 300 fold in a medium containing 3 mM non-radioactive lactose and 15 μM CCCP. The data in Figure 10 show that proteoliposomes prepared with inactivated or non-inactivated lactose carrier catalyzed a rapid efflux of radioactive lactose. This efflux was blocked by treatment with pCMBS, a thiol reagent known to inhibit transport via the lactose carrier.

A transmembrane potassium diffusion potential can be generated across the proteoliposomes and its extent can be monitored by following the quenching of diIC1(5) (see Material and methods). As shown in Figure 11, dilution of the inactivated proteoliposomes in sodium phosphate resulted in a quenching of the fluorescence of the dye, reflecting the generation of a membrane potential since it could be reversed by the addition of nigericin (CCCP could not be used because of interferences with the fluorescence of the dye). The membrane potential is stable for at least 10 minutes. Dilution of the proteoliposomes in potassium phosphate resulted in a smaller fluorescence quenching reflecting either an unspecific binding of the probe to the proteoliposomes or the generation of a small $\Delta\Psi$. A very similar pattern was observed for the non-inactivated proteoliposomes (data not shown).

Inactivated and non-inactivated proteoliposomes were assayed for active transport of lactose in the presence of a potassium diffusion potential. The results are shown in Figure 12.

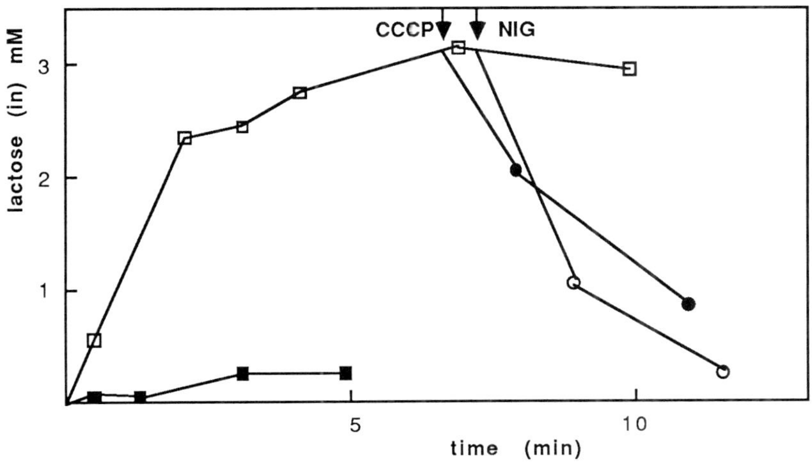

Figure 12. Active transport of lactose (0.3 mM external lactose concentration)by proteoliposomes prepared with inactivated (■) and non-inactivated (□) lactose carrier stimulated by a potassium diffusion potential. For the experimental details see Material and Methods. Where indicated, nigericin (10 μM,O) or CCCP (30 μM,●) were added to aliquots of proteoliposomes which had accumulated lactose.

Non-inactivated proteoliposomes are able to catalyze active transport. At steady-state, the accumulation ratio is about 10. Dissipation of the diffusion potential by nigericin or CCCP resulted in the release of the accumulated lactose. In contrast, under the same conditions, no accumulation of lactose could be detected in inactivated proteoliposomes.

The data indicate that inactivated lactose carrier purified and reconstituted into proteoliposomes display the same caharacteristics as in whole cells: they still catalyze facilitated diffusion but active transport is lost.

Caracterization of the solubilized inactivated lactose carrier.

We have attempted to further characterize the inactivated lactose carrier in solution. Since the carrier has a tendancy to precipitate in octyl glucoside, this detergent was completely exchanged for C12E8, another non-ionic detergent, as described in Houssin et al., (1985). It has been shown that in this detergent the carrier is essentially a dimer and that it is possible to separate dimer from monomer by HPLC (Houssin et al., 1985). We found that the HPLC elution profiles of both the inactivated and non-inactivated carrier are similar (data not shown). Thus, the inactivation does not alter the state of association of the carrier.

Finally, we have also determined the amino acid composition of the inactivated carrier. Within experimental errors, it was similar to that of the non-inactivated carrier. This does not exclude that the inactivation is the result of the removal of a few amino-acids at one end of the protein or of a change in nature of some amino-acids (hydroxylation, phosphorylation...).

DISCUSSION

The lactose carrier of E. coli is prone to an irreversible inactivation upon incubation. The extent of this inactivation depends on the experimental conditions.

The transport of lactose by the carrier strongly enhances its own inactivation. On the other hand, the transport of TDG (a competitive inhibitor of lactose) strongly inhibits the inactivation. The various substrates of the lactose carrier could be classified into two categories: those, which ike lactose, enhance the inactivation; those, which like TDG, inhibit the inactivation. Remarkably, this classification coincides with the one previously proposed on the basis of the binding (dissociation constant K(D)) and of the transport (kinetic parameter K(T)) properties of these various substrates : the lactose class of substrates displays a high K(D)/K(T) ratio while the TDG class of substrates displays a K(D)/K(T) ratio close to 1 (Wright et al., 1982). Whatever the exact reason for these differences, they indicate that each class of substrates interacts differently with the lactose carrier. We propose that there exists three different conformations of the lactose carrier with different succeptibility towards inactivation: in the absence of bound substrate, in the presence of the lactose class of substrates, in the presence of the TDG class of substrates. The susceptibility towards inactivation increases upon binding of the lactose class of substrates; it decreases upon binding of the TDG class of substrates.

The inactivation of the lactose carrier results simultaneously in the irreversible inability of the cells to catalyze active transport and in an irreversible collapse of $\Delta\Psi$. However, the collapse of $\Delta\Psi$ is not the direct reason for the loss of active transport. Indeed, we were able to reconstitute the inactivated lactose carrier into proteoliposomes. Imposition of a potassium diffusion potential across the membrane of the proteoliposomes did not stimulate active transport. On the other hand, the inactivated lactose carriers are still able to catalyze facilitated diffusion, efflux and exchange as well in whole cells as in reconstituted proteoliposomes. These data are best explained if one assumes that the inactivation of the lactose carrier leads to a structural change of the protein resulting in an uncoupling of lactose and proton transport. The inactivated lactose carrier calalyzes facilitated diffusion of lactose uncoupled to protons explaining the absence of active transport; this may also explain the difference in the kinetic parameter (K(T)) of efflux between inactivated and non-inactivated lactose carrier. The inactivated lactose carrier is also able to catalyze facilitated diffusion of protons in the absence of lactose explaining the collapse of $\Delta\Psi$. Thus, while the non-inactivated lactose carrier functions as a strict lactose/proton symport, the inactivated lactose carrier would functions as two independent lactose and proton uniports.

Inhibitors of serine proteases (PMSF, TLCK) supress the irreversible inactivation of the lactose carrier. These inhibitors have been shown in vivo to block protein breakdown in E. coli cells (Prouty and Goldberg, 1972). 8 soluble endoproteases have been isolated and purified from E. coli cells, 6 of which are serine proteases as shown by their inhibitor sensitivity (Swamy and Goldberg, 1981; Goldberg et al., 1981). More recently, 3 membrane serine proteases have been identified (Pacaud, 1982; Palmer and St John, 1987). We therefore conclude that proteolytic enzymes, and thus a protein degradation are involved in the inactivation of the lactose carrier. The proteolytic attack could be at the level of the lactose carrier itself. If this were the case, it would only involve a few amino-acids at one end of the protein as the inactivated lactose carrier migrates in SDS-PAGE like the non-inactivated carrier. On the other hand, the proteolytic attack could be at the level of a protein different from the lactose carrier, resulting in an activation of that protein. This activated protein could then in turn modify in one way or another the lactose carrier (hydroxylation, phosphorylation...) leading to the inactive form of the lactose carrier.

The stimulatory effect of external energy sources on the inactivation of the lactose carrier is the result of an increased respiratory activity of the cells. Decreasing artificially this enhanced respiratory activity by controlled addition of inhibitors of the respiratory chain inhibits the stimulatory effect of the external energy sources. The stimulatory effect is not related to the metabolism of the energy source. Indeed, we observe the same stimulatory effect using an artificial electron donnor system which injects its electron directly into the respiratory chain without prior transformation. The increased respiratory activity may modify the state of the lactose carrier in a way wich renders it more suceptible to the inactivation. This could occur either by a direct interaction of the carrier with a component of the respiratory chain (e.g., a semi-quinone) or by an interaction with free radicals generated by the respiratory activity. Alternatively, the increased respiratory activity may stimulate a proteolytic activity.

Site directed mutagenesis has been performed on the lactose carrier (Kaback, 1987). In particular it has been shown that replacement of histidine 322 leads to a protein unable to catalyze active transport but still able to catalyze facilitated diffusion. Moreover, efflux of accumulated lactose occured without protons (Padan et al., 1985). Thus this modified protein is uncoupled in lactose and proton transport and it ressembles the inactivated lactose carrier which we report here. However, the amino-acid analysis of the inactivated lactose carrier indicates that it possesses the 4 histidines normally present in the non-inactivated carrier. Thus, different type of modification of the lactose carrier can result in the same type of functional modification.

Finally, one may speculate as to the physiological implication of the inactivation of the lactose carrier. The inactivation may be part of a general process of protein degradation. In this respect we note that a similar mechanism has been advanced to explain photoinhibition in thylakoids. It has been shown that a 32 kD protein from thylakoid membrane, a structural component of PS II, becomes prone to inactivation (a proteolytic attack) under conditions of high light intensity. This inactivation results directly from the light dependent enhanced electron flow (Matoo et al., 1984; Kyke et al., 1984). In addition the degradation of the lactose carrier which we report here may serve as a control of lactose uptake by the cells under conditions of high rates of respiration, i.e., when the cells are fully energized and do not require additional sugars as energy sources.

A similar phenomenon has been reported for Saccharomyces cerevisiae by Lagunas and coworkers (Lagunas et al., 1982; Busturia and Lagunas, 1985; DeJuan and Lagunal, 1986). Sugar transport systems for maltose, glucose and galactose are irreversibly inactivated in this yeast upon protein synthesis impairment. This inactivation is considerably accelerated by the presence of a carbon source.

The experimental conditions which we have deliberately chosen here (EDTA treated cells, low external potassium concentration) amplify the inactivation to extreme levels. Under more physiological conditions a more subtle regulation may take place.

Abbreviations: CCCP: carbonyl cyanide m-chlorophenylhydrazone; C12E8: dodecyl octaethylene glycol monoether; pCMBS: para chloro mercuriphenyl sulfonic acid; diIC1(5):3,3' dimethyl indodicarbocyanine iodide; α-NPG: 4-nitrophenyl-α-D-galactopyranoside; ONPG: o-nitrophenyl β D-galactoside; PMSF: phenylmethylsulfonyl fluoride; Ph4P: tetraphenylphosphonium ion; TDG: β-D-galactosyl-1thio β–D-galactopyranoside; TLCK: N-α-p-tosyl-L-lysine chloromethyl ketone; TMG: methyl-1thio-β–D-galactoside.

REFERENCES

Booth I, Mitchell WJ, Hamilton WA (1979) Quantitative analysis of proton linked transport systems. Biochem J 182:687-696.

Büchel DE, Gronenborn B, Müller-Hill B (1980) Sequence of the lac permease gene. Nature 283:541-545.

Busturia A, Lagunas R (1985) Identification of two forms of the maltose transport system in Saccharomyces cerevisiae and their regulation by catabolite inactivation. Biochim Biophys Acta 820:324-326.

Carrasco N, Antes LM, Poonian MS, Kaback HR (1986) Lac permease of E. coli: Histidine 322 and Glutamic acid 325 may be components of a charge relay system. Biochemistry 25:4486-4488.

DeJuan C, Lagunas R (1986) Inactivation of the galactose transport system in Saccharomyces cerevisiae. FEBS let 207:258-261

Foster DL, Garcia ML, Newman MJ, Patel L, Kaback HR (1982) Lactose proton symport by purified lac carrier protein. Biochemistry 21:5634-5638.

Garcia ML, Vitanen P, Foster DL, Kaback HR (1983) Mechanism of lactose translocation in proteoliposomes reconstituted with lac carrier protein purified from E. coli. Effect of pH, imposed membrane potential on efflux, exchange and conterflow. Biochemistry 22:2524-2531.

Ghazi A, Shechter E (1981) Lactose transport in E. coli cells. Dependence of kinetic parameters on the transmembrane electrical potential difference. Biochim Biophys Acta 645:305-315.

Ghazi A, Shechter E, Letellier L, Labedan B (1981) Probes of membrane potential in E; coli cells. FEBS let 125:197-200.

Ghazi A, Thérisod H, Shechter E (1983) Comparison of lactose uptake in resting and energized E. coli cells: high rates of respiration inactivate the lac carrier. J Bact 154:92-103.

Goldberg AL, Swamy KHS, Chung CH, Larimore FS (1981) Proteases in E. coli. Methods in Enzymology 80:680-702.

Houssin C, LeMaire M, Aggerbeck L, Shechter E (1985) The lactose carrier of E. coli: evidence in favor of a dimer. Arch Biochem Biophys 240:593-606.

Kaback HR (1983) The lac carrier protein in E. coli. J Memb Biol 76:95-112

Kaback HR (1987) Use of site directed mutagenesis to study the mechanism of a membrane transport protein. Biochemistry 26:2071-2076.

Kaczorowski GJ, LeBlanc G, Kaback HR (1980) Specific labeling of the lac carrier protein in membrane vesicles of E. coli by a photoaffinity reagent. Proc Natl Acad Sci (US) 77:6319-6323.

Kaczorowski GJ, Robertson DE, Kaback HR (1979) Mechanism of lactose translocation in membrane vesicles from E. coli. Effect of imposed $\Delta\Psi$, ΔpH and $\Delta\mu$. Biochemistry 18:3697-3703.

Kyle DJ, Ohad I, Arntzen CJ (1984) Membrane protein damage and repair: selective loss of a quinone-protein function in chloroplast membranes.Proc Natl Acad Sci (US) 81:4070-4074.

Laemmli (1970) Cleavage of structural proteins during the assembly of the head of bacteriophage T4. Nature 227:680-685.

Lagunas R, Dominguez C, Busturia A, Saez MJ (1982) Mechanism of appearance of the Pasteur effect in Saccharomyces cerevisiae: inactivation of sugar transport system. J Bact 152:19-25.

Letellier L, Shechter E. (1979) Cyanine dye as monitor of membrane potentials in E. coli cells and membrane vesicles. Eur J Biochem 102:441-447.

Matoo AK, Hoffman-Falk H, Marder JB, Edelman M (1984) Regulation of protein metabolism: coupling of photosynthetic electron transport to in vivo degradation of the rapidly metabolized 32 kD protein of the chloroplast membranes. Proc Natl Acad Sci (US) 81:1380-1384.

Mitchell P (1963) Molecule, group and electron translocation through natural membranes. Biochem Soc Symp 22:142-168.

Newman MJ, Foster DL, Wilson TH, Kaback HR (1981) Purification and reconstitution of functional lactose carrier from E. coli. J Biol. Chem 256:11804-11808.

Pacaud M (1982) Purification and characterization of two novel proteolytic enzymes in membrane of E. coli. J Biol. Chem 257:4333-4339.

Padan E, Sarkar HK, Vitanen PV, Poonian MS, Kaback HR (1985) Site specific mutagenesis of histidine residues in the lac permease of E. coli. Proc Natl Acad Sci (US) 82:6765-6768.
Palmer SM, St John AC (1987) Characterization of membrane associated serine proteases in E. coli. J Bact 169:1474-1479.
Prouty WF, Goldberg AL (1972) Effects of protease inhibitors on protein breakdown in E. coli. J Biol Chem 247:3341-3352.
Püttner IB, Sarkar HK, Poonian MS, Kaback HR (1986) Lac permease of E. coli: Histidine 205 and Histidine 322 play different roles in lactose/proton symport. Biochemistry 25:4483-4485.
Rickenberg HV, Cohen GN, Buttin G, Monod J (1956) La lactose perméase d'E. coli. Ann Inst Pasteur 79:829-834.
Rosen BP, Kaschket ER (1978) Energetics of active transport. In: Rosen BP (ed) Bacterial transport, vol 4. Marcel Dekker Inc., New-York, p 559-620.
Swamy KHS, Goldberg AL (1981) E. coli contains 8 soluble proteolytic activities, one being ATP dependent. Nature 292:652-654.
Thérisod H, Ghazi A, Houssin C, Shechter E (1985a) Inactivation of the lactose permease of E. coli by the respiratory activity. Biochim Biophys Acta 814:68-76.
Thérisod, H, Ghazi A, Shechter E (1985b) Implication of proteases in the respiration dependent inactivation of the lactose permease of E. coli FEBS let 187:38-42.
Wright JK, Riede I, Overath P (1981) Lactose carrier protein of E. coli. Interactions with galactosides and protons. Biochemistry 20:6404-6415.
Wright JK, Schwarz H, Straub E, Overath P, Bieseler B, Beyreuther K (1982) Lactose carrier protein in E. coli: reconstitution of galactoside binding and countertransport. Eur J Biochem 124:545-552.

NATO ASI Series H